| DATE | | | |
|---|---|---|---|
| | | | |
| | | | |
| | | | |
| | | | |
| | | | |
| | | | |
| | | | |
| | | | |
| | | | |
| | | | |
| | | | |
| | | | |

# TECHNICAL COMMUNICATION IN THE GLOBAL COMMUNITY

SECOND EDITION

DEBORAH C. ANDREWS
*University of Delaware*

Upper Saddle River, New Jersey 07458

Library of Congress Cataloging-in-Publication Data
Andrews, Deborah C.
Technical communication in the global community / Deborah C.
Andrews. — 2nd ed.
p. cm.
Includes bibliographical references and index.
ISBN 0-13-028152-2 (pbk.)
1. Technical writing. 2. Communication of technical information.
I. Title.
T11.A515 2001
808'.0666—dc21 00-064255

VP/Editorial Director: *Leah Jewell*
Acquisitions Editor: *Craig Campanella*
Editorial Assistant: *Joan Polk*
AVP/Director of Production and Manufacturing: *Barbara Kittle*
Director of Marketing: *Beth Gillett Mejia*
Marketing Manager: *Rachel Falk*
Managing Editor: *Mary Rottino*
Project Manager: *Alison Gnerre*
Production Assistant: *Elizabeth Best*
Prepress and Manufacturing Manager: *Nick Sklitsis*
Prepress and Manufacturing Buyer: *Mary Ann Gloriande*
Line Art Manager: *Guy Ruggiero*
Creative Design Director: *Leslie Osher*
Art Director: *Anne Bonanno Nieglos*
Interior and Cover Design: *Tom Nery*
Electronic Illustrations: *Mirella Signoretto*
Cover Art: *PhotoLink/PhotoDisc*

This book was set in 10/12 New Baskerville by Publications Development
Company of Texas and was printed by Courier Westford.
The cover was printed by Phoenix Color Corporation.

Printed in the United States of America

10 9 8 7 6 5 4 3 2 1

ISBN 0-13-028152-2

Prentice-Hall International (UK) Limited, *London*
Prentice-Hall of Australia Pty. Limited, *Sydney*
Prentice-Hall Canada, Inc., *Toronto*
Prentice-Hall Hispanoamericana, S.A., *Mexico*
Prentice-Hall of India Private Limited, *New Delhi*
Prentice-Hall of Japan, Inc., *Tokyo*
Pearson Education Asia Pte. Ltd., *Singapore*
Editoria Prentice-Hall do Brasil, Ltda., *Rio de Janeiro*

*To William, Chris, and Kirsten*

# BRIEF CONTENTS

# CONTENTS

# PREFACE TO THE SECOND EDITION

The focus in this edition remains the same as in the first: to provide strategies for technical professionals to use as they reach audiences and collaborate on projects across borders of culture, language, and technology. In revising, I have reshaped some sections to reflect the rapidly changing context for technical communication. I've also incorporated new research findings and comments from students and colleagues.

The problems giving rise to communication remain muddled and multidimensional. To solve them, students must learn how to write and speak as global citizens as well as global engineers or health professionals or scientists. As in the first edition, I've taken an international context for granted. Science and technology are international activities, and an informed professional writes with the international community in mind. The pervasiveness of the Internet and especially the Web only underscores this reality.

As I reviewed the first edition, I kept in mind how current information technology aids—or hinders—good communication and looked for ways to update advice about using that technology. These updates appear overtly in the *Electronic Edge* feature boxes and are interwoven throughout the book wherever appropriate. One perhaps cynical observer notes that the Internet only lets bad information get to you faster. To avoid such risks, students will find here expanded strategies for managing information in a 24/7 world.

In this edition, too, *design* takes center stage. To adapt a line from *Alice in Wonderland* (the original is quoted at the beginning of Chapter 10): "Much of what you see depends on how *it* looks." Extensive research in academe and the marketplace confirms the central role of design in shaping the content of information products. In light of this research and my own years as an avid observer of the visual world, I thoroughly revised Part 3 on expression. The revision reverses the order of the first edition (sentences to visuals to design). In this edition, design comes first, followed by advice on composing visuals and then on composing text. Design issues also thread through the chapters on individual genres as well as oral presentations. In addition, to practice what we preach, we redesigned the look of this edition.

In other ways, too, I hope to have enhanced elements of the first edition without messing up what readers found attractive there. Several chapters have been edited for greater crispness and clarity following excellent suggestions from students and reviewers. The *Crossing Cultures* and *A Closer Look* feature boxes remain, most in updated form. So do the models, cases, exercises, and checklists readers depended on to make rhetorical concepts come alive. (Badgered by reviewers and students, however, I reluctantly let the Chapter 1 rats go.) For additional updates on information in the book, more cases and models, and expanded exercises, please visit the book's Web site: www.prenhall.com/andrews. An instructor's manual is also available.

## ACKNOWLEDGMENTS

All writing, at least the way I write, is collaborative, and I'd like to thank publicly some of the many people who guided me in improving the text.

The following reviewers for Prentice Hall gave me just the advice I needed: Genie Babb, University of Alaska, Anchorage; Daryl R. Davis, Northern Michigan University; Matt Pifer, University of Oklahoma; and Sharon K. Wilson, Fort Hays State University.

I'd also like to thank Craig Campanella, my editor at Prentice Hall, and Alison Gnerre, Production Manager.

In addition, the Consortium for the Study of Engineering Communication provided some of the best conversations about writing (and desserts) that I can remember. I'd particularly like to thank Linda Driskill, consortium convener, Rice University; Penny Hirsch, Barbara Shwom, Northwestern University; Rebecca Burnett, Iowa State University; Marj Davis, Mercer University; Karen Schriver, KSA, Document Design and Research; Dick Hayes, Carnegie-Mellon University; Steve Youra, Cornell University; Deborah Bosley, North Carolina State University-Charlotte; Margaret Hundleby, Auburn University; and Lee Odell, Rensselaer Polytechnic Institute. Colleagues in the Council for Programs in Technical and Scientific Communication also informed this book through our annual meetings in interesting places and our special roundtable in June 2000 in London. I've also profited greatly from workshops and panels at the Conference on College Composition and Communication and at the regional and national gatherings of the Association for Business Communication.

Talking about writing is not the same as *writing*, of course, but it helps. In particular, I'm grateful for the bright ideas of Stuart Selber, Penn State University; Bruce Maylath, University of Wisconsin-Stout; Paul Anderson, Miami University of Ohio; Jimmie Killingsworth, Texas A&M; Carolyn Rude, Sam Dragga, Texas Tech; Steve Bernhardt, New Mexico State University; Kathy Richardson, Capital University; Lil Rodman, University of British Columbia; Becky Worley, Martha Carothers, University of Delaware.

In the end, this book is of and for students. My students and I have spent a lot of time together—in Newark, Delaware; in London, UK; in Mikkeli and Helsinki, Finland; and in cyberspace. As a product of that time, this book represents my deepest gratitude to them.

*Debby Andrews*
Newry, Maine

# PART ONE
# Performing Your Role as a Communicator

# INTRODUCTION



"COMMUNICATION EQUALS REMEMBERING WHAT IT WAS LIKE NOT TO KNOW."

RICHARD S. WURMAN

A biologist who heads the medical writing group at a large pharmaceutical company says that he spends his days helping scientists and physicians to "think like writers." Whatever your technical or scientific discipline, this book will help *you* think like a writer. Such thinking will in turn help you understand your own discipline better by knowing the range of audiences who read about it, the purposes for which they read, the specialized language practitioners use, and the types of information products that are common. You'll learn strategies for managing and designing information for readers and listeners in a wide range of contexts. This chapter identifies the major elements of a communication context. Subsequent chapters discuss each element in detail. In addition, the text's Web site provides further examples and updates.

## THE CONTEXT FOR TECHNICAL COMMUNICATION

The *context* is the environment in which communication occurs. Narrowly speaking, *context* is the text that surrounds other text and shapes its meaning. When you "quote out of context," you take someone's words and twist them into a new meaning in a setting of your words. More broadly, *context* refers to all the components that interweave as you create an information product and as your audience reads or listens to that product. The term *information product* takes into account an important dimension of the context for technical communication: change. Although writers continue to create documents and give oral presentations, electronic technology is rapidly changing the forms in which messages reach audiences. You'll read about both traditional and evolving forms.

The context for technical communication includes these elements:

- You
- Your audience
- Purpose
- Information
- Design

Aligning all of these elements are *culture* and *ethics*. You'll read about these shortly.

## You and Your Audience

As a student and as a professional, you'll sometimes write for yourself. You record field or laboratory observations as well as notes on conversations or meetings as a way not to forget what you saw or heard. The notebook or journal serves as a holding place for your facts and concepts. In addition, you may write as a way to capture what you are thinking. Sometimes the best way to work through a problem or design a solution is just to write things down, to see where a line of text can take you.

You also write to achieve some purpose *through* other people. In doing so, you may write as an individual or as a representative of one or many organizations. You may also write as part of a team. Organizations increasingly emphasize teams as the unit of work and value collaborative problem solving. As a collaborator, you adjust your writing to an audience of your teammates. You also have to adjust your writing to other audiences your teamwork addresses.

Often, you'll write in response to a direct request from the audience. The audience may ask you to answer a question, help them make a decision, or instruct them in the use of a system or tool. At the same time, you try to achieve your own purpose within that context. In response to a job listing, you write a cover letter and resume that aim to persuade the reader to hire you. As a technical expert, you write a set of online instructions to help customers use your accounting system. You help your audience to understand and to do.

**For more about audience, see Chapter 6**

As you write, imagine a scene in which the reader is using your information product, and design your approach to make that use as easy as possible. Ask and answer some of the following questions:

***How Skilled* Is Your Audience at Reading About the Subject?** In your career, you'll write for a variety of audiences, including your technical colleagues, professionals who are not in your field, customers, clients, and the general public. Your audience will try to make sense out of a document, talk, or

Web site in light of their experiences, their values, their familiarity with terms and concepts, and their opinions about you and your organization. You will need to keep their level of expertise and interests in mind.

***What Else* Have They Read or Heard About the Topic or Situation You Are Addressing?** Rarely are documents one time phenomena. Most reflect a series of communications on the topic, and you need to write with that series in mind to control the range of possible interpretations of what you say.

***How* Will the Audience Read?** In a context of too much information, readers may resist reading or approach the task reluctantly. Visitors to Web sites arrive with mouse in hand or at the touch of a finger, ready to click off in a moment if they don't feel well served. Design units of text and visuals to motivate the audience and match their capacity for paying attention.

### Purpose

The audience will pay attention best if what they read or hear meets their purposes in reading and listening. Here are three general purposes for communicating:

- To record
- To inform
- To persuade

When you write to *record,* as in a lab notebook or insurance form, you establish the facts of the case. You describe and define as accurately as possible. Often, you use visuals and numbers more than words. The weather maps widely available in newspapers, on the Web, and on TV, for example, record information at various levels of detail.

When you write to *inform,* you want readers to understand what you know. You interpret and shape the information to meet a particular reader need. Based on the weather maps, you may draw other diagrams and compose textual explanations that help readers understand the movement of the jet stream or deviations from normal in a city's record of high temperatures over the period.

When you write to *persuade,* you want readers to act in a certain way, come to a decision you think is right, change their attitude toward something, or agree that something you think is significant is indeed significant. For example, you might use the summer's weather data to convince your manager that your building needs central air conditioning.

In assessing a communication's purpose, consider, too, how readers will use your information:

- As a reference while they operate some machinery or apply a policy?
- As a reminder of something to do that will be discarded when the action is undertaken?
- As the source of information to be filed with other documents on the topic?

Reflect the answer in your design. For example, consider designing a reference manual in a spiral binder that will lie flat next to the machine while the operator toggles between reading the instructions and operating the machine. Make sure the information, too, responds to the operator's direct needs. Perhaps use a question-and-answer format based on conversations with expert and novice users who point out key features of operation and places where things might go wrong. Include visuals. Include any necessary warnings.

For more about design, see Chapter 7

### Information

Readers expect information in a technical document, and they expect writers to provide the *right* information: not too much and not too little. As a technical professional, you will spend a good deal of time developing information from observation, experiment, and reading and then managing it for yourself and for other readers.

### Design

Information is the content of your message. *Design* is its shape. The content sometimes dictates the design. At other times, the design determines appropriate content, as in documents that follow conventional approaches or reader expectations.

**Conventions.** Learn and adhere to any conventions or standards that apply to what you are writing. Many scientific and technical documents reflect strict conventions of presentation. They are written *to spec;* that is, they conform to specifications in content, order of presentation, citation of references, expression of dates, and the like. The organization requesting the work—for example, a publication or a government agency—makes the rules you follow. These conventions ease both the recording and the reading of information.

**Genre.** Match reader expectations, too, in the genre or type of document you are writing. Many situations for writing repeat themselves, and thus a pattern has emerged for arranging documents that meet that context. An approach for reporting original scientific research in a report or article, for example, is so well ingrained that it is known by its acronym, IMRAD: Introduction, Materials and method, Results, And Discussion. The following list suggests other common genres:

- Correspondence that connects people worldwide, including e-mail postings, memos (circulated generally within an organization), and letters (circulated to readers outside an organization)
- Procedures and manuals that instruct readers about how to behave or how to operate or maintain some device or system
- Abstracts that summarize other documents and can circulate as standalone items or appear as part of a parent document

- Proposals that identify a problem or need, describe an approach to solving the problem or meeting the need, and seek approval and funds for the writer to do the work or supply the product or service
- Reports that complete a stage in a project (progress report), document a project or investigation (information report), or support a future action (decision-making report)
- Articles for publication that enhance the understanding of technical, semitechnical, or popular audiences and enrich the reputation of the author

You'll read in detail about each of these genres in Parts 4 and 5.

**Language.** Select the appropriate language and style for communication. Even if that language is English, you are likely to augment your discussion with numbers, images, and sounds. Images in particular play an important role in technical and scientific communication and may be the major language for presentations, especially on the Web and in instructional manuals that cross national and cultural borders. In addition, you may need to accommodate audiences reading English as a second (or third or fourth) language or write in an English that is easy to translate.

**Media.** Choose the best medium in which to produce your message. The range of media is expanding greatly, especially in response to innovations in electronic technology. That technology has produced a wide range of computers beyond the familiar desktop and laptop models, especially handheld ones, sometimes called *personal digital assistants,* or PDAs. Such computers are converging with cell phones and other wireless technology, fax machines, cable, television, videos, and sophisticated color printers in a vast international electronic network that is radically changing the context for communication. As *Crossing Cultures: The Electronic Highway* notes, the Internet speeds things up and fosters interaction. It gives readers even greater power to select what to read and to display information in a format and order they choose. As it provides new ways to transfer information, it changes the information itself as well as the relationship between writer and reader.

## CULTURAL AND ETHICAL DIMENSIONS OF TECHNICAL COMMUNICATION

All these elements align themselves on two other dimensions of context: culture and ethics. One way to look at culture is to see it as consisting of activities and a vision, "a way of doing things, and a way of thinking and feeling about them" (Bronowski 108). People in the same culture share ways of doing things and ways of thinking and feeling about them. An organization may have a distinct culture. So may a profession or an ethnic group or a nation.

The anthropologist E.T. Hall developed a useful system for understanding the communications implications of culture. He identifies cultures in a

CROSSING CULTURES

## The Electronic Highway

The highway of commerce, research, and community in the twenty-first century is the Internet, which attracts increasing numbers of users every day—so many that keeping track is impossible. The U.S. Department of Commerce estimates that traffic on the Internet doubles every hundred days. Americans have led in use internationally, a condition that once caused a British observer to complain that global usage figures could be broken down into only two categories: the United States and the rest of the world. But the situation is rapidly changing; as of 2000, United States users made up a minority of global Internet users.

In the early 1990s, only 22 countries were connected, but that number had risen to 217 by 1999. For example, the number of users in India was expected to grow 10 times between 2000 and 2004 (Perlez A4). Latin Americans, too, are increasingly wired to the Internet. According to the International Data Corporation, 916,000 Mexicans had Internet accounts in 1999, but the number is projected to increase sevenfold by 2004. Only 4.3 million PCs were in use in Mexico compared with 177.4 million in the United States (Petersen 1).

Internet use in Europe is also growing. In Britain, some four million new users join the Internet population yearly, according to one estimate. A 2000 survey of people in five European countries who said they had been on the Internet at least once in the past month found the following results (Cowell B1):

| Country | Percentage |
|---|---|
| U.S. | 60 |
| Greece | 4 |
| Germany | 14 |
| France | 13 |
| Britain | 22 |
| Sweden | 40 |

Another study found that Germany had more than 12 million Internet users in 1999; the United Kingdom had 14 million (*Global Reach Express,* 8 March 2000).

A limiting factor to further growth is cost: Going online in Greece, for example, is over twice as expensive as in the United States; in Germany, it's almost double. Local phone calls are metered, so browsers must pay both their phone companies and their Internet service providers. In addition, customers face a bewildering array of such providers and of portals for accessing the Web. Europeans also lag behind the United States in the use of personal computers, the traditional connection utility. But Europeans are way ahead in mobile phone and wireless technologies, and those technologies may lead to a surge in Internet usage there—very soon (Cowell B1).

range from "high context" to "low context" (Figure 1.1). In high-context situations, speakers or writers can leave things unsaid or use highly coded language, knowing that they share with the audience the prior knowledge essential for interpreting the message. You and your friends, for example, probably can use shorthand expressions that require only a few words, or just a certain kind of look, to communicate. Low-context situations bring together a more heterogeneous group of people who do not share the same common ground or base of knowledge. This lack of common ground has

**High-Context Cultures**

- Homogeneous, with fairly strong distinctions between inside and outside.
- Group oriented. Individuals identify themselves as members of the group.
- Unwelcoming of deviant behavior. The culture has high expectations for individuals to internalize group norms and behave accordingly.
- People oriented. The focus is on maintaining relationships.
- Nonconfrontational. Business requires soft bargaining and indirection to preserve face.
- Humanistic. Control is internal, preprogrammed in each individual as part of the culture.
- Averse to risk.

**Low-Context Cultures**

- Heterogeneous and generally open to outsiders.
- Oriented to the individual. Individuals seek to fulfill their own goals.
- Welcoming of a variety of behaviors; you "do your own thing."
- Action and solution oriented. The focus is on completing the task.
- Able to separate a conflict from the conflicting parties. Conflict in pursuit of a goal is positive. Business requires hard bargaining with a direct, confrontational attitude.
- Procedural. External rules govern behavior.
- Welcoming of risks.

FIGURE 1.1 **High-Context and Low-Context Cultures**
In terms of their communication strategies, cultures exist on a spectrum whose end points can be identified by some of these characteristics.
(Source: Adapted from Ting-Toomey 71–86.)

many implications for communicators, one of which is that more information needs to be recorded in explicit, detailed messages. Moreover, levels of trust in society also run from high to low. In high-trust societies, people share a vision that encourages collaboration and allows for informal and unwritten contracts. In low-trust societies, people are less able to agree or solve disputes without resorting to written contracts or litigation. Notes one expert, for example,

> Trust can manifest itself on the factory floor. High-trust Japan pioneered the use of just-in-time manufacturing, which places responsibility on factory-line workers. Germany elevates the role of the foreman, who has the flexibility to move workers around. In low-trust France, by contrast, work rules are highly codified; there is a nationally established job-classification system

that grades each worker in the hierarchy and rigidly controls his career progress. (Brooks A7)

### Organizational Culture

The distinction between high and low context can also help you understand the culture of the organizations you work for or belong to. You'll learn about an organization's culture through

- Documents: mission statements, annual reports, statements of shared values, training manuals, as well as day-to-day memos and reports.
- Rules and regulations, whether implicit or explicit. For example, how much control do employees have over their work? How much individual initiative is encouraged?
- Rewards. How are work incentives defined and allocated?
- Stories that circulate around the company about "heroes" and "heroines."
- Any rites or rituals the company observes, for example, "dress-down" days, company picnics, or teams.
- The design of the organization's physical spaces. (Deal and Kennedy)

It might also be helpful to compare old economy workplaces with new economy ones. The new workplace is

- Leaner, with fewer layers of management. People need to work more independently, with computers performing many routine supporting functions.
- More networked, less hierarchical, with an emphasis on sharing rather than withholding information.
- More flexible in its processes. Instead of a static structure of fixed and isolated jobs, it is organized dynamically to meet goals through projects.
- More flexible in its spaces, with people distributed geographically and linked electronically.
- More collaborative, with people moving in and out of teams as goals and projects demand.
- More culturally diverse.

Employees describe an old economy insurance company as "low context": It is "sluggish," "slow to make decisions," "risk averse," "bureaucratic." Its large home office building is traditional in style and filled with many walled-off spaces for individuals or groups of workers. Its cafeteria posts calorie counts next to the costs of each food item to remind workers about the need for a balanced diet. A new economy software company, by contrast, rests higher on the spectrum. It is "scrappy and aggressive," "informal," "adventuresome." It operates in a former warehouse converted into a single huge loft. Phone numbers for local pizza shops are posted next to a vending machine that offers pretzels, candy, and nuts.

Organizations and professionals also develop their own insider language and patterns of communication, their own *discourse,* that marks them as organizations and professions. Performing well within those communities depends in part on learning that discourse. You will learn strategies for designing the discourse of science and technology in this book.

### International Cultures

The discourse of science and technology is, in many ways, an *international* discourse. Similar technical training, interests, organizational roles in multinational companies, and consumer tastes are making professionals around the globe think a lot alike. But strong cultural differences still remain. For example, after a video meeting with colleagues in Japan, which has a high-context culture, one American manager noted, "You have to combine their words with their unique cultural context and then derive a meaning from those variables" (Malone 29F). To be effective, then, you have to look at things from someone else's perspective. Here's one small example: American rock climbers, or, that is, their Chinese interpreter, demonstrated such thinking when Chinese spectators frantically signaled the climbers away from a difficult route on a rock tower they had come to ascend in China. The spectators tried to point the climbers toward the back side, which contained a bamboo ladder system that made the climb easy. The interpreter saved the day by explaining the Americans' action in terms of a Chinese philosophical principle: At a hypothetical fork in the road, someone who wants to be stronger chooses the harder path. The interpreter shouted that the climbers were choosing the harder path, and the crowd immediately understood (Talbert 4).

### Ethics

The interpreter referred to a principle in explaining an action. As a technical communicator, you, too, will make choices based on principles as you work on teams and write for audiences. These principles may derive from your upbringing in your culture. In addition, you may learn these principles as a member of a profession or an employee of an organization. Increasingly, professional associations and other organizations develop statements, called codes of conduct or codes of ethics, which emphasize what a good person in that context *is* or what a good person *does.* In brief, ethical communicators respect other people. They are honest, fair, and trustworthy. They fulfill obligations. They respect the law. They are civil and polite. These attributes are easier to state than to put into practice, however, especially in the murky and complex situations professionals often face. You'll find detailed advice about ethical behavior in technical communication in Chapter 3.

A CLOSER LOOK

## Writing at Work

As a student, you probably recognize the importance of writing effective term papers, lab reports, and examinations, among other documents. You also need to demonstrate good communication skills as you conduct a career search. Your resume and letter of application, for example, represent you to a broad audience of potential employers, and your interview skills help confirm their positive impression of you. Once you enter the workplace—you may be there now while you study, too—you'll find good communication skills essential. Technical professionals spend extensive amounts of time writing and reading on the job, and their reputation depends on spending that time well. A manager offers the following typical breakdown: 20 percent research, 20 percent management, 60 percent writing. A chemist who oversees hazardous materials management for a major California corporation spends about 40 percent of her time in the field observing the work of company employees and the other 60 percent in her office, writing letters, memos, reports, protocols, and procedures.

Sharp skills lead to a good reputation and advancement; weak skills make your good thinking fall flat. Here's how one manager at a large research and development organization sums up the situation:

> The reputation of a research group, and indeed its survival, depend primarily upon the quality of the written materials—reports, proposals, and papers—that it disseminates. Failure of a researcher to write well places a burden on the editorial resources of the organization. Caught between the need to release only high-quality materials and the failure of a researcher to deliver high-quality drafts, the research manager has no alternative but to downgrade the performance rating of the researcher. Low performance ratings ultimately result in denial of advancement or dismissal. (Freudenreich, personal interview)

Researchers, then, along with physicians, engineers, technicians, chemists, lawyers, and other technical professionals, enhance their reputations by writing well.

## DESIGNING FOR THE CONTEXT

To begin to see how ethics and culture interweave with other elements in the context of one information product, let's look at an example. Here, briefly, is how Novo Nordisk, a Danish pharmaceutical company with manufacturing facilities in nine countries, developed, registered, and marketed tiagabine, an epilepsy drug (Hunter 6–8).

First, the project required teamwork. At its beginning, Novo had little experience with central nervous system therapy and no real international expertise, although it had extensive expertise in insulin and diabetes care and industrial enzymes—and it had the core idea for the new chemical entity. The company found a partner in Abbott Laboratories of the United States, which had the international experience and needed new products. The development team consisted of 50 specialists from the two companies, including pharmacologists, physicians, and chemists. Together they developed an important product, along with strong personal relationships and a team spirit.

Second, the story takes place across cultures. Each company had its own culture, that is, its own values, preferences, and way of operating. The team had to negotiate those differences in corporate culture. For example, the different sizes of the two companies led to differences in approach. Novo is much smaller than Abbott, so the Novo team members felt they had to prove themselves before they could be listened to as equals. The companies also had different research practices from which the team had to create its own procedures. In addition, the team had to negotiate differences in national cultures. For example, health care practices and standards differ between the United States and Europe. At that time, European standards were less strict than those imposed by the U.S. Food and Drug Administration (FDA), standards that Abbott insisted on upholding. European physicians resisted some of the requirements of Abbott's "Good Clinical Practice" policy in the test procedures. But keeping standards "consistent, and high, meant that Abbott could use [Novo's] studies for their file in the US, and we could use their studies to file in Europe," notes the project manager. In working with human subjects in clinical tests, then, the research team also had to confront ethical questions concerning the choice of participants, disclosure of information to them, careful treatment, and the protection of privacy in keeping records.

Third, the project emphasized speed. Because it centered on a breakthrough drug, it received priority from management and was fast-tracked. That speed served two purposes. It meant the companies could get exclusive rights to the drug and penetrate the global marketplace before their competitors. Such penetration promises rewarding financial returns. In addition, tiagabine addresses a largely unmet medical need. Physicians in the project were eager to put that option for treatment into practice. Team members worked 60 to 80 hours a week. Such dedication meant that in six years—record time—they were able to go from first use in humans to registration.

Fourth, what kept the whole project together was communication. The team had to talk with each other—through faxes and electronic mail, in the lab, over lunch, in formal meetings. They kept detailed lab notebooks. In addition, the team prepared the lengthy documentation necessary to get approval from their audience of regulatory agencies. The registration file they sent to authorities in Europe filled 79 ring binders and 40,000 pages. Then selling the product required an extensive marketing approach. A first step was an exhibit Novo developed and displayed before 3,000 visitors at an international congress on epilepsy in Sydney, Australia. All Novo activities there were bound by a communications theme, "the art of epilepsy care," with a visual logo (Figure 1.2).

FIGURE 1.2 **The "Art of Epilepsy Care"**
This logo was developed by Novo Nordisk to represent a new epilepsy drug. The logo suggests both a chemical structure and an Aboriginal dream painting.
(Courtesy of Novo Nordisk A/S. Used by permission.)

As the development of tiagabine suggests, writing is an essential ingredient in a professional career. Charles Darwin may have summed up many scientists' lament when he said, "A naturalist's life would be a happy one if he had only to observe and never to write," but Darwin recognized the need to express his ideas so others could read them. You, too, should not be daunted or discouraged by having to commit your thoughts to writing. This text provides you with strategies for communicating effectively as a technical professional—and for enjoying the challenges of such communication.

## CASE
## DISORIENTATION

To learn about methods for delivering technical information in different countries (an assignment in her technical writing class), Maria Alquinna interviewed a fellow student from Pakistan. The interview helped her write a report, but it also made her aware of a problem at the university that she felt was even more important. There was no orientation program for international students. The dean of students told Maria that the regular freshman orientation was open to all students, but many international students were older than the freshmen and had very different concerns.

So Maria proposed that an orientation program be developed. The dean liked her idea and asked Maria to develop a plan for a one-day session, along with suggestions for any handouts or electronic sites that might assist the students in taking full advantage of their time on campus. Maria and three classmates talked with several international students and recorded these notes about topics they'd like to see covered in an orientation:

**Health care.** The university health center, emergencies, methods of payment, wellness programs, prescription and over-the-counter drugs

**Social life.** Public gathering places, liquor laws, tailgate parties and other activities around sports events, fraternities and sororities, local church groups, baby-sitting services

**Banking.** Bank opening hours, types of banks for different services, paying utility and other bills at the bank, electronic banking and automatic teller machines, loans and credit requirements

**Transportation.** Public and private transportation, parking permits and regulations, car registration, driver's licenses, stated and unstated "rules of the road"

As a member of Maria's team, use information pertinent to your own campus to develop a one-page handout or fact sheet on one of these topics (or on another topic you feel should be included).

## CHECKLIST: Writing in Context

1. **Take stock of the communications context before you write and as you write**
2. **Assess your role as a communicator**
   As an individual
   As a team member
   As a member of an organization
   As the representative of an organization
3. **Analyze your audience**
   Answer their request
   Match their level of expertise and interest
   Adjust a new document to the context of other documents on the topic
   Motivate the audience to read
4. **Determine your purpose in writing and your audience's purpose in reading**
   To record
   To inform the audience so they understand
   To persuade the audience so they can decide or act as you intend
5. **Design the message to best deliver its content**
   Incorporate the conventions appropriate to the context
   Imitate the structure and approach of the appropriate genre
   Write or speak in the appropriate language and style
   Choose the best medium for the reader and situation
6. **Accommodate the cultural context for the communication**
   Organizational culture
   National and international culture
   Professional culture
7. **Communicate ethically, with respect for your audience**

## EXERCISES

1. Interview a professional practitioner or a faculty member to find out about the context of writing in your field. You might ask about the importance of writing to career advancement. In addition, find out about the people who read what the professional writes and special techniques used to accommodate those readers. Ask about the design as well. What conventions of presentation does the professional have to follow? What genres does he or she use? What electronic forms are appropriate? You might also ask about the individual's writing routine and the amount of writing done as a member of a team. At your instructor's direction, be prepared to discuss your findings in a brief oral presentation to the class.
2. You'll improve your communication skills if you understand the culture of your profession, of any organization you belong to or work for, and of your nation. Read two articles on the subject of "corporate culture" or "cultural analysis" and write a brief summary for your instructor about your understanding of the role of culture in communication.
3. Write a brief description of yourself as a reader. Take notes on how you read different documents, for example, a newspaper, a textbook, a novel, or a magazine. Take notes, too, on how you read text and images displayed on a screen. Do you, for instance, begin on page 1 with the first word and move from left to right, line by line? Do you skim first to look at the pictures? Do you skip through all the headings (titles of sections within the text) or headlines? Do you use different strategies for different reading tasks? What helps you focus your reading? Describe your reading technique in general or the approach you took to one reading assignment.
4. This chapter has introduced the topic of technical communication. As background for a discussion in class, find one information product you see as an example of technical communication. It may be a print document, a Web site, or some other form. At your instructor's direction, bring a copy or printout to class, or reproduce a segment on an overhead transparency for the class to review.
5. As you begin your course in technical communication, write down the goals you plan to achieve during the term. Consider, for example, the current strengths and weaknesses of your writing. What do you need to do to maintain and enhance the strengths and overcome the weaknesses? Review, too, any goals your instructor specifically sets for your performance. Later in the term, review your performance against this statement of your goals.

## FOR COLLABORATION

In a group of three, prepare a brief report to present orally to the class that develops this general statement:

> Electronic technology has changed radically the way we communicate scientific and technical information.

The chapter has suggested some directions of the change. For example, you might want to discuss the conventions of electronic mail or the culture of global corporations whose growth depends on electronic systems for communication.

# WRITING COLLABORATIVELY

**"WHAT IS SCIENCE ANYWAY? NOTHING MORE THAN AN INTERNATIONAL COMMUNITY OF NOSY PEOPLE LOOKING FOR ANSWERS IN ONE BIG GLOBAL LAB. IF WE ALL LEARN TO WORK TOGETHER, THOSE ANSWERS ARRIVE MUCH MORE QUICKLY."**

**DIRECTOR OF A MAJOR CANADIAN RESEARCH LABORATORY**

The unit of work in science, medicine, engineering, and business in general is, increasingly, the team. Whether you currently view teamwork as a waste of time, a necessary evil, or an effective way to get things done, you are likely to be a member of a team both now as a student and later in your career. In academic and corporate settings worldwide, teams collaborate to conduct research, develop and manufacture products, write a variety of documents, and deliver presentations. Many advertisements for job openings stress that successful candidates will demonstrate an ability to work on teams. In this chapter you will learn guidelines for effective collaboration.

## ADVANTAGES OF TEAMWORK

Why should you learn to work well on a team? Here are some advantages:

- *Division of labor.* A group can divide the work of a project to accomplish tasks that would overtax one person working alone.

- *Speed.* A global team of specialists working with electronic technology can collect data much faster than a solo worker can.
- *Expertise.* Faced with the task of solving a complex problem, a team can assemble specialists in a variety of areas to focus on the issues at hand. Such expertise is especially valuable when projects cross national borders and thus require familiarity with local conditions and the local language. For example, because legal systems differ across the world, multinational organizations often need to work with local experts to prepare such documents as contracts and operating manuals. These documents include symbols, warnings, and other cautionary statements to meet regional government standards and codes and conform to varied liability laws.
- *Check on conditioned reasoning.* Working alone, it's easy to get into a rut, to take the same approaches to problems, and to be so conditioned by past experience that you can't apply fresh insights and reasoning to a new situation. Members of a team, by contrast, can stretch each other's thinking to prevent mistakes and oversights.
- *Synergy.* As they interact, members of a team often generate ideas that none would have thought of alone. Moreover, in explaining an idea to someone else, team members clarify their own thinking.
- *Group learning.* Team members learn from the group as they also achieve the organization's goals. They hone the skills and knowledge they bring to the collaboration, learn new information and techniques from others, and become better collaborators as they work with a team. Teamwork also fosters an organizational identity as employees who have been with the company many years teach the organization's culture and procedures to newcomers.
- *Group ownership.* Teamwork can foster a sense of employee control and empowerment in organizations where individual teams manage themselves to create entire products. The team approach often reduces levels of management and may provide a more democratic, open culture than in traditional hierarchical organizations.
- *Market test.* If the team includes representatives of the intended audience for a product or document, they can provide valuable insights into its marketability and usability. Such an approach is currently unusual, but the practice is growing. Even without such representation, team members can help each other review drafts from the audience's point of view.

## DISADVANTAGES OF TEAMWORK

If you've worked on a team, you may recognize one or more of the advantages we've discussed. Or perhaps you think this list doesn't provide an accurate picture. Not all problems lend themselves to team approaches, and not all teams work. Here are some potential disadvantages of a team approach:

- *Group think.* The team may "impose a deadly uniformity and stifle the special qualities of individuals" ("Management Focus" 61). Such groupthink may be the price a team pays to achieve harmony. Unable to find healthy ways to disagree and air conflicting ideas, the team may settle on a conservative, mainstream idea no one finds objectionable but that lacks innovation. Some members may comply with the group's decision out of fear or dependency rather than from conviction that the decision is right.
- *Withholding of effort.* If members don't feel comfortable with each other or resist the authority of the team leader, they may withdraw from the discussion, content themselves with leaving things to others, and do less than they would if they had tackled the problem alone.
- *Withholding of information.* Similarly, team members may refuse to share their ideas and information with the team. That refusal may reflect a personal or cultural value in not wanting to confront others and speak up publicly, it may reflect their fear of making a mistake and their need to save face (Bosely 57), or it may derive from an unwillingness to disclose proprietary information in, for example, a discussion of a joint venture among otherwise competitive interests. If team members hold their cards too closely, the effort collapses.

## TEAM VALUES

Most of the disadvantages of teamwork result when team members fail to trust themselves, to trust others, and to trust the process of collaboration. For example, such trust is one of several overlapping requirements noted by a pharmaceutical company that sees teamwork as the essential shared value of its operation: "We can only succeed if we all succeed together." The company describes the elements of teamwork as

> Mutual respect
>
> Loyalty—support for each other in all that we do
>
> Diversity—experiential and cultural variety creating a great global company
>
> Trust—open and honest communication builds trust. (DuPont Merck)

Figure 2.1 shows the company's guidelines for encouraging productive teamwork.

In addition to following such guidelines, you also need to take responsibility—for yourself and for the group. Complete your portion of the project well and on time, prepare for meetings and attend them actively, contribute to the discussion, and see that the group's assignment is completed, even if it means assuming extra duties. You'll read more about respecting others in Chapter 3.

To put team values to work, you'll need to compose the right team, meet effectively as a team, structure and monitor the document's collaborative development, and comment appropriately on other people's writing. We discuss each of these responsibilities in the following sections.

CROSSING CULTURES

## Collaborating Internationally

The profusion of cross-border research, investment, and production means that you are likely to participate on an international team during your career. Here are some examples:

- The U.S. National Science Foundation sponsors an electronic research effort, called a "collaboratory," through which some 30 space scientists from California, Maryland, Michigan, and Denmark conduct studies based on data gathered remotely from five instruments in Greenland. The system feeds data and commentary to each researcher's terminal. Each screen has a window that displays the data in various visual forms as well as message boxes for sending and receiving text. A small menu allows researchers to shift to different instruments for gaining data. Such collaborative technology has obvious benefits, including providing an abundance of observations without the need for costly trips to Greenland. In addition, "More eyes on the data increases the chances of a 'Eureka!'event," according to one scientist (Watkins A16).

- To market a product in Japan, a Chicago-based pharmaceutical company needs to comply with multiple Japanese regulations and protocols in business and medicine. Because of physiological differences between Americans and the Japanese and differences in medical practice, the Japanese government requires additional testing of drugs beyond those necessary in the United States. For these reasons, the Chicago-based company has formed a joint venture with a Japanese company. Members from each partner collaborate to design an appropriate product and an appropriate marketing approach.

- In what one researcher calls a "reawakening of amateur science," students are becoming field researchers, collaborating over the Internet to conduct investigations that require decentralized observations. Elementary school students from Texas to Newfoundland, for example, are helping an entomologist to track the migratory patterns of monarch butterflies by catching, tagging, and releasing the butterflies and reporting any tagged butterflies on route. Other students participate in projects to track migrations of songbirds and whales. The sightings can be displayed as an interactive map that can then serve for comparisons year by year. The National Geographic Society has also developed science projects that encourage students to gather information, for example, on acid rain, and communicate and compare that information with other classrooms world wide. The availability of such data as weather maps and the reports of field expeditions on the Internet also provides an opportunity for amateurs to take part in scientific investigations (Bulkeley R16).

- Listen.
- Be open to new ideas and experiences.
- Be ready to learn.
- Respect the others as they respect you.
- Be patient.
- Don't always be assertive or competitive.
- Don't always be right.
- Give everyone special treatment.

FIGURE 2.1 **A Pharmaceutical Company's Guidelines for Productive Teamwork** (Reprinted by permission of the DuPont Pharmaceutical Company.)

## COMPOSING THE TEAM

One of the most important steps in a collaborative effort is composing the right team.

First, because communication becomes more difficult as group size increases, keep teams as small as possible. It's usually easier to exchange ideas on a team of 3 or 4 rather than 10 or 20. You'll also feel more individual responsibility for the outcome and more commitment to the team's goal.

Second, decide if you need a standing team or a special team. Standing teams who serve for limited terms attend to routine, recurring problems. Such teams often set policy in an organization or review actions to see if they comply with the policy. For example, your school may appoint a standing committee of students to advise the food service or to regulate behavior in the dormitories. In a hospital, an ethics review committee routinely responds to cases brought to its attention.

One-of-a-kind problems require the attention of a group assembled specifically to address that particular need. Here are some examples:

- A corporate team that includes two representatives of each company, a consulting financial analyst, and a translator to establish an international joint venture
- A committee selected from the membership of a professional association to promote diversity in the profession
- A team of four civil engineering students to complete a senior year design project

Third, as these examples suggest, determine the kind and level of expertise you need. If the problem crosses disciplinary lines (as in defining "wetland") or national borders (as in an international joint venture), you'll need specialists from several fields. When you are developing a product, for example, it's effective to include marketing specialists and representatives of your potential customers on the team. In addition, if those customers are located in different countries, you might include a translator who can advise you in advance about the taboos and preferences of that culture. As another example, delivering medical care often requires a team of experts in different specialties. A patient who requires knee surgery depends on a coordinated team of health care providers including a primary care physician who evaluates the problem, a surgeon who practices arthroscopic surgery, nurses, and physical therapists.

Finally, name a leader or facilitator who will assume special responsibility for seeing that the team functions properly and carries out its tasks. That person reminds others about the schedule, mediates differences of opinion, draws out members who seem reluctant to contribute, and consults with the client, sponsor, or supervisor. The leader may be

- Appointed by the person requesting the work
- Elected by the team
- Acknowledged as the technical expert by the group
- Known as the gatekeeper for important information
- More committed than others on the team to seeing the project accomplished

## CONDUCTING MEETINGS

One of the team leader's major roles is conducting the meetings that monitor the team's progress and advance the team toward its goals. This section provides guidelines for participating in meetings, either as a leader or as another member of the team.

Meetings help to build trust and monitor responsibility. They also keep everyone's attention focused on the group's work, preventing individuals from drifting too far into their own activities as ends in themselves. As you prepare for a meeting, you consolidate the work that has been accomplished and think about where the project is headed. You may meet with other team members face to face or, increasingly, you'll use electronic technology to communicate easily and rapidly.

### Face-to-Face Meetings

Depending on how many people attend and how familiar they are with each other, meetings may range from informal to formal. Figure 2.2 shows effective guidelines for formal meetings. The writer, head of a large division in an engineering firm, is addressing project teams that report to him. In point 2, he emphasizes the need to establish who is in charge. As a team, you need to decide whose word counts, where final authority rests, and how independent or dependent the team is in its work. Some teams, for example, serve as consultants to one person who is chiefly responsible for a project; others emphasize a more democratic approach.

Memo

Date: 26 July 2001
To: All Members of Project Teams
From: J. E. Rasmussen
Subject: Guidelines for Conducting a Meeting

Most meetings are enlightening and otherwise beneficial, but some take a lot of time and contribute little toward furthering our objectives. To make meetings as effective as possible, please consider the following suggestions.

1. *Purpose.* People like to know the purpose of each meeting. Is the meeting being called to monitor progress on a project, to make a decision, to hear one or more reports, or to brainstorm on an issue? State the purpose when you announce the meeting and again at the beginning of the meeting itself.
2. *Who's in charge?* Normally, the person who calls the meeting is in charge. Sometimes, I find out that I am expected to call the shots or that a vendor is running it. Please let me know in advance if you want me to take charge (or at least start the meeting).
3. *Agenda.* Always provide an agenda of items for discussion, and set a time limit for the meeting. Try to accomplish the important objectives in that time.
4. *Early punch line.* Although sometimes you need to be indirect in meetings with customers, for most internal meetings, begin with the punch line, even if it's bad news. Explain what decision is needed up front. Knowing the expected outcome helps everyone focus on the relevance of the parts of the presentation.
5. *Minutes.* Someone should take minutes to ensure that we get some substance in writing from the time spent. So that the person in charge can concentrate on making the meeting work, someone else should take the minutes. Keep minutes short; summarize the information presented, the decisions made, and any future actions decided on during the meeting.
6. *Advance notice.* Let people know the purpose and agenda as far in advance as possible. This permits them to schedule it and prepare for it.
7. *Attendance.* Pick your attenders carefully. If you don't need everyone on the team at the meeting, then send the agenda only to those who must be there, with copies to the others.
8. *Decisions and next step.* Meet until the appropriate decision is made or information exchanged. The minutes should clearly indicate action items, the responsible persons, schedules, deadlines, and next decision point. If you'll need a follow-up meeting, schedule it at the current meeting.

Without stifling discussion, let's try to maximize our time in meetings and avoid turning ourselves into a debating society. These suggestions should help improve our meetings—but ignore them if they would hinder the productivity of informal get-togethers, and certainly don't schedule a meeting just to show compliance with these guidelines!

**FIGURE 2.2** **A Memo That Provides Guidelines for Conducting a Meeting**

## Electronic Meetings

Increasingly, teams use electronic technology to enhance meeting productivity. For example, some companies use a room specially equipped with networked personal computers to conduct "meetings by keyboard." In such meetings, teams design benefits packages and employee recognition programs, develop ideas for new products and new product names, edit documents, or design objectives for different departments. Participants begin the meeting by keying in their ideas on the topic for perhaps a half hour. Everyone writes at the same time, adhering to strict limits on length. Statements appear anonymously on a large screen in the front of the room. After this initial brainstorming, participants categorize and rank the statements through a system provided by the linking software. Seeing specific text helps participants to avoid the generalizations they may use to hide their disagreements in more direct meetings. At the end of the session, the computer provides a printout of the meeting that serves as a rapid form of minutes. Because such meetings avoid grandstanding and digressions, companies are finding the technique saves both money and time (Bulkeley B1).

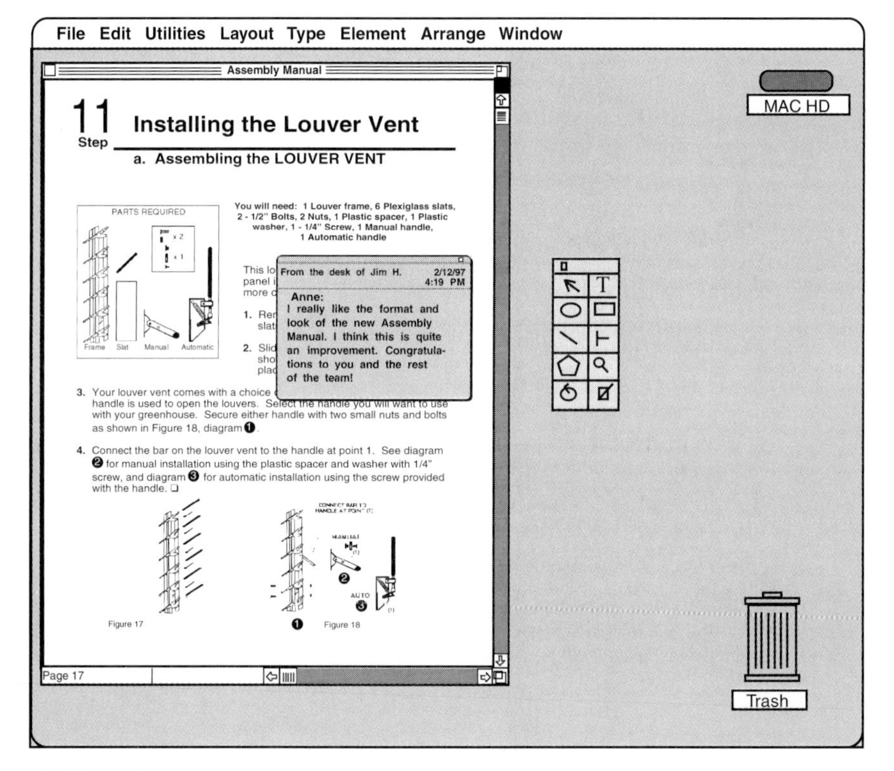


**FIGURE 2.3 Groupware**

This screen shot shows how software can help a team collaborate on a document. Each member of the team can review both text and graphics. The team can all work at the same time at remote locations or at different times, calling up the information when it suits an individual schedule. Each writer can edit the text itself or include a notation in a comment box, as on the screen in this figure.

(Courtesy of Erin Gantt-Harburcak.)

You may also use electronic technology to substitute for face-to-face meetings, especially if distances or time constraints prevent team members from meeting in person. You can circulate and respond to ideas sequentially through e-mail, or you may coordinate the team's work with more sophisticated joint-authoring software or "groupware" that allows the team to

- Circulate documents, both text and graphics, to the entire group simultaneously
- Display the design of several pages at once
- Create comment boxes separate from the text for questions and suggestions (Figure 2.3)
- Store drafts and comments to establish an explicit group memory

Teleconferencing systems that link people at remote locations through telephone or satellite hookups also enhance the opportunity for groups to discuss issues. Full-motion video conferences can be beamed worldwide among specially equipped rooms so participants can see and hear one another as well as share displays on computer screens. Personal computers can also be linked across several geographically remote locations to provide a conference-room-on-a-desk. Small video cameras photograph each participant for display in one window on the screen while data and comments appear in other windows (Figure 2.4).

## DEVELOPING A COLLABORATIVE DOCUMENT

Whether you are working with others in person or electronically, you meet to achieve some goal. When that goal is to produce a document, you'll

FIGURE 2.4 **A Conference Room on a Desk**
An advanced technology system links personal computers so that team members can talk with each other in a video display. Note the small video camera above the terminal on the left.
(Courtesy of PictureTel Corporation.)

ELECTRONIC EDGE

## Online Peer Reviews of Writing

Electronic technology provides new ways for teams to build documents as they review each others' writing across a classroom or across the world. To foster such collaboration, teams use such applications as e-mail attachments, newsgroups, and communication client software like NetMeeting, a popular freeware easily downloaded from Microsoft's Web site.

When the team shares compatible word processing applications, e-mail attachments are a simple and effective way to transfer document drafts. Redlining and comment features allow reviewers to write as much as they want (no problem keeping between the lines and in the margins), highlight the changes they are making, and provide readable comments, especially significant for those whose poor handwriting might otherwise reduce readability. The software's password feature can be activated if the reviewer wants to make sure only selected people see the comments.

Another application, a dedicated newsgroup, provides a bulletin board where team members can post, read, and respond to messages. It is available at the writer's and reader's convenience and allows all members to see all comments. The newsgroup format, however, discourages the posting of lengthy text. That disadvantage is overcome with NetMeeting, which allows writers and reviewers to use a split-screen format to display and edit text during a session. This advantage does come with one disadvantage: the need for reviewer and writer to agree on the time for review because it is a real-time application. In addition, reviewers need to learn how to organize the screen, contribute to a "chat box," and control that box if they are the leader of the review session.

Reviews may appear anonymously or with attribution. If the team is small, anonymity may not be an option. With a large group, reviewers may use pseudonyms to protect their identity. That protection sometimes leads to more honest and productive comments than surface in face-to-face reviews.

For the review to be effective, the team should agree in advance on a checklist of items by which to measure the document. Reviewers then limit themselves to those specific concerns. As with any commentary, reviewers should emphasis the positive and word their comments diplomatically. Electronic technology sometimes encourages speedy and rude responses, a tendency that reviewers have to consciously resist (MacLeod 87–94).

need to agree on your purpose and audience, assign tasks, develop a schedule, create an outline of the document, and choose or develop a style guide.

### Agree on Your Purpose

First, agree on the desired outcome of the work and the audience that your writing must address. How narrow are the limits on your approach? How much latitude do you have? Reduce your assignment to the most concrete terms possible. Listen or read well if you are given an assignment. For example, Pierre Jules Boulanger, chief executive officer of Citroen, the French car manufacturer, gave this vividly concrete assignment to engineers designing the 2CV (introduced in 1948):

> Design a car that can carry two farmers in wooden shoes and 50 kilos of potatoes at a top speed of 60 kph, and travel 100 kilometers on three liters of gas. (*Wall Street Journal* 1)

If your assignment is less explicit, you may have to meet with your group several times to state it in operational terms and iron out your approach. The wisdom of having such agreement was demonstrated recently by a multinational team designing a joint venture in the chemical industry. The two sides had different expectations for the business plan they were writing. The Americans thought they were creating a sketchy "vision" document, long on speculation and short on financial information. However, their North European colleagues thought that even an early document in their negotiations should present detailed financial information. When they couldn't agree on the form of the document, the team recognized even deeper incongruities, which caused the team to abandon the venture.

To avoid working at cross purposes, write a statement of purpose that all can agree on. The following statement directed the activities of a group of engineering students preparing a project report:

> *Purpose:* To inform the Senior Design Review Board about our investigation and to persuade them that the framing system we chose is the best one.
>
> *Problem Statement:* We were given the problem of designing the framing system for a three-story, mixed-occupancy commercial building. The exterior architecture and interior layout are fixed. We need to decide on the best framing method. To do so, we need to design several candidate systems, establish criteria for selection, and then determine which system is best. Once we make our selection, we have to complete the design of the building. Justification for that design, along with drawings and construction specifications, are then to be presented in a final report of not more than 20 pages that must be submitted no later than 16 May.

Expand your purpose statement to develop a preliminary outline of the document as a whole. Agreeing on an outline helps the team negotiate conflicting opinions, evidence, and formats before positions become hardened into prose. In addition, the outline lets a client, supervisor, or sponsor preview the work to see if it will meet requirements and correct the course of the project early on if problems develop.

## Assign Tasks

Once your team has settled on a purpose, you have many options in dividing the work. Two student groups, for example, assigned tasks differently as they prepared similar reports on the feasibility of a new road:

**TEAM A: DIVISION OF RESPONSIBILITY BY SEGMENT OF THE FINAL REPORT**

*Executive summary.* Everyone

*Geotechnical review.* Shahab. Survey the area and existing highways; research codes and compliance; write the site review and the introduction.

*Alternative alignments.* Jan. Develop alternative routes for the road; draw maps and write explanations of the criteria determined by the group.

*Construction schedule and cost projections.* Tony. Calculate schedules and costs for all alternatives; present these in a matrix and develop a more detailed projection for the group-selected alternative. Write the conclusion.

**TEAM B: DIVISION OF RESPONSIBILITY BY TEAM MEMBER SKILLS**

*Graphics.* Steve. Take aerial photographs and draw maps of the routes decided on in meetings; design the final report.

*Government research.* Chris. Review applicable codes and federal, state, and local requirements that impact on road design in the area. Digest state surveys of resident preferences and projected traffic.

*Financial.* Jeff. Calculate costs for constructing the alternative alignments.

*Technical.* Keesha. Survey the site and develop construction plans.

*Primary writer.* Ken. Pull together information from meetings and from team member notes, write a draft, circulate that to the team on e-mail, collect comments, and edit the final work.

Team B relies on one writer to lead the team in creating drafts and editing a final document. In corporations and research organizations, professional technical writers may assume primary responsibility for documents. The writers pull together information from subject matter experts like software developers or chemists.

In general, assign tasks to profit from the strengths and interests of the team members. If your group works most effectively when you collaborate on each task, then arrange many face-to-face meetings or use electronic technology to simulate such participation. Often, however, you'll find group work more productive if each member carries out a separate assignment that represents a particular skill or a different segment of the document. The team then meets to blend these products into a finished document. When dividing tasks, ensure that

- The division of tasks is fair and is perceived as fair by the members
- Everyone understands her or his individual responsibility
- The same task has not been assigned to two different people
- Group members' strengths are encouraged and not thwarted

The clearer the division of responsibility, the more effective the result—and the less energy and goodwill lost along the way.

## Develop a Schedule

To monitor the completion of each task and progress toward the goal, develop a schedule. Figure 2.5 shows the schedule for a collaborative research project. You'll read the final report on the project in Chapter 15. Agreeing on a schedule helps you reach concrete decisions about the scope and tasks in a project.

## Develop a Style Guide

Developing a style guide helps you reach concrete decisions about the form and expression in the final document. Refer to standard sources, like any guidelines published by your academic department, organization, professional publication, or professional association. Even a brief list of stylistic preferences will help save time during the final editing of your project and

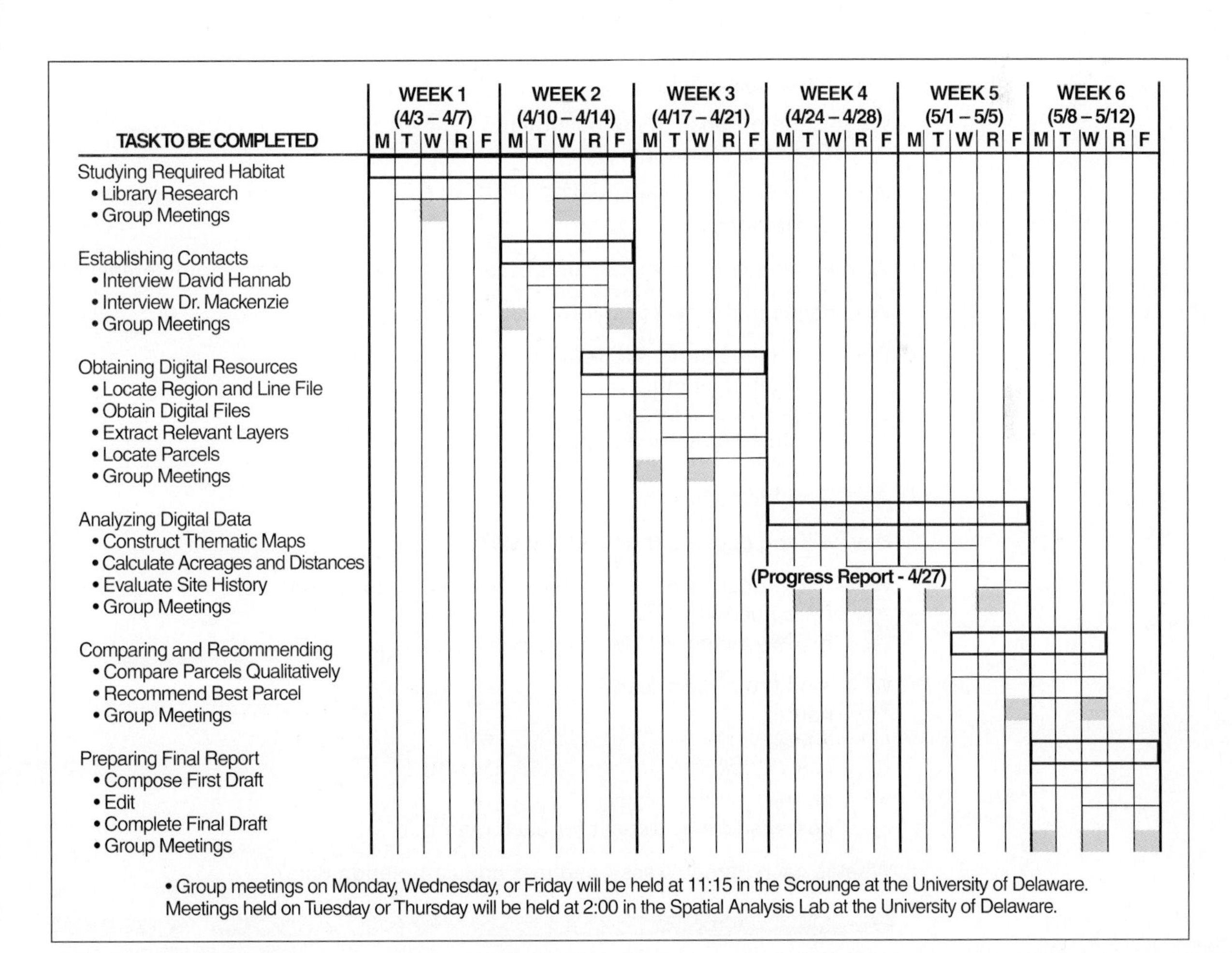


**FIGURE 2.5 Schedule for a Collaborative Student Research Project**
(Courtesy of Julie A. Fine.)

will forestall potential disagreements as team members draft their contributions. Here are some topics you might cover:

- Length of the document as a whole and of segments
- Format for introductory paragraphs to each section
- Method for citing sources
- Number and type of graphics
- Glossary of special terms and their meanings
- Format of lists and headings

Title of document:
Today's date:
Date document is due:

Who is the document for?
  Internal readers:
  External readers:

What is the purpose of the document?
  Specific technical problem:

What will the document cover?
  Scope/topics:

Who needs to review the document for technical accuracy and completeness?

Who needs to review for style?

What will the document look like?
  Estimated length:
  Page size and image area on page:
  Production method (media, binding):
  Typeface:
  Graphics:

How will the document be distributed?

What is the document's timetable? (Attach detailed schedule.)
  Date needed:
  Review schedule:

Who is on the project team?
  name:
  e-mail:
  fax:
  phone:
  postal address: (repeat for each member)

What is each person's assigned task and completion date?

**FIGURE 2.6 Document Worksheet**
(Courtesy of Emily M. Skarzenski.)

Negotiating matters of style is particularly important when you write on a multinational team. Separate that negotiation from argument about the document's content. Don't let a debate about appropriate placement of quotation marks, for example, sap energy and goodwill that should be reserved for issues of policy or scientific understanding. Establish a glossary of major terms and their equivalent in the target language for any document that will be translated. Expectations about page layout, typeface and font, and appropriate use of visuals differ among cultures. So does the technology for exchanging information electronically and for printing. Clarify explicitly the standard you will use for a particular document.

### Record Your Approach

A good control system is important when you're working individually, and it's essential when working collaboratively. The team leader or an appointed scribe should record information about the purpose, audience, tasks, team member logistics (for example, Bob can't meet on Tuesday evenings because he teaches karate), and schedule for circulating drafts among team members. Figure 2.6 shows a fill-in-the-blank form for recording the approach to a group writing project. A schedule and document outline would be attached.

## REVIEWING SOMEONE ELSE'S WRITING

Throughout a team project, you may have to comment on your teammates' writing—a situation that is often emotional and presents many opportunities to offend and to be offended. The following guidelines will help you know what to look for and how to comment effectively and politely about another's writing.

*Praise good work.* Cite places where the writer handled material particularly well. Praise is a strong motivator. In addition, sometimes writers don't really know what they've done well or badly; reinforcing their knowledge of what's good helps them not to change that—and to imitate their good skills elsewhere.

*Determine the extent of commentary desired by the writer.* Talk with the writer about the extent and direction of commentary desired. Determine how invasive the writer wants you to be:

- Proofread only? Look only for simple errors in punctuation and spelling and ignore the rest.
- Restructure? Outline the text as it stands and discuss whether that outline conforms with the writer's intentions.
- Offer suggestions for a particularly sticky section? Confine yourself to that section.

*Assess your authority as a reviewer.* For example, if you are the final authority, your word goes, and others on the team write for you. If everyone

A CLOSER LOOK

## The Landscape of Collaboration

Architects and manufacturers of office furniture are changing the landscape of research facilities and offices to support teamwork. For example, in the 1970s a researcher typically worked in one of many identical laboratory modules arranged along a corridor, each module containing a small enclosed office for the senior researcher and a bench where she or he and a few technicians performed experiments. New buildings, however, are designed specifically for collaboration. Rather than being isolated in their own modules, researchers are brought together, both formally and informally, in joint offices from which they operate remote instruments through computer links; in formal spaces for conferences, lectures, and training programs; and in alcoves with blackboards, bulletin boards, coffeepots, and mailboxes strategically placed near work areas to draw researchers into conversation (Solomon 123–24).

Office furniture, too, is being designed with flexibility and portability in mind (Bigness B1–2). Teams need flexible arrangements and portable equipment as they assemble and disband depending on project goals. Companies also save money if they can house more people in less space. Increasing numbers of organizations are maintaining wired warehouses to which "telecommuters" or "teleworkers" and field-based personnel can come a few times a month. They take over a desk the way they would rent a car in a practice now called "hot desking." One essential device of hot desking is an electronically lockable storage cabinet on wheels, dubbed a "puppy" in the United States, because it follows you around an office as you work on different teams (the British term is a "bobby trolley"). One designer predicts that offices of the future will need a mix of "'commons,' or shared work areas, and 'caves,' or private offices" (Bigness B2).

Changes in design have led to changes in terminology, as in the concept of "hot desking": a desk is a "workplane"; a storage cabinet is a "stowaway"; "perimeters" are movable screens that temporarily mark boundaries. One furniture system features private workstations "arranged around a common area—sort of like tents around a campfire—that give workers access to one another" (Bigness B2). All the furniture offers easy adjustment so that people of different sizes and shapes may use it.

on the team is a peer, then your word is equal in weight to the others. If you are the team manager, you can add weight to certain comments. You also represent the client or sponsor of the work and measure the writer's work against your perception of the reader's goals. If you are the copyeditor, your task is to measure the text against preset and primarily objective standards of format, mechanics, usage, and style. Your role determines the weight of your comments and the amount of discretion the writer has in accepting them. Note that none of these roles is "English teacher." You're not grading the writer; you're helping to ensure the quality of the final document.

*Take your role seriously.* Take other writers seriously. Be sensitive to the writer's feelings. Such sensitivity is particularly important when you deal with people who might be embarrassed about their writing skills, who are writing in their second (or third or fourth) language, or who may have been brought up to avoid open criticism and confrontation.

*Avoid making changes just to feel you've left your mark.* Modify the writer's work only to prevent real damage. You want to interfere as little as possible.

*Control the impulse to rewrite.* Commenting and editing are repair work, not construction. If you rewrite, you risk misunderstanding the writer's point and changing the substance. You may also inadvertently introduce a new style or tone that clashes with the original.

*Be specific.* Start with the big picture and work toward a word-by-word analysis (unless you are a copyeditor whose major interest lies in the word-by-word analysis). Write a summary of your criticisms for a long report. Use consistent notations.

*Don't nitpick.* The writer has only so much energy for revision. Don't waste that energy by focusing on a few minor problems in comma usage on page 1 when, for example, a lengthy discussion on page 12 should appear as a table rather than text. Just as hospital emergency rooms sequence patients for care based on the acuteness of their condition, practice a kind of triage on the writing you're reviewing. Distinguish between preferences and requirements. You may think "utilize" sounds better than "use," but generally the words are interchangeable. Avoid changing the writer's style merely because it isn't yours. Achieving a coherent voice in a document often necessitates the team's agreement on certain matters of vocabulary. That's the role of a style guide. Use the guide—not just your own taste.

Especially on an international team, creating an appropriate voice may be difficult. An American, for example, may see as praise the characterization of her prose as "hard hitting"; someone from a less confrontational culture might find that a negative. Vague or indirect language may represent the writer's polite stance toward the reader, but it may also confuse the reader. Consider the following from a British document that an American reader found frustrating:

> We are still making progress with the matter you looked at last year, but sadly we seem to be some distance from final resolution. We remain very grateful for your own help in this matter.

Establish a standard that lets you sort essential content in the document from negotiable and less critical matters of style. Documents are infinitely revisable, so fatigue alone should not determine when to call the latest draft the final one. Instead, review the assignment to make sure you have completed everything you were asked to do.

## THE PLEASURES AND PITFALLS OF COLLABORATION

Creating teams to develop new products or services has become a popular management practice. A Canadian lab director says that in his office "the majority of work gets done almost exclusively in a group setting" (Dennis and Brown). A psychobiologist at a federal research lab notes, "Group work seems to be the fastest, most efficient, and most accurate (thanks to the checks and balances system inherent in the team structure) way of doing things" (Dennis and Brown). Reflecting the advantages of teamwork, some companies may be run entirely by teams who report directly to the chief executive or the board of directors.

But such an approach can fall into difficulties. According to one team management expert,

> It's not just a question of people feeling good about each other and the company. A completely diverse group must agree on a goal, put the notion of individual accountability aside and figure out how to work with each other. Most of all, they must learn that if the team fails it's everyone's fault. (Stern F5)

To pursue a technical career, you'll need to develop a talent for working with other people and assuming responsibility for the group's actions.

## CASE
### SHOWING OFF

A wealthy alum gave the Life and Health Sciences Department at your university money to purchase two large display cases for the department's foyer. The cases measure $10' \times 4' \times 3'$ deep and have a glass top that can be locked with a clasp. They are quite handsome—and were embarrassingly empty for a year until the chair put some faculty publications in them, just to fill the void. However, the cases still look rather bleak and unattractive.

No one seemed concerned until about a week ago, when the university's president called the department chair to find out the status of the cases. His phone call was prompted by a call from the alum, who was eager to visit campus and see how her initial gift had helped "make the foyer a teaching space that would inform and entertain students while they waited between classes and would showcase the department's research" (the alum's own words). The president, well aware that the initial gift would become a final gift if the alum wasn't satisfied, told the department chair to "spruce up those cases." The department chair turned to you, in your role as president of the life science's honorary society, to solve the problem.

You decide that a "total foyer" effort is needed, including posters on the wall that will accompany whatever you put in the cases. In addition, you want more than paper in them; you need artifacts that will represent the life and health sciences to students and visitors. Those artifacts should support a theme, for example, "Exercising Your Way to a Healthy Life." The alum has broad interests, so you really can't go wrong with any topic related to the quality of life. The artifacts you choose to illustrate that theme should not be expensive, because the cases, although they lock, are not vaults. On a team of three or four, solve the cases problem. If you have access to a museum or other site that displays objects, visit it to inspire your own thinking. Then compile a team-written memo to your chair that states your theme, lists the artifacts you will include, provides the text of a label for each item, and sketches their arrangement in the cases. Justify your approach.

## CHECKLIST: Writing Collaboratively

**1. Select the right team members**

Gather the smallest number of people able to solve the problem

Determine whether a standing team or a special team is appropriate

Determine if members of the team should represent one specialty or different interests and skills

Agree on a leader

**2. Use meetings effectively to advance the team's work**

Assume responsibility for yourself as an individual and for the team as a whole

Trust yourself, others, and the collaborative process

Monitor progress toward goals

Use electronic technology to substitute for or enhance face-to-face meetings

**3. Structure and monitor the document's collaborative development**

Agree on your purpose and audience

Assign tasks to reflect individual strengths and skills and mitigate weaknesses

Develop a schedule

Develop a style guide

Document the approach

**4. Comment appropriately on other people's writing**

Praise good work and be polite

Determine the extent of commentary desired by the writer

Assess your authority as a reviewer

Control the impulse to rewrite

## EXERCISES

1. Write a brief description of a group to which you belong (a sports team, club, professional society, or class project team). In your description, note, for example, how the group is organized, how a leader is chosen, how members communicate with one another through formal and informal channels, how meetings are conducted. Comment, too, on the "culture" of the group in the terms you learned in Chapter 1.

2. This chapter said that the unit of work in science and technology is the team. To test that generalization, talk with a professional in your field about her or his work on project teams. Does the professional work with others, too, in writing documents? If so, what is the nature of the collaboration? Present a brief oral report to the class about your findings.

3. Several software programs aim to enhance the ability of writers in organizations to share documents and to write together. Review the literature on one or two such programs (you'll find information at your local computer store

as well as through the Internet). Then write a brief report describing how the software aids collaborative writing.

4. If you are working on a team project, develop strategies for monitoring and evaluating the work of the team in addition to your strategies for pursuing that work. At your instructor's request, keep an individual log of group meetings. Prepare a form that lists each person's tasks, their performance of that task, the quality of their results, and your own comments on the work. The instructor will then collect these logs at the end of the project to provide further evidence for the team's final grade.

5. Ann Thompkins was a summer intern at a consulting firm specializing in environmental assessments. Her position consisted of "shadowing" Janna Bennke, a microbiologist and member of a six-person interdisciplinary team investigating the site for a major new housing development. Bennke was very busy that summer and happy to turn over much of her work to Thompkins, who, in turn, liked to be busy and found the work exciting. So Thompkins did the extensive field analysis assigned to Bennke, represented Bennke at the one face-to-face meeting held by the group during the summer, and corresponded on e-mail every day with other members of the group, using Bennke's account, with Bennke's blessing. Thompkins rarely saw Bennke all summer, except for 5-minute meetings here and there to catch up on the work. Practically speaking, Thompkins worked on her own, answered queries from the team, and raised some of her own—all in e-mail that looked as if it came to (and from) Bennke. At Bennke's request, Thompkins also wrote (and sent via e-mail) the appropriate segment of the final report. She wanted to please Bennke, who could help her find additional employment and who had, after all, taken her on. But Thompkins also felt a bit betrayed, because her work would not be recognized as her own. The credit would go to Bennke. If you were Thompkins, would you raise these issues with Bennke? With other members of the team? With someone else in the organization? Discuss these issues of collaborative responsibility in class.

## FOR COLLABORATION

This chapter suggests that a team approach works well when you need to generate ideas and solve problems. As an exercise in deriving the benefits of teamwork, assume that a health official in the United Kingdom has asked your technical communication class to come up with a strategy for keeping British citizens from "burning themselves up," as he puts it. Residents of an often rainy country, English people often lie out in the sun with great abandon whenever it shines. They do little to protect themselves from sunburn, and because many are fair skinned, that sunburn comes rapidly. Cumulatively, those burns lead to an increased risk of skin cancer among the population. In addition, many people, especially young people, watch American television programs and movies that make tan skin a prerequisite for beauty. The health official has asked your class to come up with a strategy for warning the English about the dangers of sunburn.

Your class agrees to take on the project. For this exercise, determine the approach you would use to develop a public health campaign. Generate ideas about what information you would need to include in any warning notices along with the appropriate media for distributing the warning. You don't need to develop the warnings themselves, but you do need to determine the questions you'll have to answer.

1. Divide the class into teams. Decide, as a group, the basis for that division. Will you simply count off three or so people as you sit around the room? Should the instructor assign the teams? Should you group people by major? By skills? By interest?
2. Once the teams are formed, then each team should compose a written plan for solving the health official's problem. Include in the plan
   a. Statement of purpose
   b. Brief statement of the problem
   c. Tasks you will need to perform to achieve the purpose
   d. Team member assignments
   e. Schedule and technique for monitoring progress
3. Submit the written plan to your instructor
4. Discuss these plans orally in class and note differences among the approaches suggested by each team.

# WRITING ETHICALLY



"**TAKE CARE OF VALUES. THE REST IS SHOPPING.**"

**JOHN ASHBERY**

- In a recorded morning telephone message, a ski area spokesperson noted that the day would be sunny, all 15 lifts were "slated for operation," and conditions were "excellent." The message did not mention the wind: steady at 20 to 25 miles per hour with gusts to 50 at the time the message was composed and predicted to continue all day. In fact, none of the upper mountain lifts (11 of the 15) ran that day. Available terrain was severely wind packed, and even seasoned skiers had trouble negotiating it. It was sunny.
- As classmates in a physical education class, Jenny, Raoul, and Paul collaborated on an assigned term project to develop an exercise program for an assisted-living facility. Paul never accompanied the team on visits to the site ("I can't take old people," he said). He did gather brochures about equipment from the gym where he works and gave these to the team. He didn't write his agreed-on section of the report. Jenny and Raoul wrote the entire report and included Paul's name as a coauthor on the title sheet.
- Three Americans lived at Holburn House in London, England, a building with ten flats (apartments) occupied mainly by British, Indian, and Pakistani students. The Americans were shocked at what they saw as their neighbors' flagrant disregard for recycling. The neighbors simply tossed newspapers, cans, and bottles in the litter bins (trash containers). The Americans pulled these items out of the trash and brought them to nearby recycling bins. Later, they posted a

notice next to the lift (elevator) on each floor. The notice began, "If you don't recycle, you are what you throw away: Trash."

Each of these three brief scenarios raises an issue in *ethics,* the study of how to behave well toward other people. In this chapter, you will learn some guidelines for making ethical choices as you communicate scientific and technical information.

## TAKING CARE OF VALUES

The central value or principle in ethical communication is easy to state: Respect other people. That means you pay attention to their individuality. In doing so, you learn how they are similar to you and different from you, and you engage in dialogue with them. You think of their needs and their points of view. As environmentalist Paul Petzoldt put it, "You've got to think about your effects on the people downstream" (Daniel 10). The Quakers define ethical behavior in a similarly simple guideline: In all your dealings, leave other people at least as well off as you found them. That is like the physician's code: Do no harm. And as you avoid doing harm, be polite and civil.

Legal standards set the baseline for good behavior. In the low-context culture of the United States, detailed and comprehensive laws regulate many of the ways that people should relate to one another. In other settings, you may find fewer written guidelines and more need to understand the basic values of the culture. From these values, you determine expected patterns of good behavior.

For example, in the United States and the European Union, among other countries, legal guidelines establish liability, that is, whether advertising is deceptive and whether misrepresentations about a product or a process cause harm. Manufacturers of industrial and consumer products keep a close eye on liability laws when they write manuals and other product documentation. So do producers of software, who must ensure that they don't promise something their product can't deliver. The Product Liability Directive of the European Union places special emphasis on documentation and holds manufacturers liable for defects caused when instructions do not meet minimum requirements. Explicit guidelines also define libel, false statements that damage someone's reputation or career.

In addition, well established guidelines define "intellectual property," that is, fair use of someone else's ideas or language. In a report or article, citing sources justifies your stance, indicates what's new in your thinking and how your investigation departs from what others have done, and shows that you know the work of people the audience admires—and that you are thus in good scientific company. Being part of this community means that you acknowledge the originator of the information. Acknowledging print sources follows a rather clear routine, but guidelines for behavior in an electronic environment continue to evolve. Information in that environment is especially volatile. Texts are easily reproduced and forwarded; messages pile up in new documents created by the reader that may mask original source identities and thus defy easy assignment of ownership. As an ethical user of information technology, you need to respect the integrity of authors and clearly delineate your own from others' work.

Within a company, you may be asked to sign a "nondisclosure" agreement. Keeping information private is often essential for companies to take full advantage of their efforts. A chemical company, for example, whose costly research led to the development of a new material for surgical implants, may be reluctant to release data that would let competitors replicate and thus profit from the work. In addition, securities laws define "insider information" not in the public domain that would, if made public, seriously affect the price of a financial product (like a stock or bond) or contract.

Although legal standards carve out a good deal of territory, they do not cover all situations, and they often engender questions of interpretation and application. In the complex and murky world of professional life, you'll need to rely on basic values to pass daily tests in good behavior. Let's look at some of those tests.

## HONEST MESSAGES

One test is being honest: When you respect other people, you are honest with them. Was the ski area spokesperson honest? Did he deliberately suppress information about the wind? Or did he think sunshine was the critical factor for "excellent" skiing? Strictly speaking, he did not lie. But a caller might still feel the report misrepresented the situation. In the United States, someone who calls for a report on ski area conditions probably expects to hear the most positive news. In Europe, skiers are likely to find that resorts describe themselves in more measurable terms: snow depths at mid-mountain, temperature range, cloud and wind conditions. Marketing statements in general, like product brochures and proposals, tend to be more upbeat than, for example, technical reports and articles.

One way to determine what's honest is to look at what is dishonest or fraudulent. In 1830 an English mathematician defined fraud in three categories that are still pertinent: forging, trimming, and cooking (Hunt 42). *Forging* is completely making up data and thus easily considered dishonest. *Trimming* is manipulating data to make them look better, "clipping off little bits here and there from those observations that differ most in excess from the mean and sticking them on to those which are too small" (Hunt 46–47). *Cooking* is deliberately selecting only those data that fit your hypothesis and throwing away the rest. What matters, in the end, is your intention. If you intend to deceive, you are dishonest.

### Exaggeration and Ambiguity

Short of making things up, you may be tempted to trim and cook through exaggeration or ambiguity. Resumes, for example, may stretch the truth, describing a filing job as "managing the office information system." Here are some other examples of misrepresentation:

- Television commercials for drugs that relieve the symptoms of a cold are not allowed to say the drugs cure the user but only that they provide symptomatic relief. The visuals, however, often tell a different story; the former sufferer looks cured.
- An advertisement for a pet food notes that the product "has been through the USDA inspection process." The implication is that it passed, although a wary customer should wonder.
- Product literature sometimes exaggerates the effectiveness or ease of use of the item in ways that lead the consumer to choose it. The consumer, however, may find the promise empty. Many insect repellents, easy weight-loss programs, how-to guidebooks, and get-rich-quick schemes fit into this category.
- Proposals that seek funding for research or a contract to supply a product or service sometimes exaggerate the proposer's record of previous work, overstate the benefits of the proposed solution, embellish the credentials of the team, or manipulate the cost and time figures.
- Progress reports may paint a rosier picture of a project than the actual events warrant. You're six weeks behind your anticipated schedule, for example, but gloss that over by saying you're "virtually on target." Such misinformation hinders the reader's ability to apply proper corrective action—like bringing in someone else to do the work.

Using abstract language to hide something negative goes back a long way; for example, in the first century A.D., followers of Erik the Red, an outlaw from Iceland, named a windy island snowfield "Greenland" in order to attract others to join them. Today, a nuclear engineer might use the phrase "energetic disassembly" to downplay the force of an explosion or the description "rapid oxidation" to hide a fire. A plane crash is no less deadly when described by the official phrase "controlled flight into terrain." Such usage has given a negative connotation to the term *jargon,* which originally referred in a neutral way to any specialized language. Jargon has come to mean lengthy sentences, highly abstract terms, and automatic expressions. Although it is often harmless, it can be deceitful, disguising the author's lack of understanding and lack of results and misleading the reader.

The Lexicon of Inconspicuously Ambiguous Recommendations (LIAR) (Figure 3.1) shows how ambiguous phrases in a letter of recommendation can seem to praise a job candidate that the writer thinks is unqualified. Test your language to make sure you represent your information honestly and don't trim or cook your material in a way that intentionally misleads the reader.

### Misleading Technical Terms

Although technical terms are essential in technical communication (of course), such terms, appropriate for people who share your understanding, may violate the trust of other readers. Accurate, they may still mislead. The American Chemical Society notes that scientists may have to use "common words of lesser precision" than scientific terminology "to increase public

- To describe a [job] candidate who is woefully inept: "I most enthusiastically recommend this candidate with no qualifications whatsoever."
- To describe a person who is not particularly industrious: "In my opinion you will be very lucky to get this person to work for you."
- For a colleague who has had trouble getting along with others: "I am pleased to say that he is a former colleague of mine."
- For the person who is not even worth serious consideration:"I would urge you to waste no time making this person an offer of employment."
- For the coworker who is so unproductive that the position would be better left unfilled: "I can assure you that no person would be better for this job." (Thornton 60–62)

**FIGURE 3.1 Using Ambiguous Phrases to Deceive: LIAR (The Lexicon of Inconspicuously Ambiguous Recommendations)**

From "How to Write a Difficult Letter of Recommendation" by Robert Thornton, which appeared in *Planning for Higher Education,* Volume 23, no. 1, pp. 60–62. Reprinted with permission of the Society for College and University Planning (www.scup.org).

comprehension" ("Ethical Guidelines" 222). An expert sums up the problem neatly: "The would-be popularizer is always confronted by the dilemma of incomprehensible accuracy or comprehensible inaccuracy and the fun of [the] work lies mainly in the solution of that problem" (Slossen 819). One example of "incomprehensible accuracy" is the legal forms people encounter while conducting their everyday life. As contracts, forms such as rental agreements, warranties, and insurance statements should adhere to the common law guideline that these documents are agreements between parties who can understand the terms. But many contracts use language that disserves the less technical reader, as in Figure 3.2. Recently, both the U.S. federal government and many states have developed guidelines to encourage the use of "plain language" in such contracts. These guidelines aim to restore good faith and meet an ethical requirement on the part of the provider. See *A Closer Look: Plain English* for more about this topic.

## Misleading Visuals

Technical communicators write as much in visuals as in words; and like words, visuals can mislead, perhaps even more easily. Electronic technology only increases the opportunities for deception. Guidelines for ethical conduct in writing are fairly abundant, but such guidelines are less clear for visuals, especially for computer-enhanced images.

Some people use visuals to hide information. Architecture students, for example, pressed for time at the end of a project, may draw trees to cover sections of a facade they haven't yet developed. A company official compelled to deliver unfavorable statistics about an oil spill may display the

Either Landlord or Tenant may terminate this rental agreement at the end of said term by giving to the other party written notice thereof at least sixty (60) days prior thereto, but in default of such notice, this rental agreement shall continue upon the same terms and conditions in force immediately prior to the expiration of the term hereof for a further period of one (1) month, and so on from month to month unless or until terminated by either party giving the other sixty (60) days written notice of termination previous to expiration of the then current term; provided, however, that if Landlord shall give written notice to Tenant prior to the expiration of any term hereby created, of its intention to renew the terms and conditions of this Rental Agreement, and Tenant shall not within thirty (30) days from such notice notify Landlord of Tenant's intention to vacate the rental unit at the end of the then current term if Tenant shall pay to Landlord the rental payment required by said renewal notice, Tenant shall be considered as Tenant under the terms and conditions mentioned in such renewal notice for a further term as provided, or for such further term as may be stated in such notice. However, under no circumstances shall the term of this lease be extended beyond a term of three (3) years and six (6) months more than the initial term.

**FIGURE 3.2** **Misuse of Technical Jargon: A Two-Sentence Segment from a Rental Agreement**

numbers in small type on a slide that shows a color photograph of a sunlit bay filled with ducks. Such practices are deliberately dishonest.

Less deliberate distortions still occur if you fail to account for inherent problems in how images represent reality. A specialist on maps notes, in a statement that applies to all visuals,

> A good map tells a multitude of little white lies; it suppresses truth to help the user see what needs to be seen. Reality is three-dimensional, rich in detail, and far too factual to allow a complete yet uncluttered two-dimensional graphic scale model. Indeed, a map that did not generalize would be useless. But the value of a map depends on how well its generalized geometry and generalized content reflect a chosen aspect of reality. (Monmonier 25)

Technical professionals have to ensure that any suppression of truth does not mislead the reader.

### Lack of Disclosure

As a technical communicator, recognize and disclose any limits on your information, discussing problems as well as successes. The pressure to trim and cook is often strong, with laboratory notebooks revealing much bumpier procedures than those that find their way into published reports. Some selectivity is natural, but your frank disclosure of mistakes and wrong turns could help other researchers avoid them or assess their significance. Don't be afraid to say, "We were unable to . . ." or "We had to reduce the number of samples because . . ." Never be afraid to admit when you're wrong.

A CLOSER LOOK

## Plain English

To benefit fully from the protections offered by our federal securities laws, investors need to read and understand disclosure documents. Because many investors are neither lawyers, accountants, nor investment bankers, we need to start writing disclosure documents in a language investors can understand: plain English.

Whether you work at a company, a law firm, or the U.S. Securities and Exchange Commission, the shift to plain English requires a new style of thinking and writing. We must question whether the documents we are used to writing highlight the important information investors need to make informed decisions. The legalese and jargon of the past must give way to everyday words that communicate complex information clearly.

Thus the chair of the U.S. Securities and Exchange Commission introduces a set of guidelines for mutual companies who must conform to plain English standards in writing their prospectuses. The requirement reflects a growing movement to use language customers and citizens can understand. Government agencies as well as private individuals and groups support the movement. Companies, too, are beginning to sign on. The movement has waxed and waned over the last 25 years, but the information explosion that has accompanied expanded use of the Internet has recently provided even more incentive.

In the U.S., government officials hope that the use of such language will make their pronouncements more trustworthy. Marketing professionals hope the language will help their messages reach customers amid all the noise on the Web and in direct-mail ads. Companies hope plain English will increase workplace efficiency as employees understand corporate expectations better. In a deregulated marketplace, former monopolies, like telephone companies, need to communicate better with customers—or face losing them (Leonhardt C1).

In the U.K., too, government agencies are requiring such organizations as mortgage lenders to write in language that is "fair, clear and not misleading."

In both the U.S. and the U.K., advocates for plain English point to the financial as well as the ethical benefits of the approach. The U.S. Veterans Benefits Administration expects to save about $500,000 a year because employees trained in the techniques were able to write more effective letters to clients; the response rate on one letter rose from 35 percent to 62 percent. Chrissie Maher, who founded a Plain English campaign in the U.K. in 1979, claims, among other victories, that her campaign prompted the British government to rewrite and redesign over 58,000 official forms in the early 1980s, a process that resulted in a savings of £15 million ("Plain English Campaign").

## RESPONSIBILITY IN TEAM REPORTING

Another test of your ethical values comes as you work collaboratively to prepare a report. You have to play fair, trusting each other and correcting situations in which that trust breaks down. In the opening scenario, Jenny and Raoul included Paul as an author because he had been assigned to their team, they liked him, and they didn't want him to get into the trouble that would follow if they left his name off the report. He would fail the course. But in doing so, they felt conflicted: They were misrepresenting themselves as they were fulfilling their obligation to the team. They took responsibility for getting the work done, even though that meant extra work for both of them. In the classroom, instructors often aid students in such decisions by requiring team members to sign statements describing their own individual contribution to a collaborative report. But some instructors insist that student teams work out such problems on their own. In professional life, you'll have to work out the problems on your own. How? By talking about the team *process* as well as the team *product,* taking time, perhaps, from forward motion on solving the team's assigned problem to devote to solving the team's relationships.

## CONFLICTING VALUES IN INTERNATIONAL SETTINGS

The third scenario, about recycling in London, suggests another—and much bigger—test: carrying your ethical values abroad. The Americans saw their neighbors as "untidy disturbances," violating a principle of environmental stewardship. For the Americans, too, recycling at home was a legal mandate. Their hometowns impose stiff fines on residents who don't sort their trash and recycle. The presence of recycling bins two blocks away from their flat made them think that London ordinances might be the same. But more than a matter of mere compliance with the law, the students were offended by their neighbors' neglect of a shared responsibility for making life on the planet more tolerable. Angered and frustrated, they wrote their note. Many neighbors, however, came from places where no one even thought about sorting trash and no facilities existed to handle recycled materials. To them, good personal relations were more important than the fate of a few cans or bottles. Being called "trash" only confirmed their belief that Americans are arrogant, belligerent, and rude.

In leaving one culture to work in another, you're likely to find many situations where values conflict. Those conflicts often center on your role as an agent in the transfer of technology (Morgan). You'll find that the technology itself embeds the values of the originating society and affects the standards and roles of people in the new society (see *Crossing Cultures: Ethical Issues in Technology Transfer*). To conduct yourself ethically abroad, assess the impact of a product or process on the new culture. Observe well and listen before you take a stand. The Americans of Holburn House would have been wise to ask *why* their neighbors didn't recycle, discussed its importance with them, and perhaps established a group effort to achieve that goal if the neighbors agreed to its significance—or found a way to make themselves responsible for at least some of their neighbors' recyclables in the flats.

Instead, the Americans tried to impose their values on others. They could also have decided to join their neighbors in not recycling, especially given the difficulties of carrying all their materials on foot to the recycling

CROSSING CULTURES

## Ethical Issues in Technology Transfer

When you bring a technical process or product developed in one culture into a new culture, you raise a host of ethical issues. The following questions suggest some of these (Bernhart 140–41).

• *Does it distort social patterns?* Agricultural experts, for example, often point out the efficiency and increased yields of planting seeds in rows. In some parts of Africa and Asia, however, planting is a social process conducted by women in conversational clusters. They may be reluctant to change to a new technique that isolates individuals. Similarly, conflicts arise when a new technology increases noise levels and reduces opportunities for conversation where that has been a part of work.

• *Does it upset or distort traditional differences in status or power?* A new technical system may upset the balance between supervisors and workers. It may either reduce or enhance a supervisor's authority and perquisites in ways that cause frustration. A factory manager, for example, may value autonomy in decision making and hiring and resent the imposition of companywide standards by a new foreign owner.

• *Does it change the amount of control a worker has over the pace and conduct of work?* New technologies often require changes in attitudes toward work. For example, people unused to automated processes often fail to understand their role as monitors. One illustrative story concerns a man hired to watch the warning system of a generator. If the light came on, he was to shut down the system. He was used to working continuously with a machine. When no light came on after several days, he figured it was working fine, and just napped—with harmful results (Bernhart 140). Technologies that require workers to change their work habits—for example, to skip a traditional long lunch or meet guidelines for punctuality—raise core issues of values.

• *Does it invalidate the society's reward system?* A new technology may change the rewards people receive for their work. Workers may balk, for example, at a shift from piecework to salary. Increased automation may reduce their satisfaction in seeing an end product. Some intrinsic rewards available under the old system, like a slower pace or less supervision, may be absent under the new.

• *Does it require other technologies or skills?* Introducing one technology to a country often brings with it the need for others. For example, if you sell cars in China, those cars will create a demand for a vast increase in transportation infrastructure, more fuel, service stations, and the like. They will also intensify pollution and raise environmental issues.

bins, located two blocks away. At home, recyclables were picked up at the curb in front of their houses; if they missed a pickup, they could always carry them in a car. But they had no car in London, and they could use the difficulties of recycling, and the views of their neighbors, as reasons to justify their conduct as ethical.

When you work abroad, you may find yourself negotiating through conflicts in values. Here are a few such situations in which values conflict:

- In many countries it is perfectly legal to specify the required gender, age, even physical appearance ("an attractive, 20–25-year-old female") in hiring notices—a practice that violates U.S. law. North American companies operating in those countries can adopt that practice too. But if they do, any statements they make in North America about commitments to avoiding sexism in hiring may be seen as shallow.
- You may need to redefine your notion of "bribery" to exclude the under-the-counter payments that are common in some countries to get cargo off a ship or to secure licenses from a bureaucrat. One foreign businessman in China provides this "calculus of corruption": "To get a small contract, say $20,000, I figure it's a couple dinners out and bowling. For a $250,000 contract, it's a 'familiarization' trip for the official to the States. A contract of a million and above? Then we're talking about helping get a visa for the guy's son" ("Cracking the China Market" R1).
- Advertising practices vary in different countries. In Japan, directly criticizing a competitor or overtly soliciting a rival's clients or customers is considered unethical. In Germany and France, where memories of World War II are vivid, U.S. chemical companies often follow local precedent and avoid mentioning defense industry applications of their products—applications often cited in U.S. ads.

## BUILDING TRUST AND CIVILITY

As this discussion of the three scenarios illustrates, you need to read any communication situation with ethics in mind. In deciding what to do, follow the law as well as codes of ethics that may be circulated where you work or study and published in the journals of professional associations you belong to. We'll look briefly at a few such codes. But in the end, building trust and acting with civility come less from adherence to laws and codes than from adherence to the right values.

### Corporate Codes of Ethics

Within organizations, codes of conduct often pass from person to person orally as individuals learn an organization's culture. But organizations tend to publish a code in these circumstances:

- When they exceed the size where conversations alone can work
- When they face complex ethical issues
- When they expand their markets beyond home territory
- When they merge or collaborate with other companies or organizations

As an example, let's look at one European company's code of ethics, which includes segments typical of such a document (*Code of Ethics*). It has two main sections: "Our Values" and "Rules of Conduct for Employees." Under values, it notes the need, first, to comply with the law and contractual commitments. More than that, the code stresses the company's fulfillment of obligations to each of its constituencies:

- For customers, quality service and "commercial transparency"
- For shareholders, "optimum return" as well as "accurate and honest information" and equal treatment for all, regardless of importance
- For employees, human relations based on "mutual personal respect" and a recognition that employees are the company's "prime asset" and the company must foster a "feeling of belonging" and the "right climate for each person's professional and social development"
- For the communities where it operates, improvements in the economic and social life, in both environmental and urban affairs.

The Rules of Conduct section discusses guidelines for developing and distributing information, particularly the need for "true and honest financial information" and the need to be discrete with proprietary and confidential information. The section also discusses "commercial practices," that is, fair and honest contracts and accounting documents and avoidance of illegal payments like bribes. Finally, it defines conflicts of interests (situations which "might generate a conflict between their personal interests and the interests of the business"), and it discusses the suspension of any financing for political parties or candidates, the responsibilities of the board of directors, and the management and monitoring of subsidiaries.

## Professional Codes of Ethics

Professional associations also publish standards of good behavior, often focused on the ethical reporting of information in the association's publications and adherence to high standards in the performance of work. As an example, the American Chemical Society (ACS) directs its members to limit "narrow self-interest" and at the same time do good for others, that is, maximize "perceived benefits to society." The ACS begins a discussion of the obligations of editors and authors in writing for publication as follows:

> An essential feature of a profession is the acceptance by its members of a code that outlines desirable behavior and specifies obligations of members to each other and to the public. Such a code derives from a desire to maximize perceived benefits to society and to the profession as a whole and to limit actions that might serve the narrow self-interests of individuals. The advancement of science requires the sharing of knowledge between individuals, even

though doing so may sometimes entail foregoing some immediate personal advantage. ("Ethical Guidelines" 217)

## Ethical Use of Electronic Technology

Corporations, colleges and universities, professional organizations, and advocacy groups are also establishing benchmarks for the ethical use of information technology. These guidelines are works in progress. They evolve as the Web expands not only as a delivery system for e-mail and research but also as a site for e-commerce, both business to customer and business to business. Bad behavior breaks down the trust of a community of users. Because it's hard to police cyberspace, good conduct often depends on personal integrity. The expansion of e-commerce, for example, means that consumer privacy is increasingly under attack. Monitoring software can track a consumer's movement across many Web sites and collect and sell that information, without the consumer's knowledge. The World Wide Web Consortium, which designs standards for the Web, is developing ways to help users keep their personal information private and share with the system only what they want to share. Companies also publish policies that help them keep customer information private (Figure 3.3).

**For more about reviews of literature, see Chapter 21**

Respecting privacy is a key concern in ethical computing. Networks "let people indulge in cyberspace snooping and electronic eavesdropping" (Hays R22). To forestall abuses, organizations may monitor messages circulated on their systems and impose sanctions on those who enter accounts without authorization or otherwise disrupt the functioning of the system.

## Respecting Others

Ultimately, good behavior centers on respect for others, the respect that builds trusting relationships. In low-context societies, especially in the U.S., legal guidelines attempt to mandate such respect. For example, the Americans with Disabilities Act of 1990 tries to level the playing field for people with disabilities. Equal employment opportunity and affirmative action directives also try to redress traditional discriminatory practices in hiring or in academic admissions, although some affirmative action guidelines in admissions have been overturned in the courts. The attempt to legislate respect faces difficulty in application. It is also an action that is hard for many people outside the U.S. context to understand.

**Inclusive Language.** Whatever the difficulties of *legislating* respect, the core value is right, and you should take care of that value whenever you communicate. Show respect, for example, by using inclusive language. Bias against someone of a different gender, age, or ethnic background can surface in your writing, although you may not mean any harm. In their defense, for example, the Americans at Holburn House did not mean any harm in calling their neighbors "Trash." They exaggerated for effect and liked the style of the thought. The neighbors thought less about style and more about being insulted. Such insults are becoming even more common

/ SEARCH / SITE MAP / CONTACT US /

/ ABOUT DuPont / SCIENCE / PROUCTS & SERVICES / FOR INVESTORS / COMMUNITY / CAREERS

LATEST NEWS

**DuPont Company Privacy Policy**

The people of DuPont respect the privacy of visitors to our web sites. You are able to control what personal information, if any, you share with us. You are not required to tell us who you are or to reveal any information about yourself.

Listed below is the information we do collect and how we use that information:

- Our web server collects your domain name, but not your e-mail address.
  **Used how?** To measure the use of our site's content.
- Forms you fill out to interact with our business units (to request product literature, place an order, track an order, or otherwise correspond with a DuPont business contact) may request personal information such as your e-mail address, telephone number, and specific product/materials interests, for example.
  **Used how?** To conduct business with you. At no time do we sell or share your information outside the DuPont Company.
- Some of our pages may use *"cookies"* -- data that a web site can send to your browser, which in turn is stored on your computer system.
  **Used how?** To allow our web site to recognize you and provide custom information when and where appropriate. You can configure your browser to notify you when you receive a cookie or you can instruct your browser to refuse cookies altogether.

Periodically we administer online surveys to gather your views on our web site content, structure, and overall experience. The information you provide in these surveys is used to improve our site. Within the survey itself, we will let you know exactly how we will use the information collected.

Note that our Web site contains links to nonDuPont Web sites. Although we try to link only to Web sites that share our respect for customer information privacy, we are not responsible for their content or their privacy practices.

Individual DuPont business units may have more detailed privacy statements on their pages that explain any further use of your information within their section of www.dupont.com. If you have any questions about our privacy policy, please write to us at info@dupont.com, with the subject line, "privacy."

/ HOME / DuPont WORLDWIDE / TRADEMARKS / PRIVACY /

If you have questions or comments, please contact us by e-mail or by phone.
Copyright © 2000 E. I. du Pont de Nemours and Company. All rights reserved.

FIGURE 3.3 **A Statement from the DuPont Company's Web Site concerning How the Company Protects the Privacy of Visitors to Its Site.**
Source: *www.dupont.com/corp/privacy.html.* Used by permission.

in electronic messages where the coolness and impersonality of the medium and ease of sending a posting can lead to rudeness and *flaming,* responses that can offend the recipient. It's not difficult to avoid discriminatory or derogatory language if you think about it. To avoid *sexist* language, eliminate nouns and pronouns that indicate gender:

> ***Sexist:*** An average engineer spends twelve hours of his day at his terminal.
>
> ***Inclusive:*** An average engineer spends twelve hours a day at a terminal.

Or use the plural:

> ***Sexist:*** A scientist should be aware of the implications of his work as he decides what experiments to conduct.
>
> ***Inclusive:*** Scientists should be aware of the implications of their work as they decide what experiments to conduct.

ELECTRONIC EDGE

## Irritating Gadgets

In addition to behaving well when you use electronic devices, you need to recognize when good behavior requires that you *not* use them. The easy portability of cell phones and computers today means that norms must be established for when and where they are appropriate. A sign in the lobby of a London hotel makes explicit the proper behavior there. Under the heading "Irritating Gadgets," the sign reads:

> In deference to the wishes of the majority of our guests, the use of portable telephones and laptop computers in the public areas of The Cadogan is not permitted.

In other settings, you'll have to judge the appropriateness of these devices. One such setting is an airplane. Because of possible interference with navigational instruments, most airlines prohibit their use on landings and takeoffs. In addition, several companies require that employees avoid work on computers when sensitive or proprietary information would be displayed. Similarly, as the person next to someone at work on a laptop, you need to resist any temptation to read that screen. Moreover, in the tight spaces of a plane, your right to type at the keyboard may interfere with your seatmate's right to be free from disturbances. Telephones can also be disruptive.

At a much broader ethical level, company-purchased mobile phones and computers that at first seem to be a benefit may violate your privacy in ways that raise serious issues. They make you reachable wherever you are, contribute to a 24/7 work week, and apply not-so-subtle pressure for you to work harder and longer—at home, on planes, in cars. Increasingly, the problem you'll face is not *connecting* to a corporate system—but *disconnecting*. The speed of these devices also exerts pressure for you to speed up your own response to messages, sometimes to the detriment of your ability to think things through and develop a suitable response. After early enthusiasm about computing and cellular technology, debates about the personal and social costs of such devices are percolating in both the business and the popular press.

Or recast the sentence:

> ***Sexist:*** Profile a client so that you can write your proposal to meet his needs.
>
> ***Inclusive:*** Profile the needs of your client; then write your proposal to match those needs.

It's fairly easy, then, to avoid calling attention to gender in nouns and pronouns. It's more difficult to avoid nouns that carry connotations derogatory to women, the elderly, or people of a different race or color from yours. The context for your remarks is critical. Sometimes authors match terms in derogatory ways, as in a reference to the *men* and the *girls* at the office; *men* should match *women; boy* matches *girl.* Gender-free titles help: *flight attendant* rather than *stewardess.* But the main point is being sensitive and polite, attributes that would have prevented a court case, for example, in which a man claimed he was being harassed by a waitress who kept calling him "Honey."

In addition, avoid using yourself as a standard. For example, *Asian* is preferable to *Oriental* because the Orient is east only from a Eurocentric perspective; *European* is preferable to *Western.* Similarly, *nonwhite* implies a norm of "whiteness"; *flesh colored,* meaning beige or pink, ignores other colored flesh. A U.S. publication calls its January issue the "Winter Edition"; that designation is valid in the United States, but not, for example, in Australia, where January is high summer. The author of an article about problems in the computer industry dated the events as "during the Thanksgiving weekend." Countries celebrate Thanksgiving at several different times (often related to harvests in their geographic region)—and some countries don't have such a celebration—so the designation may be misread.

Sometimes, inclusive language goes overboard and thus fails to achieve its purpose. One problem is euphemisms, positive-sounding words that describe something considered negative. Although it might be appropriate to use the term *nonreader* rather than *illiterate* to talk about someone who is intelligent but unable to read text on a page, using the phrase *visually challenged* for someone who is blind may do that person a disservice in making the condition seem exotic.

**Politeness.** Show respect, too, by being polite. A dry cleaner attached this small notice to a shirt being returned to a customer with a small stain still on it: "We have done our best to remove the stains on this article but regret that we have not been entirely successful. Further efforts could cause irreparable damage to the garment."

The notice uses personal pronouns (we) to express personal responsibility, two complete sentences (one rather long), the term *regret,* and an explanation about further damage, to treat the customer with respect. It is polite, an attitude and a voice that the American customer of the British cleaner found engaging.

Simply being polite may help overcome misunderstandings and foster goodwill, especially in a communications environment that is often either impersonal or strident, especially with electronic communication. The French have a term—*les règles de politesse* (the rules of politeness)—that suggests the extent to which politeness doesn't just happen but is the result of cultivation. Politeness requires both the right attitude toward others and careful attention to making that attitude explicit in the language one chooses. That's one reason why French for many years was the international language of diplomacy.

Every day as you write and as you talk with people in person, online, on the phone, or in a video conference, keep in mind the central value of ethical communication: Respect those other people. To show that respect, take time to think of the other, to build trust, and to be polite.

# CASE
## GOING METRIC

You are an intern at the U.S. subsidiary of a large British supermarket chain. When selling at home or in Europe, the company must comply with a regulation of the European Community that all retailers use metric measurements rather than imperial units (like pints, pounds, and inches). Nutritional labels on food packages and gas at the pumps, for example, follow the metric system. Many of your customers, however, object. "Metres are something you see on a wall, to gauge the gas or electric you've used," notes one. Older customers in particular find it hard to think in metres and kilos. Conversions are also difficult: You multiply one pint by 0.56825 to get litres or divide a kilogram by 0.454 to get pounds. Small shopkeepers object because they must buy new scales and other equipment and think that metrication is just another ploy by big business to run them out. Shoppers fear that such measurements really misrepresent items and provide an opportunity for stores to round out their measures and charge higher prices, because customers won't have an intuitive sense of what the prices should be. Moreover, U.K. customers are reluctant to become European citizens; preserving British units helps preserve their national identity. Metrication (or "metrification," as the process is generally called in the United States) thus raises emotional as well as practical concerns. Metric units seem to undermine firmly held values and identities. Attempts to impose such units in the United States have stalled for years.

European Union rules include some compromises. For example, Britain can keep the mile, and pubs can sell beer and lager by the pint. Measures classified as "descriptive," like the "quarter-pounder," a 9" pizza, and dress and shoe sizes, are acceptable, although most consultants advise that labels include both imperial and metric measurements. Prepackaged food has to conform to metric standards (grams or kilograms), but customers can buy loose items by the pound. During a transition period, notices posted in supermarkets can include both systems but must give more emphasis to metric than to imperial.

As a global company, your employer will save money by standardizing all its products in the metric system common everywhere but in the United Kingdom and the United States. Most of what they buy, for example, is already measured in the metric system, and thus they don't need to repackage and convert those measurements. Once learned, it's a simpler system than imperial units. In Britain, most young people learn the metric system in school, and science students in the United States usually measure their work metrically. For a short period, people usually need to use a conversion table, just as new learners of a language are tied to a dictionary. But after a while, metric units become second nature.

Your company circulates a quarterly newsletter to its customers. Your supervisor asks you to write a short article encouraging customers to recognize that the metric system won't harm them, doesn't misrepresent weights and measures, and isn't an evil attempt to impose a foreign culture or kill small businesses. The newsletter has traditionally circulated only to customers in the United Kingdom, but as a pilot study, this issue will also circulate in the United States.

Using the information in this case, write the article. You may also add information you find in journals at your library or in online sources.

## CHECKLIST: Writing Ethically

1. **Treat your audience with respect and civility**
2. **Be honest**
   Avoid deceitful exaggeration or ambiguity
   Avoid misleading technical terms
   Disclose the limits on your information
   Acknowledge your debt to others
3. **Take responsibility in team endeavors**
4. **Carry your ethical values abroad**
   Recognize the values inherent in technology transfer across borders
   Work through conflicts in values by listening to and respecting your audience
5. **Adhere to any codes of ethics that apply to your organization or your profession**
6. **Use inclusive language**

## EXERCISES

1. Some instructors request that students write an independent report on the conduct of a team they participated on. Let's suppose that one person on the team did nothing: never attended a meeting, never completed an assignment. Should you convey that information to the instructor? Should you cover for him or her? Why or why not? Or suppose that you think the rest of the team ignored your ideas and wrote a report that was effective but didn't reflect any of your work. Should you still take credit for it? Be prepared to discuss your response to these questions in class.

2. In Chapter 2 you saw an example of a "shared values" statement a company used to codify its corporate culture and encourage compliance with standards of good behavior. If an organization you work for or belong to has such a statement, obtain a copy and write a brief report analyzing, at least from your point of view, whether the organization lives up to its statement or how it could improve either its statement or its behavior.

3. To make ethical decisions, you need to clarify your own values and codes of conduct. Some of the values you hold are probably critical, for example, being honest and making good on promises. They are bedrock values. Others may range from highly significant to current preferences, for example, loyalty to your family or university, being financially secure, taking risks, practicing religious beliefs, saving time for privacy, maximizing leisure time, or eating a vegetarian diet. Sometimes, values conflict: You may value wilderness, for example, but value comfort, too, and want to build a house with flush toilets, telephones, and electric service in a wilderness area. Create a visual, perhaps in the form of a pyramid, that suggests the hierarchy of your values and could thus help you make decisions when values conflict.

4. Collect examples of jargon that causes confusion and may do harm to the reader. As you saw, insurance forms and rental agreements are two potential sources. Translate the two-sentence segment from a rental agreement in Figure 3.2 to communicate its message clearly to a lay audience.

5. Try your translation skills on these other examples of jargon. Not all jargon harms the reader, of course. But when jargon is used only to impress or to make information obscure, it may slow down the reader, and that in itself is a disservice. Strip down the sentence to say what the author probably meant:

   - It has been suggested by various engineering groups throughout this industry that departmental coordination could be enhanced by employing the use of additional departmental functional controls in order to facilitate familiarization with any malfunctioning in present routine.
   - The hope for future returns through the use of natural breeding is definitely one of the primary sparks that keep the purebred breeder's candle glowing in hope of attaining future progress and success in the dairy industry.
   - This project intends to develop musicality in elementary school children through innovational instructional approaches utilizing multisensory channels through the natural elemental avenues of speech, drama, movement, rhythmics, singing, instrumental playing, and symbolization.
   - On the basis of personal judgment, founded in past experience conditioned by erudition and disciplined by mental intransigence, the incumbent integrates the variable factors in an evolving situation and, on the basis of simultaneous cogitation, formulates a binding decision (irreversible) relative to the priority of flow of interstate and intrastate commerce, both animate and inanimate. (From the U.S. Civil Service Commission)
   - Travel which is incident to travel that involves the performance of work while traveling. Simply stated, travel which is incident to travel that involves the performance of work while traveling means travel to a point at which an employee begins to perform work while traveling or travel from a point at which an employee ceased performing work while traveling. (From a U.S. Department of Justice publication titled "Travel Time Policy and Regulations")

6. Find examples of ambiguous or otherwise misleading language in a daily newspaper, professional journal, advertising brochure, or other publication. Bring the examples to class and explain why you think such language obscures or deliberately falsifies meaning. Hint: A source of extensive examples is the *Quarterly Review of Doublespeak,* published by the National Council of Teachers of English.

7. Review the syllabi you've received in your classes to note any codes of conduct or classroom behavior the instructor includes. How extensive are such guidelines? What attitude toward students do the guidelines reflect? What do the guidelines suggest concerning the level of trust among students and between students and the teacher in that classroom? Is it appropriate for

students in one class to hand the same report, essay, or paper in to the instructor of another class? Why or why not?

## FOR COLLABORATION

1. One excellent way for you to show respect for others is to contribute time as a volunteer in your community. Several colleges and universities support programs for volunteerism. These programs may offer formal internships or course projects called "Service Learning" or "Community Service Placements." On a team, develop a project that aids your community. For example, if your school does not have a volunteer program, propose one, through the appropriate offices. If your school does have a program, investigate its offerings and propose that your group work in one of them. Or develop a service learning project on your own. Consider opportunities working with people of all ages and in situations that promote wellness, a sustainable environment, fair housing, access to scholarships or social welfare programs, and the like.

2. Gather a team to update the code of ethics for appropriate use of electronic technology on your campus—or to write such a code if one is not available. What topics need to be covered? What emphasis needs to be given to each topic? What form of delivery is best for the code? Paper? Online? As its own Web site? As a page on the computer center's site?

3. Prepare a review of literature on a topic in the ethics of technical and scientific communication. This chapter should suggest some areas for further study. You might want to visit Web sites devoted to case studies in ethics and the home pages of professional societies and publications that have developed guidelines for appropriate presentation.

PART TWO

# Managing Information for Readers

# COLLECTING EMPIRICAL INFORMATION



"YOU CAN SEE A LOT BY OBSERVING."

YOGI BERRA

Professionals conduct research to find information that will help them understand something they didn't understand before or to solve a problem:

- The team from Novo Nordisk and Abbott Laboratories, as you read in Chapter 1, conducted research to develop a new drug to treat epilepsy. First in the laboratory, and then in clinical trials, they tested the effectiveness and safety of their new therapy.
- Computer specialists gather information to create new software.
- Foresters examine a stand of trees to determine the best time and plan for cutting them.
- Industrial engineers look for ways to streamline a manufacturing process.

In this chapter and Chapter 5, you will learn guidelines for gathering information efficiently and effectively.

## PLANNING YOUR RESEARCH

Although approaches to research can be complex, two broad categories are generally recognized. A manager at a large insurance company neatly labels these categories "on the ground" research and "book research." After suggesting a strategy for research in general, this chapter provides guidelines for "on the ground," or empirical research. The manager's empirical research consists of visiting Europe and Japan, where his company aims to

For more about collaboration and research proposals, see Chapters 2 and 14

enter the market for a highly specialized form of insurance. During his visits, he interviews local insurance agents, talks with businesses that might buy the insurance, and observes local customs and practices. The epilepsy drug team used an empirical approach as they designed and ran experiments and observed clinical tests. When you observe phenomena in the laboratory or field, conduct experiments, and gather information through interviews and surveys, you use an empirical approach. Chapter 5 discusses "book" research: gathering information from documents and Web sites. A research project may require one or both approaches. These approaches, too, sometimes overlap as you use electronic technology in an investigation.

You'll discover the right approach if you spend some time at the beginning of a project setting up your strategy. Figure 4.1 provides an overview of the research process you will learn about in this chapter and Chapter 5. Informally, you may simply make notes for your own use. If your investigation is complex, and particularly for any collaborative project, write the plan in a document you circulate to other team members, the sponsor of the research, a supervisor, or a professor for approval:

- Divide your approach into specific tasks.
- Assign a deadline for each task so you avoid sinking under the weight of the endeavor as a whole.
- Set monitoring points at which you'll determine if you need some midcourse corrections to keep the whole project on target.

Without such a plan, as one authority states, your life will be a miserable series of emergencies. This section will help you define your research objective and develop a list of tasks.

## Focusing the Subject

How do you get started on a project? On the job, you often respond to a direct request:

- You are a summer intern at a computer manufacturing company. Your supervisor asks you to prepare a guide to regulations concerning recycling of paper and other materials in the six countries in which you do business.
- You are a nutritionist in a nursing home. The director asks you to plan guidelines and menus for the food service.
- You are a senior architecture major on a team whose design project is to retrofit an existing home to meet the needs of a disabled client.

As a student, however, you may need a preliminary step: finding something to write about. Where should you look for inspiration?

1. *Play to your strengths.* What are you interested in? What do you know well? What have you seen or read recently that captured your attention? That invites investigation? Browse the library or the Web. Talk with a friend.

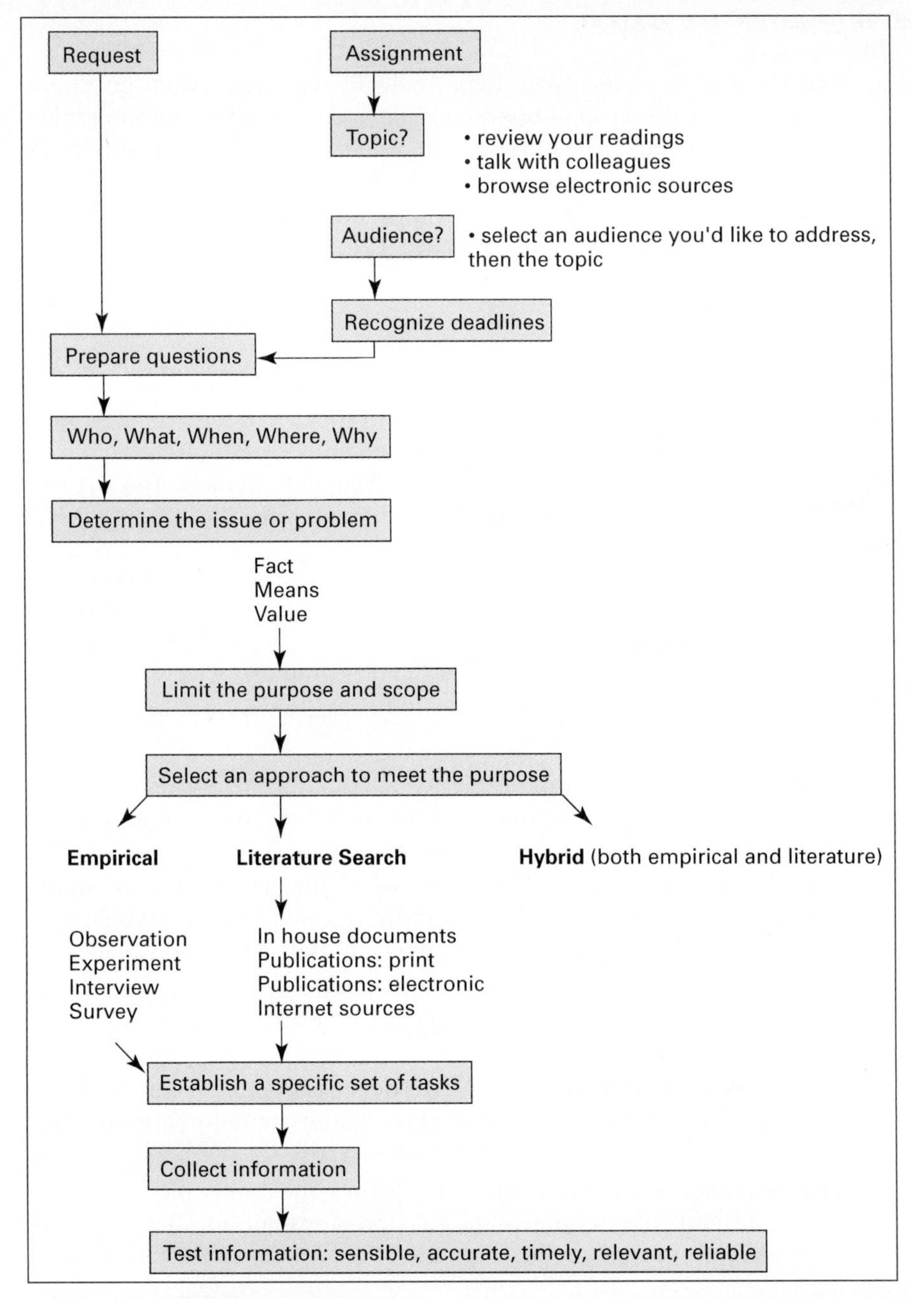


**FIGURE 4.1 The Research Process**
If someone has directly requested your research, begin right away to develop the questions that the request requires you to answer. If you have been given an assignment in class to "write a report," you'll need to decide on a topic first.

2. *Think about your audience.* To get a handle on a topic, think about who would like to read about that topic, and let your image of a reader help shape what you will write about.
3. *Be realistic.* Choose a topic you understand and one on which current information is readily available to you—where you are. Letters of

inquiry sent to people you don't know, for example, often go unanswered. Interlibrary loan books and journals may take weeks to arrive at your library. Stretch yourself to do something worthwhile, but don't make your life unnecessarily difficult.

4. *Keep your deadlines in mind.* If your report is due in two weeks, avoid an experiment that will take three weeks. Let the deadline determine the scale of your approach.

### Preparing Questions

Once you have settled on a subject (or one has been settled on you), convert your general topic into questions that will frame your research. "The outcome of any serious research can only be to make two questions grow where only one grew before," notes the economist Thorstein Veblen. To start the process, think of the journalist's five Ws and an H: "who," "what," "when," "where," "why," and "how."

For example, assume that shipments to the sporting goods store where you work are processed only haphazardly. Boxes remain unopened in the stockroom, crowding the aisles and making movement impossible. Employees don't know what inventory is available and couldn't find what they need even if they did know because it remains in boxes. To explore the problem, consider each of the five elements:

**For a proposal developed from this line of reasoning, see Chapter 14**

*Who.* Consider who is responsible for unpacking. All employees? People on certain shifts? No one in particular? A failure in assigning (or assuming) responsibility may cause the problem.

*What.* Decide what can be done to move shipments out onto the shelf. Divide the work into four discrete tasks: receiving, opening, verifying content against the shipment manifest and the order, and storing.

*When and Where.* Examine the setting. Should shipments be processed in the storeroom? On the loading dock? Every day? On shipment days only? At night?

*Why.* Look at the short- and long-term benefits of effective processing. Is it just a good thing on its own? Does it contribute to better morale at the store? Can it improve customer service?

*How.* Develop a management plan. Within the limits of the store's budget, hours of opening, and total staff, consider how people could be deployed to process the shipments.

The summer intern interested in recycling composed the following questions (among others) to focus her research:

1. What is the definition of *recycling* in each country? Does the term occur in their language?
2. Is there legislation concerning recycling? If so, is it local, provincial, or national?
3. To whom does any legislation apply?
4. What products (if any) does the legislation cover? Paper? Metal? Glass? Aluminum? Used oil? Plastics?

5. Are there any penalties for violations or noncompliance?

The nutritionist might ask such questions as these:

- What is the recommended caloric intake for a largely elderly and sedentary population?
- Are certain foods less or more tasty to an elderly palate?
- What interactions are there between diet and specific drugs used by individuals in the nursing home?

The scale of your questions varies from the simple to the complex. For example, finding information about required courses in your major should be routine and easy, while finding the causes of cancer involves more complex questions and answers.

## Defining the Problem

Your questions help you probe systematically into the issues or problem you need to solve. As George Bernanos, the French novelist and political writer, said, "The worst, the most corrupting lies [are] problems poorly stated." Novices often cut short the process of setting up the research problem (Flower 3–36). Compared to experts they

- Define smaller problems.
- Overlook ramifications of a particular issue.
- Leap to solutions.
- Seek fewer sources outside their own minds.
- Think about academic assignments rather than problems.

Experts spend more time generating possibilities and honing research strategies. To define problems like an expert, identify what type of problem you are working on. Here are three useful categories.

**Problems of Fact.** One type is a problem of fact. Such problems are the core of "basic science": Researchers observe phenomena and suggest explanations of how, and sometimes why, things occur. Physicists engaged in identifying stars, naturalists examining the feeding habits of quail, geologists studying the substructure of the Yucca flats—all deal with problems of fact.

How big is the polar ice cap?

What was the lobster catch last year in Casco Bay?

These are problems of fact.

**Problems of Means.** These "how-to" problems occupy the attention of engineers, technicians, mechanics, and other applied technologists and businesspeople.

- What procedure should veterinarians use to insert metal replacement joints in sheep?
- What capacity pond is necessary to water 300 cattle?

The building of devices, machines, industrial plants, bridges, software, interstate highways, and information highways—all present problems of means. The scale of such problems ranges from the invisibly small, for example, microsurgery on a knee, to the huge, like building a self-supporting space colony. One form of a problem of means is the development of sophisticated instruments with which to solve problems of fact, like the Hubble telescope for examining the solar system.

"How come?" problems also fit into this category, that is, determining why engineered structures fail. Bridges and highway overpasses fall and planes crash (fortunately, not too frequently). Someone has to assess the circumstances of the failure, determine the possible causes, select a probable cause, and recommend what to do to get the item back in action—and to prevent such failures in the future.

**Problems of Value.** Problems in this third category, as the name implies, lead to an evaluation. You have to rate something, or rank priorities, or determine policy. Sometimes, such problems are closely related to problems of fact. In other words, you may need to measure whether something meets predetermined, objective standards. Here are some examples:

- Does aspen bark contain compounds with medicinal properties?
- Does the substructure of the Yucca flats meet the Department of Energy standard for storage of nuclear waste?
- Does the company's procedure for recycling paper conform with state requirements for waste treatment?
- How much of the recommended daily requirements for nutrition do these fruit drops provide?

Other such problems require that you rank items, or assess priorities, or determine trade-offs in a given situation. If your purpose is to purchase a computer system, then you might ask these questions:

- What is the prevailing system at your college?
- What is the easiest interface for you to use?
- How much can you spend?
- Could you get along by using publicly available sites?

You look for criteria against which to measure your choice.

On a broader scale, organizations and nations also make such choices. One major form of report in the United States, for example, the environmental impact statement, flows from the problem of determining the likely effect of some proposed human intervention on the landscape. Such reports lay out alternatives for decision makers who aim to mitigate negative impacts. Here are other problems of value:

**For more about environmental reports, see Chapter 15**

- What level of funding from the National Science Foundation is necessary to help ensure an adequate supply of scientists and engineers in 2020?
- Should smoking be banned in all public locations in the city?

Sometimes the opportunity for observation that will enhance your work comes unexpectedly. For example, a container ship from Hong Kong spilled 7,200 rubber duckies, along with some other tub toys, into the Pacific Ocean during a storm. The toys then drifted to shore. Oceanographers were pleased with the opportunity the duckies provided for studying the dynamics of currents and winds. "In our field, it is kind of like the Big Bang of drifters," noted one researcher who plotted "daily ducky sightings" and has published his work (Carlton B1). The ducks helped researchers

The native groups, however, don't get fooled by "which-line-is-longer" illusions like those here (Miller). The mind makes a box that seems to suggest line (a) is farther away, and thus shorter. (They are the same length.)

More recent studies have pointed out cultural biases in Hudson's approach. The point, however, is clear: Culture shapes how one sees (Miller).

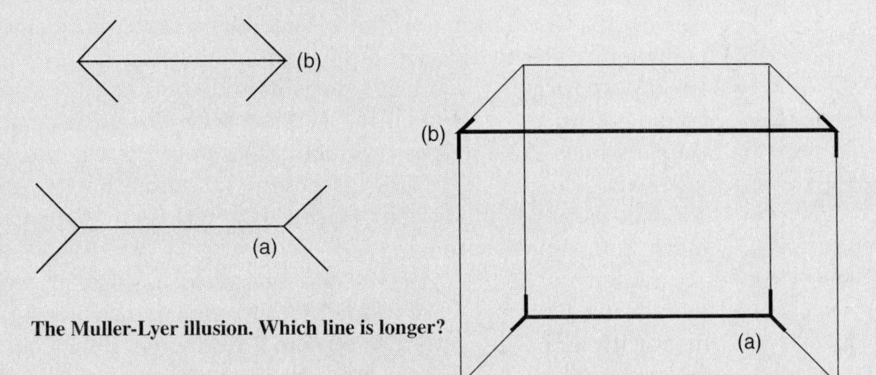


The Muller-Lyer illusion. Which line is longer?

What the Muller-Lyer illusion suggests.

predict the movement of oil spills and helped fishermen find the best sites for fishing.

- Is the 10-acre tract near Red Clay Creek a wetland?

Based on your understanding of policies governing the designation, you walk out on the tract and take field notes, make drawings, perhaps photograph plant species.

- Do bacteria cause ulcers?

You set up controlled experiments with animal and perhaps human subjects to test that hypothesis.

Some observations are processed automatically by instruments that extend our vision, as with computer analyses of the weather and observations from space probes. You may also record observations in notebooks or on preset forms. Often, visuals are more useful than words. To solve in particular problems of fact or means, then, you often need to observe well either in the field or in the laboratory and to record those observations accurately. For more about how culture shapes such observations, see *Crossing Cultures: Learning How to See.*

## INTERVIEWING

Another method for gaining firsthand information is to talk with people in an *interview*. Interviews take many forms. Conversations with other students or researchers constitute an informal interview. Through conversations you find out what others are doing about the topics or problems that interest you, you test your own thinking, and you glean current, often suggestive, information that isn't otherwise available. Tap into the grapevine in an organization and you may learn better information than what's provided in official statements. Join an electronic newsgroup to enter conversations (in writing) with people who share your interest in some topic. Recognize cultural settings that favor face-to-face meetings rather than documents for conveying information and set up such meetings. A wide range of problems may require you to talk with others about how to proceed.

More formally, you may interview an expert in a prearranged meeting conducted in person or online (Figure 4.2). For such occasions, develop your instincts to know when to relax, when to push, when to listen, when to talk—and when to call it quits. Watching talk shows and the nightly news on TV may help you hone your skills, although their emphasis is entertainment, and yours is gathering information.

A form of group interview is a *focus group* discussion, a technique common in marketing research. In a focus group, six to 10 customers or clients meet with a facilitator to discuss some product or policy currently in a development stage. Groups are selected on the basis of similar characteristics; to confirm the accuracy of the findings, two groups per characteristic are usually required. The discussions last about an hour or two. As clients listen to each other, they are reminded of thoughts or memories that can enrich

**Before the interview**

1. Select carefully whom you want to interview. Don't waste your time—and that of the person being interviewed—by approaching someone who cannot tell you what you want to know.
2. Do your homework. Know the basics about the subject and the person. You'll be resented if you inquire about facts you could easily have learned before. Such inquiries also consume valuable time better spent at a higher level.
3. Outline the form you want the interview to take; prepare a list of questions.
4. Request the interview well in advance. In your request
   - Suggest several times to suit the person's schedule
   - Establish your own credentials
   - Clarify how you will use the information
   - Explain why you have chosen this individual
   - Suggest the topics you'd like to cover, perhaps with sample questions.

Write or e-mail the request if the interview involves potentially controversial or proprietary information.

**In the interview itself**

1. Arrive on time.
2. Take a few minutes to establish a cordial relationship and encourage trust. Let your own nerves settle. Reiterate the reason for the interview.
3. Be ready to adapt questions to a new situation. Be flexible. Let the person being interviewed talk. Provide gentle reminders if the person wanders from the topic but don't indicate boredom or impatience. Be ready to pursue a new line of questioning if the person suggests a desirable alternative to what you had planned. Allow a little slack. Interrupt only if necessary for clarification.
4. Either take notes in a notebook or, having obtained permission, record the session on tape. Avoid extensive note taking. Record statistics and other precise data, but don't try to capture every word.
5. Verify key or controversial statements before you end the interview.
6. If you are conducting a series of interviews, ask questions rigorously in the same order and avoid offhand remarks that might prejudice the comparability of results.

**After the interview**

1. Compile the notes or transcribe the recording while the interview is fresh.
2. Thank the person being interviewed in writing (if the request was in writing) or informally. You might send along a copy of the final product.
3. In your report, refer to the interview by a citation that includes the name and title of the person interviewed, the place, and the date. The date is needed to establish a context; it will protect you and the interviewee should his or her opinions subsequently change.

FIGURE 4.2 **Guidelines for Conducting an Interview**

the data. The Program for Appropriate Technology in Health (PATH), for example, uses focus groups to develop health care materials for low-literate and nonliterate audiences, mainly in developing countries. The discussions generate evidence clients will find persuasive, identify myths or beliefs about a product or practice, and help information developers compose questionnaires and design and evaluate drafts of brochures and other materials. Political campaigns use focus groups to test how a candidate's platform will play out with voters. Producers of consumer products use focus groups to check on the appeal of their product and correct any problems the product may pose.

For more about focus groups, see Chapter 6.

## CONDUCTING A SURVEY

While interviews usually center on one person or a small group of people, a *survey* allows you to gather information more broadly. A survey often precedes proposals for changes in practices, educational programs, or service delivery. Surveys are also important marketing tools. Recreation supervisors survey the users of national parks to determine priorities for development. Marketing specialists survey eaters' preferences for certain foods before full-scale production of new lines. Students often use surveys to solicit opinions on university policies and practices. Such educational research with human subjects is probably exempt from formal review, but it would be wise to make sure you've complied with your ethical obligations to those subjects. In addition, conducting a survey requires you to

- Develop an effective questionnaire, the tool of the survey.
- Select an appropriate group to respond to the questionnaire.
- Determine the best method for distributing the questionnaire to the group.
- Properly interpret what the respondents say.

A survey must meet two criteria: reliability and validity. A reliable survey is predictive; that is, its results can be replicated among a different group of respondents. Reliability is a measure of the sample population. A valid survey is one that provides the information you need. Validity is a measure of the questionnaire. This section briefly discusses how to choose a sample population and compose a questionnaire.

Survey research requires special training in the social sciences and statistics to select the best approach and interpret results accurately. If you are not familiar with statistics, for example, you might want to collaborate with someone who has such skills. In addition, conducting surveys across cultures presents special problems. Americans are generally comfortable responding to questions. But you may find resistance or fear when you administer a survey elsewhere, particularly if you ask about such subjects as health practices and personal values that are not generally discussed in public. More than merely translating the words in a questionnaire designed to elicit health information, for example, you may need to work through a local agent who is trusted by your target population and elicits information indirectly.

### Reliability

To make sure your survey is reliable, select carefully the sample of people you'll survey. Make sure they are representative of the larger group whose behavior or attitudes you seek to identify and that you haven't stacked the deck to achieve the result you want. If your goal is a survey of campus attitudes toward fraternities, you bias the result if you question only fraternity members—or only those who decided not to join a fraternity. Similarly, in seeking opinions on restricting use of a wilderness preserve, you skew the outcome if you mail a questionnaire only to those who hold hunting licenses or only to members of a nature conservancy. A survey on the drug-taking habits of the elderly is biased if only those in nursing homes are questioned. In addition, make sure the sample population is large enough to warrant the interpretations and trends you ascribe to it.

### Validity

Design the questionnaire carefully to avoid bias or misleading language that will invalidate your results. For example, politicians may structure questionnaires in a way that elicits answers favorable to their candidate or cause. In an even more ethically suspect practice, they may use the survey form, especially in a phone call, to disguise a "push poll," a deliberate attempt to change voters' minds. In addition, survey respondents are often highly suggestible. In one classic test, students offered their detailed opinions about three nationalities that (unknown to the students) don't exist (Goleman C1). The options given in the questions sometimes bias the results. For example, on the forms he provided to students for evaluating his course, one professor listed a range of answers whose lowest point for each question was "Very good" and the highest "Outstanding."

In addition to bias, another problem that reduces accuracy in surveys is misleading phrasing. To be effective, keep questions as concrete as possible. Responses are less likely to be valid and comparable when readers have too much room to interpret. Instead of asking, "Do you eat regularly," which leaves open the meaning of "regularly," ask specifics: "How many times a day do you eat? Do you eat in the morning? At noon?" Instead of asking "over the last few years," which might mean any number from two to seven or more, provide a specific time frame: "over the last three years."

Avoid using jargon or bureaucratic language that will be meaningless or ambiguous to respondents. For example, a standard question in a U.S. Department of Labor survey of unemployment reads: "Is there a job from which you are on layoff?" To the agency, the term *layoff* refers to a specific category of those who will be called back to work and are not then counted in the same way as those who are otherwise out of work. But researchers discovered that respondents thought the term was simply a nice way of saying "fired." The question now adds the phrase: "and expect to be called back" (Goleman C11). Similarly, in a National Health Survey, people failed to recall visits to a doctor when asked simply, "In the past 12 months, about how many times did you see or talk to a medical doctor?" When their responses were checked against medical records, however, people were found to have forgotten 60 percent of the visits. A question that offered cues to recalling specific visits, especially the most recent one, improved the accuracy of results (Goleman C11).

Figure 4.3 provides guidelines for composing an effective questionnaire. Figure 4.4 shows a covering memo that meets these guidelines. Figure 4.5 provides examples of questionnaire segments that explain, instruct, and ask questions.

In a cover note or at the top of the page, explain your purpose, your role, the intended use for the information, and the deadline for returning the completed form.

- Limit the number of questions. A long list may frustrate potential respondents before they even begin.
- Number each question for ease of reference.
- Place the easiest questions first (or the most interesting) to build motivation to respond.
- Include only one idea in each question.
- Make the questions easy to answer: yes/no, multiple choice, or fill in the blank.

A Likert Scale also provides for easy responses—and easy tabulation of those responses; for example,

7 = strongly agree
6 = agree
5 = somewhat agree
4 = neutral, neither agree nor disagree
3 = somewhat disagree
2 = disagree
1 = strongly disagree

Just be sure that you clarify whether a high number indicates agreement or disagreement—and repeat those numbers on each page of a multipage questionnaire.

- If your project has multiple goals, make a separate questionnaire to match each goal.
- Pretest the form with a small group of potential respondents to weed out ambiguities and unproductive questions.

FIGURE 4.3 **Composing an Effective Questionnaire**

MEMORANDUM

Date: 10 January 2001
From: Roger P. Curtis, President
To: All Employees
Re: Commuting Survey

We say we are an environmentally friendly company.

builds motivation by evoking a shared value

The federal government has provided us with another occasion to prove it. Under the provisions of the Clean Air Act, we must develop a plan to ensure that each vehicle used to commute to our worksite will have an average of 1.5 or more occupants. Our plan must also suggest a variety of commuting options, including carpools as well as other forms of transit.

explains the questionnaire's purpose and the writer's need

The enclosed Employee Commute Survey is a critical first step in formulating our plan. We need to hear from you. Please complete the survey and return it to your supervisor no later than 1 February. Your responses should reflect actual trips to the workplace between 6 and 10 January. If you were traveling offsite during that time, please return the form with that notation. Information gathered in this survey will tell us how employees get to work and help us develop a computer model to assist in planning future transit routes.

notes the deadline for returning the form

The survey asks you to identify yourself, but only as a method for the survey coordinators to get in touch with you if they need to confirm or clarify information. All responses are strictly confidential.

encourages compliance by promising confidentiality

Thank you for your assistance in providing information about your commuting habits. Your responses will help us continue to maintain our role as an environmentally friendly corporation. If you have questions about the survey or about the Clean Air Act, please send e-mail to Amy Tsai, coordinator of the compliance effort, at Tsai@insco.com.

reiterates the intended use for the information

FIGURE 4.4 **Covering Memo That Accompanies a Questionnaire**

## TESTING INFORMATION

The goal of your research is information, that is, facts and interpretations that will help you and your reader solve a problem or make sense of a situation. The information serves to plug a hole in your understanding of a concept, phenomenon, or event and to make things more certain for yourself and for the people you write to. But not all the information you gather will be useful or right. You'll need to screen and evaluate what you find to make sure it meets five criteria: Your information should be sensible, accurate, timely, relevant, and reliable.

### Sensible

First, information should make sense. When you observe, experiment, listen, or read, don't lose your common sense. Of course, revolutions in scientific

**Explanations**

Responding to the following questions will help you review the current status of your health and determine how your life style may influence your health in the future. The information you supply will be kept in strictest confidence by your physician. Please complete the questionnaire before your first appointment.

**Question**

6. **When do you normally arrive at work? (Circle the number that corresponds with your response.)**
   - a. **Work rotating shifts**
   - b. **Between 10 A.M. and midnight**
   - c. **Between midnight and 6 A.M.**
   - d. **Between 6 A.M. and 10 A.M.**

**Question**

9. Please rate our technical support staff on ...
   - **a.** Its overall performance
   - **b.** The length of time between your dialing our number and your receipt of a satisfactory response
   - **c.** The staff's willingness to help customers
   - **d.** Accuracy and usability of the information you received

**Question**

**2.** How often do you eat the following foods?

Never 1-3 times/week 4-7 times/week 2 times/day
3+ times/day

Fried food
Fresh fruits and vegetables
Ice cream, cheese, butter, sour cream
Red meat (hamburger, pork, lamb)
Pasta

**Instructions**

1. Mark one box for each item with a small "x," not with a checkmark.
2. If you don't know the answer, or if the question doesn't apply, mark an "X" in the "Don't know" box.
3. Base your answers on your activities over the last three months.

**Instructions**

We would like your opinion of our technical support help line. Please rate our performance over the last three months. Use a seven point scale, where 1 means "poor," 4 means "good," and 7 means "excellent."

**FIGURE 4.5** **Sample Segments from Effective Questionnaires**

thinking like those associated with Galileo, Newton, and Einstein didn't make sense against the backdrop of prevailing theory at the time. But the numbers, images, and words you deal with day by day should be sensible. Be wary of the odd, the bizarre, the too-easy explanation, the spike in an otherwise normal curve. Check the arithmetic. Look for corroboration. Sometimes things that are too good to be true are too good to be true.

**For more about accurate interpretation, see Chapters 3 and 11**

### Accurate

Second, information should be accurate. Every time information is relayed from reader to reader, the possibility for inaccuracy increases. When data flow digitally from the instruments recording it into the computer systems of those who need it, that direct flow enhances accuracy. In addition, you need to interpret information properly. Answer the following questions about your own research or someone else's research on which you are relying:

- Did you or another researcher use the proper technique and appropriate instruments?
- Is the information subject to direct observation? If so, was it derived from such direct observation?
- What is the degree of accuracy with which the information can be recorded—and was it indeed recorded accurately?
- Can the information be verified? Was it verified?
- Is the information complete or "representative"? If "representative," how were the examples chosen?
- Did you clearly separate facts from interpretations?
- Is the reasoning sound? Does the information support the conclusions drawn from it? Could alternate conclusions also be supported?

### Timely

Third, information should be timely. Unless you specifically seek historical references, work from the latest data in the field. Use electronic communication to cut down on "information float," the time lag between when information is developed and when it is available to others. Old information, accurate in its day, may now be wrong. On the other hand, resist the temptation to publish information too early in an investigation before it has been adequately verified. In another meaning of "timely," make sure the information gets to the reader when the reader needs it.

### Relevant

Fourth, information should be relevant to your purpose. No matter how hard it was to find, if your information doesn't fit the purpose of your project, eliminate it—or save it for another project.

### Reliable

Finally, information should be reliable. People should be able to trust your information, and you need to determine the reliability of information from people you talk with or whose documents you read. To establish that reliability, ask the following questions:

- What are the source's credentials: education, employment, chief interests, publications, professional activities?
- Is the source speaking in her or his own field and up to date?
- How reliable have the source's other reports proven to be?
- Did the source observe the data directly, or are they secondhand?
- Does the source seem to know the topic?
- Does the source face pressures to suppress or distort information in order to raise profits, to provide (or avoid) opposition, to support a prejudice, to exploit a situation?

As a technical professional, you'll spend your day dealing with information. You'll take information from one context to use in the new context of a document or oral presentation you are preparing. In this chapter, you've learned some strategies for gathering information firsthand, by observing, experimenting, and conducting interviews and surveys. As you take notes, look for inconsistencies and holes. Keep searching until the inconsistencies are resolved—or well explained as inconsistencies—and the holes are plugged. You've also seen that you need a critical attitude toward the information you gather through these methods. Don't take everything you see or hear on faith, but weigh your material so you are not fooled by the evidence and you don't fool others when you convey your findings to them.

## CHECKLIST: Collecting Empirical Information

1. **Create a detailed plan for your research**
   - Clarify the request and focus the subject
   - Sort what you know from what you need to know
   - Prepare questions
   - Define the problem or issue
     - Fact? Means? Value?
   - Write a specific statement of purpose
   - Determine your approach to gathering information
2. **Observe phenomena or events and record them well**
3. **Interview experts**
   - Select the interviewee strategically
   - Know appropriate background before the interview
   - Prepare questions, but be prepared to change tack if needed
   - Verify key or controversial statements before closing
4. **Conduct a survey**
   - To make the survey reliable, avoid bias in the sample and include a large enough sample
   - To make the survey valid, compose an effective questionnaire
5. **Test your information to make sure it is**
   - Sensible
   - Accurate
   - Timely
   - Relevant
   - Reliable

## EXERCISES

1. Develop a plan for researching a topic that could develop into a report in your technical writing class. Define the problem you'd like to investigate,

narrow the approach if necessary, and prepare a set of questions. In addition, make a list of techniques and sources you would use to conduct your investigation. Briefly profile the intended reader of the report. Create a schedule of your work in visual form. Write up this information in a memo to your instructor.

2. Write a five-question questionnaire concerning a policy of your college or university on, for example, holidays, class schedules, open hours of the library or computer center, parking privileges. If the policies of your school allow, distribute the questionnaire to ten people. Create a table to display your results (see Chapter 8) and explain them in a brief message to your instructor.

3. Find the answer to these questions:

   What is the cost of an oil change at local garages?

   What is the cost of a hamburger in town?

   How far is it to the nearest airport?

   What is the best route to the airport?

   How many people are employed by your school or university?

   What are the periods of peak traffic on nearby roads?

   In class, compare individual answers and test them for the qualities of good information you read about in this chapter.

4. Using "Electronic Edge: The Ethics of Using Electronic Postings in Research" as a departure point, analyze ethical issues in using materials from online discussions in research. Who should have access, for example, to student postings in a class chat room or Listserv? Who *owns* those postings? Should students be notified if a teacher is using those postings in a research project?

## FOR COLLABORATION

On a team of two or three students in the same major, prepare a brief, informal oral report (see Chapter 22) on the role of communication in your field. You'll need to narrow the topic to some aspect of communication, for example, typical documents written by professionals in the field or commonly used channels of communication. To gather information, interview professionals and a sample of graduates you know. If you have access to computer newsgroups in your field, you might want to "lurk," that is, read the postings without adding your own messages, for a week or two. Note topics of discussion and prevailing attitudes and issues. Provide a summary of your results in a report for the class.

# COLLECTING INFORMATION FROM DOCUMENTS AND WEB SITES



"EVERYBODY GETS SO MUCH INFORMATION ALL DAY LONG THAT THEY LOSE THEIR COMMON SENSE."

GERTRUDE STEIN

This chapter continues the discussion of techniques for gathering information begun in Chapter 4. You learned some guidelines for empirical research in Chapter 4, and in this chapter you will learn about another kind of research: finding information in documents, broadly defined, and on the Web.

## PRINT AND ONLINE SOURCES

The amount of print and online information is staggering. According to one account, more information has been produced in the last 30 years than in the previous 5,000, and a weekday edition of the *New York Times* holds more information than most people came across in a lifetime in seventeenth-century England (Wurman). Another researcher predicts that by the year 2018, information will double every eight hours. To measure the expanding volume of online messages, researchers have developed new terms for expressing the number of bytes or information units: A terabyte, for example, is a trillion bytes and a petabyte is a quadrillion bytes. One of your major challenges as a professional will be dealing with this stream of information, because almost all research projects at some point require you to read what other people have written.

*Where* will you find information? Look around. You may find information in such archives as the library on campus; the file cabinets at your home or office; provincial, state, and federal archives; the archives of national and international agencies; and the archives of professional organizations (see *A Closer Look: Documents and Documentation*). The list of places where information is stored is long. You may also pull information from these and other sources at your desk or in the field on your own computer or personal digital assistant linked to the Internet. As you have probably discovered, the Web is the fastest growing archive around.

Wherever it is stored, information is increasingly stored *electronically.* Such storage vastly decreases the amount of space needed for an archive, facilitates the organization of documents, and speeds access for a variety of purposes. A slim CD-ROM can hold the equivalent of 250,000 pages of text as well as sound and video; double-sided disks hold even more. Documents can also be stored online; the folders in your e-mail account are one small example.

### In-House Documents

Company documents will probably be your main sources of information on the job. These include

- Annual and quarterly reports
- Financial statements
- Personnel records
- Project reports
- Catalogs
- Memos
- Policy handbooks
- Manuals

Some companies produce vast quantities of such documents, especially professional information traders, including investors, consultants, distributors, and economic planners. For example, Mitsubishi Corporation, the largest of such traders (called *sogo shosha*) in Japan, transmits information equivalent to 7,000 pages of the *New York Times* within the company every day (Pitman 17).

You may access in-house documents electronically from a single database through a local area network or shareware program. Select what you need from old documents to create your new document. Use information management software to track your work, store different versions, and sift through your personal files. Merging a database of names with a database of all the memos you wrote last month, for example, allows you to create a file of all memos sent to one person. Take advantage of formatting techniques that let you prepare information for flexible reuse. Many companies are enhancing their documentation through multimedia programs. Videotape, for example, is often better than a textual description to record how to operate a one-of-a-kind instrument and to teach someone how to perform a task.

### Publications

In addition to in-house documents, you may need to consult a variety of publications to find the information that will meet your purpose. These publications are referred to collectively as "the literature." A review of such literature often forms an early segment of a report or proposal, and review articles provide excellent summaries of the literature on a particular topic or problem. Print books and journals continue to be a major source of information: For example,

**For more about reviews of literature, see Chapter 21**

- The total number of scientific and technical articles published in the world in 1973 was 271,512; in 1991 it was 405,554 (NSB 421); in 1994 it was more than 1 million (Broad C1).
- In 1966 the National Library of Medicine indexed about 157,000 articles from medical journals; by 1993 the number had more than doubled to 400,000 (Goad 1B).
- At least 40,000 scientific and technical journals are published worldwide (Broad C1).
- The Chemical Abstracts Service, which publishes abstracts of scientific literature, produced 3,437 pages of material in 1907; 48,713 in 1970; and 119,000 in 1993 (Weiss 410).

The Government Printing Office is the largest publisher in the United States. The government issues technical reports, brochures, policy statements, regulations, maps and charts, census data, manuals, speeches, records of Congress, and periodicals. Similarly, other governments and international agencies such as the United Nations and the World Bank produce publications. Governments worldwide maintain patent records. Corporations issue such publications as quarterly and annual reports, reference manuals and other instructions for users, and catalogs. Professional associations publish proceedings of meetings and reports as well as journals and newsletters, although some highly specialized journals are available only online. Electronic technology enhances specialization that would be otherwise uneconomical; such journals encourage new forms of interactive argument.

### The Web

The database of choice for many students and professionals is the Web. The number of Web pages available continues to grow at a staggering pace. By one estimate, there were a billion such pages in 2000 (Specter 88); another authority says the number doubles about every 80 days ("Who Can Measure the Net?" 61). Increasingly, government agencies, universities, research institutes, individuals, and commercial vendors post information on the Web. It is not, however, a comprehensive database. Many topics are not covered at all, and coverage of some includes only very recent data.

## INDEXES AND LINKS

The task of finding the information you need in this maze of publications and Web sites may seem daunting. Take heart. Sophisticated tools can help you find your way, and as they increase in their searching power, such tools are also becoming easier to use. Through electronic technology, librarians and other information professionals are developing structures for storing information that will speed you on your way to what you need. These structures fall into two general approaches: access through an index and direct access through hypertext links.

### Indexes

Technical professionals and students are served by a wide range of indexes:

- The catalog of a library
- Published indexes that survey literature on a particular field
- Yearly indexes to individual publications
- Web directories and search engines

Finding information through an index is a two-step process. First, you look up your search term in the index, which provides a *citation,* that is, an address for where information on that topic is stored. You'll read more about search terms shortly. Then you go to the address and find the information.

**Catalog.** Use the catalog of a library to find out where (and whether) a book or other separately published item may be found in this specific collection. Through the Internet, you can enter the catalogs of many libraries across the world to find the works you need.

**Published Indexes.** Both publishers and professional organizations issue special indexes to information published serially, that is, in magazines, journals, and newspapers. Such indexes are the workhorses of research. Each index usually lists citations under subject headings as well as author and title. Horizontal (or subject) indexes survey the literature on a single subject or field, for example, *Biological Abstracts, Engineering Index,* and *Chemical Abstracts.* Vertical, or publication, indexes survey a single publication (for example, the yearly index to *Science* or *Gourmet*). Some indexes are issued weekly, some monthly, some yearly; frequently issued ones are usually compiled annually.

For more about abstracts see Chapter 13

Some indexes also include an abstract of each entry. An abstract is a short summary of the contents of an original document. The abstract may be particularly useful if it presents in English the major findings of an article published in a language you don't understand.

One variant index is *Science Citation Index.* It traces, in effect, the offspring of an article by indicating all the articles in which a known article is cited. It indicates the clusters of research around a particular topic. It is thus a useful tool in showing collaborative patterns in science as well as the influence of one article on other work (Figure 5.1). *Science Citation Index* is one of many available electronically as well as on paper. Because your library pays a onetime cost for a CD, you can usually use such indexes for free, as you would use a print index. You may have to pay a fee to access

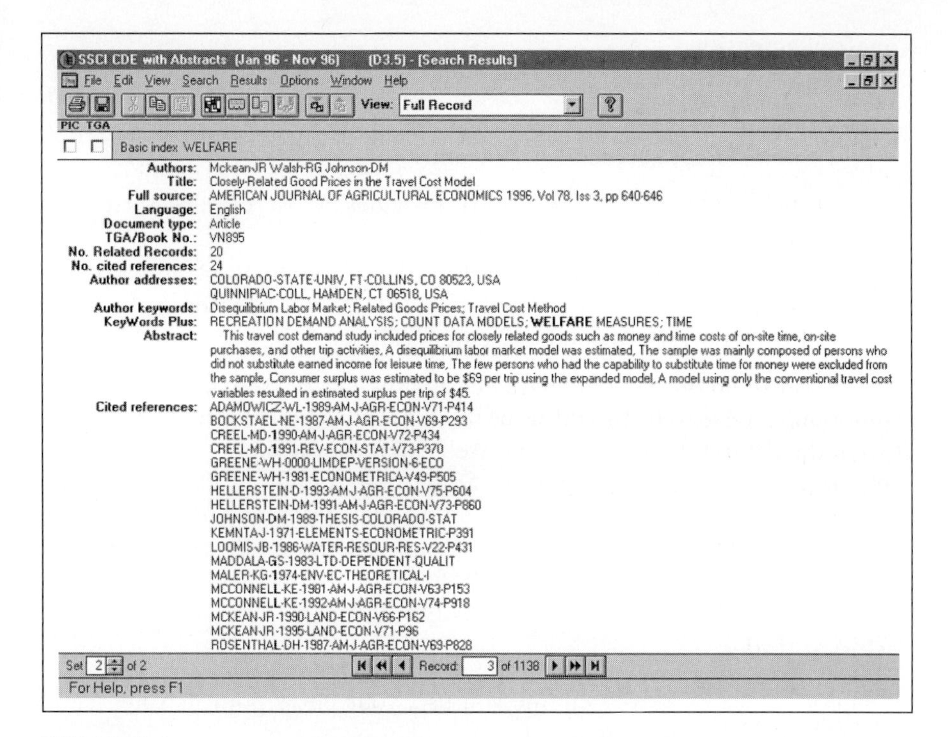


**FIGURE 5.1** **Screen from *Social Science Citation Index***

A sample entry that provides information about one article in the *American Journal of Agricultural Economics* and about other articles that cite it.



an online, commercial bibliographic service, but you may be willing to pay that for its more frequent updates.

**Web Searches.** Many programs and strategies help you search the Web for information:

- A *browser* is a program that lets you navigate the Internet—like Netscape's Communicator or Microsoft's Internet Explorer.
- A *search engine* is a program that works within your browser to help you find the information you need. You type a keyword or keywords into a dialog box; it looks for Web pages that contain those words and provides a list of documents (*hits*). Popular search engines include AltaVista, Excite, Northern Light, Google, HotBot, InfoSeek, and Yahoo; by some accounts, there are more than 1,500 search engines.
- A *directory* classifies, organizes, and prioritizes information. Browsers and search engines provide such directories for their customers; typical topics include recreation and travel, business, finance, arts and leisure, and the like. Within each topic, you can move down several levels:

Recreation
Skiing
Snow or waterskiing
Snow skiing: extreme, cross country . . .

Beaucoup (www.beaucoup.com) provides a list of directories and search engines.

- A *portal* is a gateway to the Web, increasingly focused on a specific topic or cluster of topics, like the science portal Ingenta or the financial news portal provided by Reuters (Figure 5.2).

Because no one search engine covers the entire Web, use several for a comprehensive search. In addition, don't limit yourself to the first page or two in the list of hits. The more the Web expands, the more likely you are to get irrelevant or useless responses. Some search engines also accept fees to prioritize listings. So try several and use a metasearch engine that combines other engines, like Metacrawler and AskJeeves. These present one long list of hits, usually ranked by relevance.

## Using an Index

Whatever index or search engine you use, your entry point is your search terms. To find appropriate terms

- Read a few articles on the topic.
- Review textbooks to see what words they use to refer to the subject.

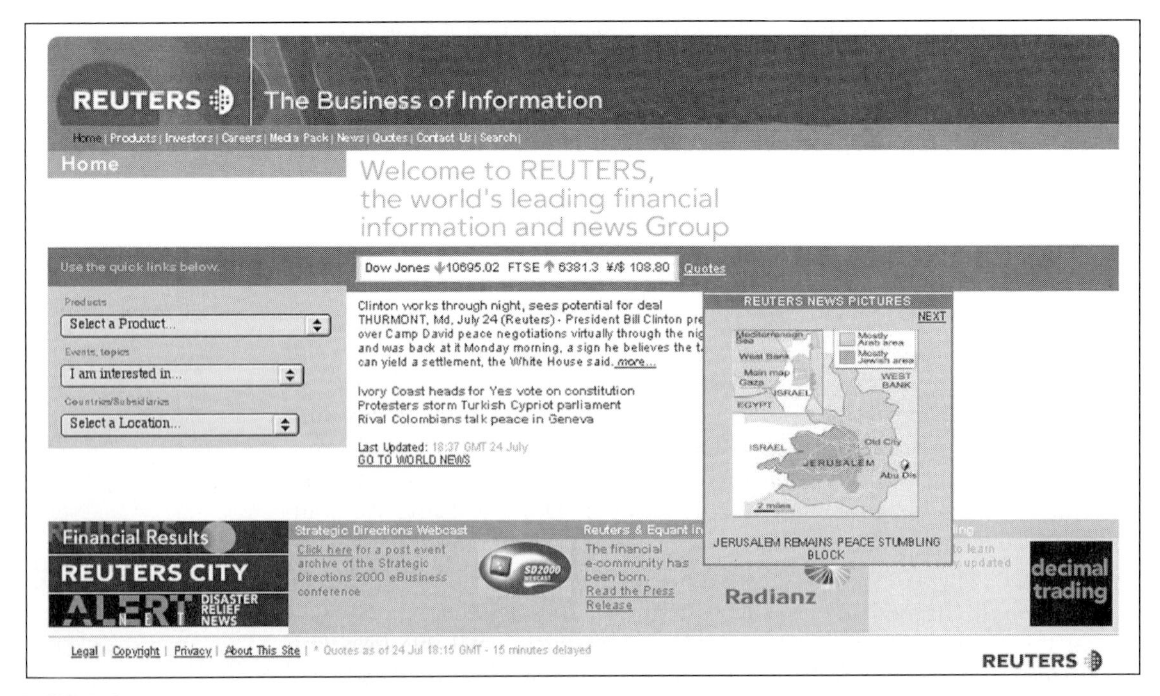


**FIGURE 5.2 Home Page from Reuters, a Source on the Web for Up-to-Date Financial Information**

- Talk with a reference librarian.
- Consult specialized dictionaries, the thesaurus published with most indexes, and the Library of Congress Subject Guide.

Start with the more unusual term; add a more general term later, if necessary. Think of synonyms if your initial terms produce no results. Because terms also change meaning over time, you may have to enter new categories in extensive searches.

Your goal is to find publications that contain your search terms in them. Each match is a hit, and your hit rate will increase in a computerized search if you incorporate specialized terms called "Boolean operators" that help you specify the range of items you are interested in. For example, one operator—AND—allows you to find articles that combine two or more key concepts:

- Entering the terms "Nutrition AND women" will locate articles in the database that discuss the nutritional needs of women.
- "Nutrition AND women AND athlete AND California" will locate discussions about nutritional issues among women athletes in California.
- To limit the search to articles that don't refer to pregnant women, use the "NOT" term: Nutrition AND women NOT pregnant.
- To account for synonyms or other similar terms or spellings that may appear in the literature, use "OR": Nutrition AND women OR Female.
- Plurals sometimes cause problems. If you want to retrieve articles that use terms in both singular and plural form as well as in different parts of speech, add an asterisk to the root word: female* or athlet*. That tells the computer to find anything with that root and any kind of ending (*females, athletes, athletic*).

These operators also work with search engines, although the default settings may vary from engine to engine. For example, some use "or" rather than "and." In one, a + functions like "and"; – functions like not:

+ heart + disease + women locates sites with material about women and heart disease

+ pets – dogs locates sites about all pets *except* dogs

With some search engines, you can also search for phrases, placing quotation marks around the words you want to find cited together in some documents. For example, entering "I have a dream" in AltaVista brings you directly to the URL for the full text of Martin Luther King, Jr.'s speech.

Although it is a powerful tool, Boolean searching is limited to the presence of the right keywords in the right documents. A U.S. government project aims to remove this limit. A researcher is devising a way to map keywords in the neighborhood of other words whose company they often keep. Thus, even if the original term doesn't show up, your search will still spot the neighbor. For example, an article about the Boston Red Sox baseball team may not include the term *baseball,* but it might include *a three-hitter.* That neighbor term would be enough to achieve a match if you begin your search with the term *baseball.* When sports terms are put on a map, the words *baseball* and *three-hitter* would be neighbors (Miller B1).

Searching an index electronically has significant advantages over a manual search. It is

- Faster.
- More convenient; you can sit at a terminal, perhaps even in the comfort of your own home, and log onto indexes that might be widely spaced in the library—or not available at your particular library.
- More dogged and comprehensive; computers don't tire in the way humans do.
- More current; online indexes can be updated more frequently than print ones, and thus information becomes available faster.

But electronic searching also has a downside:

- Commercial online services can be expensive.
- Many electronic indexes cover only the last ten or so years of publications; you'll need to use a print index for items published earlier than that.
- Currently, only a small percentage of Web pages are indexed.
- Electronic systems are also subject to power outages and malfunctions that don't affect print.
- You may see something as you browse a page or the library shelves that an overly focused computer would miss.

For the present, it's probably best to use both forms in your work.

### Hypertext Links

Indexes continue to serve as a major tool for finding information. In addition, electronic technology allows information to be structured so you can bypass indexing tools and go directly to the material you need. The concept underlying this structure is *hypertext,* the storing of information in marked modules that can be linked in multiple ways. Clicking on a key term in one document takes you directly to related discussions in other documents. Once your search engine has identified a match between your search term and sites on the Web, you can click on the title of the Web page in the index to go directly to the page itself or to a fact sheet about the source of the page, if one is available. Some browsers also provide a translation function, which you can invoke if the site is not in a language you understand. When you reach the selected Web page, you may find it has hot links to other sites appropriate to the search you are conducting.

A hypertext structure also underlies many reference tools on CD-ROMs. One large category is encyclopedias. Entries include audiovisual essays that combine narration, music, and photography and provide a rich network of links to each other. A CD-ROM on the periodic table allows you to click on an element of interest and obtain information on the element's history, spectra, physical properties, and the like. Professional societies like the American Heart Association produce compact versions of journals in cardiology that allow you to enter search terms and receive the full text of current journal articles that contain those terms.

Online services bring current information in hypertext format into your home or office. One such service, MEDLINE, is a database of 3,700 journals that helps physicians focus in on diagnoses, select effective test procedures, and generally keep abreast of the latest information in their specialty. For example:

- A physician consulted MEDLINE when he suspected that a patient's pancreatitis (inflammation of the pancreas) might be associated with a drug she was taking. He typed the name of the drug and "pancreatitis" into the system. That combination soon produced the abstracts of four articles, one translated from Japanese and two from French. Those sources, which he would not otherwise have encountered, provided the confirmation needed for his diagnosis (Winslow B1).

A CLOSER LOOK

## Documents and Documentation

The shape of information is changing rapidly, especially in response to electronic technology. In this setting, familiar terms like *document* and *documentation* are assuming new and sometimes confusing meanings.

On paper, a *document* is a separately issued item, anything from a single-page memo to a magazine, a book, or a multivolume report or encyclopedia. Online, a document may be a digital equivalent of any one of those. But an additional feature of online presentation is the permeability of the borders between documents. An author assembles a document in one form. As a reader at the screen, you can take segments from many documents to create your own personal one in another form. Documents merge into a database on the topic.

Let's turn to the term *documentation.* It has a familiar meaning to students and other researchers: citing sources in a report. In a technical or business context, the term also means "a written record of a process or system":

- When you *document* a software program, you write down all its features and all the steps a user must perform to make it work.
- When you *document* the accounting process at your company, you observe and survey everyone who contributes to the process, you record what they do and the forms they use, and you create a standard for the performance of those tasks. This standard documents the "best practice" in a process.

Documenting processes helps companies ensure quality in their work and certifies to others that the company knows what it's doing. The certification reflects a growing emphasis on meeting standards in the global marketplace. As they attend to their processes, companies record that attention in documentation.

### Finding Images and Sounds

Multimedia systems are solving many problems in structuring information for easy retrieval. But they are creating at least one other: the problem of retrieving items from the vast archive of images and sounds that can now be stored in the system. Sophisticated programs that find keywords easily locate information in text. But images and sounds resist such techniques. You can create a separate database of text that describes the images, but that description may miss the features you need. Information scientists are working on techniques for compressing, storing, and indexing digital images in ways that preserve their essential meaning (Wilson A20). One approach is to codify simple characteristics, like colors or patterns, as a search procedure. The merging of computer and entertainment technologies means that videotapes, films, network television, music—all may be accessed through a common system.

## TAKING NOTES

However you get there, at some point you have to read the text, numbers, and images on the screen or the page:

For more about reviews of literature, see Chapter 21

- Begin your reading with current documents and work backward. The latest articles offer clues to earlier work.
- Look in particular for review articles on the subject. These summarize publications on a topic and list extensive bibliographies.
- Take notes to create a holding place for information you'll need to incorporate in whatever document you are writing.
- Filter the information you read to make sure it's sensible, accurate, timely, relevant, and reliable.
- Evaluate all publications:

  Could the source be biased?

  Is material reviewed before it is published or posted? By whom?

  What is the reputation of this publication in the field?

  Who quotes this publication? Are they reliable, and thus do they consider the source reliable?

If you read online, you can download documents or segments into your own personal computer to read later. You can keep your notes online as well. If you write between the lines of what you're reading, include some icon or symbol to mark your comments or create your comments in a different typeface or font. A better plan is to develop a separate file of notes. With your preliminary plan in mind, arrange units of information into modules, each keyed to a topic heading. The goal is to limit each module to an item that will not have to be further divided in your report: one fact, a brief summary, a paraphrase of a section, or a direct quotation (in quotation marks).

If you prefer the feel and flexibility of paper, you can put each module on a note card. Maintain a separate file that lists complete citations for all your sources, and write an abbreviated version of the citation for the source within the module as well. That may be the author's last name and the date of publication or a number that corresponds with a citation in the list.

When you take notes on material from the Web or other online sources, be careful to mark the beginning and end of any text you download. Once it's in your word processing system, that text looks on the screen just like what you write. But you need to clearly indicate, whether your notes are on a screen or on a card, what information is in your own words and what is quoted matter. Make sure, too, that your note will be adequate for understanding when you reread it a week or a month later. Writing good notes also means you have already written a module of your report when you start a draft. Good notes ease the composition. Take time while you have the source at hand to be complete. Your notes will probably take one of three forms: quotation (textual or visual), paraphrase, or summary.

## Quotation

Early in your writing, you may be tempted to quote everything because the topic is new. But once you are familiar with the field, you can resist that temptation and take confidence in using your own form of expression. Save direct quotation for rare occasions. The quotation should be especially economical, memorable, and authoritative. Visuals you seek to reproduce should bear up under reproduction; a color original, for example, may not be as informative in a black and white reproduction. Change figure numbers and captions to fit the sequence in your own report.

**For more about quotation, see Chapter 21**

## Paraphrase

When you restate information from a source in your own words or visuals, you paraphrase. Use paraphrasing

- To test how well you understand a passage.
- To highlight an author's main point.
- To condense a wordy or roundabout original.

In doing so, make sure you are faithful to the author's context and intent. Name the source as part of the paraphrase and provide a source note to mark its end.

> ***Original:*** It is significant to note that the estuary has exhibited a dramatic recovery in water quality, especially in oxygen levels, compared to the conditions that prevailed 20 years ago. This is a result of wastewater treatment facility construction. It also results from the reduction of industrial discharges that has followed the enactment of such legislation as the federal Clean Water Act.
>
> ***Paraphrase*** (with source named as part of the sentence and citation at end): To paraphrase Guerreo's testimony, water quality in the estuary, especially

oxygen levels, has improved greatly in the last 20 years because of enhanced wastewater treatment and federally mandated reductions in industrial discharges (Guerreo 10).

## Summary

Use a summary to communicate all the main points of a whole document or a segment in a nutshell. To summarize

- Read the original carefully, probably more than once.
- Sort out the main point or points from the examples or illustrations in which they are embedded.
- Look for key terms.
- Economize on words.

**The original has a leisurely style, with more qualifiers**

***Original:*** The soils of the Pinelands have developed on sandy geologic deposits and are unusually porous and acid. Since Pinelands soils for the most part have a larger proportion of coarse particles than fine particles, they drain rapidly, have little ability to retain moisture or nutrients, and are highly erodible. Major Pineland soils series and their characteristics are listed in Table 2.21 of the CMP Part 1 (see Appendix G). Characteristically, the surface ("A" horizon) of an unlimed Pinelands soils is extremely acid (pH 3.6–4.0). The subsoil and substratum ("B" and "C" horizons) tend to be either extremely acid or very strongly acid (pH 4.2–5.0), with an average pH of about 4.6. The seasonal water table is generally high. There is great potential for nitrate, ammonium, pathogenic viruses, phosphorus, and certain organics to move into Pinelands groundwater, since the soils are generally low in clays and organic matter which can hold such materials. (Source: *Final Environmental Impact Statement, Proposed Comprehensive Management Plan for the Pinelands National Reserve.*)

**More details and references**

**The summary style is more compact: "mostly" replaces "usually" and "for the most part"**
**Fewer details; the focus is on main characteristics**

***Summary:*** Pinelands soils are mostly coarse, sandy particles that drain rapidly, leach readily, and erode freely. The topsoil is generally extremely acidic, and the underlying soil is only slightly less acidic. The acidity along with the seasonally high water table leads to dangerous leaching of fertilizer components, pathogens, and other organics into local groundwater (Julie A. Fine).

## Annotation

Unlike a summary, which takes a neutral stand on the source, an *annotation* is a critical and evaluative statement, sometimes as a form of advertising. You may be asked to prepare an "annotated bibliography" of sources to be used in a report or as a free-standing document.

An annotation may be short and informal ("must read," "don't bother"). A longer version assesses the reliability of the book, article, or report and discusses its significance and importance to a potential user. It may discuss the author's credentials, the subject and purpose of the work, its format, its strengths, and any errors or weakness. Some annotations are written in full sentences, but this practice is not always followed. The following annotation shows an appropriate approach:

*Envisioning Information.* Edward R. Tufte. Cheshire, CT: Graphics Press, 1990. 126 pages. This is a highly regarded publication by an internationally known expert on information design. Tufte is a polymath who teaches political science and statistics at Yale, as well as lecturing in the university's law school and reviewing students' graphic design work in the School of Art. The thread that pulls these different activities together is his study of how information is organized and conveyed visually, and the excellent fruits of that study appear here. The book includes both contemporary and historical examples, all integrated seamlessly into his discussion, in accordance with his

CROSSING CULTURES

## "The Internet Speaks English"

Although the Internet connects people across the globe in ways not even dreamed of ten years ago, until recently, it has made that connection mostly in English. The language of air traffic control, scientific publications, and business transactions, English, particularly American English, has dominated the Net. For some Internet users, the American bias raises ethical issues. One Internet provider in Russia called it an "act of intellectual colonialism" that reinforces American cultural powers and further divides the world into haves, who can access information electronically and in English, and have-nots cut off from information sources ("World, Wide, Web: 3 English Words" 5E).

But even now, and especially over the long term, English is losing its monopoly. A prominent reason is the explosive growth of e-business. According to one marketing research company, nearly two-thirds of e-commerce projections ($1.6 trillion) will be accounted for outside the U.S. in 2003. In that year, Americans will constitute no more than one-third of those online (against one-half in 1998) ("The U.S. no longer has the monopoly on the Internet . . . and even less in e-commerce by 2003").

Companies are springing up to aid Internet users in escaping English. One free e-mail service site, for example, allows you to compose a message in 28 languages, including Vietnamese and Urdu. In Japan, Internet use is increasing, especially among young people (those under 34). Some of these new users are comfortable with English, but others are fueling a market for more Internet sites in Japanese. Various companies are developing systems to translate computer data into many languages, even Chinese characters and the Cyrillic alphabet. E-business organizations in particular are finding that customers click onto a competitor when they find a site only in English, with product prices calculated only in U.S. dollars. In a 1998 study, Forrester Research found that people spend twice as much time on sites in their own language and are three times more likely to buy from them than from a site only in English. In addition, customer service costs are lower when the site offers instructions in the customer's native language (*Global Reach Express* 23 November 1999). English will probably remain a dominant language of the Internet, but customer-oriented sites in the U.S. are dimming their American accents.

belief that "tables and graphs are really paragraphs about data." His guidelines for achieving good design are at times idiosyncratic, but following them would help an author spark a reader's imagination. He also gives compelling examples of "chartjunk," illustrations that fail his tests; they distort information and mislead and annoy readers. The *Boston Globe* called this book a "visual Strunk and White," a comparison that gives a good sense of its practicality and wit.

## CITING SOURCES

Good notes will help you store, and thus remember, the information you need. Good notes should also help you remember where that information came from. You'll acknowledge your sources as a way of recognizing ownership—of words, of pictures, of numbers, of facts, of ideas, of points of view. *Why* and *how* you acknowledge sources in a document depend on the context in which you write and, more broadly, on an understanding of the obligations you as a writer have to the people you write for and the people whose ideas and language you write about.

### Why

Appropriate citing of sources is a matter of honesty; it also serves to establish your own authority as you cite sources whose credentials are well established. Let's look at what's appropriate in several situations.

- Readers of some genres or types of documents expect explicit and detailed accounting about who owns what in the material presented (Jameson). Your instructor, for example, is probably such a reader, and any assignment you write in class must explicitly acknowledge the source of such proprietary information as the following:

  Direct quotations
  Opinions and predictions
  Statistics derived by the original author
  Visuals in the original
  Author's theories
  Case studies
  Unique research procedures

  Failure to acknowledge sources may bring severe disciplinary sanctions.
- Professional journals require authors to let their readers know whose work informed the article at hand. The "literature review" confirms the thoroughness and validity of the author's research and points readers to original sources from which they can pursue another line of inquiry.

For more about literature reviews, see Chapter 21

- More broadly, in professional circles, anonymous information is often suspect. To be credible, information must bear the mark of the person responsible for creating it, the conditions under which it was developed, and the context for its interpretation.

- Within an organization, documents stored in paper or electronic databases are often fair game for reuse by others in the organization, to the point where no one can say with certainty who the original author is. In effect, the organization is the author of a collaborative work. One person may write something signed by another person; coworkers blend their texts in composite documents that list no authors' names. Such practices are considered perfectly acceptable.
- Newspaper and magazine writers routinely give less than a full accounting of all their sources. Their readers trust that the situation is being fairly and ethically represented; they avoid the publication if it proves untrustworthy; and they do not expect the same hefty apparatus of documentation as in a term paper.
- Outside the United States, with its intensely legalistic and individualistic view of property rights, standards for citing sources are generally more relaxed, except in those scientific and technical publications that address U.S. readers as part of their international audience.

### How

Knowing *how* to acknowledge sources is simply a matter of finding the right rules and applying them. The right rules are those your audience (an editor, a professor, a supervisor) expects you to follow. Your documentation should be both unobtrusive and complete. Every documentation system consists of two components: a brief citation next to referenced material in the text itself, and a full citation or "address" for each source in a list of references at the end of the text. In-text citations may be indicated by a number keyed to a numbered list of references or, more commonly, by an author's name. The name may appear as an agent in a sentence. Or it may appear in parentheses, accompanied by a page number, as in this textbook. The references may be listed in alphabetical order or in the order of citation. They are numbered if numbers are used to key citations to the text. In doing your research, make sure you record all the publication information for a source, even if some information (for example, the author's full first name or the volume and issue number of a journal) proves not to be appropriate in a particular style. It's easier to eliminate an element than to recover a missing one.

Different organizations, government agencies, professional societies, and journals publish their own style guides (see Chapter 12). These are frequently updated to reflect changes in publication practices, especially the increase in the number and types of electronic sources. Read the rules that apply to your document, and then imitate examples of citations that conform to those rules. Note particularly

- What information is included
- What punctuation separates the elements

- How capital or lowercase letters are used
- Which items are underlined

Apply only one style in any one report or article, and apply it consistently throughout.

As you prepare a list of references, conform to your reader's stipulations for its content as well as its form. To do so, you may also need to prepare a *bibliography*, which is an alphabetical list of all the sources you consulted in writing the document, whether or not you refer to them directly in the text. A bibliography is rarely necessary in company reports and journal articles, but as a student you may find that an instructor requires both references and a bibliography to confirm your search was thorough.

**ELECTRONIC EDGE**

## Documenting Electronic Information

Electronic information requires new approaches to documentation. Such information is volatile and often changeable. The postings on Web sites, for example, come and go, and a source note needs to account for such change. In addition, you compile information by linking easily from source to source, creating a single text that, unless you are careful, may mask the identities of individual contributors. To avoid problems in citing electronic sources, one authority suggests taking these steps (Arnzen):

1. *Save all electronic documents you intend to cite.* To avoid a potential problem with not finding again a source you bookmarked for a later reading, download documents to your own disk or save them in your mailbox. That copy provides a text for you to use even if the Internet source deletes it. You can also cut and paste direct quotations without retyping (but emphatically *with* proper credit). And you have a text to authenticate the source in case your instructor requests it.

2. *Cite original print sources, if appropriate.* Many electronic documents summarize or duplicate ones also available in print. Because one purpose of citing sources is to help readers find them, you may be able to help the reader in that task if you also find and cite the original print version.

3. *Find signed documents whenever possible.* When a document is posted, linked, or forwarded electronically, the author's name is sometimes stripped or included only at the end. But knowing that name, and that person's credentials, is an important factor in judging the document's validity. So avoid giving much weight to anonymous postings, and instead locate the name of an author. Sometimes an e-mail message to a Web master or Listserv owner will help you find that information. Validate sources against each other.

The differences in approach to documentation may seem both overwhelming and perverse, but a little practice will help you adjust to the reader's preferred style. Three guides commonly used by technical students and professionals are those of the American Psychological Association (APA), the Council of Biology Editors (CBE), and the Modern Language Association (MLA). Other popular guides are published by the University of Chicago Press and the U.S. Government Printing Office (GPO). Links on your textbook's Web site will take you to detailed instructions for each of these styles.

## CHECKLIST: Collecting Information from Documents and Web Sites

1. **Exploit the wide range of information available in print and online sources**
2. **To do so, use *indexes***
   Develop keywords to search for appropriate information in a database
   Apply them to print and electronic indexes
   Use them to launch search engines on the Web
3. **Follow *hypertext links***
   On the Web
   In a CD-ROM database or in online journals
4. **Take notes**
   Work backward from current documents, especially review articles
   Indicate clearly what is quoted material and what is in your own words
   Avoid long quotations: summarize and paraphrase
5. **Cite all sources**
   Recognize the ownership of words, ideas, images, and points of view
   Establish your own authority by citing experts
   Apply the rules for citation style expected by the audience in the document's context

## EXERCISES

1. Compile an annotated bibliography on the topic you have chosen for an individual or group report. In your annotations, note the intended audience, scope, and special features of each source and describe briefly the contribution you expect that source to make as you write your report.
2. Find the following information (and note how you found it):
   - The names and manufacturers of drugs used to control blood pressure
   - The names and addresses of the editors of two journals: the *Japan Journal of Opthamology* and *Corrosion*
   - The size of a soccer field
   - The current price of an ounce of plutonium
   - The names of people who have cited the article by Watson and Crick about DNA in *Nature*

3. Examine at least one index and one abstracting service in your field available in both CD-ROM and print. Compare the print and the CD-ROM versions. Which is easier to use? Why? Then compare the two different sources. How do they differ:

   - Scope (What do they include? What's missing?)
   - Purpose (For whom are they prepared? What kinds of inquiries should they satisfy?)
   - Type of information given (What details are provided in each entry? What accessing terms do you need: author, subject, keyword, other?)

4. Locate an article on some aspect of science policy in the United States or abroad. You might, for example, examine rulings or actions by the U.S. Food and Drug Administration, a United Nations organization, or the World Bank. Summarize the article.

5. Develop ten keywords that would be useful in a literature search on one topic, for example, "nutrition." Conduct a Boolean search with those terms in several indexes available on CD-ROM and online. Record any cross-references or other terms that seem relevant as you proceed. Note which terms are most effective in helping you gain the right material.

6. Write a memo to fellow students in your major field evaluating for them the five most significant journals in your field. For each journal, include the following information:

   - Complete citation, including availability in print or electronic form (or both)
   - Readers: who, how many (circulation) (see Chapter 6)
   - Coverage: the general topics of the articles
   - Reliability: how do you know you can trust it?

   Justify your choice of the journals as "significant."

7. Be prepared to discuss in class any differences you discover in the content of the leading journal in your field within the last five years. Then expand your review to cover differences in the last ten years.

## FOR COLLABORATION

### A Workshop on Plagiarism

At the extreme, *plagiarism* is a term easy to define: presenting as your own original work something which is not. When you copy a source word for word, and don't tell your reader the text is a copy, you plagiarize. But as the discussion in the chapter suggests, that simple definition may not accommodate all the decisions you need to make to use sources ethically. The following collaborative exercise, adapted from Jameson, should help you fine-tune your skills at ethical documentation.

Compose six teams of three persons each. If the class is large enough, two teams may work on the tasks described here. Or you may investigate additional genres, for example, internal corporate documents, handbooks, and advertising material. After completing its assigned task, each team will report its results orally to the class for general discussion and provide a brief written summary (with full documentation) for the instructor. Throughout the exercise, pay particular attention to how your sources use direct quotation, paraphrasing, and summarizing.

*Task 1.* Collect definitions of plagiarism. Sources include your department or university as well as professional societies. Such societies publish their definitions in society-sponsored journals or in separate brochures available from the society's office. Look for definitions in style guides like those listed in Chapter 12 and on the Web. Summarize the definitions and comment on any inconsistencies.

*Task 2.* Using the term *plagiarism* as your team's keyword, conduct a search of the *New York Times Index* and the *Periodicals Index* to locate articles about people who plagiarized, the form of their plagiarism, and the consequences of their acts. Each team member may report on one major case of plagiarism or each may detail several occurrences in one category, for example, politicians, scholars, or scientific researchers.

*Task 3.* Research the distinction between *plagiarism* and *copyright infringement,* two actions that both overlap and differ. In particular, look at the electronic implications of the terms.

*Task 4.* Collect examples of how popular writings about science represent their sources. Look at newspapers as well as magazines. As a reader, what kind of accounting do you expect? Are your expectations fulfilled? Do the articles give generalized source statements ("According to a recent article in the *New England Journal of Medicine,* . . .") or more detailed ones? Do practices differ in different popular publications? In general, what are the ethics and mechanics of documentation in this genre?

*Task 5.* Apply the questions listed in Task 4 to a sample of professional journals.

*Task 6.* Apply the questions listed in Task 4 to a sample of manuals for consumer products, for example, for computer software and hardware.

CHAPTER 6

# MANAGING INFORMATION FOR READERS

"INFORMATION IS NOT COMMUNICATION. IT IS PERCEPTION."

PETER DRUCKER

When you conduct research, you enhance your understanding of a topic or problem. You develop a warehouse of information related to your research questions. That warehouse is the starting point in managing information for a reader. It's not the end point, although it's easy to make that mistake. Instead, think of your reader as a *customer*. Select information from the warehouse, and design your presentation as a showroom that attracts and holds the customer's attention. This chapter will show you how to analyze your audience so you can meet both their needs in an information product and your own needs.

## FROM WAREHOUSE TO SHOWROOM

Customer-oriented thinking underlies good print documents; it's also especially important in oral presentations and on the Web. Understanding your readers as customers is the first step in designing an information product, the subject of extensive discussion in Part 3.

### Taking the Customer's Perspective

When North Americans look at the moon, they traditionally see a face in the dark spots—the man in the moon. When people in India and parts of China look at the moon, they traditionally see a rabbit; Australians see a cat; Fiji Islanders see a rat (Miller 87). The next time you look at the moon, look for the image you don't usually see. People do see things differently depending on their expectations, their prior knowledge and experience, and their purpose. Astronomers studying the moon, for example, see details of structure and composition that have nothing to do with faces, rabbits, or rats. To begin the shift from warehouse to showroom, learn how your readers see things. Here are some examples:

1. A technician tests a soil sample taken from a site formerly occupied by a gas station to determine the presence of any toxic products. She records the findings and includes them in a quarterly report to the Environmental Protection Agency, which monitors the site.
2. A U.S. team and a Swiss group meet to see if their companies might form a joint venture. One Swiss member takes the minutes of the meeting, which both groups approve, and then summarizes the critical issues in a report that recommends the joint venture to the teams' supervisors in each country.
3. A technical writer interviews the developers of her company's new business management applications as well as five customers who used the company's earlier products to record the features of the new and any problems with the old. She then develops a prototype for a Web site that will present the new features as a way of marketing the product. The site also has links to customer support organizations, an interactive checklist for customers to determine which specific combination of applications is best for them, and the capability to be accessed not only in English but also in German and French.

Each writer selected from a warehouse of information those facts or interpretations needed for this particular document. Each document addresses a different audience: a regulatory agency (1), corporate decision makers (2), and users of a product (3). Each has a different purpose: to inform (1), to recommend (2), and to sell and instruct (3). Documents 2 and 3 also required the writers to communicate across cultures—corporate cultures and national cultures.

## Selecting from the Inventory

To see how writers convert their stock of information to meet the needs of different readers, look at the five passages that follow. All are about the same topic: the development of a theory to explain color perception. Each introduces an article in a publication, and they are in order from most to least technical.

*Journal of the American Chemical Society*

The chromophoric unit of visual pigments is known to consist of 11-cis-retinal covalently bound in the form of a protonated Schiff base to the c amino group of a lysine in the apoprotein opsin [1]. Protonated Schiff bases of retinal absorb at ~440 nm in polar solvents while various salts formed in nonpolar solvents absorb at somewhat longer wavelengths (~440–180 nm) [2]. The visual pigment bovine rhodopsin has an absorption maximum of ~500 nm

> while other 11-cis-retinal-based visual pigments have maxima as far to the red as 580 nm. The mechanism through which the protein shifts the absorption maximum of the chromophore from its solution value to wavelengths ranging from 440 to 580 nm has been a question of major interest. In this communication we present the first experimentally based model which accounts for the absorption properties of a specific pigment, bovine rhodopsin. (Excerpted with permission from Barry Honig, Uri Dinur, Koji Nakanishi, Valkeria Balough-Nair, Mary Ann Gawinowicz, Maria Arnoboldi, and Michael G. Motto. "An External Point-Charge Model for Wavelength Regulation in Visual Pigments." *Journal of the American Chemical Society* 101 [1979]: 7084–7086. © 1979 American Chemical Society.)

In this passage from a professional journal, the authors, who are the scientists who did the work, write in a "high-context" environment to colleagues familiar with technical terms and concepts like *chromophoric unit, Schiff base, 11-cis-retinal.* They move through a swift recap of what is known, noted in the references to the literature ([1], [2]). That review introduces their new information: the mechanism of absorption. The discussion is theoretical and dense and serves the authors' purpose of establishing through publication that they developed the "first experimentally based model."

***Chemical & Engineering News***

> A single molecule, that by itself has sensitivity to ultraviolet light, serves the eye by proving sensitivity to the broad spectrum of visible light. The way that the molecule, 11-cis-retinal, presents the brain with a rainbow instead of mere shadowy images now has yielded to precise chemical explanation. The effort to develop that explanation "took many years" and involved about a dozen scientists, principal among them organic chemist Koji Nakanishi of Columbia University in New York City and biophysicist Barry Honig now at the University of Illinois, Urbana. The project has depended on the synthesis of a family of highly unstable compounds closely related to 11-cis-retinal and subsequent analysis showing how that one compound can serve several biochemical masters to give broad spectrum visual perceptions. (Excerpted with permission from Jeffrey L. Fox, "Chemical Model for Color Vision Resolved." *Chemical & Engineering News* 12 November 1979: 57(46) 25–26. © 1979 American Chemical Society.)

A staff writer, not the researchers, wrote this news item in a journal that serves a broad base of scientists and engineers. The explanation of light sensitivity that begins the article is more concrete than theoretical. The name of the molecule (11-*cis*-retinal) is presented as new information the reader is not expected to know and is thus defined. The author focuses on both the researchers themselves as news makers and on the findings. The first paragraph sets up a mystery that motivates the reader to move to the second. Shorter paragraphs than in the first passage also ease reading.

***Science News***

> Each light-sensitive cell of the human eye responds to a particular wavelength of light. Some sense red, some green and others blue. Yet the same chemical component is involved in detecting each hue. A molecule called 11-cis-retinal absorbs light in every receptor cell, but the large protein molecule to which the retinal is bound determines what wavelength of light it best absorbs. Now Koji Nakanishi of Columbia University and Barry Honig of the

> University of Illinois report just how the protein influences retinal's light absorption. Precisely located negative charges, probably on the amino acids of the proteins, are responsible for color discrimination. (Reprinted with permission from *Science News,* the weekly newsmagazine of science © 1979 by Science Service Inc.)

This explanation is even more concrete and assumes less prior knowledge of chemistry. The author uses fewer technical terms and includes transitional words (*yet, but, now*) to guide the reader. The last sentence provides a summary of the mechanism that allows the reader to stop there and move on to another article—or to continue to read the details, if interested.

***The Chronicle of Higher Education***

> How and why human beings, monkeys, freshwater fish, and a few other animals see colors has been explained for the first time by Koji Nakanishi, an organic chemist at Columbia University. For years, scientists have known that the body gets "11-*cis*-retinal," a light-absorbing molecule that governs perception of color, from fish and dark green vegetables that contain vitamin A. Once absorbed into the body, the vitamin-A derivative travels to the eye's retina, where it binds with one of four "visual proteins," known more commonly as pigments, three of which are involved in color vision. (© 1979, *The Chronicle of Higher Education.* Excerpted with permission.)

This passage, from a weekly newspaper for university administrators and faculty, assumes little background in the topic. It aims to entice the reader. To do this, the author moves to the level of the picturable and the familiar—human beings, monkeys, freshwater fish—and away from the level of molecules and protons. The author also stresses the news value: "for the first time." The names of the researchers are generalized as we move to more broadly popular accounts. Technical terms are enclosed in quotation marks, the author's acknowledgment that the reader is not expected to be familiar with them.

***The New York Times***

> Working with highly sensitive chemicals in a red-lit laboratory at near-freezing temperatures scientists at Columbia University have performed experiments enabling them to answer a hundred-year old question about color vision.
>
> Their new understanding of normal color perception may also point the way to future practical applications in the treatment of color blindness.
>
> Prof. Koji Nakanishi and his collaborators have demonstrated how a single substance, called retinal, can be responsible for perception of all four types of color messages: red, green, blue, and black and white. (From "Scientists Discover Answer to Color Perception" by Dava Sobel, *The New York Times,* 11/28/79. Copyright © 1979 by The New York Times Co. Reprinted by permission.)

In this account from the *New York Times,* emphasis shifts from the research to the researchers, particularly those at a local institution, Columbia University. The tale is told as a story of happenings in a "red-lit laboratory" at "near-freezing temperatures." The sense of drama and discovery motivates the reader. Practical applications are also stressed—applications not mentioned in the other passages. They answer the nonscientist's question,

"What's in it for me?" The molecule is here generalized to "retinal." This audience does not require more precision in the name. One-sentence paragraphs keep the reader's eye moving.

The authors of these passages selected, arranged, and expressed information to match their image of what readers of that publication would look for—or should look for—in an article on the topic. Often, you, too, will create an "ideal reader" to write to, especially in such documents as reports and articles that address several readers at once. At other times, you'll need to incorporate details and devices that accommodate one individual.

## PROFILING THE READERS

Whether your readers are ideal or real, establish a profile of them. The checklist in Figure 6.1 provides a good starting point for any document you need to write, whether in print or online. Then adjust your approach to match that profile. Knowing their personal characteristics may also help you engage their attention: age, gender, handicaps, hobbies, interests, socioeconomic status, political leanings, and memberships.

### Purpose

Determine why the reader or readers need the document—its *purpose* or *purposes:*

- To keep up to date?
- To learn how others have solved a problem?
- To inform themselves before making a decision?
- To accomplish some task?

Find or create the overlap between your purpose in writing and your readers' purpose in reading. For example, if you are selling home systems for solar heating, you'd need to create one kind of document, with extensive engineering drawings, to convince a zoning board that your system conforms with codes. You'd write another document for a home owner who would like to add a solar system and would be interested in the difficulties of such an addition, potential savings in energy costs, any trade-offs in comfort compared with the current system, ease of addition and maintenance, and the like.

Your readers will use your document as a tool to get their work done. To return to our earlier examples:

- A regulator at the EPA reads the soil toxicity report to ensure that the company owning the site is in compliance with agency regulations.
- Division managers at the U.S. and Swiss companies read the recommendations of their project team to decide whether to go ahead with the proposed joint venture.

What is the document's purpose?
Why should the readers read this document?
What do they already know about the topic? The situation?
What do they expect from *you?*
How credible are you in their eyes?
Do you need to *establish* credibility?
What standards or conventions apply to this kind of document?
What *genre* is appropriate in this context?
How will they use the document? Where?
What delivery system do they prefer: On paper? Online?

**Addressing Multiple Readers**

Do you need to address multiple readers with one document?
If so, what is the relationship between or among the readers?
Who is the primary reader?
Are there also secondary readers?
Are there immediate readers in your organization who must approve the document?

**Writing Across Cultures**

Is the document form you're used to for this purpose different from what your readers might expect?
What do your readers consider polite behavior? What are safe (and not safe) topics?
What elements of style and format, what rules and norms, apply internationally in this situation?
Do you need to teach your readers about the context for the document?
How can you collaborate with your readers before writing and as you write to help ensure effectiveness?

FIGURE 6.1 **Profiling the Reader or Readers**

- Customers and potential customers visit the Web site to learn about features of the business management program and determine its applicability to their tasks.

To avoid frustrating your readers, determine early on how they will use your document.

## Prior Knowledge

Base your approach and selection of information on the readers' prior knowledge about the topic, and thus the new information they need. In the passages on color vision, you read how different writers pitched their discussion of color vision to a range of levels from experts (who are well informed before they read) to novices (who know little). The writers took into account what information their readers already knew and what they needed to explain and elaborate.

Writing for expert colleagues, who share a high-context environment with you, encourages shorthand. When writing for novices, however, you may need to motivate them and fill in some gaps in understanding. For novices, you spend more time on the background details and preliminary steps (how to locate an on/off switch, for example) that would tax the patience of an expert, and you avoid technical terms that might be off-putting. As you saw in the passages from the *Chronicle* and the *Times,* you often need to capture reader attention through such interest-getting devices as analogies, anecdotes, stories, and startling statistics. You're more likely to focus on practical rather than theoretical dimensions and to present concrete rather than abstract evidence. For any document, sort through the information you need to present to determine what will be new information for the reader and what the reader already knows. Begin with a review of the known as a framework for the new.

A reader's prior knowledge can speed communication and foster understanding, but it may also impede comprehension if the information conflicts with the presentation being made. People often resist the new. In addition, ingrained habits of thought may make learning difficult. As Nobel laureate Roald Hoffmann notes,

> When I try to explain chemistry to outsiders, I have three main audiences: the person in the street, fellow academics in the humanities and physicists. All three audiences are equally ignorant of chemistry, but the most difficult audiences are the physicists, because they think they understand, but they don't. (Browne C1)

## Expectations

In Hoffmann's terms, the physicists let their prior knowledge stand in the way of new information. Readers also interpret information to conform to what they expect from you and from a particular kind of document. If they expect to see a face—or a rabbit—in the moon, that's what they'll see. Determine what readers expect in your document and then meet that expectation. Or devote an opening section of the document to teaching them to expect something new.

**Your Persona.** First, establish yourself as someone whose documents should be read. Readers are more likely to read a message from someone they know and trust than from the many others who write to them. So create a professional persona that establishes your authority as a writer who can be depended on. Write individual messages when you need to gain special attention, especially in an e-mail system that highlights such messages. Readers are more likely to respond to a message sent to them personally than as a member of a group.

For more about professional persona, see Chapters 7 and 21

**Conventions and Genre.** Second, arrange your message to conform to the conventions and the genre appropriate for your context. Use your knowledge of the literature in your field to ask the questions readers would ask, to select acceptable forms of evidence for answering the questions, and to sequence that information in the right way. Each genre is a kind of recipe, and following the recipe helps you meet readers' expectations about what kind of information will be presented and where it will be located. Parts 4 and 5 provide detailed guidelines for composing different genres of scientific and technical communication. The first passage you read on color vision, for example, begins with a brief review of the literature and sets the stage for the discussion in a pattern typical in the scientific article.

**For more about scientific articles, see Chapter 21**

Following a recipe also makes it easier for you to write. Figures 6.2 and 6.3 show some of the diverse genres necessary to document the development of a product. At each stage, a document or an oral report summarizes what has been done, aids decision makers in determining if the product should be taken to the next stage, and provides salespersons with evidence to sell the product to customers.

Knowing the kind of document readers expect at each stage will help you meet their expectations. Figure 6.2 arranges the documents on a scale that measures their relative emphasis on the record or the reader: Those at the bottom are warehouses; those at the top, showrooms. The base of

Speech to a local group (boy scouts, Kiwanis) — showroom/ for the reader
Oral presentation to customers
Paragraph in the company's annual report
Announcement in company bulletin
Product brief
Sales guide
Sales brochure
Trade journal article
Speech at professional conference
Professional journal article
Formal memo to board of directors
Oral presentation to company management
Operation manual
Maintenance manual
Installation manual
Market assessment report
Progress report
Patient disclosure
Laboratory notebook — warehouse/ for the record

**FIGURE 6.2.** **Genres of Communications About a Product: From Warehouse to Showroom**

**INTERNAL**

Laboratory notebook entry. Progress report. Presentation to company management. Formal memo to board of directors. Product brief. Announcement in company bulletin. Paragraph in annual report for stockholders.

**EXTERNAL**

**Marketing.** Market assessment report. Manuals: installation, operation, training. Magazine article. Trade journal article. Sales brochure. Product labels and packaging information. Sales guide.

**Professional.** Article in a technical or professional journal. Paper at a meeting that is then published in the meeting's proceedings.

**Other.** Patient disclosure or filing with a regulatory agency. Specifications for a supplier or manufacturer.

FIGURE 6.3. **Genres of Communication About a Product: Internal and External**

information remains roughly the same, but a rise on the scale is accompanied by a decrease in information density and an increase in the devices needed to appeal to the audience. Figure 6.3 takes a different look at the genres by dividing them into those that address audiences within the company (internal) and those written to audiences outside (external). Marketing documents address customers directly or aid salespersons in approaching them. Professional documents review the development process and address technical colleagues. Other documents address such audiences as government agencies and suppliers.

**Method of Reading.** Third, imagine how (and maybe where) your readers will use the document. Simplify whatever elements of the document you can to reserve the reader's energy for conquering information that is inherently difficult. Readers are often resistant and inattentive, tossing what seems unsuitable, selecting material worth a second glance, and then reading the selections only in bits and pieces. They scan and skim to find enough information to solve their problem or perform their task. To get readers to settle into your text, make the information they most need most accessible to them:

- Structure information in segments.
- Label segments clearly.
- Provide indexes, tables of contents, glossaries, and other supplements that aid readers in retrieving information.

In addition, accommodate the direction in which your audience reads. Arabic and Hebrew, for example, are read from right to left (the reverse of English); Japanese, from right to left and top to bottom. That bias affects how people read visuals as well. Japanese charts, for example, sometimes include the vertical axis on the right side; Japanese readers of

**For more about design, see Chapter 7**

a diagram that included a "man/auto" (for "manual/automatic") switch called it an "auto/man" switch as they read from right to left (Stevenson). When you arrange a series of pictures that show a process, consider using vertical arrangement to avoid the confusion of what is the beginning and what is the end. Most people read vertically arranged material from top to bottom.

**Media.** Finally, in this brief list, decide which medium your reader would prefer. The scientists who wrote the original article on color vision selected the *Journal of the American Chemical Society* because it reaches the expert audience they wanted to address, and its prestige lends special authority to their findings. You'll need to select from an increasingly large range of print and electronic channels for delivering your material. As you do so, be careful to avoid incompatibilities in technology. Computer systems are increasingly able to talk to one another. But you and your reader may not share the same version of a software program, and a printer abroad may not use the same paper size as yours does—seemingly small problems that may become big ones if you depend only on electronic delivery.

## ADDRESSING MULTIPLE READERS

Your reader profile will help you to determine your approach as you convert information from the warehouse of your investigation to the showroom of your document. For example, each passage on color vision accommodated a composite profile of the audience for that publication.

In addition, in an organizational setting, you may need to write one document that addresses readers who have different interests reflecting their roles in the organization and their differing responsibilities. As shown in Figure 6.1, you can identify these differences by thinking about three types of readers:

- A *primary* reader who is probably the person who requested the document and who will be its major user
- An *immediate* reader, someone (or several people) in your organization who must approve the document
- *Secondary* readers who will be affected by the decisions or information the document contains (Mathes and Stevenson 40–50)

The following examples will help you identify these three types.

- A computer specialist documents a computer system for a bank. Her audiences are the head of information systems (her supervisor), the managers of the bank, and ultimately the programmers and clerks who will have to learn the new system.
- A chemist in an independent research laboratory reports to the National Institutes of Health (NIH) the results of tests that the NIH is sponsoring on certain drugs. Other readers are his supervisors at the lab and the drug manufacturers who may use the results in developing their products.

- A consulting engineer recommends a water storage system for a city. Addressed to the city engineer, the report also needs the approval of managers at the consulting company before it is sent to its destination. In addition, it may be read by voters.

The bank managers, the NIH, and the city engineer are, in each instance, the primary readers, whose needs should take priority in your writing. But before your document reaches them, you may have to meet the needs of immediate readers, that is, colleagues or supervisors who will approve its content and approach before sending it along to the readers. In the examples here, the computer programmers and clerks, the drug manufacturers, and the city residents are secondary readers.

The path through such multiple readers can become complicated. You may need to create a different document for each. You may have to negotiate with immediate readers if they contradict each other in reviewing your document. Such negotiation is common in many organizations as supervisors take on the role of coauthors of your document. The patient record in a hospital, for example, serves as a major way for health care providers to communicate. But some hospital administrators insist that the record be written for another audience, that is, lawyers, insurance companies, and review boards who monitor the hospital's accountability. When you address multiple readers, you may find it helpful to draw a chart that indicates the relationships between them. Figure 6.4 shows the relationship among immediate, primary, and secondary readers in a large corporation. An engineer writes a report whose primary reader is a real estate supervisor. The immediate reader is the supervising engineer, who may ask other engineers to look at it as well. Secondary readers are real estate agents who may be interested in the content of the report. Depending on their roles and responsibilities, some readers focus on the technical discussion, others on financial considerations, still others on management issues or policy implications. Figure 6.5 provides a different example: the readers of a student's resume. Here, the primary audience is potential employers. Immediate readers who provided advice are friends and other people who agreed to write letters of reference.

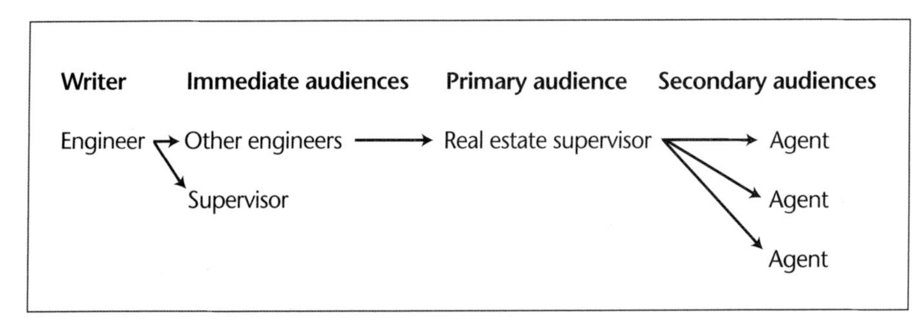


**FIGURE 6.4 Readers in a Large Corporation**

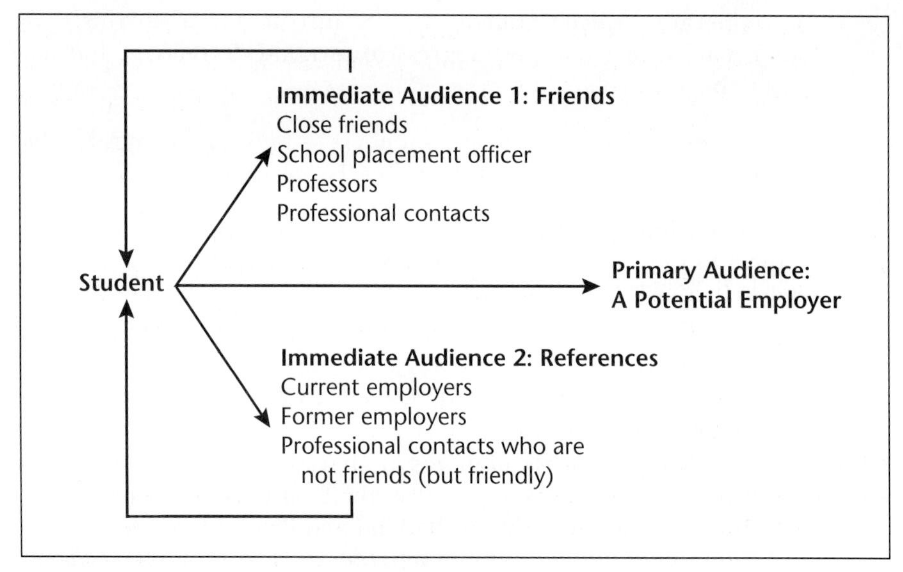


**FIGURE 6.5** **Readers of a Student's Resume**

## WRITING ACROSS CULTURES

Increasingly, you'll find yourself addressing readers who are different from you. To do so effectively, determine the values and forms that guide communication in that setting. Differences in both corporate and national culture often show up as differences in genre, that is, in the form for presenting information:

- In one company, internal proposals for action always appear on green paper, one page only, with segments clearly blocked out by preprinted headings that limit the text to be included. Only proposals that meet these requirements will be read.
- A computer company successfully adapted its instructions from the genre of the manual familiar to U.S. readers into the genre of the *manga,* or comic book, familiar to Japanese readers.
- Health care professionals in Mexico were effective when they cast their instructions in the form of *fotonovela,* soap opera comics that attract a wide readership there (Forslund).

More broadly, determine the values that guide communication in the culture you are addressing. For example:

- An American computer company, acting on its awareness that the Japanese value the artful packaging of food, presents, and flowers, created instructions with a glossier look, including bright colors and photographs, than those for the American market (Lombard).
- Several researchers remark that the Japanese have a higher tolerance for dense page design than Americans, who require more white space.

- The imperative voice common in U.S. instructions ("do this," "do that") may seem harsh and aggressive to non-U.S. readers and may need to be avoided.

Match visual messages about time and numbers as well as warnings to what your audience's culture leads them to expect. For example:

- Researchers found that the symbol of the moon is appropriate in some cultures to indicate the concept of month. A picture instructing non-readers in Sierra Leone to breast-feed their babies for six months after birth includes six moon-and-star symbols over a drawing of a mother and child.
- But in Mexico, the symbol of the calendar was better understood as depicting a month (Forslund).
- Some cultures do not use numbers at all and some count differently from you. In Nepal, for example, as in the United States, people count each finger as one unit, but in Thailand and Bangladesh, people consider each joint of the finger as a separate number, so a picture showing three raised fingers would mean nine units, not three.
- To warn a reader that something is harmful or not allowed, graphic artists in Euro-American societies often use a circle with a slash through it or an X over an object. Nonreaders in Kenya, however, did not understand the meaning of a picture in which an X was superimposed on a drawing of a pregnant woman carrying a heavy load. Instead, publication developers used the more positive image shown in Figure 6.6. The pregnant woman's load is smaller than her companion's.
- An icon of warning common in U.S. software is the palm-up "stop" sign. That icon may be misread, however, in an international context. A raised hand has an opposite meaning in Buddhist iconography where the gesture (*mudra*) means "fear not." Moreover, in some cultures a raised hand has pornographic connotations.

You meet your readers' needs by understanding how they see things and presenting information in their preferred form, even if it's not yours. But in some circumstances, that approach may not be the best one. A Swedish academic tells a story that suggests why. A Swede who was an upper-level manager in a technical company traveled on business to Tokyo. Well aware that Japanese businessmen dress formally, he packed a suit in his carry-on luggage and changed into it just before arriving. His Japanese hosts, expecting their Swedish colleague would dress informally, changed from their suits to casual wear to meet him at the airport. In the global economy, it's not always easy to tell in each situation whose rules and norms apply.

So another approach to dealing with differences is to develop in your own mind a picture of your ideal reader and write for that audience. What should a reader of the document want to know? How should the reader respond to what you have to say? What should the reader's attitude be? Teach your reader to take on that role and feel comfortable. Explain, probably in

**FIGURE 6.6 A Positive Approach to Conveying a Visual Warning**
A pregnant woman should not carry a heavy load.
(Source: Zimmerman et al., 34. Used by permission.)

the document's introduction, your reasons for choosing the approach you take in the document. You'll use such an approach particularly in low-context situations where every element of the communication needs to be spelled out.

Finally, you may accommodate diverse professional audiences by writing in what might be called an international style. Similarities in education and consumer habits make this option attractive. Many professionals worldwide have been educated in the United States, collaborate with U.S. researchers, and use U.S. software for communication. So standards for international communication reflect many U.S. practices, especially the use of English, making writing abroad easier for North Americans. Electronic messages in particular often reflect U.S. norms. But an effective international style isn't blindly American. Instead, you need to incorporate an approach that recognizes, for example, your audience's expectations about polite behavior and safe or inappropriate topics for discussion.

CROSSING CULTURES

## Medical Instructions for the Nonliterate

Writing across cultures requires collaboration between writers and their audience and a deep understanding of cultural differences. A good example is the approach used by The Program for Appropriate Technology in Health (PATH), a Seattle-based group, to prepare health-related materials. PATH works with other international development agencies to improve "the safety, availability, and acceptance of health products and technologies worldwide" (Forslund 2). PATH responds to invitations from host countries and collaborates with representatives of the target audience in developing the materials. First, PATH holds "focus group" discussions that include a leader (usually from the host country), randomly chosen representatives of the target audience, and publication developers. Early discussions elicit the health care issues as well as the attitudes and visual literacy of the audience. In later sessions, sample materials are tested. Then, in-depth interviews with the target audience help to confirm a final design.

In one project, PATH developed visual instructions that demonstrate for nonreaders in Sudan how to prepare a packet of oral rehydration solution (ORS) to be given to a child with diarrhea. The figure shows four successive versions that incorporate changes recommended by the target audience.

Commentators interpreted draft A as a picture of a woman in an office because of the large table and because the woman wears a *tobe,* a long piece of cloth women in Sudan typically wear over their dresses when they go outside the home. The audience did not notice or could not identify the sun or the ORS packet. Nor did the audience perceive the woman as preparing something for a child because no sick child was shown. Thus draft B was drawn to include a father and child as well as a shorter table. The mother still wears a *tobe* because, during other pretests, the audience mistook the woman without a *tobe* for a man; a woman would be responsible for child care. The audience noted two problems with draft B: The child looked too active to be sick, and the scene did not show what time the ORS packet should be prepared.

In response to these comments, draft C was created. It includes a clock whose face shows the time, a mosque, and a lamp—all indicating the evening. The father is shown praying (another sign of night) and the child looks sick. But the target audience could not understand what the clock meant because it was an unfamiliar symbol. Although they recognized the mosque, its presence did not improve their comprehension. Pretests of draft D showed that respondents could recognize the picture's components and could derive the intended message. The lamp replaced the mosque to indicate night, and the

## THE READER AS COLLABORATOR

Although this chapter has emphasized your role as a writer in creating documents that work for your readers, readers don't just take what you give them. They play a dynamic role in creating meaning based on their

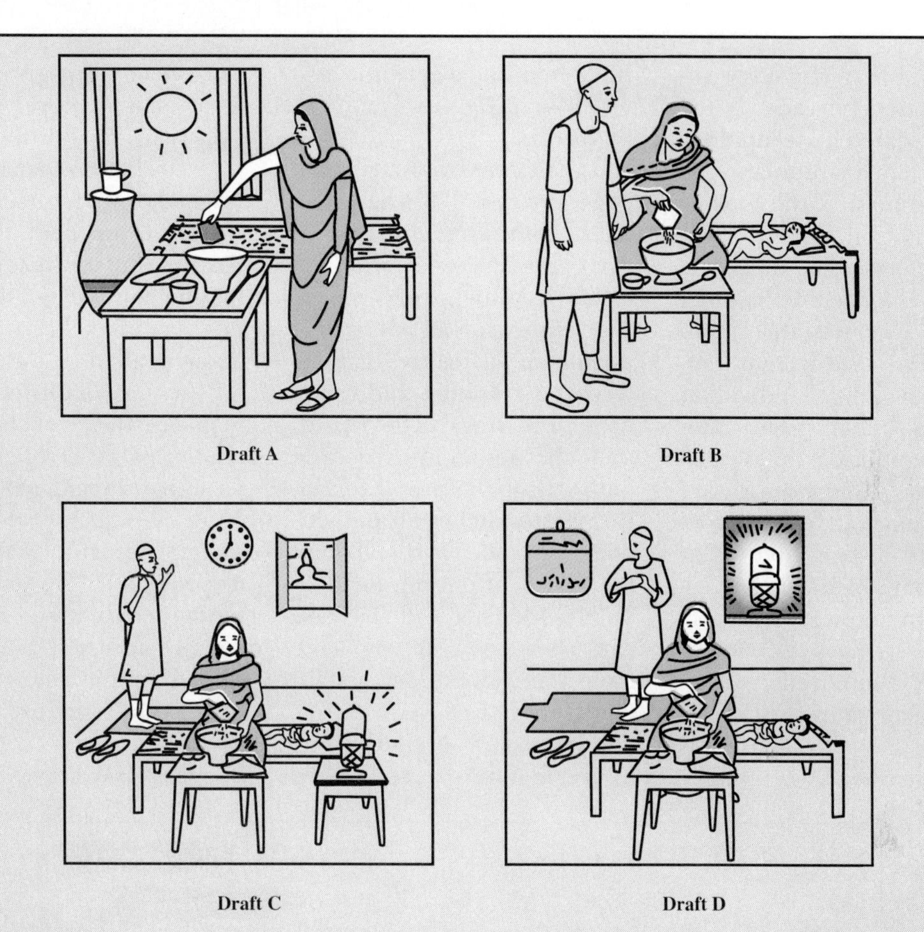


Four Drafts of a One-Frame Message About Preparing an Oral Rehydration Solution.
(Source: Zimmerman et al., 36–37. Used by permission.)

wall calendar was added. Even though many viewers could not actually tell what date or day the calendar indicated, they still understood it as adding a time element.

own purpose, prior knowledge, and expectations. That role is particularly significant when you deliver your information electronically (see *Electronic Edge: Reader-Created Text*). If possible, talk with your reader(s) before you write and while you write to confirm your approach and to develop

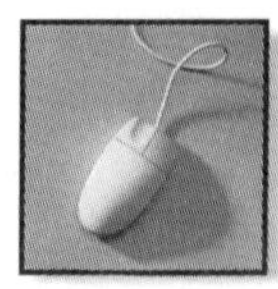

ELECTRONIC EDGE

## Reader-Created Text

Electronic technology often puts the reader in the driver's seat. The technology lets readers manipulate the organization and the look of a presentation. In hypertext and hypermedia, designers package information in independent modules that readers can link in different ways to match their individual preference and pace. The technology fosters "mass customization," that is, easy ways to make many individual documents from one large database. From what one designer wrote into an original hypermedia presentation, readers can create hundreds of different texts to reflect their own interests. Newly developed "markup" languages further allow electronic text to cross different computer platforms.

The author of a printed page creates a linear document that aims to control what readers see and the order in which they see it. The author of an electronic document writes the text, selects the graphics, and establishes the links. The reader then chooses an idiosyncratic path through the information that matches an individual's purpose and capacity for paying attention. Readers on the Internet and the Web can also assemble elements from the databases established there by several different writers or organizations to create a unique composite document. That gives readers a lot of power over the texts they read. In addition, a manufacturer who sends paper manuals to customers knows who receives them. But when that manufacturer places a manual online on the Internet, readers can remain anonymous.

It's difficult to show on a page the structure and reader paths through an interactive document. But several designers developed what they call a "Z diagram" to provide a "frozen view of hidden content and routes of content access" (Wurman 5). The diagram below demonstrates the structure of a hypertext program (courtesy of Rebecca B. Worley).

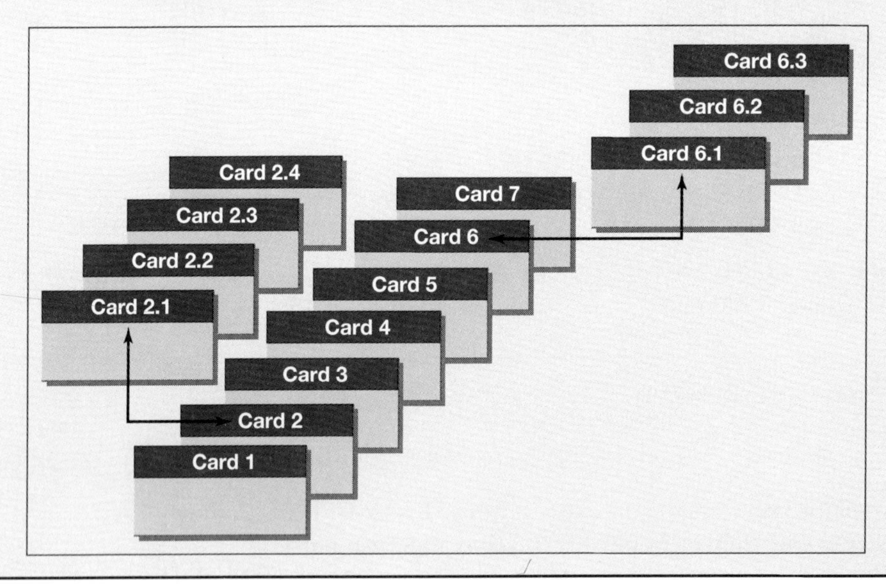

## CASE
## GETTING A RESPONSE

John Fiestra supervised the assembly room of a television manufacturing plant. He oversaw the mounting of the sets in cabinets and installation of the fittings: knobs, handles, and labels. He noticed that out of every dozen or so sets that came down the line, one had screw holes on the bottom that were misaligned enough so that someone had to jiggle the set to match it up with the cabinet holes for final mounting. This process took extra time and thus reduced the productivity of Fiestra's group. He also feared that the rough handling required for alignment would damage the sets and result in rejection by the quality control department to which the sets were shipped when they left his assembly room.

Fiestra decided to call the engineering design department, which had final authority over the design of the sets. A junior engineer to whom he spoke told him to write a memo about the problem to the senior design engineer. Fiestra did so, but two months later, he had had no response. He discussed the problem at lunch with a fellow supervisor in the chassis area, from which the sets came into the assembly area, and this person was surprised that Fiestra had even bothered to contact engineering design at all. He said that Fiestra should simply have told him and he would have followed up directly by adjustments in the assembly room. Engineering design, he said, never liked to deal with production people. Even if they got involved, it would take months of study before they agreed on a design change; he'd just modify it on his own to correct the problem.

Use your understanding of audience and channels of communication to write a one-paragraph answer to each of these questions about this situation:

**1.** Whom should Fiestra have contacted when he spotted the problem? Should he have communicated orally or in writing?

**2.** Should the junior engineer in design have told Fiestra to write the memo? Why or why not?

**3.** Should Fiestra have waited two months for a response to his memo? What alternatives did he have?

midcourse corrections when necessary. If discussion is not possible, try to take on the readers' role, asking, as you write, "Will they understand this term?" "Will they need more explanation here?" "Could they take this the wrong way?" Try to control the text so you avoid misunderstandings and never give readers the opportunity to say, "So what?"

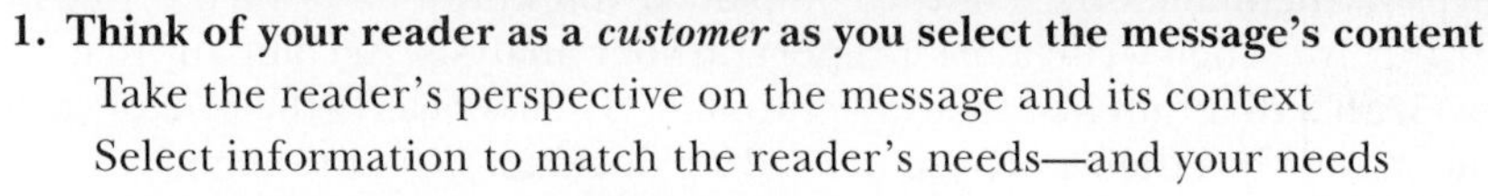

## CHECKLIST: Managing Information for Readers

1. **Think of your reader as a *customer* as you select the message's content**
   Take the reader's perspective on the message and its context
   Select information to match the reader's needs—and your needs
2. **Profile your reader or readers**
   Determine their purpose for reading and use for the information product
   Assess what they already know about the topic and situation

Identify what they expect in the message and from you
Imagine how (and maybe where) they'll use the document
Select the appropriate medium

3. **Accommodate multiple readers**
   Rank your readers to know whose needs are most important
   Chart the sequence of readers
   Make sure you meet the primary reader's expectations

4. **Write to be understood across organizational and national cultures**
   Use the structure and style the reader expects when they're different from yours
   Or collaborate with the reader to develop a new structure and style
   Use an international style

## EXERCISES

1. To prepare a report for your technical writing class, focus your purpose by selecting an audience who would need to read about the topic. For example, if solar energy interests you, consider who might need information on that topic as part of their work:

   Mechanical engineers and materials engineers
   Architects
   Manufacturers of hardware for solar conversion
   Designers who develop computer control systems
   Economists interested in energy cost-benefit analysis
   Plumbers
   Home owners
   Publishers of popular magazines
   Home financing institutions
   Zoning boards
   Tax accountants
   Engineering students

   Discuss in class the kinds of information each audience would require.

2. If you hold a job in addition to your work as a student, analyze the reporting relationships of people you work with, that is, the multiple audiences for documents written in your company. You may need to ask about audiences in the home office of the company if you work in a branch office or a franchise. Then write a brief report on your analysis. Accompany that report with a diagram like that in Figure 6.4. Discuss briefly the role of each reader and thus their purpose in reading documents sent to them.

3. Assume that you have written a resume to apply for a job (perhaps you really have written such a document). Draw a diagram like that in Figure 6.5 to indicate the people you would send that resume to (including their names and

titles). If your instructor requests, write a brief note explaining why you would choose these readers.

4. Read an article in your local newspaper or in a national newspaper like the *New York Times* or the *Wall Street Journal* that concerns some issue in science or technology. Then check in your library or online (see Chapter 5) for other discussions of the same topic in journals aimed at a more technical audience. Compare the strategies for addressing at least three different audiences on the topic. Your reading of the passages on color vision should help you construct your analysis. Attach a photocopy of one segment from each article as evidence for your analysis.

5. Write a brief profile (about 500 words) of someone who reads you: a supervisor at work, a professor, or colleagues on a work team.

6. The British and the Americans are two cultures "separated by a common language," notes American writer H. L. Mencken (the British attribute that quotation to G. B. Shaw, a British writer). Here is the abstract of an article written by a British artist. Can you detect any "Britishisms" in it that would be unfamiliar or at least uncommon for American readers?

   Consumer product instructions can often seem to be somewhat anachronistic. Whilst products, their packaging and advertising, are the subjects of intense scrutiny in the light of market trends and drives towards excellence, the humble instruction leaflet is left to fend for itself. Often poorly produced, confusing, inaccurate, or just plain ugly, the much maligned instruction booklet is frequently the butt of humour. "Destructions" rather than instructions is a term widely used to describe the obligatory post-sales bumph accompanying even the most respectable of products. (Williams 4)

7. The letter reproduced on page 120 from a concerned citizen in a London neighborhood was circulated to other residents requesting their participation in a meeting to stop the development of a restaurant in their block. As in Exercise 6, analyze differences between British and American audiences and political systems as suggested by this letter. In addition, note how the letter tries to persuade the reader. What is the letter's goal?

## FOR COLLABORATION

In a team of three, collect a range of guides to a particular city, region, or country. Then analyze how those guides appeal to different reader interests. For example, you may find guidebooks oriented especially to students (often with a theme of doing things as cheaply as possible), to people interested in literary sites, to families interested in sites to visit with small children, and to shoppers. You may find such guidebooks in bookstores or in your college or public library. In addition, review information available on the Web concerning your chosen region. Note the topics included for each audience, the level of language and thus expectations about the reader's prior knowledge of the region, the use of visuals, the amount of detail, and the form of the book. Will it, for example, fit in a tourist's pocket? Is it meant to be read at leisure and remembered? At your instructor's request, present your analysis as a talk to the class or a written collaborative report.

26th September 2002

MAJOR RESTAURANT DEVELOPMENT
10/12 Andrews Close

Dear Resident

On the 17th October the planning committee of the Royal Borough will be recommended by their officers to grant planning permission for a major catering development submitted by Andrews Estates. Whatever you may be told about the quality of the outlet, once permission has been granted further variations can be granted and within a short space of time a restaurant club could be developed with late noise etc.

I have organised a meeting on 4th October at the Danforth Hotel. The meeting will start at 6.30 PM and Shab Gupta, our MP, will be present together with Councillors and media.

I hope you have written formally to the council and more importantly will attend this meeting. Full attendance is essential to show the council and Andrews Estates that the proposal is unacceptable. Please also let your neighbours know about the meeting.

Kind Regards,
Barbara Scott-Phillips

PART THREE

# Designing Information for Readers

CHAPTER 7
DESIGNING INFORMATION FOR READERS

CHAPTER 8
COMPOSING VISUALS

CHAPTER 9
COMPOSING TEXT

CHAPTER 10
EXPLAINING

CHAPTER 11
PERSUADING AND PROVING

CHAPTER 12
REVISING

CHAPTER 7

# Designing Information for Readers



"DESIGN BRINGS ORDER OUT OF CHAOS."

Thinking of your reader as a customer helps you manage the content of a document, oral presentation, or Web site. Take a customer-oriented approach, too, in *designing* that information product. Pages and screens draw reader attention when they

- make complex information accessible.
- encourage action and remembering.
- look good.

To draw and hold reader attention, you'll need to design your information well. Good design also enhances your own credibility in readers' eyes. In this chapter, you'll look at options and conventions in information design. Integrating design principles will help you make sense while you also make a good impression.

## ORGANIZING

The first principle of design for technical documents is making the information *work* for the reader. To achieve that working order, you need to focus on a main point and plan the best strategy for supporting it. The context for your document may prompt you to find the point and the plan:

- Follow any explicit request from the reader concerning the form and substance of the document.
- Observe the conventions and rules associated with the genre of document you are writing, for example, a letter, memo, proposal, report, or article.

- Meet reader expectations about conventional patterns for presenting information, such as narrative, comparison/contrast, cause and effect, or elimination of alternatives.
- Meet expectations about where in a document the reader will find certain kinds of information (Figure 7.1).
- Deviate from the conventional and the expected when such deviation may suit your purposes, but be prepared to explicitly defend a different approach.

To supplement these prompts, try *brainstorming,* that is, thinking by association. Teams find this technique particularly effective. Jot down on a sheet of paper or a blackboard the phrases that represent the issues, and then start looking for connections in the jumble. Sort ideas into groups and identify the common thread that brings them all together (Figure 7.2). As you sort the groups, apply the principles of *hierarchy* and *parallelism.*

## Hierarchy

A hierarchy represents the *inequality* of items through a pattern of main points, subpoints, and their support. One problem with the original text in Figure 7.3 is that everything looks as if it's at one level; the writer didn't

Introduction
- Welcome the reader
- Review the problem or request that necessitated the document
- Establish a general framework for your topic and approach
- State any limits in the scope or coverage of your document, if needed
- Define key terms, if needed
- State the purpose of the document
- State the main point of the document
- State the plan of the document

Middle
- Provide the major subpoints of your argument or description
- Elaborate on or focus the meaning of those subpoints with evidence
- Arrange that evidence in a pattern the reader can recognize
- Periodically summarize what's been said and forecast what's to come

Ending
- Remind the reader of the main point
- Recommend future action, if appropriate
- Reinforce the reader' s good impression of you

**FIGURE 7.1 Document Segments: Each Segment Serves a Different Purpose for the Reader**

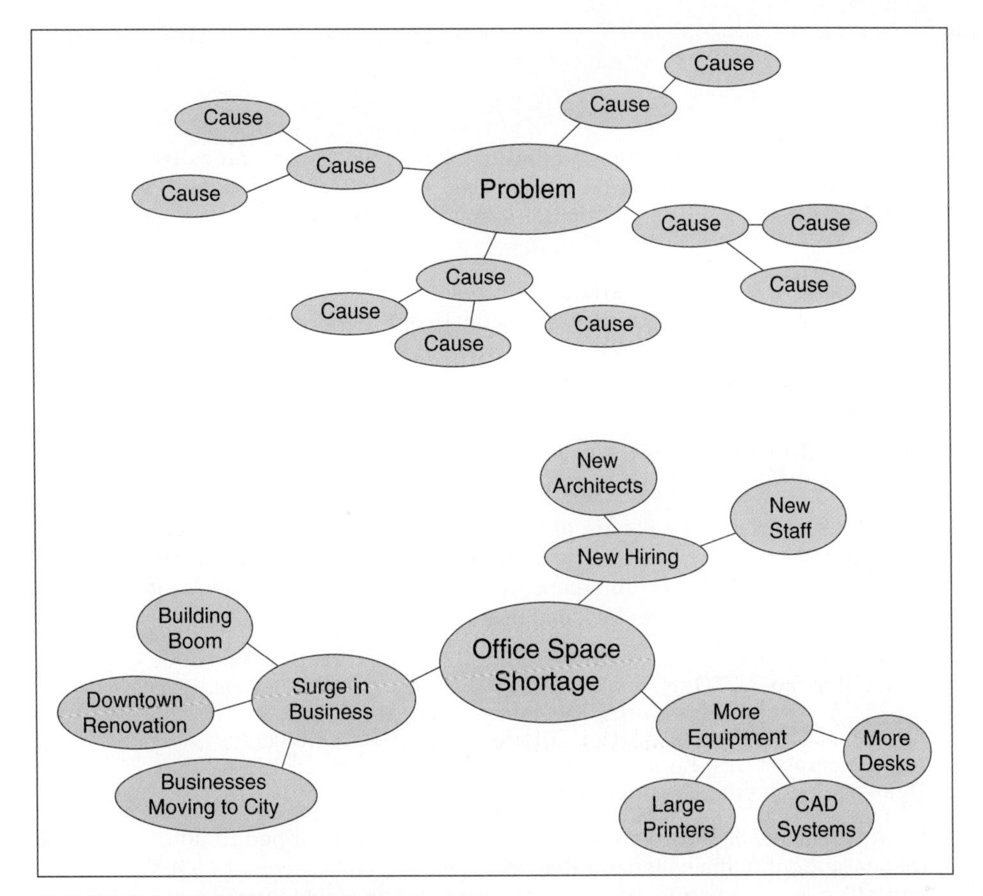


FIGURE 7.2 **Mapping a Brainstorm**
To capture the essence of a brainstorming session, try a visual technique called "mapping." A blank problem/causes map is at the top of this figure; at the bottom, you see how a team faced with a shortage of office space at an architectural firm used this map form to brainstorm causes of the problem. Clustered closest to the central circle are the major causes. Linked to these are the minor causes of each. Similarly, "solution" could occupy the central spot, and you could then record your thinking about options to solve the problem. (Maps reprinted courtesy of Robert Johnson-Sheehan.)

designate that pattern. The redesigned page reflects this hierarchy of information, with the definitions, for example, subordinated to a general statement about the types of intercepts. For both print documents and Web sites, establish the levels of your information and design your presentation to make those levels clear to the reader. Avoid creating more than three or four levels.

## Parallelism

Parallelism is a design strategy that reinforces the *equality* of items. Similar items are expressed in similar form. It operates at every level in a document: individual words and numbers, paragraphs, sections, chapters, pages,

**Original**

Running text is bland and hard to skim

Basic Intercept

Intercept calls are calls which are intercepted at the far end office because the called number is no longer valid. These calls are routed to TOPS to provide information about the called number, such as the new number.

Some reasons why the dialed number may be intercepted is the number has recently been changed or disconnected, the number is unassigned or vacant, there is temporary trouble with the number, or the number is associated with a split referral. A split referral is when one number maps to two or more different numbers.

The end office only sends the called number to the TOPS switch for intercept calls.

The types of intercepts appear before the reader sees why this information matters

The types of intercept calls are automatic intercept (AUTO INT), Intercept operator number identification (INT ONI), automatic number identification fail (INT ANIF), intercept cut-through (INT CUT), intercept recall (INT RCL), or intercept special (INT SPL).

If the called number is transmitted to the SYSTEM, then the call can be handled as an AUTO INT call. The SYSTEM communicates directly to an LSDB. The LSDB returns the information about the intercepted number to the SYSTEM. The subscriber is then connected to an Announcement System which auto quotes the intercepted information. Operator involvement is not needed.

Intercept ONI calls occur when the end office is not equipped to automatically identify the called number. Intercept ANI calls occur when the end office is equipped to automatically identify the called number, but fails to do so. The call routes to an operator terminal and the operator queries the subscriber for information on the called number and launches a query to the LSDB database. The operator then quotes the requested number or releases the call to an Announcement System. ONI and ANIF intercept calls are also known as operator handled (OH) intercept calls because the calls must go to an operator to obtain additional information. Note that the missing called number causes intercept calls to be ONI or ANIF.

**FIGURE 7.3 Before and After: How Design Makes Complex Information More Digestible**
(Courtesy of Robert Kelton.)

**Redesigned**

**BASIC INTERCEPT**

**What is an intercept call?**

An intercept call is a call that is rerouted from the end office to TOPS because the called number is no longer valid. When such a call is rerouted, TOPS offers information about the called number, which often includes a new, valid number.

**What causes an intercept response?**

The intercept response is caused when a number

- has been changed recently
- has been disconnected recently
- is unassigned or vacant
- is associated with a split referral
- is experiencing temporary trouble of some sort

Note that the end office only sends the called number to the TOPS switch for intercept.

**What are the types of intercept?**

There are several types of intercept depending upon the difficulty posed by the called number. The following section describes each type; these descriptions are summarized in the table.

**AUTO INT = automatic intercept**

Automatic intercept transmits the number to the SYSTEM, which uses an LSDB to return information directly through the Announcement System. An operator is not involved.

**INT ONI = intercept operator number identification**

If the end office cannot automatically identify the called number, the call becomes an Intercept ONI call. This is also known as an "operator handled" (OH) intercept call. An operator quotes the requested number or releases the call to an announcement system.

| Call Type | Description | Occurs when | And . . . |
|---|---|---|---|
| AUTO INT | automatic intercept | call transferred to SYSTEM | |
| INT ONI (OH) | intercept operator | End office can't identify number id | OH call |
| INT ANIF (OH) | auto number id fail | Call did not route automatically | OH call |

**Question-and-answer format builds motivation and eases comprehension**

**List form adds white space and aids skimming**

**Definitions appear when they are needed**

**Table format reduces prose and eases comparison**

**FIGURE 7.3** Before and After: How Design Makes Complex Information More Digestible (Continued)

screens, and visual display. As a technical writer, you'll find parallelism one of your most useful techniques in designing information. In Figure 7.3, for example, observe how the revised version makes the information about an intercept call more accessible as a series of parallel questions. In addition, within the section on types of intercept, the writer pulls out a series of terms and their definitions, all expressed in parallel form. These are also summarized in the table, which forces information into parallel rows and columns. The typeface for the questions is also parallel (all in boldface) as is the typeface for the terms being defined (capital letters and boldface). Similarly, the alignment of text on the page provides parallel placement for all questions. We'll look at more examples of hierarchy and parallel design throughout this chapter.

## ESTABLISHING AN APPROPRIATE PERSONA

More than just information, you convey something about yourself in each document you write. How much personality you reveal, and the voice you choose, depend on the context. In a formal report you'll probably take a relatively impersonal approach. But a memo that is soberly serious may well strike your reader as off-putting and fail to establish the authority of your words. In any document, design your information in a way that establishes a professional persona, a positive image of you that crosses the page or screen.

The persona may be yours personally, or you may represent the image of your organization. In a report to colleagues, for example, use the technical terms of your discipline and formal expression. Technical terms are essential for expressing information concisely and precisely. Such terms have only one meaning, agreed on by the experts, and no synonyms. Formal language, too, recognizes the seriousness of the occasion, contributes to clarity, and addresses the needs of audiences who come to English as a second or third language. That means that you generally avoid contractions (*don't, you'll, we've*) as well as any colloquial or conversational expressions typical when you talk with friends or want to signal that the occasion is a conversational one.

For more about London Underground visuals, see Chapter 8

Organizations often spend heavily to project their character in a design program that begins with their logo, the stylized trademark or signature that represents their name. One excellent example is the logo of the London Underground (Figure 7.4). It suggests rails running through a tube (the familiar name for the subway system is the *tube*). The design is clean, efficient, and modern; it works well both in large scale to mark stations and in small scale on stationery. It underlies choices about the typeface, visuals, and text in all Underground messages. See *A Closer Look: Branding* for more information on about corporate identity.

## SIMPLIFYING

Documents and maps of the London Underground achieve an effective simplicity. Most people on the job have too much to read and read in an

**London Underground Limited**
Group Station Manager's Office
Gloucester Road
London SW7 4SF
Telephone 0171 918 5538

October 2001

Dear local resident

**Sloane Square station**

The escalators leading from the two platforms at Sloane Square Underground station are amongst the oldest on our system and are overdue for a major overhaul.

I am writing to let you know that on 15 October work will start on the modernisation of the first of these; that which leads up from the westbound platform. i.e. trains coming *from* Victoria.

FIGURE 7.4 **Design Persona: The London Underground**
The Underground's distinctive design approach carries across its logo, system diagram, correspondence (illustrated here), and all posters and publications. The distinctive sans serif typeface provides a consistently modern, easily recognizable look.
(Reprinted by permission of London's Transport Museum.)

environment of distractions. Your design helps lessen information overload and focus reader attention. Especially in designing Web sites, avoid the temptation to dazzle. Graphics-oriented programs have brought the tool kit of a designer and a printer to a technical writer's workstation. But they have not necessarily brought the designer's wisdom. In their restless pursuit of variety, many writers create pages and screens that convey chaos more than content. Instead, simplify. The more complex the information, the simpler your design should be. Look for the underlying bones of an argument. Establish and repeat key elements rather than distracting the reader with unneeded variety.

**The London Underground Diagram is reproduced in Chapter 8**

## CHOOSING THE RIGHT TYPE

A major element in the success of the London Underground's design program is the distinctive typeface used in all their presentations. To design well, choose a typeface and size that express your information and convey the appropriate design persona. Computer text production offers a wide choice.

A CLOSER LOOK

## Branding

The *Financial Times,* the "world business newspaper," is printed on paper of a distinctive color, a salmon pink. That color makes it instantly recognizable at newsstands and in offices and libraries. The color served as the background for print ads in other newspapers, like the *New York Times,* when the *Financial Times* began strenuous marketing of a U.S. edition. It's the color of the banner heading on the newspaper's Web site *(www.ft.com).* And it's the color of the dumpster (or "skip") behind the Financial Times Company offices on the South Bank in London. That dumpster could only belong to one company.

A distinctive color is an obvious mark of the *Financial Times.* It contributes to its corporate identity, its visual look. More than that, it helps establish the newspaper's *brand.* Once only a process for burning a mark into the hides of cattle to identify ownership, *branding* today refers to business strategies for developing world wide recognition and the ownership of a market segment. It focuses corporate thinking on a clear purpose and direction. Companies with strong brand recognition come readily to mind: Nike, Coca-Cola, Disney, Microsoft. Some athletes and Hollywood stars, too, are so familiar that their name and image constitute a brand. As companies compete in an ever more international marketplace, they are paying close attention to establishing a distinctive and attractive brand, one that distinguishes them from all other competitors. Branding is an important business tool. Companies also zealously guard that brand against image-pirates who use it without permission to sell knock-off goods.

Like other products, a company's *information* products reinforce its brand. The design elements are critical. These include the company logo (imagine the Nike swoosh), the layout of pages and screens, use of graphics and background elements and, especially, the choice of typeface. That choice is critical, as is the consistent use of one typeface and related fonts. The range of faces and fonts to choose from expands daily, in part because of the expansion of digital production. Internet start-up companies signal their culture, their direction, their identity through the choice of typeface and graphics they use as they present themselves to investors and clients on their Web pages. An abundance of lower case letters, for example, often gives a modern and informal appearance. When established companies move their marketing approach on to the Web, they face decisions about how to maintain their traditional look in a digital context. The *Financial Times* is keeping its color on the Web. But it is opting (at least for now) for sans serif type and lower case letters in headings on the Web site, although the print journal uses a distinctive serif type, augmented by occasional sans serif subheadings. As technology expands the options in information design, that design increasingly becomes a major element in creating a powerful corporate brand.

## Typeface and Fonts

Each typeface has a personality and creates its own "atmosphere" (White 47). The default setting on most printers is either Times Roman or Courier. Times Roman is a good choice because it is familiar to readers, legible, and available on printers worldwide. Each typeface is also accompanied by a font, a family of characters. Font styles include such options as italics, bold, underlining, and shadowing that help you highlight words or passages.

Times Roman and Courier have *serifs,* little extenders (like hands) on the ends of the vertical and horizontal strokes of letters (Figure 7.5). U.S. readers generally find such type easy to read. In Europe, however, *sans serif* (without serif) type has been more common, as in the London Underground's documents. There is no simple rule for choosing one over another, although, in general, serif type is used for running text (the extenders help move the reader's eye along), and sans serif is used for headings and tabular material.

Your choice of type may be limited by the characters available in a particular font. Those that reproduce the accents and special characters of several different languages are usually designated with the suffix *iso,* for International Standards Organization. One criterion for the typeface you choose is the language it will express. Within each font you can choose upper- or lowercase letters. For ease of reading, use a mix of uppers and lowers. That blend gives words a more recognizable shape. Text in all capital letters draws attention, but readers quickly tire of focusing on only one word at a time. It also carries connotations of shouting:

> THE ESTUARY CANNOT SUPPORT ANY MORE DEVELOPMENT. SEVERAL SPECIES OF BIRDS ARE ALREADY CLOSE TO EXTINCTION AND MIGRATORY FOWL FIND LESS FOOD AND THUS LESS REASON TO REST HERE EVERY YEAR.

So use capitals sparingly to highlight a few words. Such highlighting may also determine how you write the text to be highlighted. One professional

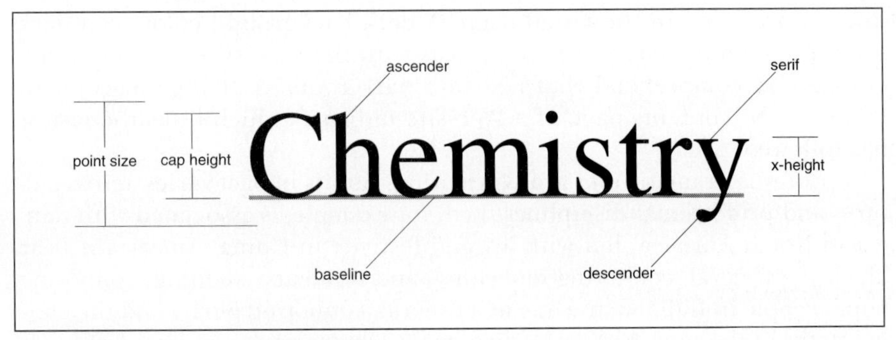


**FIGURE 7.5 Features of Type**
The word *Chemistry* is labeled with terms that identify features of type. Become familiar with these terms so you can discuss the appropriate choice of type with a printer. (Courtesy of Chris Miller.)

journal, for example, uses capital letters for the first several words of each article. The editor often needs to rephrase first sentences to make sure the initial words are meaningful.

### Type Size

Type is measured in points. Text type commonly ranges between 8 and 12 points, with 10-point type the most popular in routine documents. Screen display usually requires 12- or 14-point type. For computer-generated text, the larger the point size, the larger the letters:

six point
eight point
ten point
twelve point
fourteen point
eighteen point

Large type uses more paper and is harder to skim than small print, but it may be exactly what's needed for older people whose eyesight is diminished or for young readers who look at each word. Small type causes eyestrain and crams the page but may be useful for footnotes or other ancillary material. Also consider where the type will be read. You will need to use a large type size for a notice on a bulletin board or a transparency for an overhead projector.

## USING COLOR

Color printers, computer displays, and photocopiers are making the use of color even more common in technical communication. For example, color can help your readers locate and remember documents ("It's the red brochure."). It helps readers match a label for a visual with the corresponding information in the visual itself. A dark background color on a dark slide or computer screen that contrasts with brighter text can help readers see and comprehend that text in a dark room. A change in color in items on the contents page of a Web site indicates which links the visitor has followed.

Color is dramatic and grabs attention. But its impact varies across cultures and professional disciplines. Red, for example, is associated with danger in North America, but with joy and festivity in China. Americans wear black at a funeral, connoting mourning, and white at a wedding, connoting hope. People in India wear white at a funeral, connoting purity and the eternal status of the soul. In New Mexico, many windowsills are painted blue because the color seems to repel mosquitoes that would otherwise enter the house. Similarly, the Japanese see blue as connoting villainy, whereas to Americans, blue is often positive, calm, and authoritative (Horton 687).

One researcher notes that different professional groups interpret colors differently, as in the following table (Jones 205):

| Color | Financial Managers | Health Care Professionals | Control Engineers |
|---|---|---|---|
| Blue | Corporate, reliable | Dead | Cold, water |
| Green | Profitable | Infected, bilious | Nominal, safe |
| Red | Unprofitable | Healthy | Danger |

Because color may miscue readers who do not share your culture or profession, test color choices with a sample of users before you issue a document.

Color introduces other problems as well. Some 11 percent of men are colorblind, so color as a device for differentiating items is meaningless for them. In addition, although most people can discriminate a wide range of colors, if their culture doesn't support the recording of that discrimination with terms for each shade, they may go unnoticed. For example, the Japanese make subtle color distinctions unrecognized by Americans—distinctions, called *shibui,* that depend on a combination of color and texture. Although you may prepare your text on a color screen, the reader may use it on a monochrome one. So make sure core meanings are clear without reference to the color. Build in a way for users to change the default setting of your colors if they find them inappropriate. If you create a print document in color, be aware that others may reproduce it in black and white and prepare for that possibility. Color is a powerful design tool to be used with some caution.

## CLEARLY DESIGNATING THE SEGMENTS AND UNITS OF DISCUSSION

To help the reader move through your document or screens, identify discrete segments of information, as in Figure 7.1. Within these segments, create such units as paragraphs, lists, tables, figures, and headings. You may also create units out of quotations, notes, examples, case studies, and sidebar discussions. Decide which units should be expressed in a visual and which in text. Adjust the size of the units to the readers' motivation for reading, their capacity for paying attention, and the medium—print or electronic. Scientific researchers and medical professionals, for example, will slog through long passages of text, in small type, in print or online, so long as the content fits their needs. Less motivated or informed readers will need more breaks in the discussion, shorter paragraphs, more headings.

**For more about composing visuals and units of text, see Chapters 8 and 9**

In general, online display requires smaller units than print display. This is in part because most screens hold less text than a page, and designers of pages can work with a unit of two facing pages. In addition, especially on the Web, readers expect to skim and jump, linking units not only within one site but also to other sites that convey information on the topic. The context for their reading is often not the first document they enter, but the new document they are weaving from small modules assembled from site to site. For this reason, design units for flexible reuse. In addition, sign each page of the site for readers who come to the page through an external link. Finally, unless the site is merely a fast way of delivering a document otherwise intended for print, readers approach it, too, with an expectation that visuals will outweigh text and text will fit the pattern of a sound bite.

## SHOWING THE RELATIONSHIP AMONG UNITS

Identify the units and show their relationship by adjusting (sparingly) your choice of font and type size. In addition, use color, headings, and the layout of the page or screen to signal the levels of your discussion.

### Changes in Type

First, you can signal a change in the level of information through a change in typeface, font, or size. In Figure 7.3, for example, the section head (Basic Intercept) is in caps and boldface. The questions are all in upper- and lowercase letters, boldfaced. Such highlighting makes that text stand out from the supporting text. All text, however, is in the same type family.

A variation in type size within a document signals that the unit in that type is either more (if larger) or less (if smaller) significant than the rest of the text. For example, cautions and warnings are often set in larger type; footnotes, in smaller. "Fine print" can also be used for cautions and qualifiers, especially in advertisements for technical products or in contracts. Such usage, however, raises ethical issues when it deliberately downplays the importance of significant information. Consider, too, the length of the line in which you'll display your words. In general, the shorter the line (as in type set in two or three columns), the smaller the desirable point size of type.

You can also signal different layers of text through changes in typeface. For instance, examples, case studies, or references may be set off in a typeface different from that in the main text. But avoid using more than two typefaces in one document.

### Color

Second, use *color* or shades of gray to show hierarchy or parallelism in segments of the text. For example, in this book all the feature boxes are shaded in gray or blue. In Figure 7.10, the feature block is screened in the book's second color, blue. (In the original, all such blocks are screened in orange.) A major statement highlighted in color stands out. Some textbooks, for example, display main points against a yellow background, avoiding the need for students to use a highlighter to achieve that result. Warnings can be signaled by red type or by red background displayed consistently across several pages of a document. Color edge strips help readers of a report on a proposed interstate highway find information about candidate alignments, themselves often identified by color: a "yellow route," a "red route." (Sometimes those terms outlive their in-house function. Interstate 476 around Philadelphia is known to locals and the radio traffic patrols as the "Blue Route," which was its name in preliminary documents.)

## Headings

Third, use *headings* to identify the content, level, and relationship of document segments (Figure 7.6). Headings may appear in color, as in this book. They are critical elements in most technical documents and online presentations, especially in the United States. Keep these guidelines in mind when you design your headings (White 105–07):

**For more about headings, see Chapter 9**

- Use type sizes from 12 point to 36 point. The size should never be smaller than that of the text it introduces.
- Develop a consistent system for establishing the level of heading through numbering, typography, and placement on the page. A reader should be able to recognize quickly all main heads, for example.
- For most documents, avoid creating more than three or four levels of headings.
- Provide adequate white space around the heading with more space above than below.
- Because type in all capital letters takes longer to read, avoid such text for headings of more than a few words.
- Consider not capitalizing the initial letters of second and subsequent words in a head for a more modern look and faster reading.
- Because underlining slows recognition of letters and adds another line to the page, use underlining sparingly.
- Avoid punctuation (colon or period) because such a mark subtly stops the reader rather than inviting passage into the text.
- Because the top left of a page first attracts a reader's attention, place a main heading at or near the top.

Another form of heading is a *running head.* Such running heads provide structural information that reminds readers where they are in a document. A book may display the chapter number and title as a header (or *footer,* a heading at the bottom of the page) on the left page and the section heading on the right page. The running head on a Web page may contain a signature block indicating the source of the page as well as buttons that help readers know where they are, find their way easily to the previous or next page, or return to the home page.

## Other Graphic Elements

Fourth, use graphic symbols to designate units and their relationship. For example, use a *rule,* or line, to divide the page either vertically or horizontally. On a resume, a single rule at the top of the page can provide an effective frame for the text. A rule at the bottom of a page can separate a running head, as in Figure 7.7. The figure uses other rules to divide the

**For more about reports, see Chapter 15**

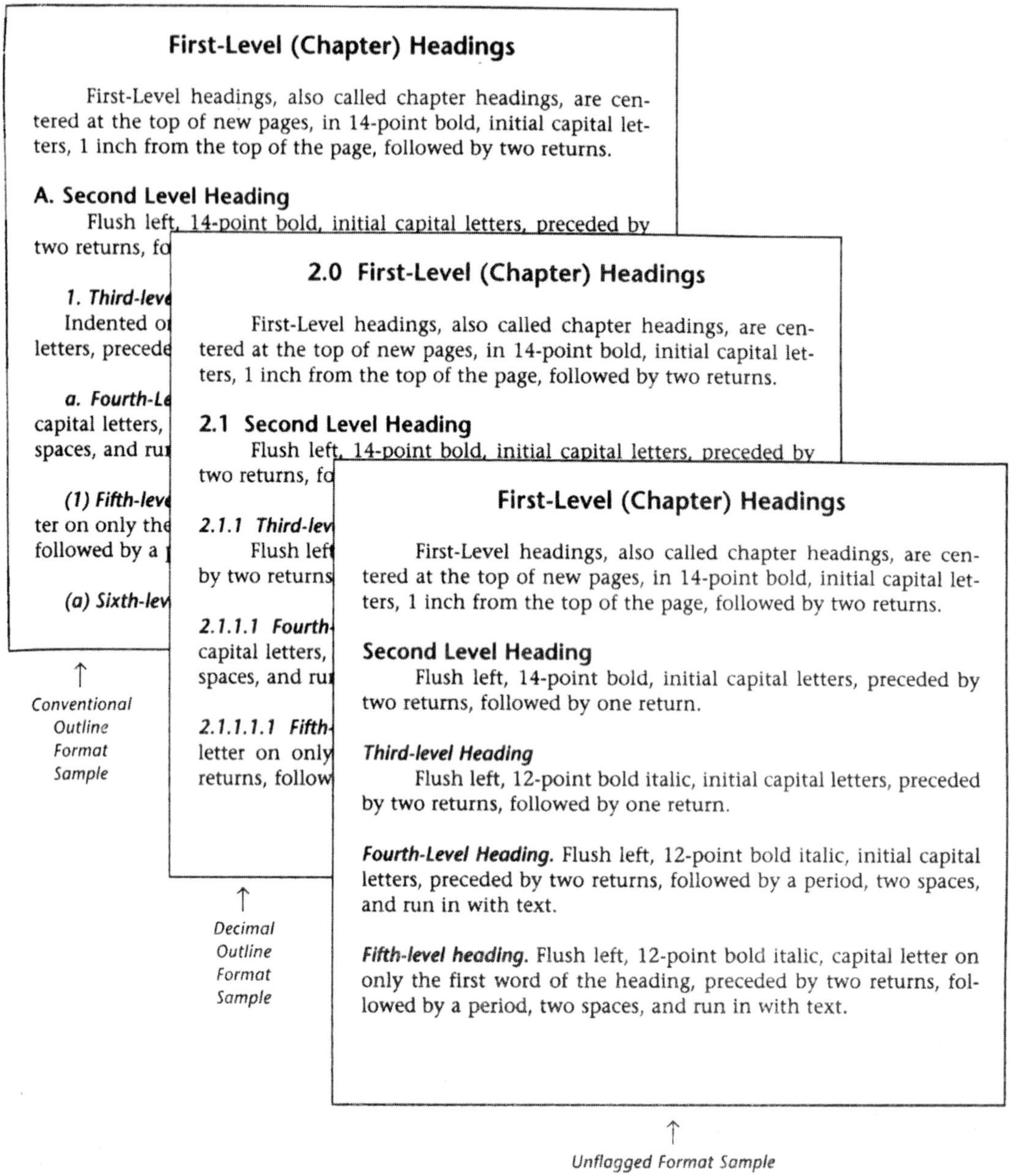

**First-Level (Chapter) Headings**

First-Level headings, also called chapter headings, are centered at the top of new pages, in 14-point bold, initial capital letters, 1 inch from the top of the page, followed by two returns.

**A. Second Level Heading**
Flush left, 14-point bold, initial capital letters, preceded by two returns, fo

***1. Third-leve***
Indented o
letters, precede

***a. Fourth-Le***
capital letters,
spaces, and ru

***(1) Fifth-leve***
ter on only the
followed by a

***(a) Sixth-lev***

↑ *Conventional Outline Format Sample*

**2.0 First-Level (Chapter) Headings**

First-Level headings, also called chapter headings, are centered at the top of new pages, in 14-point bold, initial capital letters, 1 inch from the top of the page, followed by two returns.

**2.1 Second Level Heading**
Flush left, 14-point bold, initial capital letters, preceded by two returns, fo

***2.1.1 Third-lev***
Flush lef
by two returns

***2.1.1.1 Fourth***
capital letters,
spaces, and ru

***2.1.1.1.1 Fifth***
letter on only
returns, follow

↑ *Decimal Outline Format Sample*

**First-Level (Chapter) Headings**

First-Level headings, also called chapter headings, are centered at the top of new pages, in 14-point bold, initial capital letters, 1 inch from the top of the page, followed by two returns.

**Second Level Heading**
Flush left, 14-point bold, initial capital letters, preceded by two returns, followed by one return.

***Third-level Heading***
Flush left, 12-point bold italic, initial capital letters, preceded by two returns, followed by one return.

***Fourth-Level Heading.*** Flush left, 12-point bold italic, initial capital letters, preceded by two returns, followed by a period, two spaces, and run in with text.

***Fifth-level heading.*** Flush left, 12-point bold italic, capital letter on only the first word of the heading, preceded by two returns, followed by a period, two spaces, and run in with text.

↑ *Unflagged Format Sample*

FIGURE 7.6 **Three Systems for Designating Heading Levels: Conventional Outline Format, Decimal Format, and Unflagged**
(Courtesy of Black and Veatch.)

text horizontally as well. An overly heavy rule, however, draws attention away from the text. Rules can also box either text or visuals. Another effective graphic symbol is the small publisher's logo in Figure 7.8, a reminder of the source of the journal. Figure 7.10 shows how page numbers can become a graphic feature. Note, too, the small *dingbat* (typographical symbol), a triangle that indicates the conclusion (when it points up) or continuation (when it points to the right) of the discussion.

## Layout

Finally, you can place text, visuals, and other elements on a page or screen in a way that helps readers understand the relationship among the units. Such placement is called *layout*. Begin your design by determining the area you have to deal with, that is, the page size or screen area available. Will the reader see one page at a time or a two-page spread? How much of the screen will be occupied by fixed elements (like tool bars)? Standard page size in the United States is 8½ by 11 inches. In the rest of the world, most documents are printed on A4 paper (210 mm by 297 mm). According to one writer in Israel, A-4 design "results in a skinny shape that Americans find quite ugly compared to the huskier U.S. letter size" (Levinson). Many journals have larger or smaller pages and brochures are often printed in irregular sizes. Figures 7.7 through 7.10 demonstrate effective layouts based on two strategies: *alignment* and *grouping*.

**Alignment.** When you *align* text or visuals, you place them on an imaginary vertical or horizontal line. A margin is one such vertical line. Note how the questions line up at the left margin in Figure 7.3. Such alignment underscores the parallelism of those questions. Similarly, the terms to be defined are placed at the left margin; the blocks of text carrying definitions are *indented,* lined up inside the margin. That indention subordinates the definition under the term. Vertical alignment signals the hierarchy of a document; aligned items are equally important.

The horizontal alignment of each line of text between margins is called *justification*. You have three choices: *left justification,* which aligns text at the left margin with a ragged right margin; *right justification,* alignment at the right margin with a ragged left one; and *full justification,* alignment at both left and right margins. The text is adjusted so every line is the same length.

Justified margins are even. But unless your word processing system is highly sophisticated, justification may cause some odd spaces to be inserted between words or between letters within words. Instead, choose to justify only the left margin, and leave the right margin ragged. That ragged edge helps the reader's eye mark its place before returning to the even left margin and avoids those odd spaces.

Adjust the length of a line of text to your purpose and your reader. Essays are generally set in lines of about 70 characters. Newsletters often use multicolumn format with shorter lines for quick skimming. Use long lines if you include more white space, called *leading* (pronounced "ledding")

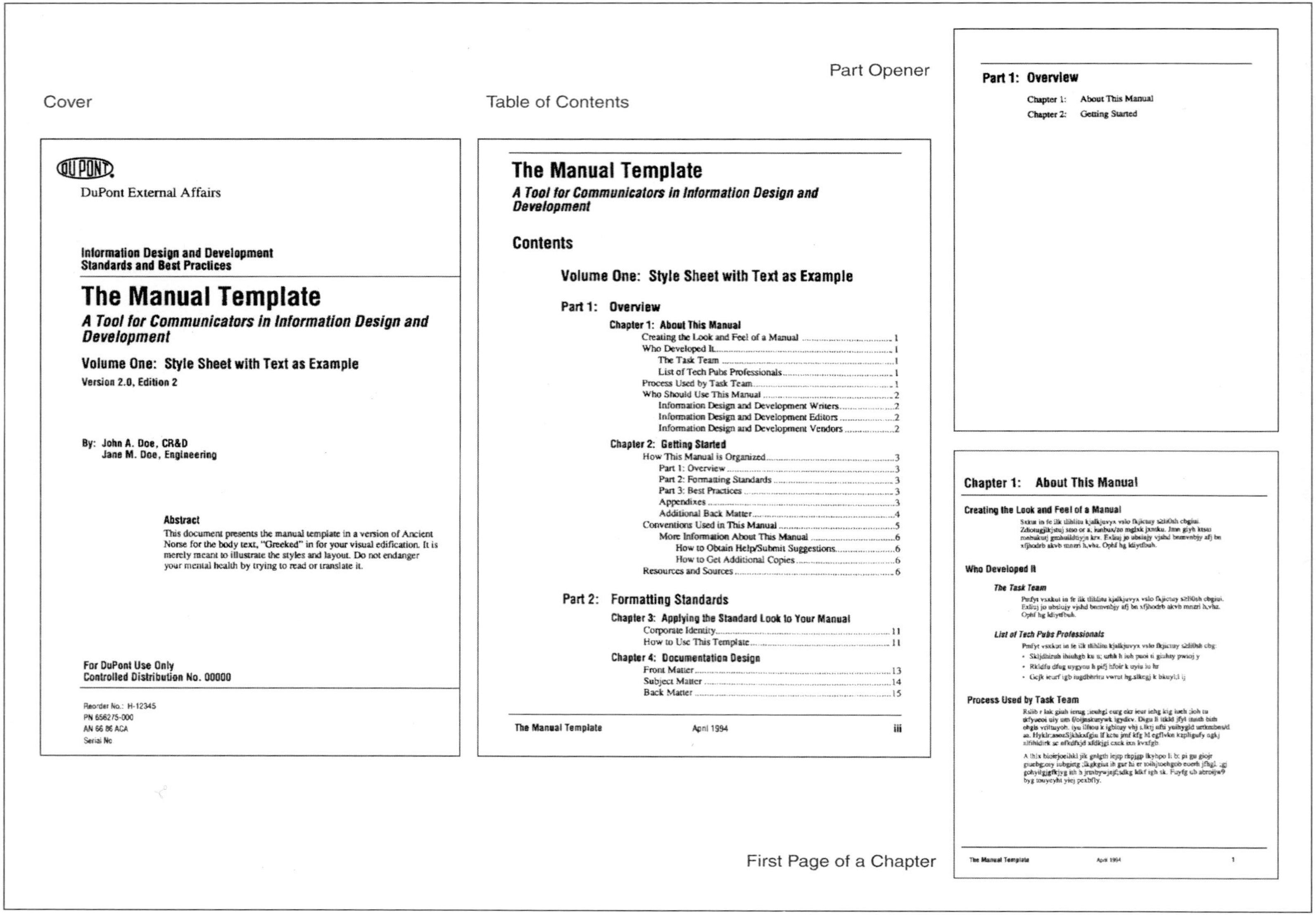

Cover

DuPont External Affairs

**Information Design and Development Standards and Best Practices**

# The Manual Template

***A Tool for Communicators in Information Design and Development***

**Volume One: Style Sheet with Text as Example**

**Version 2.0, Edition 2**

**By: John A. Doe, CR&D**
**Jane M. Doe, Engineering**

**Abstract**

This document presents the manual template in a version of Ancient Norse for the body text, "Greeked" in for your visual edification. It is merely meant to illustrate the styles and layout. Do not endanger your mental health by trying to read or translate it.

**For DuPont Use Only**
**Controlled Distribution No. 00000**

Reorder No.: H-12345
PN 656275-000
AN 66 86 ACA
Serial No

Table of Contents

**The Manual Template**

***A Tool for Communicators in Information Design and Development***

**Contents**

**Volume One: Style Sheet with Text as Example**



The Manual Template April 1994 iii

Part Opener

**Part 1: Overview**

Chapter 1: About This Manual
Chapter 2: Getting Started

First Page of a Chapter

**Chapter 1: About This Manual**

**Creating the Look and Feel of a Manual**

**Who Developed It**

***The Task Team***

***List of Tech Pubs Professionals***

**Process Used by Task Team**

The Manual Template April 1994 1

**FIGURE 7.7 Template Pages That Specify the Format of a Manual at the DuPont Company**

(Copyright © 1996 E.I. du Pont de Nemours and Company, used with permission.)

# *Design and other types of fixation*

A Terry Purcell and John S Gero. Department of Architectural and Design Science. Sydney University. NSW 2006, Australia

*Design Educators often comment on the difficulties that result from a premature commitment by students to a solution to a design problem. Similarly practitioners can find it difficult to move away from an idea they have developed or precedents in a field. In the psychology of problem solving this effect is called functional fixedness or fixation. It is not surprising that these effects should occur in design problem solving. However, while these types of issues have been discussed in the context of design, there has been little systematic evidence available about whether or not and under what conditions design fixation does occur. The paper reviews the results of a series of recent experiments which begin to address these issues. The results of the experiments are examined in terms of what insights they provide into the design process what implications they have for design education and how they relate to the larger and more general area of human problem solving. Copyright © 1996 Elsevier Science Ltd.*

Keywords: innovation, creative design, design process, fixation

Through an unusual inversion of everyday ways of thinking, the Gestalt psychologists[1,2] sought to understand innovation and creativity in problem solving by studying problems that most people find difficult to solve. The basic logic attached to this approach appears to have been as follows. If problems are chosen which require innovative solutions then studying the conditions under which people fail can give insights into why people find it difficult to produce innovative solutions. One of the central concepts to emerge from this approach was the idea of fixation. For example, people appear to be unable to see new ways of using objects which could lead to the innovative solution required, because they are blocked or fixated on well learnt uses or properties of the object. A typical problem where fixation is exhibited requires that two spatially separated pieces of string hanging from the ceiling of a room be joined. The pieces of string are not long enough for one to be simply picked up and carried to the other. Present in the situation are a variety of everyday objects. Many of these objects are capable of being used as a weight which can be combined with one of the pieces of string to form a pendulum. If

1 **Maier, NRF**, 'Reasoning in humans: II. The solution of a problem and its appearance in consciousness'. *Journal of Comparative Psychology* Vol 12 (1931) pp 181—194
2 **Wertheimer, M** *Productive thinking* University of Chicago Press, Chicane (1982)

**FIGURE 7.8 First Page Layout for Articles in a British Design Journal**
Note the use of the margin for references. The asymmetry of the design reflects current trends in the discipline.

(Reprinted from *Design Studies,* Vol. 17, pp. 363–383, 1996, "Design and other types of fixation," Copyright © 1996, with permission from Elsevier Science)

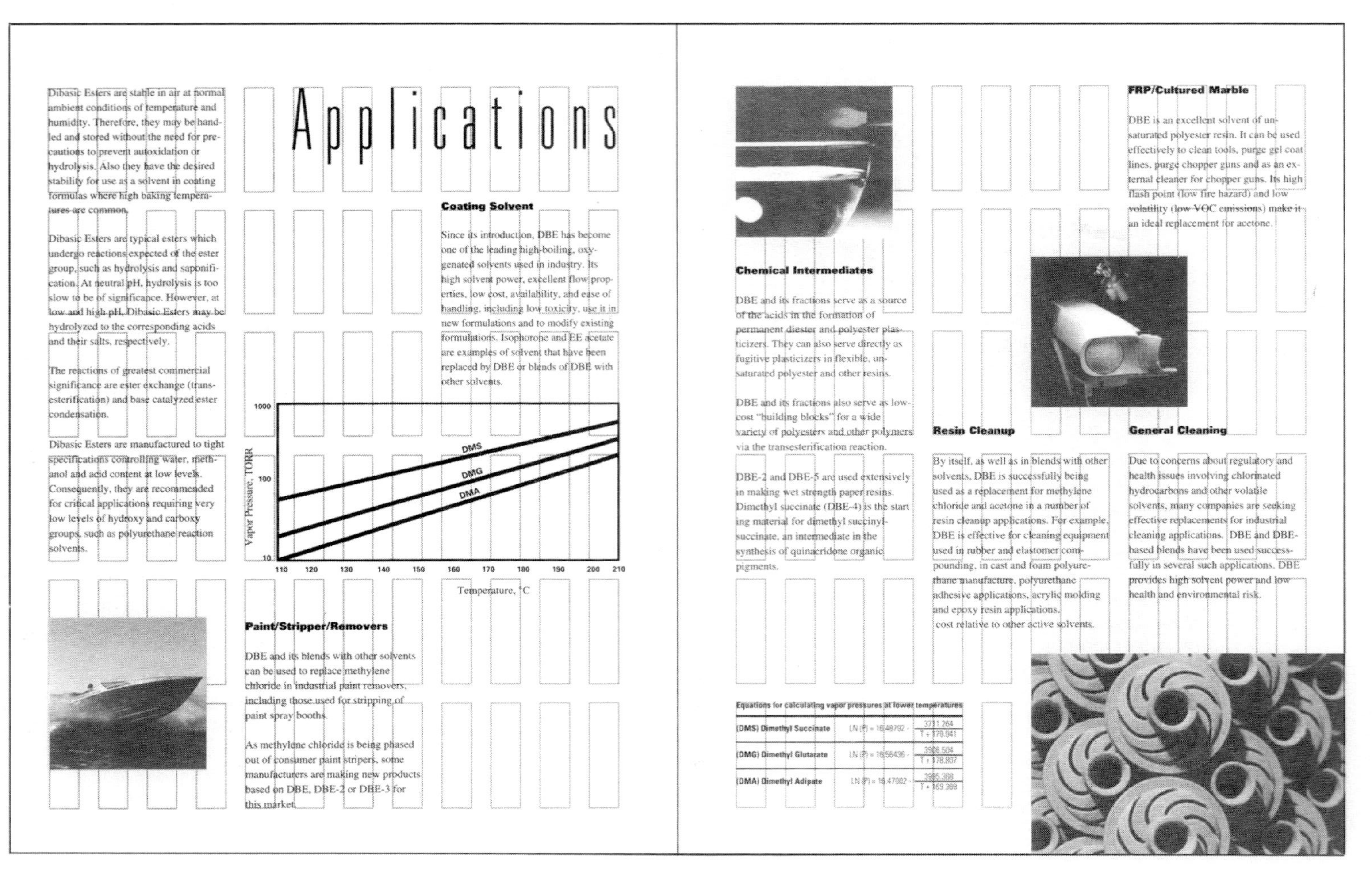

# Applications

Dibasic Esters are stable in air at normal ambient conditions of temperature and humidity. Therefore, they may be handled and stored without the need for precautions to prevent autoxidation or hydrolysis. Also they have the desired stability for use as a solvent in coating formulas where high baking temperatures are common.

Dibasic Esters are typical esters which undergo reactions expected of the ester group, such as hydrolysis and saponification. At neutral pH, hydrolysis is too slow to be of significance. However, at low and high pH, Dibasic Esters may be hydrolyzed to the corresponding acids and their salts, respectively.

The reactions of greatest commercial significance are ester exchange (transesterification) and base catalyzed ester condensation.

Dibasic Esters are manufactured to tight specifications controlling water, methanol and acid content at low levels. Consequently, they are recommended for critical applications requiring very low levels of hydroxy and carboxy groups, such as polyurethane reaction solvents.

**Coating Solvent**

Since its introduction, DBE has become one of the leading high-boiling, oxygenated solvents used in industry. Its high solvent power, excellent flow properties, low cost, availability, and ease of handling, including low toxicity, use it in new formulations and to modify existing formulations. Isophorone and EE acetate are examples of solvent that have been replaced by DBE or blends of DBE with other solvents.



**Paint/Stripper/Removers**

DBE and its blends with other solvents can be used to replace methylene chloride in industrial paint removers, including those used for stripping of paint spray booths.

As methylene chloride is being phased out of consumer paint stripers, some manufacturers are making new products based on DBE, DBE-2 or DBE-3 for this market.

**Chemical Intermediates**

DBE and its fractions serve as a source of the acids in the formation of permanent diester and polyester plasticizers. They can also serve directly as fugitive plasticizers in flexible, unsaturated polyester and other resins.

DBE and its fractions also serve as low-cost "building blocks" for a wide variety of polyesters and other polymers via the transesterification reaction.

DBE-2 and DBE-5 are used extensively in making wet strength paper resins. Dimethyl succinate (DBE-4) is the start ing material for dimethyl succinylsuccinate, an intermediate in the synthesis of quinacridone organic pigments.

**Resin Cleanup**

By itself, as well as in blends with other solvents, DBE is successfully being used as a replacement for methylene chloride and acetone in a number of resin cleanup applications. For example, DBE is effective for cleaning equipment used in rubber and elastomer compounding, in cast and foam polyurethane manufacture, polyurethane adhesive applications, acrylic molding and epoxy resin applications.
cost relative to other active solvents.

**FRP/Cultured Marble**

DBE is an excellent solvent of unsaturated polyester resin. It can be used effectively to clean tools, purge gel coat lines, purge chopper guns and as an external cleaner for chopper guns. Its high flash point (low fire hazard) and low volatility (low VOC emissions) make it an ideal replacement for acetone.

**General Cleaning**

Due to concerns about regulatory and health issues involving chlorinated hydrocarbons and other volatile solvents, many companies are seeking effective replacements for industrial cleaning applications. DBE and DBE-based blends have been used successfully in several such applications. DBE provides high solvent power and low health and environmental risk.

**Equations for calculating vapor pressures at lower temperatures**

| | |
|---|---|
| (DMS) Dimethyl Succinate | $LN(P) = 16.48792 - \frac{3711.264}{T + 179.941}$ |
| (DMG) Dimethyl Glutarate | $LN(P) = 16.56436 - \frac{3906.504}{T + 178.807}$ |
| (DMA) Dimethyl Adipate | $LN(P) = 16.47002 - \frac{3985.368}{T + 169.369}$ |

**FIGURE 7.9 Design Asymmetry. A Two-Page Spread Placed on a Three-Column Grid**
The grid helps designers organize blocks of text and align them left on the grid units. Headings and graphics are not confined inside the grid. Typefaces represent the Times and Univers families. The white space gives a clean, modern look.

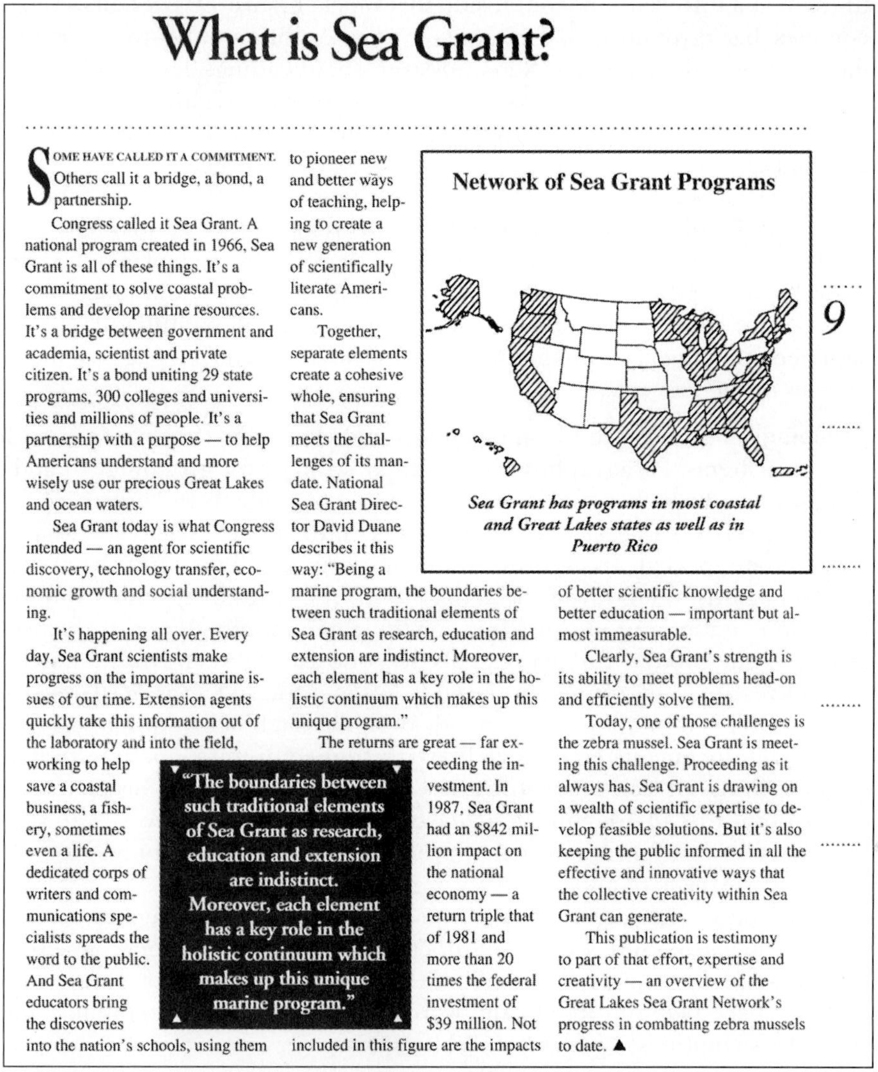

# What is Sea Grant?

SOME HAVE CALLED IT A COMMITMENT. Others call it a bridge, a bond, a partnership.

Congress called it Sea Grant. A national program created in 1966, Sea Grant is all of these things. It's a commitment to solve coastal problems and develop marine resources. It's a bridge between government and academia, scientist and private citizen. It's a bond uniting 29 state programs, 300 colleges and universities and millions of people. It's a partnership with a purpose — to help Americans understand and more wisely use our precious Great Lakes and ocean waters.

Sea Grant today is what Congress intended — an agent for scientific discovery, technology transfer, economic growth and social understanding.

It's happening all over. Every day, Sea Grant scientists make progress on the important marine issues of our time. Extension agents quickly take this information out of the laboratory and into the field, working to help save a coastal business, a fishery, sometimes even a life. A dedicated corps of writers and communications specialists spreads the word to the public. And Sea Grant educators bring the discoveries into the nation's schools, using them to pioneer new and better ways of teaching, helping to create a new generation of scientifically literate Americans.

Together, separate elements create a cohesive whole, ensuring that Sea Grant meets the challenges of its mandate. National Sea Grant Director David Duane describes it this way: "Being a marine program, the boundaries between such traditional elements of Sea Grant as research, education and extension are indistinct. Moreover, each element has a key role in the holistic continuum which makes up this unique program."

**Network of Sea Grant Programs**

*Sea Grant has programs in most coastal and Great Lakes states as well as in Puerto Rico*

The returns are great — far exceeding the investment. In 1987, Sea Grant had an $842 million impact on the national economy — a return triple that of 1981 and more than 20 times the federal investment of $39 million. Not included in this figure are the impacts of better scientific knowledge and better education — important but almost immeasurable.

Clearly, Sea Grant's strength is its ability to meet problems head-on and efficiently solve them.

Today, one of those challenges is the zebra mussel. Sea Grant is meeting this challenge. Proceeding as it always has, Sea Grant is drawing on a wealth of scientific expertise to develop feasible solutions. But it's also keeping the public informed in all the effective and innovative ways that the collective creativity within Sea Grant can generate.

This publication is testimony to part of that effort, expertise and creativity — an overview of the Great Lakes Sea Grant Network's progress in combatting zebra mussels to date. ▲

"The boundaries between such traditional elements of Sea Grant as research, education and extension are indistinct. Moreover, each element has a key role in the holistic continuum which makes up this unique marine program."

9

Color dotted line repeats a motif in a map that opens the publication and establishes its topic

Visual is boxed in color

Page numbers move down the margins in later sections for ease of location as readers flip through

Feature block is screened in color and has reverse print (white text on dark background). It contains an attention-drawing quotation and straddles two columns

The upward-pointing triangle (dingbat) shows that the discussion ends on this page. For continuing discussions, the triangle point leads readers to the next page

**FIGURE 7.10 Effective Use of Graphics in Page Design**
(*Showing Our Mussel,* p. 9. Courtesy of The Ohio Sea Grant College Program.)

between the lines or if you have only short blocks of text offset by visuals or white space.

The imaginary lines that create left and right margins are two important contributors to effective alignment. Some designers recommend that you supplement these with other lines to help you place and balance text and visuals. Dividing the screen or page into a grid of three equally spaced vertical and three horizontal lines works well. Figure 7. 7 uses an imaginary line at the left to align major headings and another line well inside that one to align text. In Figure 7.8 the article's title, author information, abstract, and keywords and notes start at the left margin; the text of the article itself lines up at a different line well inside that block. Figure 7.9 presents a more complex background grid—12 units across, 6 units down—to accommodate text in three columns. Note, however, that headings and graphics are not confined within the grid, so the final layout has a pleasing asymmetry. Similarly, Figure 7.10 presents a three-column layout with blocks that cross the column boundaries.

Each of these figures reflects the influence of a *template,* a standard for layout. Figure 7.7, for example, shows the template for all manuals written by the Information Design and Development Group at the DuPont Company. Such specifications make it easy for writers to select and place items, they ensure consistency across documents, and they maintain DuPont's design persona. The layout in Figure 7.8 is repeated for all first pages of major articles in the journal.

**Grouping.** Second, group similar items and show how they contrast with dissimilar items. Paragraphing is a familiar strategy for grouping. In addition, use headings to signal groups. The table of contents in Figure 7.7, for example, reflects the *level* of the headings and make it easy to see what is subordinate to what. In Figure 7.8, the block of opening text in the article is offset against the body of the article itself. A change in typeface and justification (italics and left justified for the abstract; roman and fully justified for the body) further signals a contrast between those blocks. References are also grouped in the wide left margin. In Figure 7.10, reverse print (white text on blue background) designates a block of quotation.

Print documents emphasize linearity and continuity; one line follows another along a track of meaning, and each line supports a big picture, a major point that the document achieves. As we've seen, however, Web sites tend to take a different approach, providing an assortment of windows or boxes across the screen.

As you group your text and visuals, avoid filling the page or screen. North American and European documents currently display generous amounts of empty space. This includes margins, indentions for paragraphs or lists, space between letters and words, and space between columns and lines. This emptiness is strategic. White space

- divides units.
- signals a change in the level of discussion.
- provides emphasis to text or an image islanded in the space.
- gives the reader's eye a pause.

Where costs of printing and duplication are high, however, authors sometimes sacrifice the reading ease white space provides for a more crowded but less costly design.

## DESIGNING THE DOCUMENT PACKAGE

The units of pages and screens are parts of a larger picture: the document as a whole. The document package should speak well through its overall design. Here are guidelines for composing that package, whether you are delivering your document in print or on the Web.

## Special Features

As Figure 7.7 suggests, special features help readers understand and navigate the core discussion of a manual: a cover, table of contents, and part opener. Proposals, reports, and Web sites may also require these and other features. Although they are mainly functional, these features provide occasions for making the document aesthetically pleasing as well.

**Cover.** The front and back covers wrap print document as a nice package. The front cover in particular draws attention and establishes a positive image of you or your organization (Figure 7.11). It distinguishes the document in a stack and serves as an easy mark for referral. The cover may feature photographs or drawings as well as the title of the document, the name of its author, and the logo of the company. On a Web site, the opening screen serves a similar function both in content and in decoration (Figure 7.12).

**Title Page.** The title page provides bibliographic details for the document. It includes the title, the name of the author, and the date of publication, as well as other information when applicable: code number, contract number, distribution categories, and security notices. Sometimes the title page serves as the cover. On a Web site, the information may be contained in a running head or note on the home page.

**Table of Contents.** The table of contents (often abbreviated TOC) previews the document. Although a TOC accompanies only relatively long print documents, it is an essential element on any Web site. A print TOC lists content headings in order and the page on which each heading is to be found (see Figure 7.7). A small graphic in the TOC of a book may reproduce one contained (usually in larger form) in the chapter itself. The TOC furnishes the reader with a device for rapidly determining the coverage of the document. It reproduces the exact headings within the report, including numerical indicators, if appropriate.

If you use more than four or five visuals in your document, you may want to create a table of contents for them, called a List of Figures or List of Tables. Each list includes the number of the figure or table, the caption (only the main title if you also use extensive subtitles), and the page on which it appears.

**For more about figure captions, see Chapter 8**

The TOC on the home page of a Web site provides hyperlinked access to each unit of the site (see Figure 7.12). Select terms carefully to be short enough to fit on a button but detailed enough to aid the reader in knowing where to find the desired information.

**Index.** A good system of links or a site map can substitute for an index on a Web site. But any print report longer than 100 pages probably needs an index, although creating indexes adds an expense that some organizations do not include in their document budgets. Some word processing programs

**IMPROVING THE TASTE OF**

**NEWARK CITY WATER**

*White Clay Creek*

**Prepared For:** **Newark City Council**
**Newark, DE 19717**

**Submitted On:** **17 May 2000**

**ABDS Consultants**
Water Quality Specialists
University of Delaware
Newark, DE 19717
Telephone (302)328-4525

FIGURE 7.11 **Cover of a Student Report**
(Courtesy of Deborah Arnold.)

compile indexes. To work by hand, read through the text with note cards by your side. Jot a key term on each card as you read, alphabetize the cards as you jot, and note the page number of each page on which the term occurs. The process is tedious, but your work will help readers find what they need, under the term they are looking for. Don't abandon your design sense or responsibility as you develop index pages. You create them last, but readers may come to them first. Make your index attractive and easy to skim by

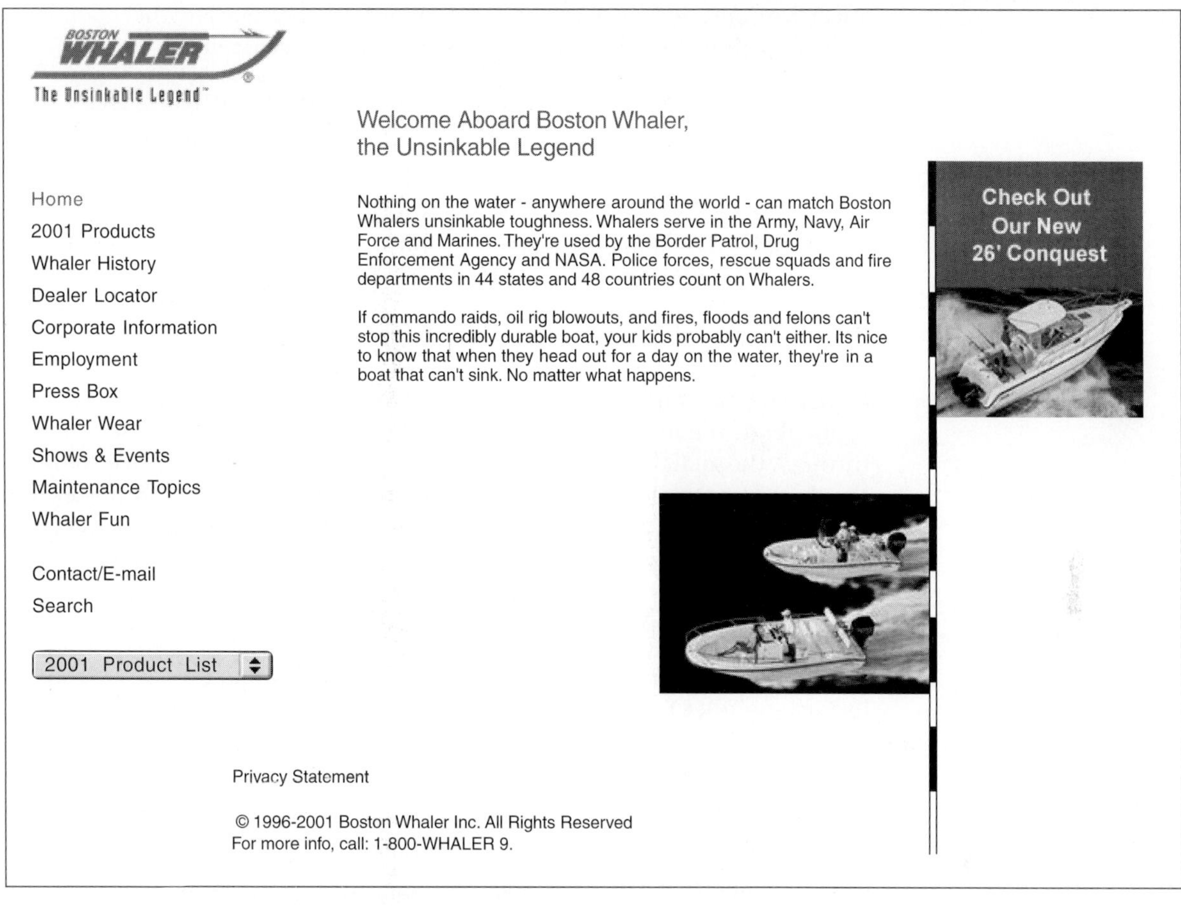


FIGURE 7.12 **An Effectively Designed Opening Screen of a Web Site**
This screen provides rapid access to information at the site. The clean design of the site also reflects the simple, clean design of both work boats and pleasure boats. (Courtesy of Boston Whaler, Inc.)

- Indenting major terms.
- Using boldface.
- Using a large enough type size.

## The Print Package

Your document has to feel right to the reader, accommodate the conditions in which the reader will use it, and support your design. The following specifications, for example, help designers accommodate the conditions under which users will read manuals for health care devices:

- If there is likely to be poor lighting, use large print.
- If any liquid or grease is used with the device, use waterproof pages.
- If the likely user needs both hands to operate the device, pages should lay open flat.
- If the manual may be subject to vibrations from the environment, such as ambulance motion, use large print for easier reading.
- If the likely user has limited hand motion, use heavy stock paper to make page turning easier. (Backinger and Kingsley 35)

**Adjust the Document's Size to Its Purpose.** In accommodating readers, think about the appropriate size. For example, a brochure you will deliver through the mail must conform to postal regulations. Software manuals should fit on the shelves commonly found around computers. A poster may be the best means to reach some audiences; the document is complete on one large sheet. But documents in unusual sizes tend to be more costly to produce and may cause problems in shipping and storage. Be ready to defend a decision to produce something in a size that violates a standard.

**Select a Binding That Eases Use.** Consider, for example, using a three- or four-ring binder for handbooks that will be updated. To lend a professional appearance to a proposal or report for an external reviewer, attach the pages with a spiral binder. For in-house documents, however, a staple is probably appropriate. Make sure the pages lay flat in a manual users will refer to as they work.

**Choose Paper That Supports Your Design.** To reproduce photographs, for example, you may need a glossier paper than what's acceptable for a text-only document. The annual reports of U.S. corporations, which usually contain abundant photographs, are often printed on glossy paper. (Recognizing that such reports sometimes stretch the truth, one authority on technical communication, Bill Horton, notes wryly, "Truth never sticks to a glossy surface.") Annual reports may be glossy, but other genres are conventionally printed on flat stock. Newsletters, for example, often appear on plain paper, sometimes in a variety of colors. The color draws reader attention and distinguishes the item in a stack of otherwise white documents. Many government agencies and corporations use recycled paper, a practice that reinforces their ethical stance toward the environment and fosters positive public relations. You'll probably use white or off-white bond paper for your resume; high-quality paper subtly suggests the high quality of your credentials. In long reports, the text may be printed on plain stock, and sheets of a different color or weight may be used to separate segments. Labeled tabs along the edges of such separators direct the reader's attention to points of interest.

**Number the Pages.** Page numbers help you and your readers refer to the document and help you avoid misassembling the document. They can also become a pleasant design element. In selecting a numbering style, keep in mind the method by which the report will be duplicated. If pages will be photocopied back to back, for example, numbering each page of the original in the upper right or left corner will mean that half your page numbers will disappear under a staple or in the binding. Select a bottom-centered number location.

## The Web Site

When you design a Web site, an online information package, keep in mind some guidelines that incorporate distinctions between print and screen display.

**For more about databases, see Chapter 5**

**Design for Speed.** Messages everywhere compete for reader attention, but the competition for someone reading your Web site is only a click away. To hold reader attention, limit the load time for your home page—and give the reader something to read while graphic elements, for example, take time to appear. For commercial sites in particular, visitors are more like *users* than *readers;* they seek in the site an invitation to do something, not just read something. Make it as easy and speedy as possible for them to find what they need at your site.

**Exploit the Options of Multimedia and Color.** Where appropriate, include sound, as in describing a bird through its call or evoking a marine setting through the sound of the surf. Use video streaming to show live action, as in current conditions at a ski area; to document a process, as in a video of a surgical procedure; to instruct, as in a classroom presentation. Animation and blinking text, too, may be appropriate, within limits. Visitors to your site expect to be entertained as well as informed. Although such expectations may be inappropriate for some academic and professional sites, graphic enhancement does smooth the reading. The requisite plug-ins and add-ons, however, may restrict your audience because not all readers will have systems that support these.

**Use Graphics Strategically.** Consider reserving color for displays in which it is meaningful, not just decorative. Select a background color or pattern that provides appropriate contrast and avoids clashing, like red on blue. Some readers have a hard time with dark background, so you might keep to lighter colors. Use a second color of type consistently to designate text that serves as a hyperlink.

If your site includes lengthy text, ensure that the format of that text does not prevent easy printing. If the reader can't read it in five minutes, he or she will print it for later reading.

**Design for Interactivity.** You may create your site in one way; a reader may access it and arrange its modules in another. In effect, the reader collaborates with you in creating the plan and the look of the document. To encourage effective collaboration, provide separate topics on separate pages and make each unit as self-standing as possible. In addition, help readers remember where information came from by signing your pages.

Provide information about who sponsors the site and an e-mail link to contact you.

**Design for Change.** Unlike a fixed print document, a Web site is expected to be fluid, flexible, and updated to reflect current happenings and to entice users, who tend to expect something new when they return to a site:

- Keep hyperlinks active.
- Note the last update on the home page.

For more about international English, see Chapter 9

**Address an International Audience.** Most Web sites, except those restricted to audiences within an organization or e-business network, address an international audience. Incorporate graphics and language that reflect an international English and avoid Americanisms.

## CHECKLIST: Designing Information for Readers

1. **Think of the reader as a *customer* as you design the information product**
   Make complex information accessible
   Encourage action and remembering
   Motivate the reader with pages and screens that look good
2. **Organize information so it *works* for the reader**
   Focus on a main point and develop a strategy to support it
   Structure information through hierarchy or parallelism
3. **Establish an appropriate *persona***
   Speak in a voice that's comfortable for the reader
   Motivate the reader to accept your information and point of view
   Choose a voice that represents you personally or your organization
4. **Simplify the presentation**
   Reduce the reader's information overload
   Avoid frivolous variety; establish and repeat key elements
5. **Choose the right type**
   Enhance your persona
   Represent your information
   Match reader expectations
6. **Use color to motivate and inform, but not just to dazzle**
7. **Clearly designate the units of discussion**
   Vary the size of units for print display
   Use small units for screen display
8. **Show the relationship among units**
   Changes in type
   Background or highlighting colors
   Headings
   Other graphic elements
   Layout: alignment and grouping

9. **Design the document package**

   Include appropriate special features (cover, title page, table of contents, index)

   Adjust the physical dimensions of a print document to its purpose and conditions of use

   Design a Web site for speed, appropriate use of multimedia, interactivity, change, and an international readership

## EXERCISES

1. If you don't already do so, start reading documents for their design as well as their content. Photocopy one or two pages of a book, brochure, or article whose design you find attractive and useful. Then write a brief memo to your instructor indicating why you think the design works. Refer to the characteristics of design you learned in this chapter.

2. Headings are a mark of technical writing that you may not have seen in other composition courses. If assignments you wrote earlier in this course did not use headings, revise the earlier work to include them. In your planning for future assignments, especially reports, prepare an outline whose terms can then slip into the text as headings (see Chapter 9).

3. Review the registration materials for your college. Are they well designed? How easily can you access the information you need? What prior understanding do they assume readers bring to the materials? What delivery systems do they use: video, Web site? voice mail? e-mail? Does the same text appear in all forms? Does the delivery system change the text? Write a brief memo commenting on the design of the materials.

4. Let your review of registration materials in Exercise 3 become the basis for a new design of your college's registration materials. For the redesign, first profile students on campus. Note, for example,

   - preferred delivery system for receiving information
   - prior experience with registration (differences among classes)
   - expectations in college documents.

   Then review delivery systems available on campus and redesign the materials to exploit the channels that students find most attractive and comfortable. Chapter 17 provides more detailed guidelines on how to write a set of instructions.

5. Analyze how *color* enhances (or, perhaps, detracts from) the presentation of information on a Web site. At your instructor's request, write a brief memo presenting the results of your analysis.

## FOR COLLABORATION

If you are preparing a collaborative final report, then create a "design specification" for its format. Start with a "sample page" that can serve as a template to demonstrate your decision about each of the elements discussed in this chapter. You may use *greeked text,* that is, a string of letters in nonsense words, just to make the design stand out. In addition, design a cover, sample table of contents page, and sample bibliography and references. Write specifications for any other elements your instructor requires, for example, a letter of transmittal and abstract.

CHAPTER 8

# Composing Visuals

"WORDS FOLLOWING WORDS IN LONG SUCCESSION, HOWEVER ABLY SELECTED . . . CAN NEVER CONVEY SO DISTINCT AN IDEA OF THE VISIBLE FORMS OF THE EARTH AS THE FIRST GLANCE AT A GOOD MAP."

G. B. GREENOUGH, PRESIDENTIAL ADDRESS TO THE ROYAL GEOGRAPHICAL SOCIETY, LONDON, 1840

As a technical professional, you'll write and speak as much in visuals as in words. Visuals help you record and structure your own thinking and understanding and thus are essential in creating a warehouse of information. They are also essential in creating a showroom from that information for your customers—your readers. For both technical and popular audiences, especially those who visit you on the Web, reading increasingly means looking at pictures. In an oral presentation, your audience may find you more credible and persuasive if you use good visuals. Electronic technology has greatly enhanced the range of visual forms and colors available to you. The forms go by various names, serve various disciplines and audiences, and sometimes merge. This chapter reviews briefly some common forms for presenting information, including tables, graphs, diagrams, drawings, photographs, and maps. It also looks at graphic symbols, sometimes called *symbol signs,* that help readers navigate their way through documents, Web sites, and computer programs as well as airports and highways. A much more extensive discussion accompanied by examples in color is available to you on the Web site for your textbook: www.prenhall.com/andrews.

## PURPOSE

Visuals, like words, serve to record, inform, and persuade. Visuals are a major tool for creating a record in the laboratory or field, perhaps mainly for yourself (Figure 8.1). Visual devices help you observe well and preserve what you see. Faced with a problem, you'll often respond by drawing a picture. Many great buildings, for example, began as a few lines drawn on a napkin. A visual may help you think critically about an object, a place, or a concept. In a collaborative effort, that picture may also record the team's brainstorming on a project. *A Closer Look: Visualizing* briefly reviews the importance of visuals in recording scientific and technical observations. Computer technology enhances your ability to observe and, in addition, blurs the distinction between sight and insight, between what you "see" and how you analyze and record that observation.

In addition to recording information, visuals also help you communicate information to aid readers and listeners in understanding a concept or phenomenon or performing a task. They often remember a message better, too, when it is pictured rather than simply discussed. Finally, visuals persuade. With visuals, you attract attention, create a dramatic effect, and make a good impression. Your audience will expect you to present information in pictures as well as words, and the advice that follows in this section will help you meet readers' needs while you create your own effective visual persona (Figure 8.2).

### Choosing Between Visuals and Text

What information lends itself to visual form? When you write a series of paragraphs, you can develop an argument one step at a time while the reader pays

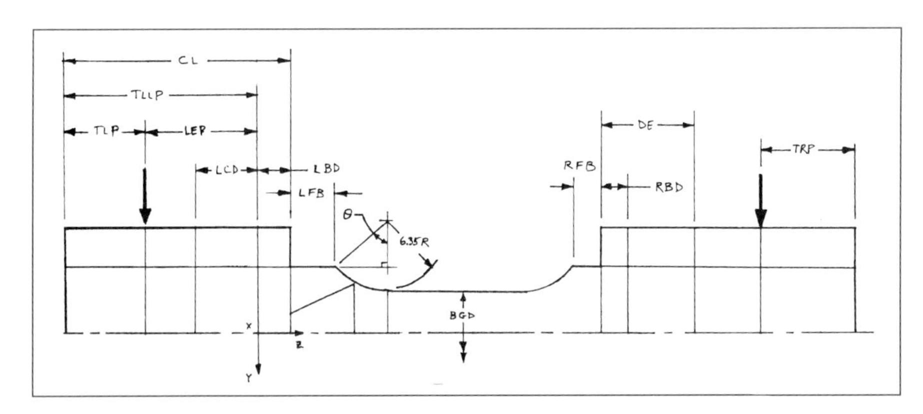


**FIGURE 8.1 Visual Problem Solving**

This sketch helped a graduate student in biomechanics think about finite element models. It shows one way to divide up a bone specimen, which is attached at each end by end caps, into elements for his computer simulations. A close colleague might be able to understand this sketch, but its main purpose was to help the student *see* what he was thinking, and in performing that function, it worked.

(Courtesy of David M. Selvitelli.)

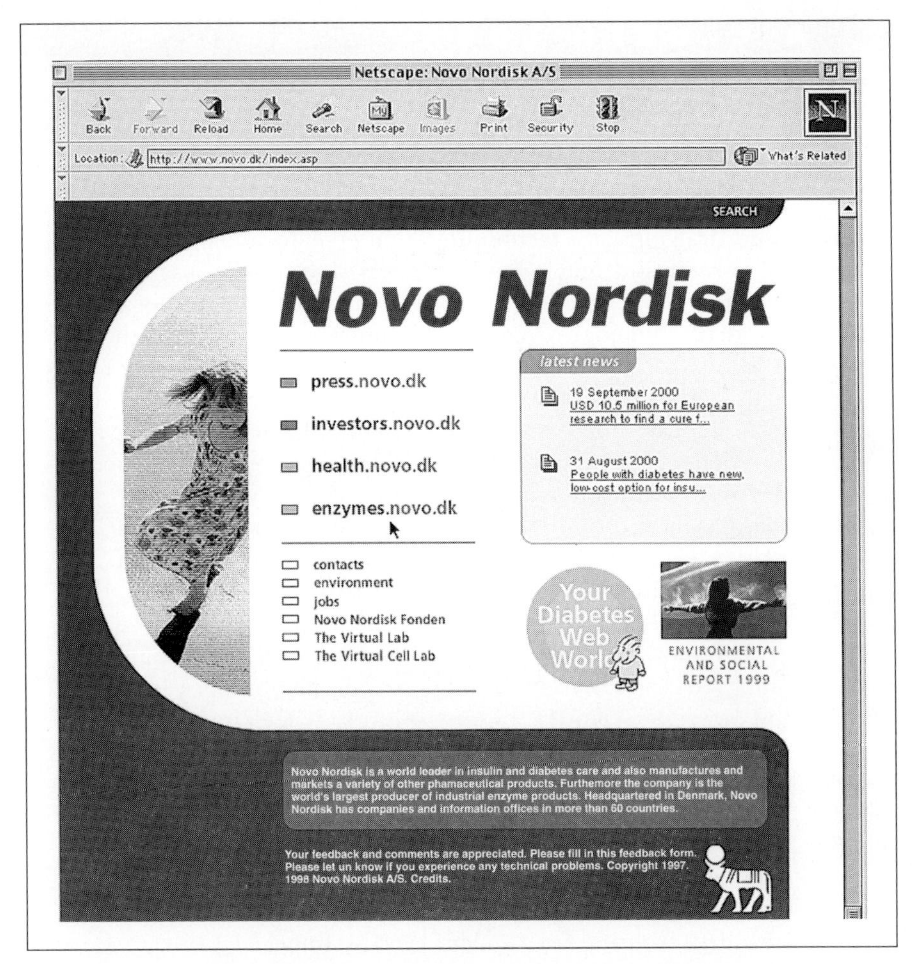


**FIGURE 8.2 A Home Page That Creates an Effective Visual Persona**
(Courtesy of Novo Nordisk A/S © Novo Nordisk. Used by permission.)

attention to that progression. A visual, however, does its work all at once. It is multidimensional rather than linear. Moreover, as one expert notes, "concepts evoked by words often need to be narrowed; concepts evoked by visual media usually need to be broadened" (Williams 671). If you're defining the concept of "waste," for example, you'll need modifiers, examples, and explanations to make sure the reader's concept matches yours. If you use the graphic symbol of a wastebasket to stand for waste, however, the concept evoked may be too limited; it is limited to being indoors, its shape varies in different cultures, it represents household as opposed to industrial trash, and the like. You need to add features that clarify or extend the meaning of the visual to explain or teach a concept or process. You'll often control the meaning a reader derives from your document better in text. Most visuals require the reader to construct a meaning from what you present rather than simply receive what you say (Figure 8.3) on page 156.

A CLOSER LOOK

## Visualizing

*Visualizing* is a process with deep historical roots. Recognition of the need for precise visual observation, for example, revolutionized anatomy and physiology in sixteenth-century Europe. The medicine of the Middle Ages was based on theoretical systems dating to antiquity. Occasional diagrams of the body reflected generalized verbal descriptions rather than clear observation of nature. It was Leonardo da Vinci whose anatomical drawings from life showed the power of careful observation, recorded so that other anatomists could compare the record to what they saw with their own eyes and thus validate the information. The drawings provided "virtual witnesses" to the reality. Later, naturalists used drawings as a technique of observation and memory. Carefully reproduced as woodcuts and engravings, the drawings of plants and animals represented the building blocks of scientific knowledge. John Ziman, a physicist, argues persuasively that observations are the primary means of documenting and validating data in all scientific disciplines.

Today, computer technology has greatly enhanced the power of scientific observation. Remote sensing devices like LandSat satellites, for example, observe the world from space and create detailed maps of the physical environment. Robots deep within a volcano or under the sea send back photographs to researchers' computers. Technology like this lets researchers observe from impossible-to-reach locations. Information arrives at the scientist's screen in digital form that can be enhanced and manipulated to display objects and events of particular interest. Simulations let chemists conduct "virtual" experiments and pattern the structure of molecules in a fraction of the time necessary without such technology. Electronic microscopes let metallurgists see into the deep structure of materials. (Photo at right shows dendrites in a nickel-based superalloy produced by electron beam melting. Courtesy of Malcolm McLean, Department of Materials, Imperial College, London, and M. Halali.)

But encouraging your audience to construct that meaning may be just what you want to do. In addition, before using a visual, assess your audience's ability to read it because visuals, like words, demonstrate different levels of technicality (Figure 8.4 on pp. 158–159). When you address colleagues, for example, you can probably base your presentation on a selection of the visuals you created in your research, and then add words that tie the visuals together. To address an audience unfamiliar with the information, however, you may have to draw new visuals that explain concepts and enliven the text. Your original visuals, appropriate for a technical report or article, may be too complex, dense, or abstract.

Engineers use computer-aided design and computer-aided manufacturing programs (CAD-CAM) to develop and then control processes on the factory floor. Architects use such programs to draw buildings and then rapidly view changes in the design. Industrial designers can test configurations of work spaces to see, for example, if a pilot can reach all the controls in an aircraft. Physicians use scanning devices that allow them to see both organs and processes within a patient's body.

Scanning devices and other computer-enhanced processes help researchers develop a form of visual that has taken on new dimensions at the frontiers of science and technology: a *map*. As early explorers developed maps of new continents, researchers are using mapping techniques to show the interior "landscape" of the human body or to visualize relationships in statistical information. Specialists in information systems talk about creating maps of document flows or communication patterns. The Human Genome Project at the U.S. National Institutes of Health has created a "genetic map" that marks chromosomes to indicate the probability that one segment will be inherited with another (Wheeler A8). By structuring known information, maps create relationships that become new information.

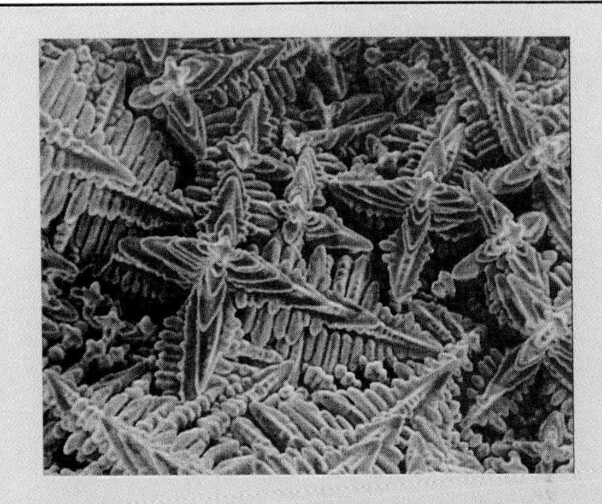

## Compressing a Discussion

Visuals, then, do things that text cannot. For example, you can compress certain kinds of discussions into a visual form that presents information more effectively than extended text. Consider the following paragraph:

> On Monday, the high temperature was 30 degrees Fahrenheit, and we had 4 inches of snow. We had 7 inches on Tuesday, when the high was 25°. There was no snow on Wednesday. The high temperature that day was 28°, followed by 32° on Thursday (with just an inch of snow). Temperatures then continued to rise, with Friday's high coming in at 35° and Saturday's at 37°. No snow either day. On Sunday, the temperature dipped again, to 30°, with 2 inches of snow.

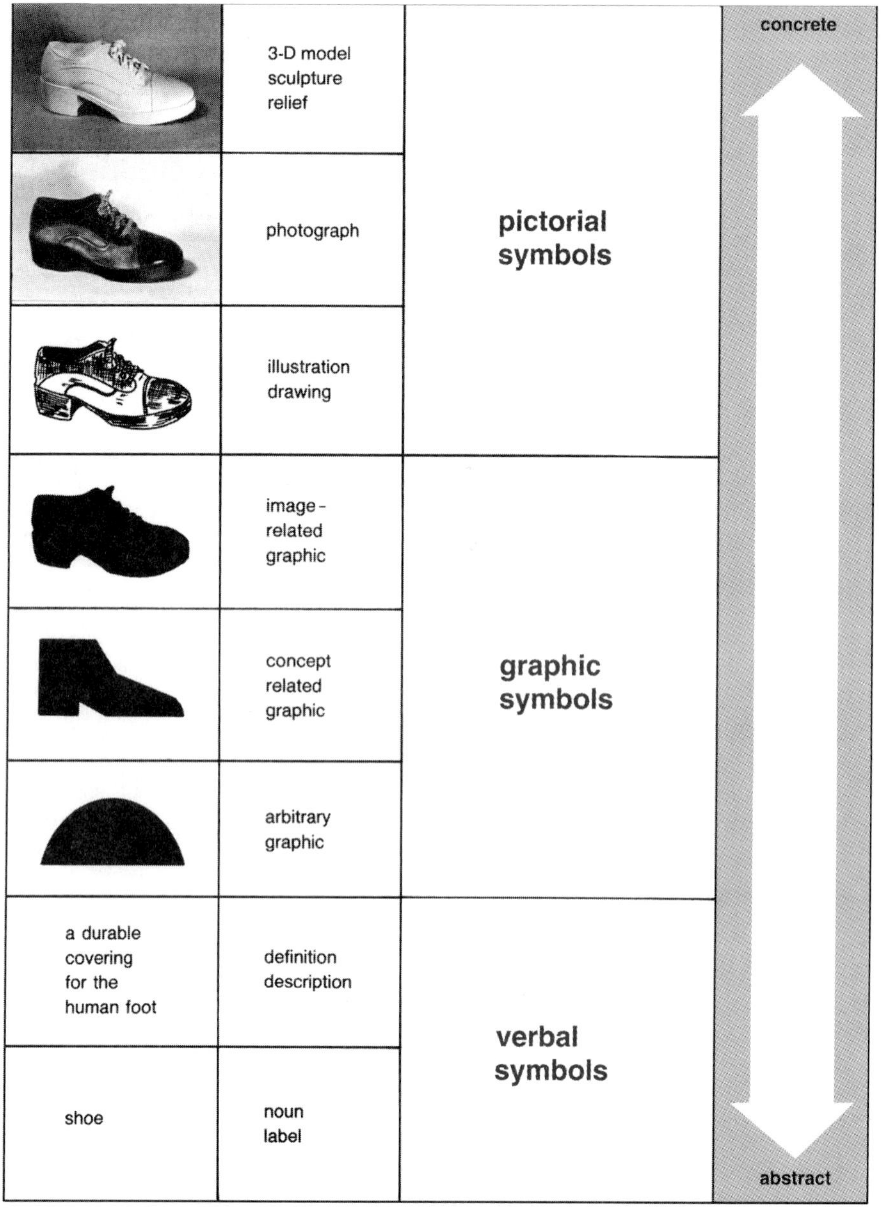

| | | | concrete |
|---|---|---|---|
| | 3-D model<br>sculpture<br>relief | pictorial symbols | |
| | photograph | | |
| | illustration<br>drawing | | |
| | image-<br>related<br>graphic | graphic symbols | |
| | concept<br>related<br>graphic | | |
| | arbitrary<br>graphic | | |
| a durable covering for the human foot | definition<br>description | verbal symbols | |
| shoe | noun<br>label | | abstract |

**FIGURE 8.3 Levels of Abstraction in Graphics**
Like words, visuals can be concrete or abstract. This chart shows the range from models and photographs, which are the most concrete, through graphic symbols. Such symbols are often not self-evident but require that the reader be initiated into the code.

(Courtesy of Ralph E. Wileman.)

The paragraph has a nice casual voice to it, but the information becomes more accessible when classified and when presented as a table. Visual display reduces redundancy and facilitates comparisons.

**TEMPERATURE AND SNOW REPORT, WEEK OF 9–15 JANUARY**

| Day | High Temperature (°F) | Snow (inches) |
|---|---|---|
| Monday | 30 | 4 |
| Tuesday | 25 | 7 |
| Wednesday | 28 | 0 |
| Thursday | 32 | 1 |
| Friday | 35 | 0 |
| Saturday | 37 | 0 |
| Sunday | 30 | 2 |

**For more about classification, see Chapter 10**

When you need to display a wide range of quantitative information, even for a general audience, you'll often find visual devices helpful. An excellent model is *The Historical Atlas of Canada,* which uses a series of visual essays to tell significant stories about Canada. (Figure 8.5 on pages 160–161).

## Clarifying Complex Information

Second, use visuals to simplify information that would be hard to explain in words alone. Figure 8.6 on page 162, for example, makes complex information about recycling more accessible. Recent advances in information mapping also allow you to display multiple layers of data in a relatively simple form, as in Figure 8.7 on page 163.

## Fostering Effective Decision Making

Third, use visuals to help clarify the issues in making a decision. As a manager or concerned citizen, you'll use visuals to help you decide about resource management, zoning, and the like. According to one expert, "Eighty to ninety percent of the decisions people make have a geographic component" (Moore 31). The map in Figure 8.7, for example, helped the manager of an environmental monitoring project in Casco Bay, Maine, track changes in land use patterns, water quality, and shellfish beds.

## Crossing Borders of Language and Culture

Fourth, use visuals as an international language. One authority notes, "A major portion of engineering information is recorded and transmitted in a visual language that is in effect the lingua franca of engineers in the modern world" (Ferguson 41). As a neuropathologist remarked, in his discipline of identifying the chemical constituents of cells and tissues, "the images are the data" (Wheeler A19). Professionals have developed highly sophisticated codes and conventions for visual presentation of information.

(a)

(b)

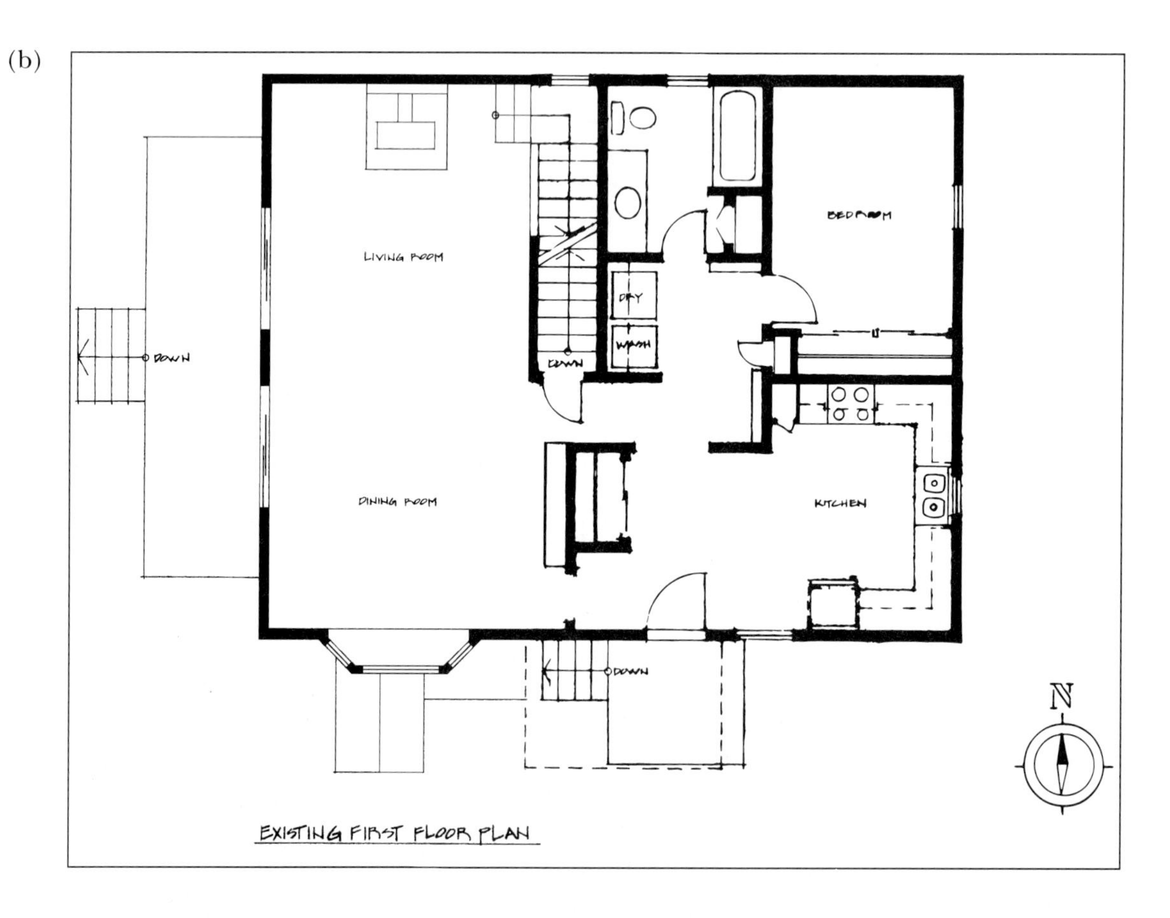
LIVING ROOM
DINING ROOM
BEDROOM
KITCHEN
DRY
WASH
DOWN
DOWN
DOWN
N
EXISTING FIRST FLOOR PLAN

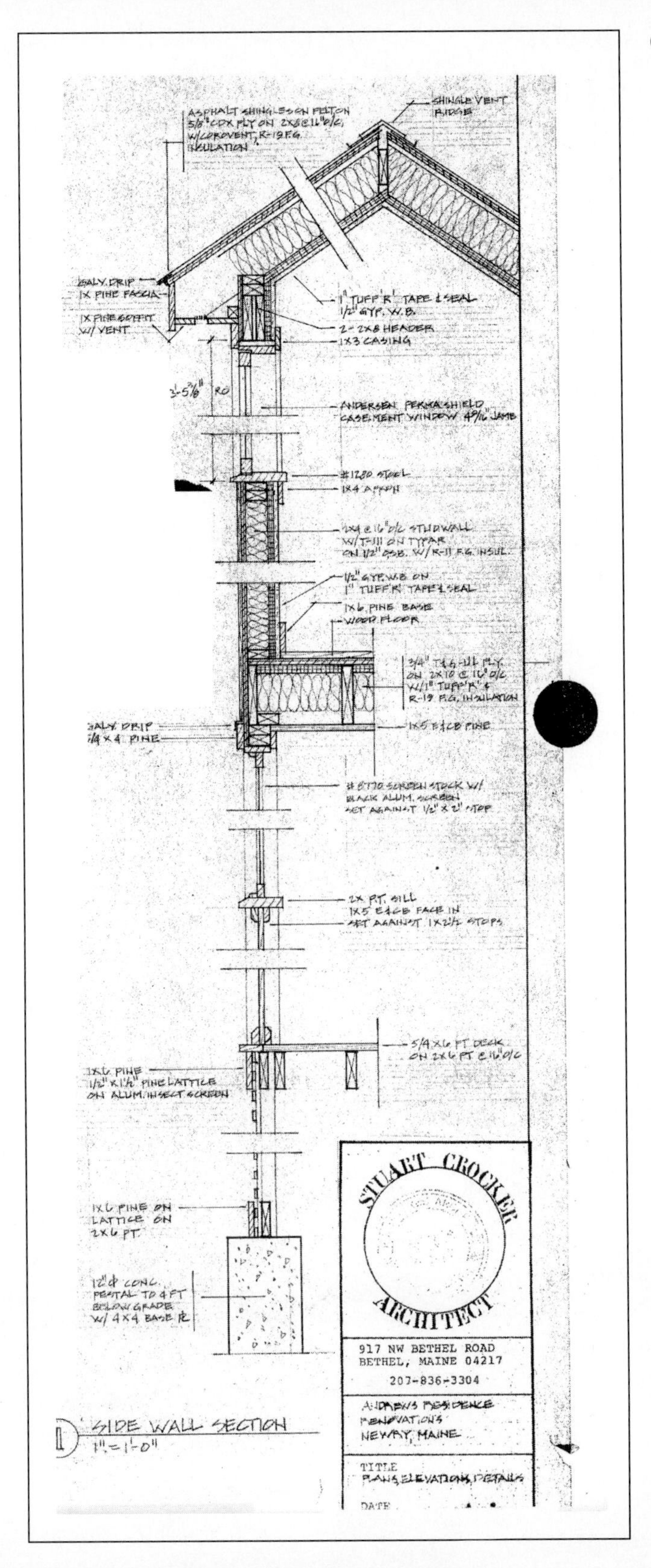


(c)

**FIGURE 8.4 Levels of Technicality in Graphics** Architects use visuals to sketch their own thinking about a new building and to communicate their ideas to various audiences, including clients, zoning boards and other regulatory agencies, suppliers, and contractors. In doing so, they show the building from various points of view and for various purposes. For example, a client may be interested in how a new addition to a house will *look.* A realistic drawing, created either by hand or, increasingly, on a computer, helps fulfill this purpose (a). If the client understands the conventions of a floor plan, then a depiction from that point of view (b) is also helpful. *Plans* are more abstract and use visual codes, like an arc to suggest the pattern in which a door swings and indicators of windows within walls. More abstract, and at the same time more technical, is a *specification* for a contractor, here shown in an *elevation* or *section* (c). The purpose of this drawing is to instruct the contractor and to provide a basis for ordering materials.

(Drawings of Andrews Residence Renovations, Newry ME, by Stuart Crocker, architect. Reprinted by permission.)

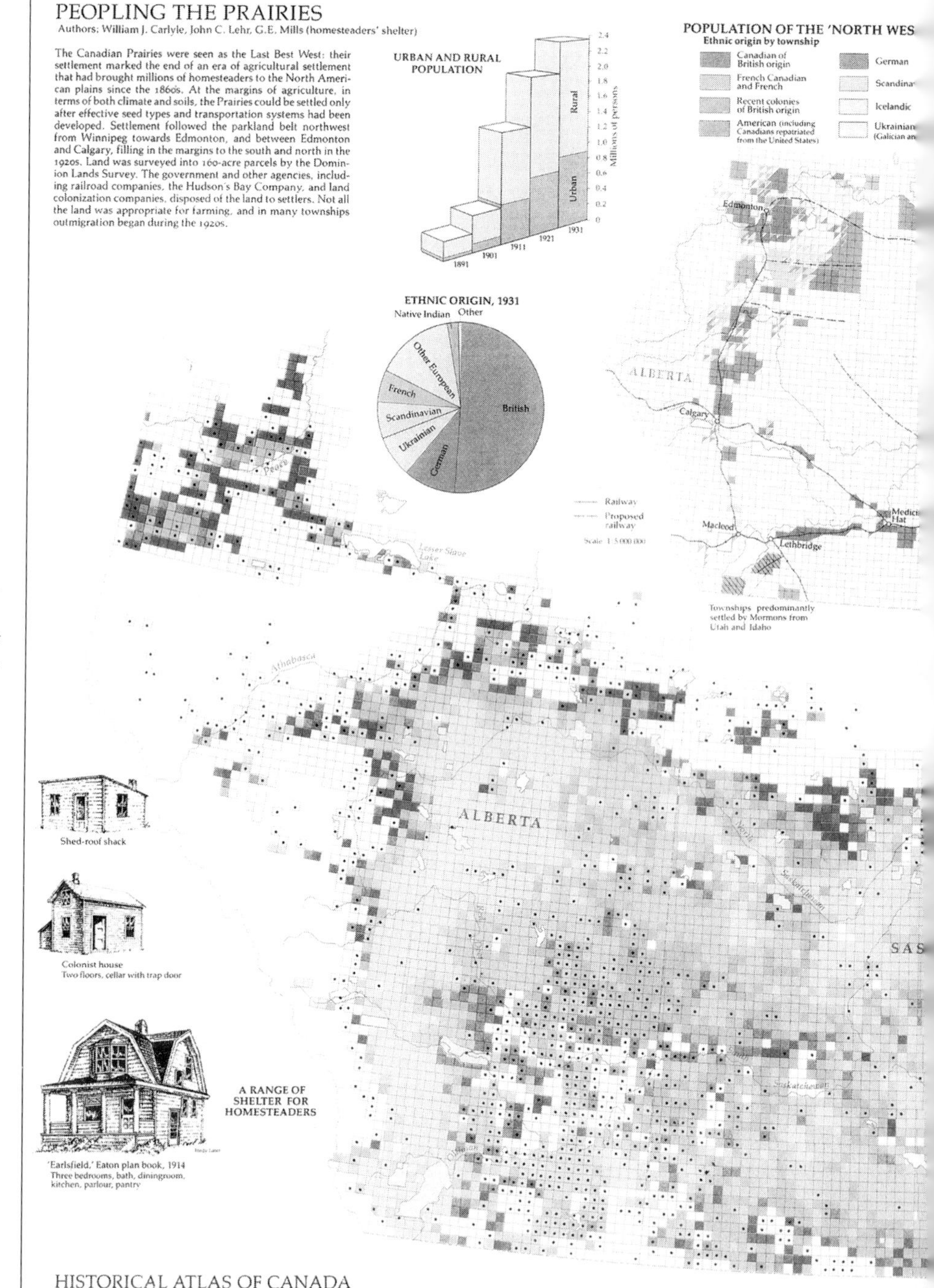


**FIGURE 8.5 Compressing a Discussion**

*The Historical Atlas of Canada* tells the history of the country through visuals. Each two-page spread is a highly compressed visual essay on a particular topic. It brings together many of the forms you are reading about in this chapter, for example, a pie chart on "Ethnic Origin, 1931," a stacked bar chart on "Urban and Rural Population," information maps (i.e., "The Progress of Prairie Settlement"), and drawings of various house types. Color (not reproduced here) is used to encode meaning in the keys to the various charts and maps. The visuals are connected through a brief textual introduction and captions.

(Reprinted by permission of the University of Toronto Press Incorporated © University of Toronto Press, 1990.)

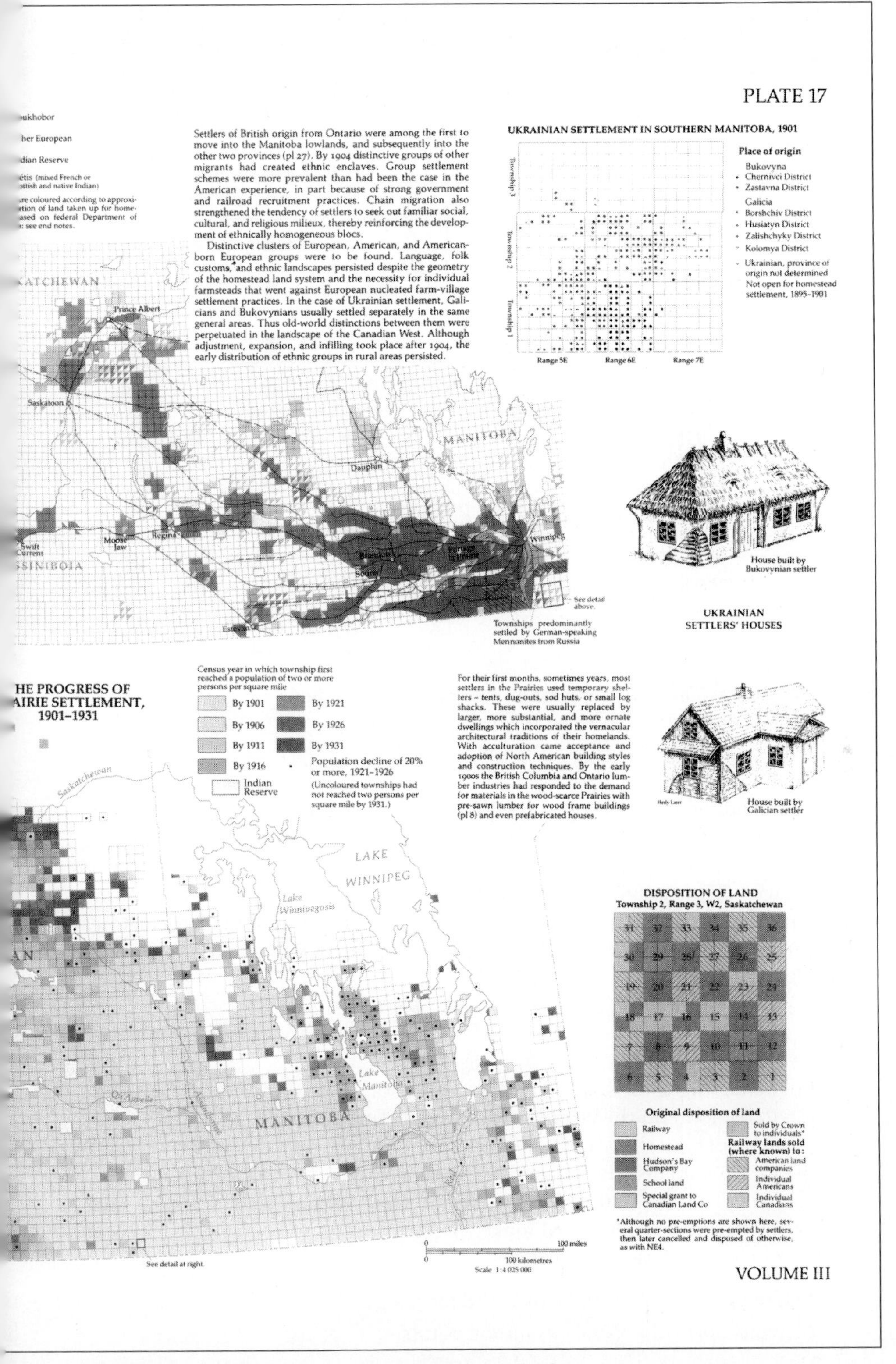

PLATE 17

Settlers of British origin from Ontario were among the first to move into the Manitoba lowlands, and subsequently into the other two provinces (pl 27). By 1904 distinctive groups of other migrants had created ethnic enclaves. Group settlement schemes were more prevalent than had been the case in the American experience, in part because of strong government and railroad recruitment practices. Chain migration also strengthened the tendency of settlers to seek out familiar social, cultural, and religious milieux, thereby reinforcing the development of ethnically homogeneous blocs.

Distinctive clusters of European, American, and American-born European groups were to be found. Language, folk customs, and ethnic landscapes persisted despite the geometry of the homestead land system and the necessity for individual farmsteads that went against European nucleated farm-village settlement practices. In the case of Ukrainian settlement, Galicians and Bukovynians usually settled separately in the same general areas. Thus old-world distinctions between them were perpetuated in the landscape of the Canadian West. Although adjustment, expansion, and infilling took place after 1904, the early distribution of ethnic groups in rural areas persisted.

UKRAINIAN SETTLEMENT IN SOUTHERN MANITOBA, 1901





House built by Bukovynian settler

UKRAINIAN SETTLERS' HOUSES

For their first months, sometimes years, most settlers in the Prairies used temporary shelters – tents, dug-outs, sod huts, or small log shacks. These were usually replaced by larger, more substantial, and more ornate dwellings which incorporated the vernacular architectural traditions of their homelands. With acculturation came acceptance and adoption of North American building styles and construction techniques. By the early 1900s the British Columbia and Ontario lumber industries had responded to the demand for materials in the wood-scarce Prairies with pre-sawn lumber for wood frame buildings (pl 8) and even prefabricated houses.

House built by Galician settler

THE PROGRESS OF PRAIRIE SETTLEMENT, 1901–1931



DISPOSITION OF LAND
Township 2, Range 3, W2, Saskatchewan



*Although no pre-emptions are shown here, several quarter-sections were pre-empted by settlers, then later cancelled and disposed of otherwise, as with NE4.

VOLUME III

| STATES | AL | AK | AZ | AR | CA | CO | CT |
|---|---|---|---|---|---|---|---|
| Labeling Laws | | | ● | | ● | | ● |
| Truth in Advertising | ● | ● | ● | ● | | ● | ● |
| Plastics Resin Coding | | ● | ● | ● | ● | ● | ● |
| Paper | | | | | ● | | ● |
| Metal | | ● | | | ● | | ● |
| Glass | | | | | ● | | |
| Plastic | | | | | ● | | |
| Used Oil | | | ● | ● | ● | | ● |
| Recycling Targets | | | | ● | ● | | ● |
| Recycling Violation Penalty | | ● | ● | | ● | | ● |
| Toxics in Packaging | | | | | | | ● |
| Recycling Tax Credits | | | ● | ● | ● | ● | |

**FIGURE 8.6 Making Complex Information Accessible**
(Adapted from data compiled by Khesha T. Jennings.)

This shared visual jargon fosters communication among colleagues worldwide, especially as the capacity for electronic delivery of visuals increases. In more popular usage, too, graphic symbols convey information across languages (Figure 8.8). Keep in mind, however, that people from different cultures may have to be initiated into that visual vocabulary (Horton). *Crossing Cultures: International Signs and Symbols* provides examples of international differences.

## Avoiding Clutter

But in achieving these four purposes, apply some restraint. Resist the temptation to overwhelm the reader with visuals (Tufte). Often, one photograph of a failed specimen, for example, has more impact than many photographs of similar specimens; the images blur in the viewer's mind. Make sure that each visual, too, communicates clearly. Overcrowded or murky visuals frustrate and mislead readers and make them think less of you (Figure 8.9). To keep your presentation uncluttered:

- Use visuals to highlight only the most significant information.
- Avoid devoting visuals to minor points.

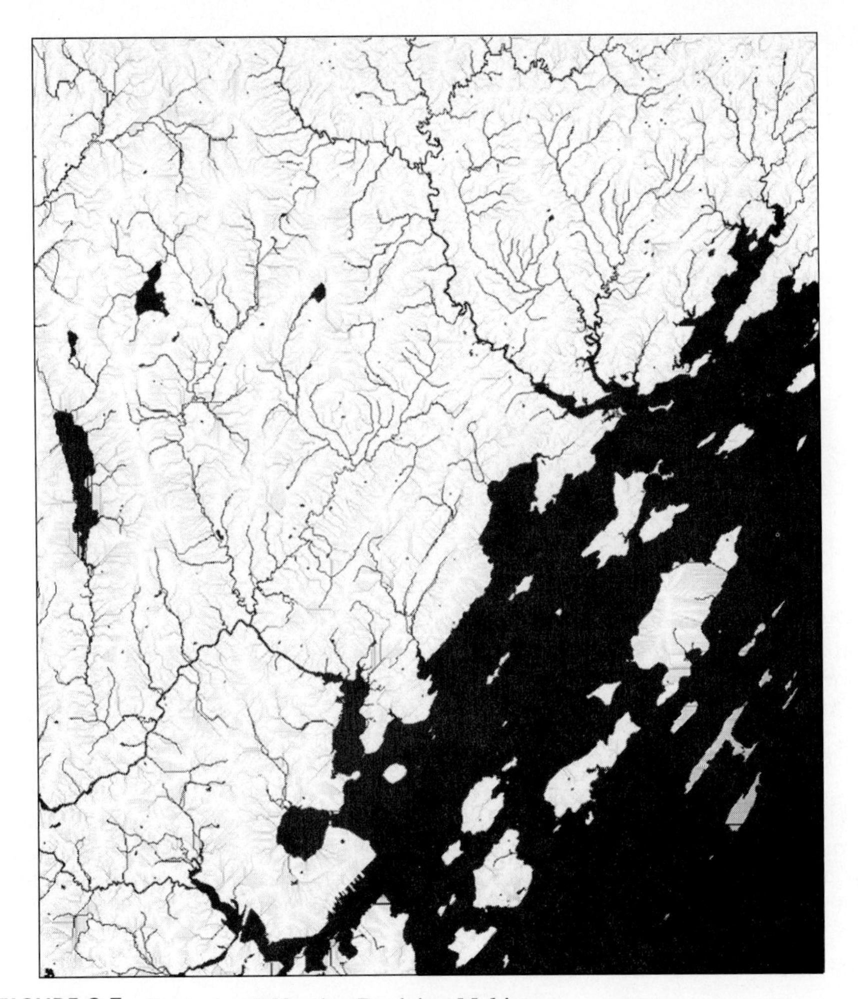

**FIGURE 8.7 Fostering Effective Decision Making**
This map, generated by a Global Information System, shows the drainage patterns of a portion of Casco Bay, Maine. When combined with maps that show soil types, land cover, and land use, it helps resource managers for an environmental monitoring project locate sources of pollution and make decisions that protect the environment.
(Reprinted by permission of Thomas R. Burns and GIS Mapping and Analysis.)

- Anticipate the final form in which the audience will see the visual. The image may be reduced in print and may need to be enlarged for an oral presentation.
- Simplify visuals to be used for oral presentations.

## Creating Effective Labels

When you use visuals primarily to enhance the atmosphere of a document or presentation, you don't need to label them. But label any visual aimed at

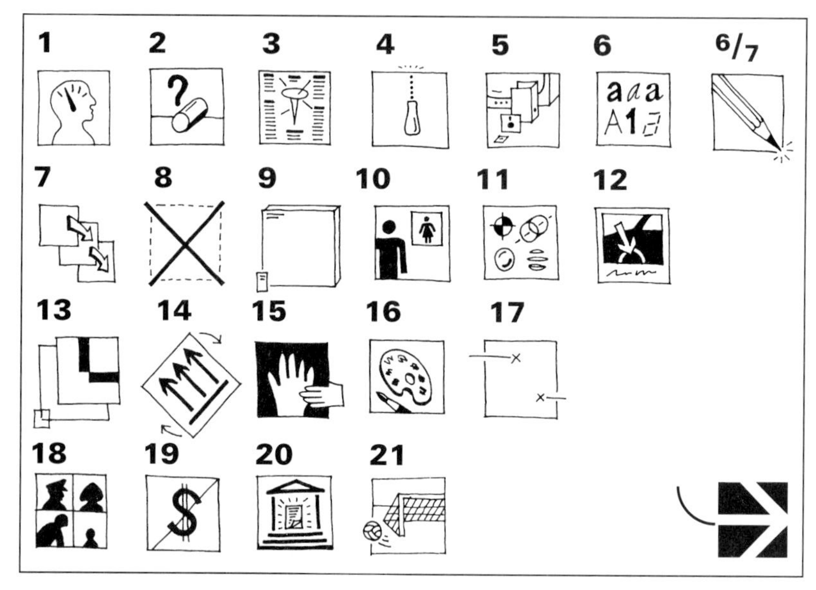


**FIGURE 8.8 Crossing Borders**

This is the table of contents of a brochure that accompanied a Dutch exhibit of consumer product manuals. The brochure, which measures 15 cm × 10.5 cm, provides text in English, Dutch, and German. To accommodate all three languages, to keep the size small and thus the cost of the publication low, and to guide readers around the many manuals displayed in the exhibit, the brochure and the exhibit itself rely heavily on graphics. Number 1, for example, notes that life today is a "continuous intelligence test" because everyone has to operate so many machines in the course of the day. Number 8 represents "don'ts": approaches to conveying product warnings to audiences who speak a variety of languages. Number 21, a soccer ball missing the net, signals the exhibit of manuals that miss their mark. The arrow in the lower right corner tells the reader to "turn the page."

(Courtesy of Drs. Piet Westendorp and Prof. Paul Mijksenaar.)

recording data or, especially, informing an audience. The extent of the label or caption (captions tend to be longer than labels) depends on your purpose, the audience's familiarity with the form of the visual and its content, and the genre of the presentation. In some documents, the visuals are the main text, and captions provide commentary that connects the images. In a newsletter or popular magazine, an informal label may identify the people or objects in the picture or add a humorous twist. In technical articles and reports, a label or caption is usually more extensive and consists of three parts: a number (or, more rarely, a letter) that shows the sequence of that visual in the report, a title, and an explanation.

The *number,* for example *Figure 1,* helps you and the reader refer to the visual. In a long document, visuals may be numbered by chapter (e.g., Figure 12.10 is the tenth in Chapter 12). The *title* helps you transform your thinking as you recorded the information into evidence that enlightens or persuades someone else. Delete any internal codes or identifiers that are meaningless to the reader, for example, "Run ll," "Alloy 60," or the date of

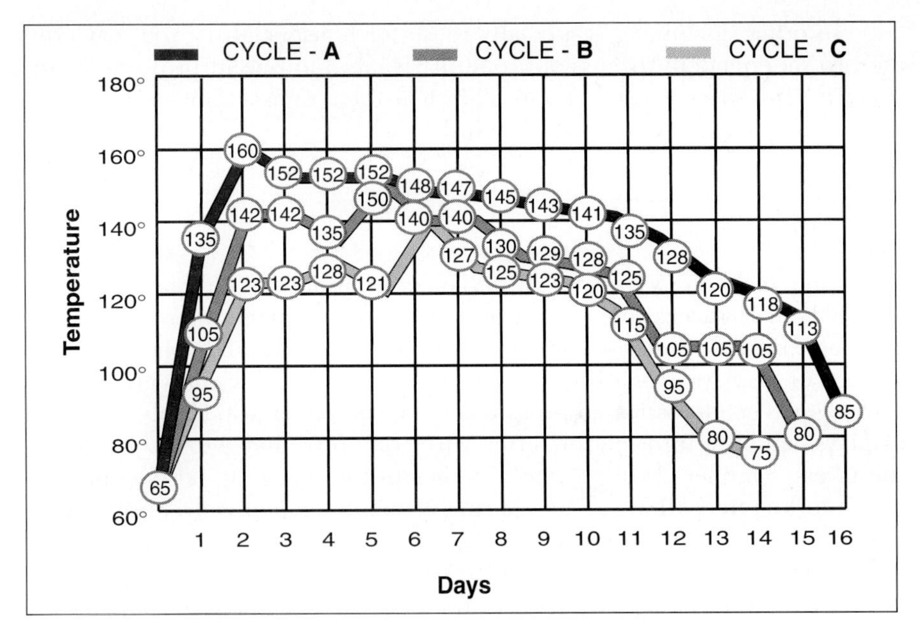


FIGURE 8.9 **Chartjunk**
An excess of symbols and lines makes this graph unreadable.
(Courtesy of Erin Gantt-Harburcak.)

the run. The *explanation* depends on the audience's sophistication and the complexity of the information you are presenting. If the image is a copy of someone else's work or uses data someone else compiled, your explanation should also include a reference to that source:

> Source: Science Resources Studies Division, National Science Foundation, *Research and Development in Industry: 2001* (Washington, DC: NSF, 2001).

You may place a reference notation on the title consistent with notations used elsewhere in the report:

> Figure 3. Lobster Landings in Casco Bay, Maine 2001–02. [6]

In a report, introduce every visual in the text; none should enter unannounced. The least invasive mention is a parenthetical one:

> The beginning of each heartbeat is detected as a large, positive spike in the differentiated pressure waveform (Figure 2).

To draw even more reader attention, use a full sentence that instructs the reader. Avoid simply pointing to the visual:

> ***Poor:*** Figure 2 gives a typical waveform.

Instead, highlight it:

> ***Better:*** As shown in Figure 2, the beginning of each heartbeat is detected as a large, positive spike in the differentiated pressure waveform.

In other documents, especially those for nonspecialists, you may create two independent discussions, one in visuals and one in prose, without specific references between them. The clear focus on one topic and the design of the page or screen make the link.

## FORMS

Electronic technology has greatly enhanced your ability to create graphics and make them accessible to your audience both in print and on the Web. You enter information into the program; it creates a visual display, perhaps a range of displays. Not all results are effective, but such visuals are a good starting place. Sophisticated programs allow you to develop highly complex displays that layer information, often with color introduced to discriminate the layers. Scanners bring a world of images into your digital documents, and color printers allow you to add that element inexpensively. Multimedia applications help your information product *move:* A viewer at a screen can fly through an "information landscape" at the angle of a low-flying aircraft (Figure 8.19 on p. 181). In that way, the user can access, for example, several different charts arranged in different planes and intersecting one another to create another informative dimension. Such technology *animates* information. Underlying these applications, however, is the basic vocabulary of six forms described briefly in this section.

### Table

One simple, if powerful, form is a *table.* In a table, you arrange words or numbers in rows (horizontal) and columns (vertical). The Periodic Table of the Elements is a classic example (see Figure 10.3 in Chapter 10). Spreadsheets are a common management and financial application. Tables serve several purposes:

- Record numerical information consistently, as in a price list
- Identify holes in your information where cells (the intersections of rows and columns) are empty
- Show many discrete units of information in a small space
- Make it easy for readers to compare items
- Present a series of parallel terms as a multidimensional list

Most tables read vertically, that is, in the same direction as the text (in "portrait" position). Long tables, however, may run horizontally ("landscape" position). Place such tables on a right-hand page with the top toward the inner margin. To show a network of interrelationships among items, either words or numbers, use a matrix, another form of table (Figures 8.6 and 8.10).

In composing a table, follow these guidelines:

1. Place units of measurement in the title or a footnote, not after each entry. Carefully note any multipliers (for example, "in thousands").

**Table 4. Examples of measured ambient and oil temperatures for seven forest machines.**

| Machine | | Temperature (°C) | | |
|---|---|---|---|---|
| | | Ambient | Oil | Difference |
| Forwarder | 1 | 2 | 53 | 51 |
| | 2 | -2 | 35 | 37 |
| | 3 | 40 | 110 | 70 |
| Grader | 4[a] | 30 | 65 | 35 |
| | 5[a] | 20 | 110 | 90 |
| Delimber | 6[a] | 25 | 100 | 75 |
| Harvester | 7 | -5 | 47 | 52 |

[a] *Data from unpublished 1995 FERIC field study; other data from Table 21 in Makkonen (1994). Due to the variability of the data, every machine should be monitored to determine its unique temperature characteristics.*

(a)

**FIGURE 8.10 Tables**

Example (a) has a number (4), a title, and an explanation keyed by the footnote letter *a*. The temperature designator appears in the column heading. Readers would be familiar with the names of the machines and need no further explanation. No vertical lines are needed to separate the columns. Example (b) is a *matrix* from the same publication, a "Technical Note" from the Forest Engineering Research Institute of Canada (FERIC) that tells practicing foresters how to convert forestry machinery from using hydraulic fluids based on mineral oils to new environmentally compatible oils. It uses highly technical codes to indicate cleanliness levels. Readers, too, would be familiar with the acronyms for the two major organizations that set standards for operation: ISO and NAS.

(Source: Ismo Makkonen, "Converting a Forestry Machine to Use Environmentally Compatible Hydraulic Oils," *Wood Harvesting,* Technical Note TN-234. November 1995. pp. 5 and 6. Courtesy of FERIC.)

**Table 3. Minimum recommended hydraulic system cleanliness limits[a]**

| Component or type of system | Recommended cleanliness limits according to ISO and NAS | | | | | | | | |
|---|---|---|---|---|---|---|---|---|---|
| ISO | 13/10 | 14/11 | 15/12 | 16/13 | 17/14 | 18/15 | 19/16 | 20/17 | 21/18 |
| NAS | 4 | 5 | 6 | 7 | 8 | 9 | 10 | 11 | 12 |
| Servo valves | | | | | | | | | |
| Regulator valves | | | | | | | | | |
| Proportional valves | | | | | | | | | |
| Pumps (p > 16.0 MPa) | | | | | | | | | |
| Pumps (p < 16.0 MPa) | | | | | | | | | |
| Low-pressure hydraulics | | | | | | | | | |
| Required filterability[b] | | | | | | | | | |
| bx = 75 | — | — | 3 | 5 | 10 | 20 | 25 | 25–40 | |
| b | 3 | 3 | — | — | — | — | — | — | |

[a] *These standards are undergoing revision, and will likely be tightened.*

[b] *"x" is the required particle size (mm) above which a 75 or 100 times reduction in particle count is required.*

(b)

2. Arrange items in a logical order: alphabetical, geographical, quantitative, or chronological.
3. Place numbers in columns rather than rows for ease of comparison and carryover to an additional page. Rows are also limited by the width of the page.
4. Align all numbers along the decimal point.
5. Place long tables that serve as references in an appendix.
6. Provide adequate space around a table for easy reading and good design.
7. Box the table if that accords with the visual style in the rest of the document.
8. On most occasions, use spacing to separate the columns and avoid vertical lines.
9. Note the source of data or of the entire table. Use a system of notation consistent with that of the document as a whole.

## Graph

A second basic form of visual is a *graph*. Use a table to classify and assemble numbers or words so those discrete units can be easily compared. Use a graph to turn numbers into pictures that show relationships among such units. Graphs indicate, for example, trends and rates of change. We discuss a few common graphs next.

**Line Graph.** In a line graph, you highlight trends and relationships in numerical data (Figure 8.11). Such graphs, for example, record the temperature or other characteristics of hospital patients, show customer preferences revealed in a market survey, detail how a vehicle performed in a series of engineering tests, or summarize a company's financial picture. Computers that monitor experimental data often indicate results directly in line graphs. Some numerical values lend themselves to expression in both a table and a graph. For an audience who needs the details, use a table; to show a trend to an audience who needs an overview or needs to be persuaded, use a graph.

In composing a line graph, follow these guidelines:

1. Show the independent variable (often a measure of time) along the horizontal axis (the x-axis).
2. Show the dependent variable on the vertical axis (y-axis).
3. Choose a scale that well represents the importance of the changes shown (Figure 8.12).
4. For ease of reading, use multiples of 2, 5, 10, and so on.
5. If you need to, include an insert that shows an important segment at magnified scale to supplement a line that covers a longer time. For example, a plot of how a specimen's weight changed over a two-day period might include an insert showing significant changes over the first few hours.

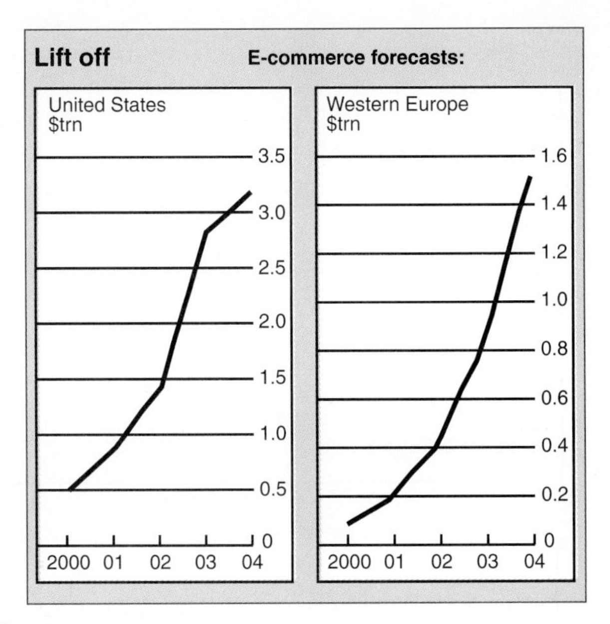


FIGURE 8.11 **Line Graph**

The units of measurement in the y-axis are identified at the right, which supports the pun in the title ("lift off"). Most U.S. graphic software places the units at the left, but non-U.S. publications vary the placement on either the right or the left. The abbreviation "$trn" (trillions of U.S. dollars) would be clear to readers of *The Economist,* where this graph was published.

(Source: "A Survey of Government and the Internet: The Next Revolution." *The Economist.* June 24 2000: 4 © 2000 The Economist Newspaper Group, Inc. Reprinted with permission. Further reproduction prohibited. www.economist.com.)

6. Use a series of graphs with parallel scales if you need to show more than two or three lines.
7. Present information as it was originally recorded. If the recording was continuous, don't show individual data points.
8. Keep each line equal in width, because one that is wider than the others appears to be more important.
9. Label the vertical axis to the left to keep the area within the grid as free of text as possible.
10. Limit explanatory material on the graph. Save descriptive discussions, like test conditions, interpretations, or apparatus, for the text.

**Bar Graph.** Use a line graph to show a trend or a continuous progression over time; use a bar graph to compare the relationship among discrete items. For example, connect each month's statistics about births in a community in a line that shows at a glance fluctuations in the birthrate over a year. To emphasize the comparison from month to month, use a bar graph. The length of the bar indicates a value (for example, 80 percent) or amount (for example, 10). You can arrange bars either vertically or horizontally. Vertical bars often represent changes over time, with the baseline acting like the horizontal axis of a line graph. Horizontal bars commonly show

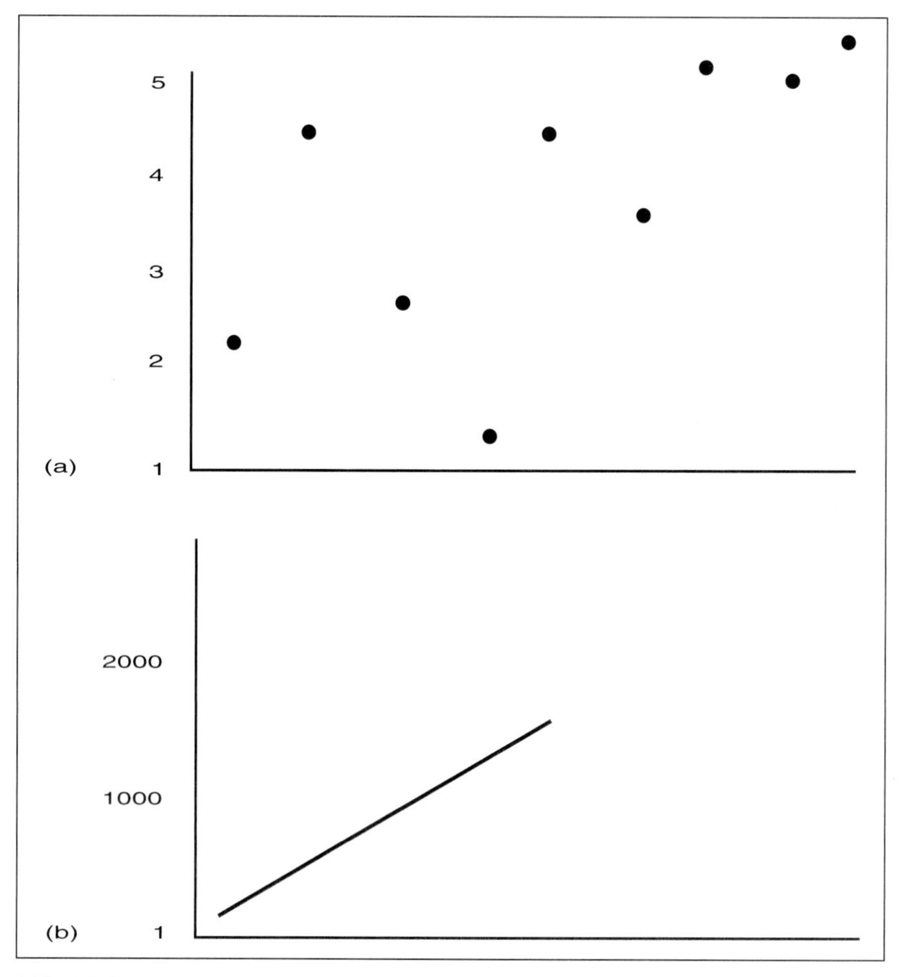


**FIGURE 8.12** **Distorted Line Graph**
At small scale, the data points fail to line up (a). But to present a prettier picture—to force the data into a straight line—researchers sometimes just enlarge the scale, which blurs distinctions (b).
(Courtesy of Erin Gantt-Harbucak.)

acceleration or percentages. Bar graphs (Figure 8.13) appear frequently in corporate annual reports, general interest magazines and newspapers, and other publications aimed at a broad readership, because these graphs are easy to understand. They can be misleading, however. You can create attention-getting graphs that cause the reader to come to the wrong interpretation (Figure 8.14 on page 172).

In composing a bar graph, follow these guidelines:

1. Use solid bars or differentiate bars with various colors, hatching, or perspective.
2. Stack information within one bar to compare data from different years in one category, or cluster the bars (Figure 8.13).

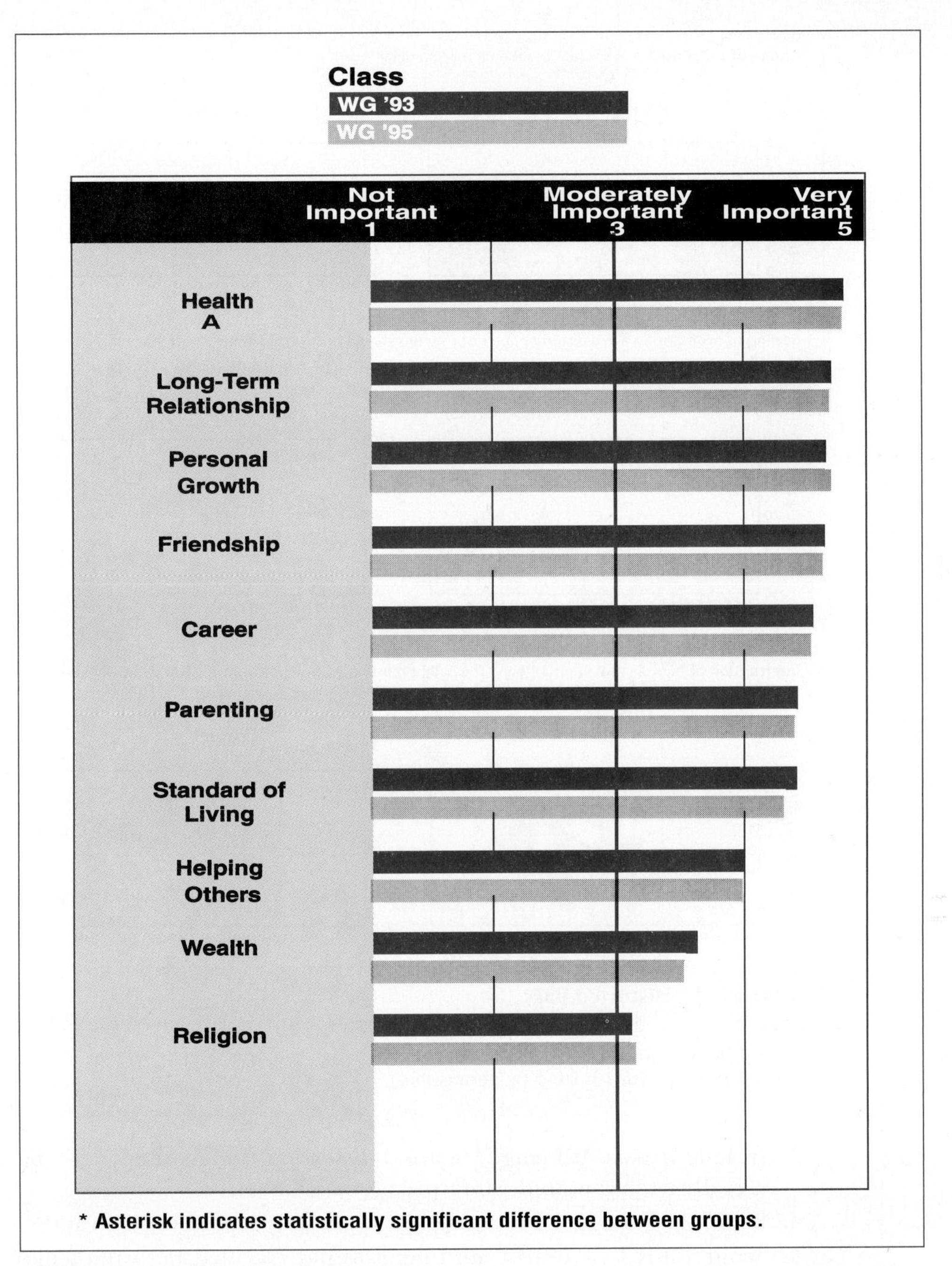


FIGURE 8.13 **Bar Graph**
Bar charts provide easy comparisons at a glance. One bar can represent one value, or several values can be aggregated into multiple bars, as in this figure, which summarizes responses to a yearly questionnaire circulated to graduates of the Wharton School of the University of Pennsylvania. Color coding helps designate the bars.
(Courtesy of the Wharton Life Interests Project, University of Pennsylvania.)

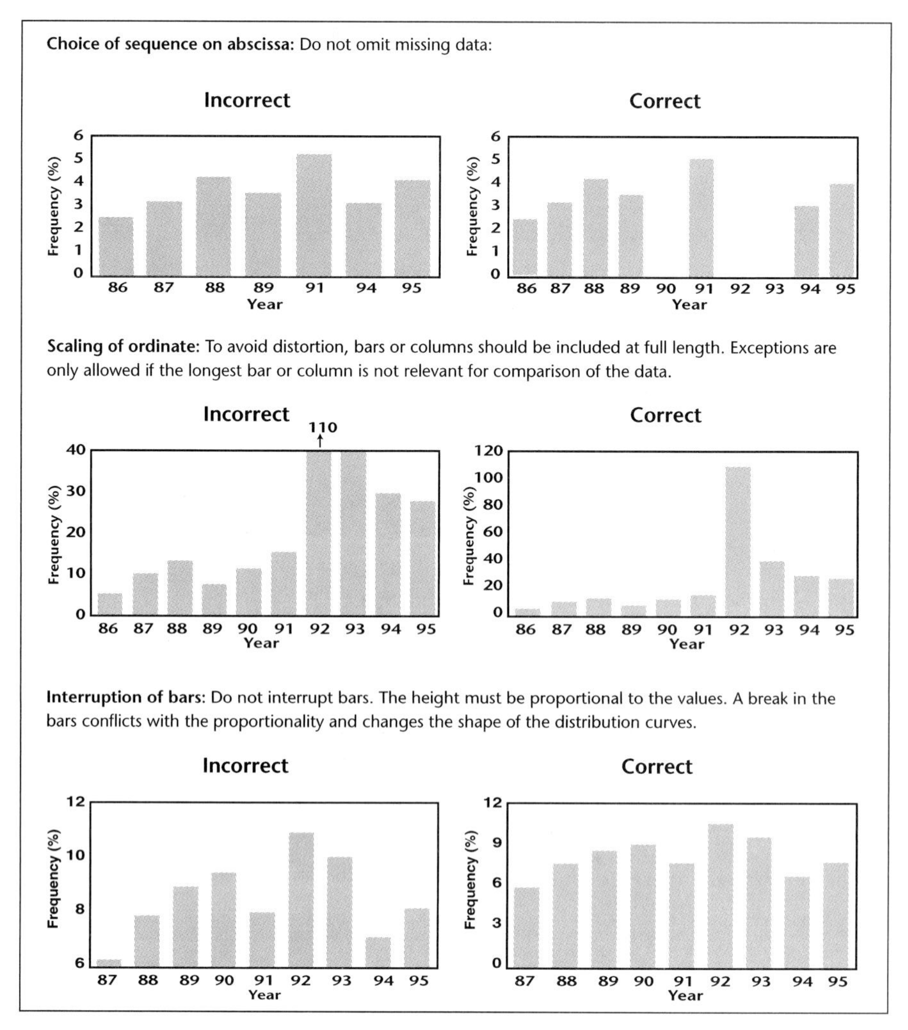


FIGURE 8.14 **Distorted Bars**
The bar graphs on the left can be misleading; correct forms are shown on the right.
(Source: Max Rüegg, "Graphical Presentation of Data," *European Science Editing,* no. 56, October 1995, pp. 16–18. Used by permission.)

3. Include bars at full length to avoid distortion (unless the longest bar is really irrelevant to the interpretation).
4. Begin the scale at "0" so the height accurately represents the value.
5. Avoid a 3-D look unless you intend to indicate that both the length and the volume of the bar are meaningful. Otherwise the reader will compare volume instead of length.
6. Similarly, maintain equal bar width and spacing so the reader doesn't think the width is as informative as the length.

**Pie Graph.** Use a pie graph, as the name implies, to divide up a total (the pie) into pieces. That total may be, for example, a sum of money or a sample population—all students at your university, all civil engineering majors, all residents of a particular city. Pie graphs are common in articles and reports for nonspecialist audiences but less common in technical discussions. They often take up more room on a page than the information deserves, so use them sparingly.

In composing a pie graph, follow these guidelines:

1. Begin the division of the pie at the top center ("noon"). Most graphics software will not let you begin anywhere else.
2. Arrange sectors by size, with the largest at the top.
3. Place labels either inside or outside the circle—and without leader dots that tie the label to the pie.
4. Label slices with percentages, not just words.
5. Try to show no more than six segments.

## Diagram

A third visual form is a *diagram*. A diagram may be based on numerical information, like a graph. But a diagram often represents other kinds of relationships, some of them relatively concrete and some relatively abstract. A diagram may be both descriptive, showing what happens, and prescriptive, showing what ought to happen. Prescriptive diagrams accompany instructions for assembling devices, operating equipment or systems, scheduling a project, or reinventing a management structure. A diagram may also be called a *chart*. Here are some common forms.

**Organizational Chart.** An organizational chart shows how people or parts relate in a structure. The most common form is the organizational chart of a corporation or agency (Figure 8.15).

**Schedule.** A schedule helps individuals and teams plan and monitor activities. Your class schedule, for example, both describes when you'll be in class and prescribes your activity for the academic quarter or semester. Contractors use schedules to ensure that the right people and equipment arrive at the job site at the right time. Researchers include a schedule of planned project activities in a proposal seeking funding to support those activities. The most common form of schedule is probably the Gantt chart, a cross between a table and a figure (page 29). Tasks are listed in the rows; they are measured against time units designated in the columns.

**Flow Diagram.** A flow diagram, as the name implies, shows a flow or process. The diagram may use text and drawings to describe something concrete, like the production of a medical device (Figure 8.16b). Or it may represent abstract thinking, as in a decision tree (Figure 8.16a). In such diagrams, connect symbols with arrows to indicate the direction of flow. Such arrows are especially important internationally, because people who read from left to right (as in English) tend to see a sequence as beginning at the

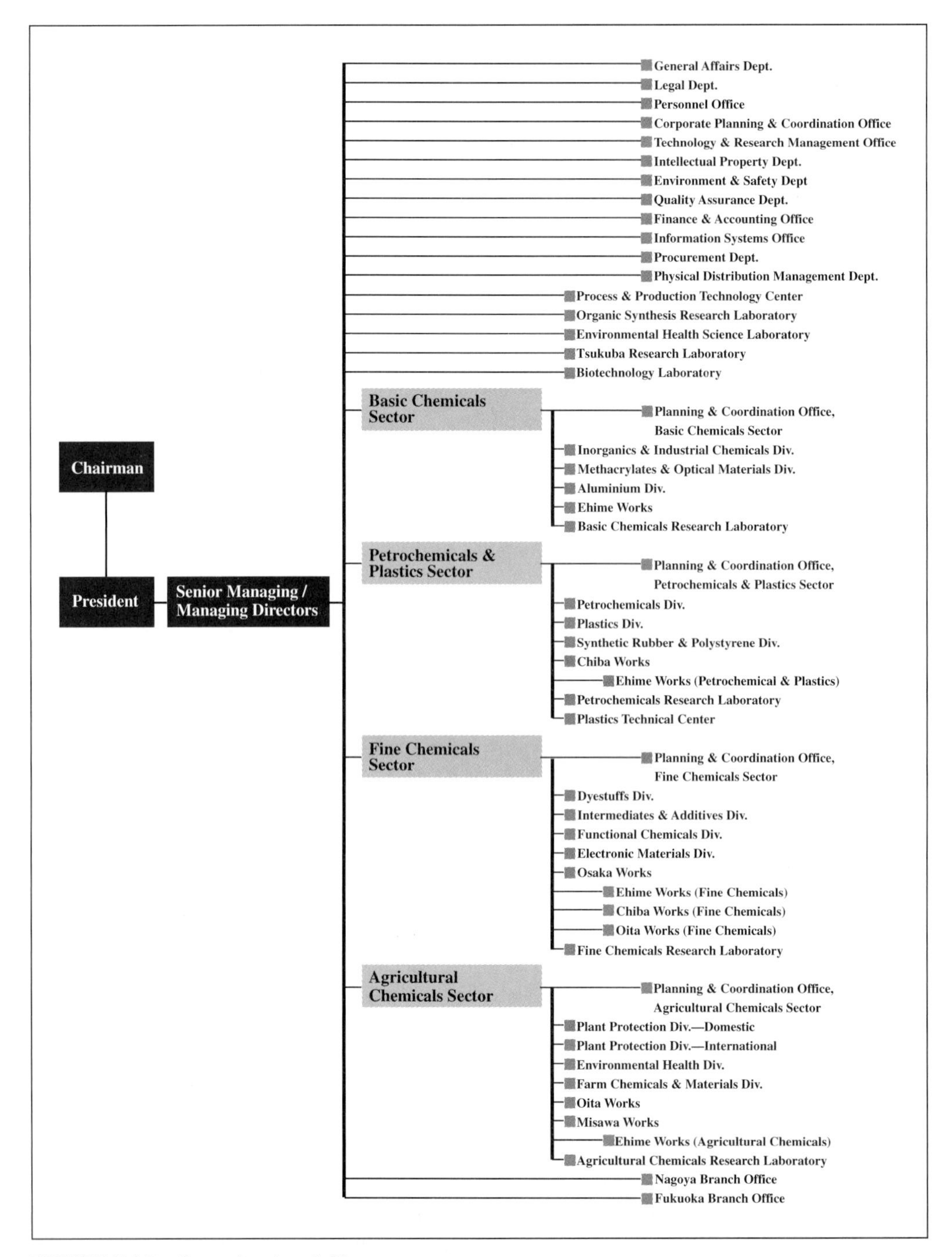


**FIGURE 8.15 Organizational Chart**
This chart presents the management structure of a large Japanese chemical company. In its horizontal orientation it differs from most U.S. charts that place the chief executive officer at the top and arrange other positions in a hierarchy below. This horizontality suggests, perhaps, a more collaborative approach.
(Courtesy of Sumitomo Chemical America, Inc.)

left and moving right; people who read right to left depict flow in that direction too.

**Schematic Diagram.** A "schematic" demonstrates electrical or mechanical connections in a series, as in a circuit or computerized system. An overview schematic may be supplemented by more detailed diagrams for manufacturers or field crews. Figure 8.20 on page 185 reproduces the London Underground Diagram, which represents a subway system by adapting the conventions of a circuit diagram.

**Pictogram.** A *pictogram* enhances the pictorial possibilities in any diagram (or graph) to motivate or entertain a popular audience. The bars of a bar graph, for example, may suggest the topic under discussion: people, bags of money, hockey sticks, or computers. To persuade patients to complete the course of an antibiotic even when their symptoms disappeared, for example, a British physician drew a chart that showed spoons running to a finish line across hurdles representing each dose. The spoon that crossed the finish was the winner. Several patients, however, objected to seeing their health practices as competitive. Such misreadings are a potential problem with pictograms. So is bias or regionalism in the symbols, for example, in the shape of electric plugs, which differ globally.

## Drawing and Photograph

Two related forms, a drawing and a photograph, provide a fifth and sixth way to show some piece of reality. Each form has advantages (and disadvantages) in depicting such items as a site, a mechanism, an organism, or an event.

**Drawing.** A drawing provides flexibility. For example, Figure 8.17 on page 180 illustrates the structure of a coffee maker. It shows both the surface and what's beneath. A photograph would not let you see inside in the same way. Readers of drawings need to understand the conventions that underlie them, however. One convention is representing the whole by a part. When you show only the head to represent a whole animal, for example, be aware that some readers may not understand that convention. Drawings can look realistic, although some people have trouble finding an image in a series of discrete lines.

**Photograph.** A photograph shows how something actually appears, in its context (Figure 8.18 on page 181). Many readers find a photograph easier to understand than a drawing. Through a photograph, for example, a researcher can examine a laboratory setup that she may want to duplicate. Photographs of authors often accompany articles in journals and magazines. They help readers visualize who is speaking to them in the article and remind them of colleagues whose faces they may recognize but whose names may be forgotten. A computer-generated image can also show things that are too big, too small, or too remote to be seen by the unaided observer. With such a photograph you can help your reader see into cosmic space, into the brain, or into the microstructure of an atom.

(a)

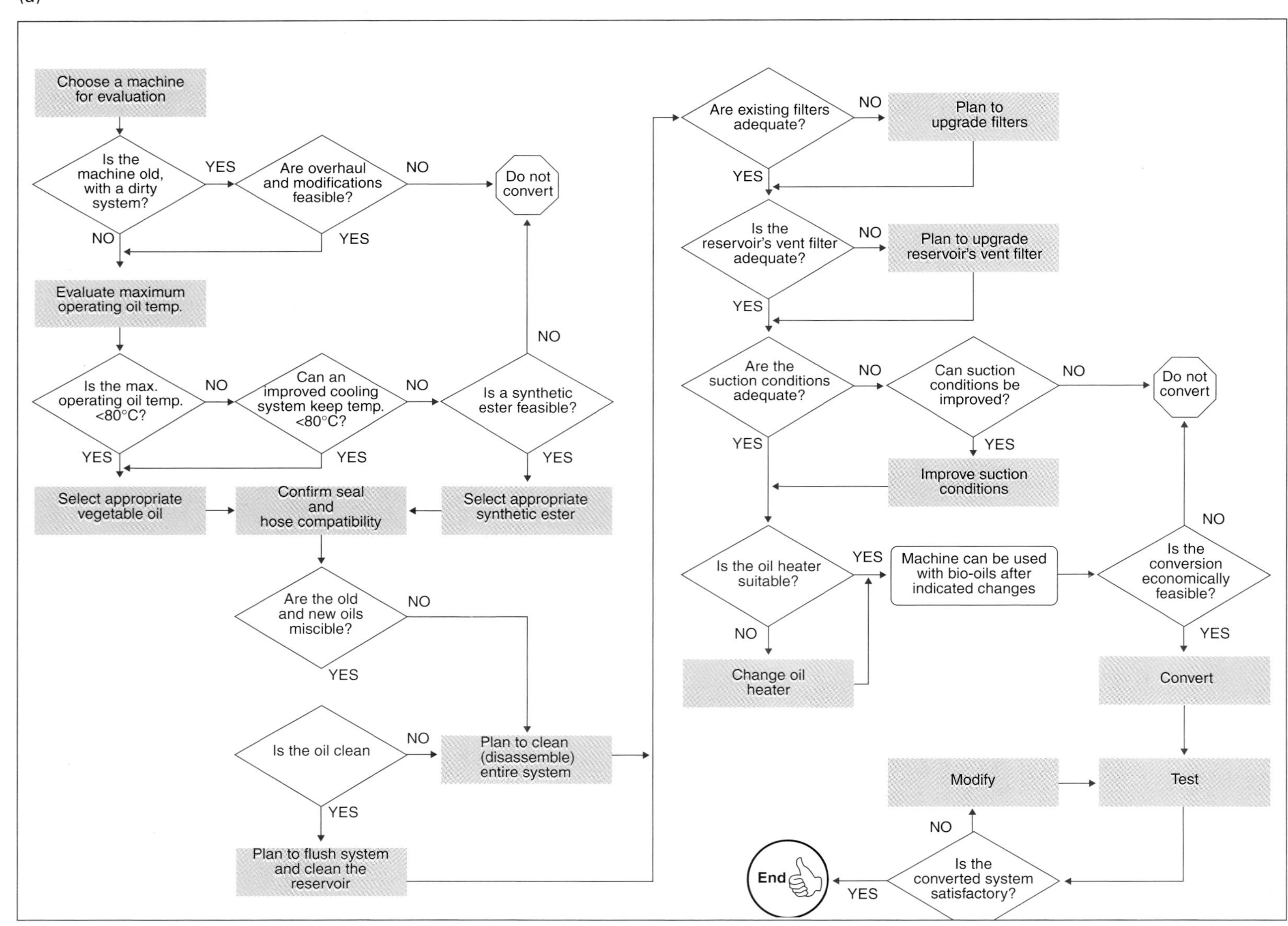
Choose a machine for evaluation
Is the machine old, with a dirty system?
YES
NO
Are overhaul and modifications feasible?
NO
YES
Do not convert
Evaluate maximum operating oil temp.
Is the max. operating oil temp. <80°C?
NO
YES
Can an improved cooling system keep temp. <80°C?
NO
YES
Is a synthetic ester feasible?
NO
YES
Select appropriate vegetable oil
Confirm seal and hose compatibility
Select appropriate synthetic ester
Are the old and new oils miscible?
NO
YES
Is the oil clean
NO
YES
Plan to clean (disassemble) entire system
Plan to flush system and clean the reservoir
Are existing filters adequate?
NO
YES
Plan to upgrade filters
Is the reservoir's vent filter adequate?
NO
YES
Plan to upgrade reservoir's vent filter
Are the suction conditions adequate?
NO
YES
Can suction conditions be improved?
NO
YES
Do not convert
Improve suction conditions
Is the oil heater suitable?
YES
NO
Change oil heater
Machine can be used with bio-oils after indicated changes
Is the conversion economically feasible?
NO
YES
Convert
Test
Modify
Is the converted system satisfactory?
NO
YES
End

(b)

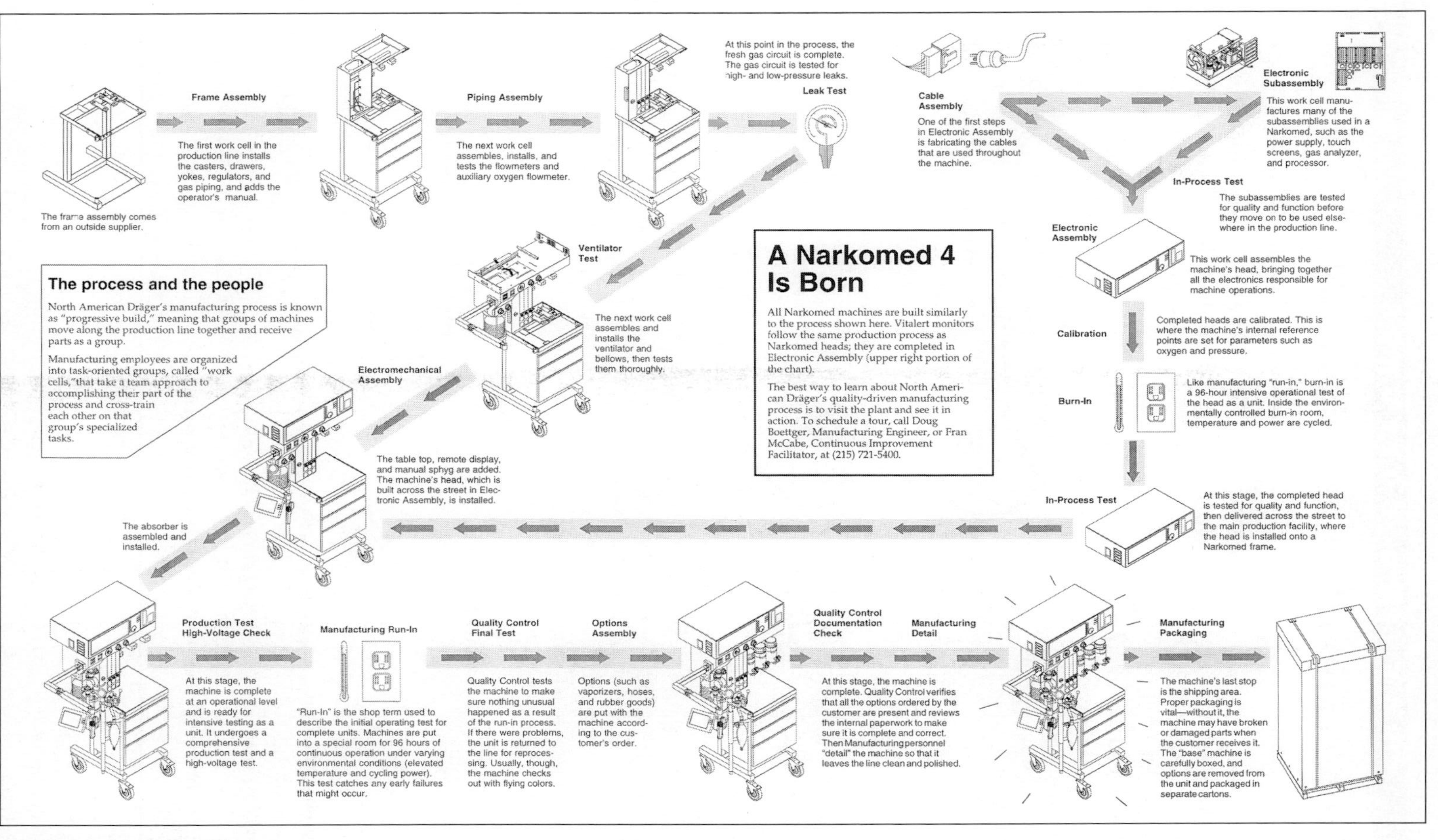


**FIGURE 8.16 Flow Diagrams**

(Example a) One form of a flow diagram is a *decision tree,* as in this chart that provides an overview of the process for deciding if a forestry machine is a suitable candidate for conversion to environmentally compatible hydraulic oils. (Source: Ismo Makkonen, "Converting a Forestry Machine to Use Environmentally Compatible Hydraulic Oils," *Wood Harvesting,* Technical Note TN-234, November 1995, p. 12. Courtesy of FERIC.)

(Example b) Flow diagrams also show production processes, as in this overview of the manufacturing of a medical device. It is aimed at a broad audience—the salespersons and customers for the device—and thus includes drawings to enhance concreteness. The line with arrows is yellow in the original, in imitation of such lines that designate areas on the factory floor. Extended captions at each step provide further details to explain the process.

(© 1994 North American Dräger [Telford, Pennsylvania]. Designed and produced by Emily M. Skarzenski and Leo Lynott.)

CROSSING CULTURES

## International Signs and Symbols

Road signs and packaging symbols must cross barriers of language in an ever more international economy. In the early 1900s, European automobile clubs tried to establish a standard system. Here are two signs proposed by the Swiss (Helfman 143):

These didn't work, in part because people read the snail as representing slime more than slowness.

The pedestrian crossing signs at right (Helfman 147) show men wearing European clothes—not necessarily the standard in other parts of the world. The Saudi Arabian figure is headless because the Muslims don't permit human images (a headless one, however, is acceptable).

International

Great Britain (no hat!)

Southern Rhodesia

Australia

Japan

Saudi Arabia

Graphics are often useful packaging symbols; thermometers, for example, that show the range of proper storage temperatures and arrows that show "This side up." But the following sign caused problems (Helfman 165):

It was meant to indicate "fragile." But a dockworker in India thought it indicated the *content* of a box, not how to handle it. Figuring the box was full of broken glass, he tossed it on the back of a truck (Helfman 143–66).

Research is being conducted to correct these problems. A committee of the American Institute of Graphic Arts (AIGA), for example, tested the effectiveness of passenger and pedestrian symbol signs. It measured them on three dimensions:

***Semantic:* Relationship of the visual to a meaning**

- Is it easy to understand—or *mis*understand?
- Is it tied closely to one culture?
- Is it tied closely to one age group?
- Is it difficult to learn?
- Is it already widely accepted?
- Does it contain unrelated elements?

***Syntactic:* Relationship of one visual to another**

- How well do the symbol's parts relate to each other?
- How well does the symbol relate to other symbols?

- Are its most important parts recognized first?
- Does it contradict existing conventions?
- Is its construction consistent in the use of format, scale, color, typeface, and the like?

***Pragmatic:* Relationship of a visual to its user**

- Can it be easily seen?
- Is it affected by poor lighting, odd viewing angle or other "visual noise"?
- Is it vulnerable to vandalism?
- Can it be successfully enlarged and reduced? (adapted from AIGA 20)

The committee, for example, looked at the candidate visuals below to represent "Exit." The word is well recognized in the United States as meaning the "Way Out," a phrase used frequently in the United Kingdom. But "Exit" as a term is less used outside the United States (unlike, for example, "Hotel" or "Taxi"), so the term alone is not enough. The idea seems to be difficult to show visually because it is abstract. Groups 1 and 2 are ambiguous. To people familiar with diagrams, groups 3 and 4 seem to show "exit from a closed space," not always the meaning. The arrow also indicates exit to the side, although usually you need to go *ahead.* The group 5 symbol, used at London Heathrow Airport, is like the "no entry" symbol. But that symbol is red, indicating "stop," and with a horizontal bar indicating a barrier. This symbol, in green, indicates a raised barrier that says "go" and that also completely bisects the green disk, indicating a passageway. Because the symbol is abstract, it probably needs to be combined with the appropriate word (exit, way out, *salida, sortie*) until it is learned (AIGA 120).

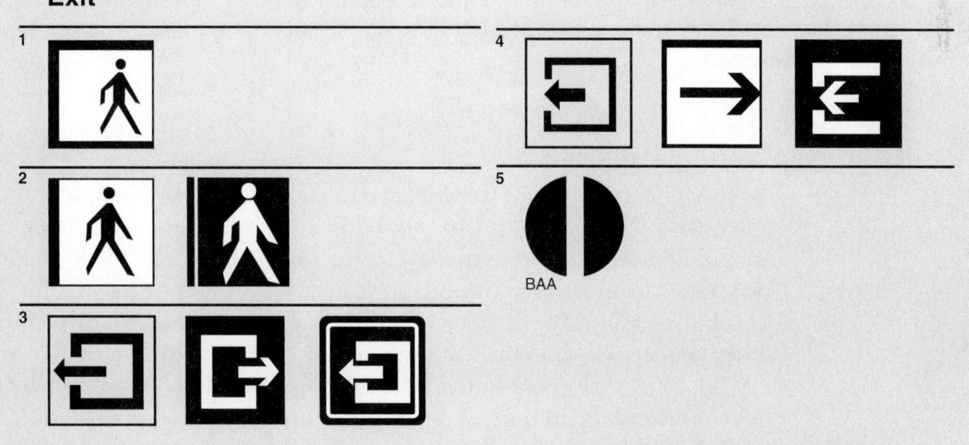

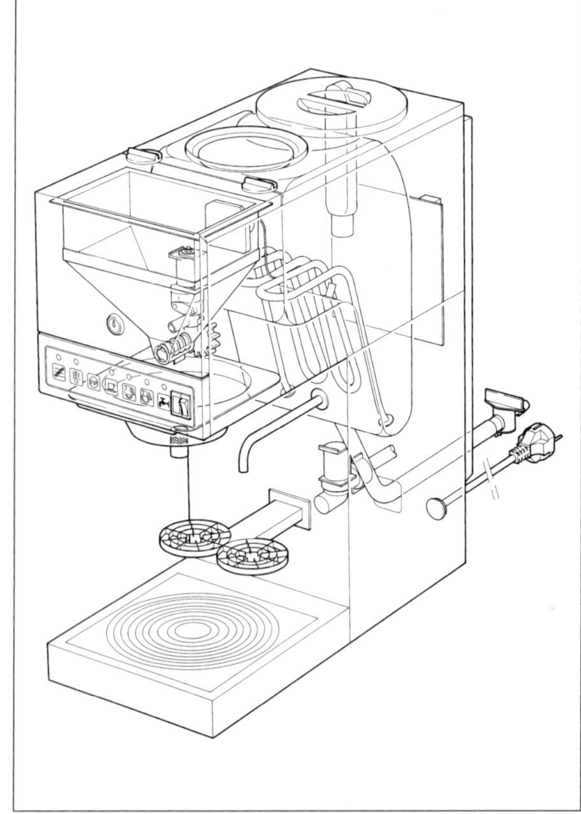

FIGURE 8.17 **Drawings**
These two drawings of a European coffee maker help explain its operation. The left drawing is a realistic depiction with both the coffee pot and a cup in place. The right drawing shows a *phantom* view that relates inside devices to the outside dimensions. Whereas a photograph might replace the left drawing, it could not achieve the inside view.
(Courtesy of Tedopres International bv.)

A photograph is often dramatic and draws attention, attractive characteristics when you want to persuade a reader. But a photograph has disadvantages. It may be expensive both to produce and to duplicate, and it may take a long time to load on a networked computer system. A photograph may also show too much detail; the main object dissolves in a field of other objects. Such details may also raise unanticipated issues in brochures or Web sites that cross cultural or national borders. A brochure for a German conference hotel, for example, contains a picture of its swimming pool. In the background are two nude women, a detail that might offend more conservative Americans. Computer-generated photographs can cross ethical borders when they are manipulated digitally to provide false evidence. Viewers tend to believe that photographs show something real, but the technology can create images that metamorphose ("morph") one person into another, eliminate unsightly features from a landscape, show a

**FIGURE 8.18 Photograph**
Photographers need to move in on their subjects, whether people or equipment. Equipment looks best at work, as in this photograph of a machine for cleaning out a forest.

(Source: Mark Ryans, "Evaluation of the Timberjack (FMG) 0450 Plantation-Cleaning Machine," *Silviculture,* Technical Note TN-224, November 1994. Courtesy of FERIC.)

building where none exists, and otherwise create a reality that may mislead a viewer.

## Map

Finally, in this brief survey, a *map* is an increasingly significant form for recording and displaying information about a landscape, either a real one or one that uses "landscape" as a figure of speech (Figures 8.7 and 8.19). In traditional usage, a map depicts landforms; a *chart* depicts water or sky. But that distinction is becoming blurred, except for mariners.

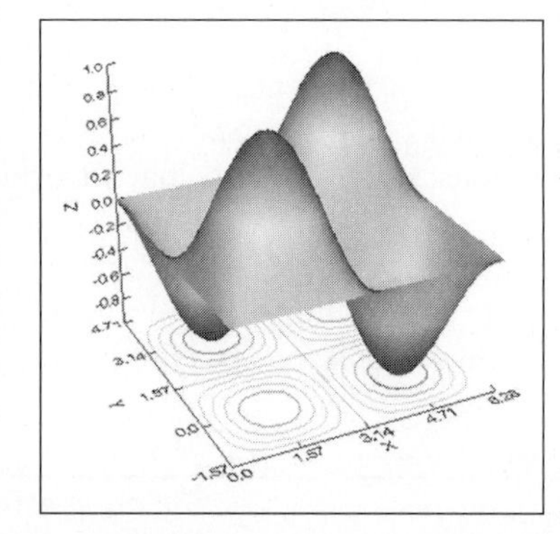

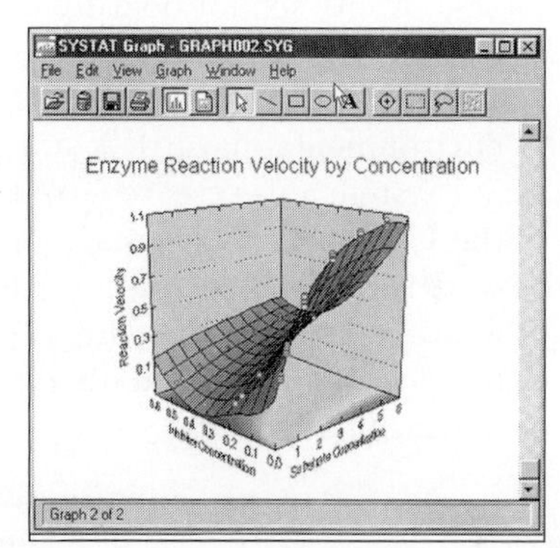


**FIGURE 8.19 Mapping Statistics**
Software developers are creating new ways to show statistical data derived from scientific and technical research. The programs place information in a computer-created landscape that adjusts dynamically to a user's request. As in these examples, the presentation offers a three-dimensional view that helps unravel complex results in graphics that both explain and persuade. Such programs give new meaning to the cliché about "speaking volumes."

(Courtesy of SPSS, Inc.)

**Attributes.** All maps display three attributes: scale, projection, and graphic symbols (Monmonier). You'll read more about graphic symbols in the next section.

- *Scale* shows how much smaller the map is than the reality it represents. Paper maps adhere to fixed scales indicated in words or numbers on the map itself. With computer-generated maps, users can vary the scale to zoom in on a point of interest in a dynamic flight through the landscape.

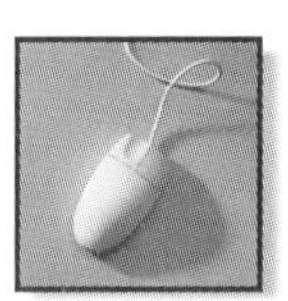

ELECTRONIC EDGE

## Geographic Information Systems

Computers are greatly enriching the process and product of mapmaking. One technology that is becoming increasingly popular—and gradually less expensive—is Geographic Information Systems (GIS). Now mainly restricted to large computers, it will soon be available on personal computers too. The market, according to one U.S. company, was worth about $75 million in 1995. GIS allows you to enter many forms of information that can then be layered as maps to highlight relationships and foster good decisions.

A city school board can use GIS to locate all schools within a mile of a potential environmental hazard. A similar system aids researchers at the University of Delaware in studying agricultural and weather trends and the effects of environmental disturbances on wetlands.

In the United Kingdom, GIS technology is also popular, especially because the necessary groundwork has been laid in the highly accurate digitized maps from the government's Ordnance Survey. North East Water, the largest water-only utility in Britain, uses GIS to monitor its 8,000 km piping network and let customers know about any emergencies, planned maintenance, or other special conditions. Marks and Spencer, a major retailer, uses GIS routinely in support of decisions about locating new stores, defining fuel-efficient delivery systems, and profiling customers and customer trends. Notes one Marks and Spencer researcher, "A map backdrop certainly helps us present complex scenarios in an easily digestible form" to aid decision making (Ireland 11). More important, it helps with detailed financial analysis through links with other software. A region in northern Spain is using GIS technology to deliver information to tourists and potential investors over the Web.

Another important—and complementary—technology in mapmaking is global positioning. With a handheld device that receives satellite signals, a field worker determines her exact location. To map a wildlife habitat, she walks along the boundary, logging points into the system, which stores the points in sequence. The information is then entered into a GIS that generates a map of the area, calculates its size and coordinates, and is ready for other manipulations, like a comparison with other maps showing housing patterns and earlier maps of the habitat to determine effects over time.

- *Projection* refers to the way a three-dimensional territory is represented on a flat plane. Projection techniques always distort, as in the Mercator projection world map that creates an overly large Greenland. But that map system worked well for Europe and North America in Mercator's time.
- *Graphic symbols* represent features on the map.

**Reader-Oriented Mapping.** As with all visuals (and words, too), when you select information for a map, consider the reader's needs. A subway map, for example, should emphasize the linkages among the lines and the sequence of stations more than the exact geography of the route. A passenger won't be aware of that geography (Figure 8.20).

Recognize, too, the limits of the information behind the map. Some of that information may be unreliable, as in aerial surveys that don't work well in mountainous areas. Moreover, maps are often used as political tools, sometimes with unethical consequences. Government leaders use maps as power plays to claim territory or to show the riches of the territory they already possess. Military agents may use maps to disinform their enemies.

## GRAPHIC SYMBOLS

Within these visual forms you may use *graphic symbols* to convey information. Such symbols also deliver messages in other contexts discussed in this section.

For more about visual ethics, see Chapter 3

**Map Graphics.** Graphic symbols represent features on a map. Three types are common: point, line, and area. Point symbols, for example, note the location of cities. The relative size of a dot or circle by the city's name can also indicate the relative size of the city being designated. Pictorial points can add even more information as in a stylized skier representing a ski area. Line symbols show, for example, roads and rivers (adapted from Helfman 125).

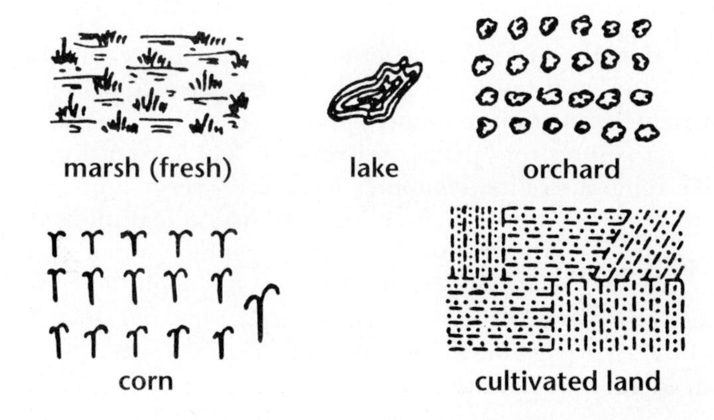

**FIGURE 8.20 (opposite page) The London Underground Diagram**
This diagram of the subway system, the "tube," in London, England, is an often-imitated model of excellence in the visual presentation of information. Earlier depictions of subway routes were more like *maps,* tracing the ins and outs of particular routes, but Henry C. Beck recognized in 1931 that there was a better way. An engineer, he adapted a circuit diagram to show a horizontal baseline with routes depicted as verticals and diagonals. The only topographic feature is the stylized River Thames that runs along the bottom edge and provides an excellent reference point. Riders needed to know about *stations* and *connections,* but not about the surface geography of their routes. The diagram helps to simplify the city itself as well as its transportation. London is not a grid city, like New York, or a radial city like Paris, and its somewhat chaotic aggregation of several smaller "cities" could hopelessly confuse a visitor. The map removes some of the confusion, something that most people find welcome. In doing so, however, it exaggerates the central area. That exaggeration also makes outer stations seem closer to the center than they are and thus may mislead some users. But it does serve well the information needs of millions of commuters who require more detail. The diagram is color coded, and those colors for each line dominate the graphic scheme at appropriate stations. The diagram is now in digital form and is easily adapted to reflect new lines and station improvements. It is reproduced in pocket brochures and on large posters as well as on T-shirts, mugs, underwear, and other consumer products as an icon of graphic design.

(Source: Text, Ken Garland, *Mr Beck's Underground Map* [Middlesex, U.K.: Capital Transport Publishing, 1994.]; diagram reprinted by permission of London Transport Museum. Reference Number 97/E/827.)

More abstractly, lines can represent contours. On topographic maps, the spacing of the lines indicates the relative steepness of slopes. Area symbols show boundaries.

**Symbol Signs.** Graphics also serve to describe processes like package handling, laundering, household maintenance, or toy assembly; to designate the locations of such facilities as hospitals, rest rooms, swimming pools, and eating places; and to provide road directions. But as *Crossing Culture: International Signs and Symbols* notes, creating graphics that serve different cultures is not easy.

**Icon.** Traditionally, an *icon* is a venerated religious image, but the term has taken on a different meaning in graphic design. On a computer screen, an icon represents a function to be performed: a file folder with an arrow for "open file"; a printer for "print"' scissors for "cut." According to one expert, "Like religious relics, computer icons are energy units, which focus the operative power of the machine into visible and manipulable symbols" (Bolter 52). The term icon also refers to a graphic symbol repeated in a document. The arrow at the bottom of Figure 8.8 might be considered an icon. The globe viewed from space is an icon repeated on part-opening pages and the first page of each chapter of this textbook. An icon also appears with each of the feature boxes.

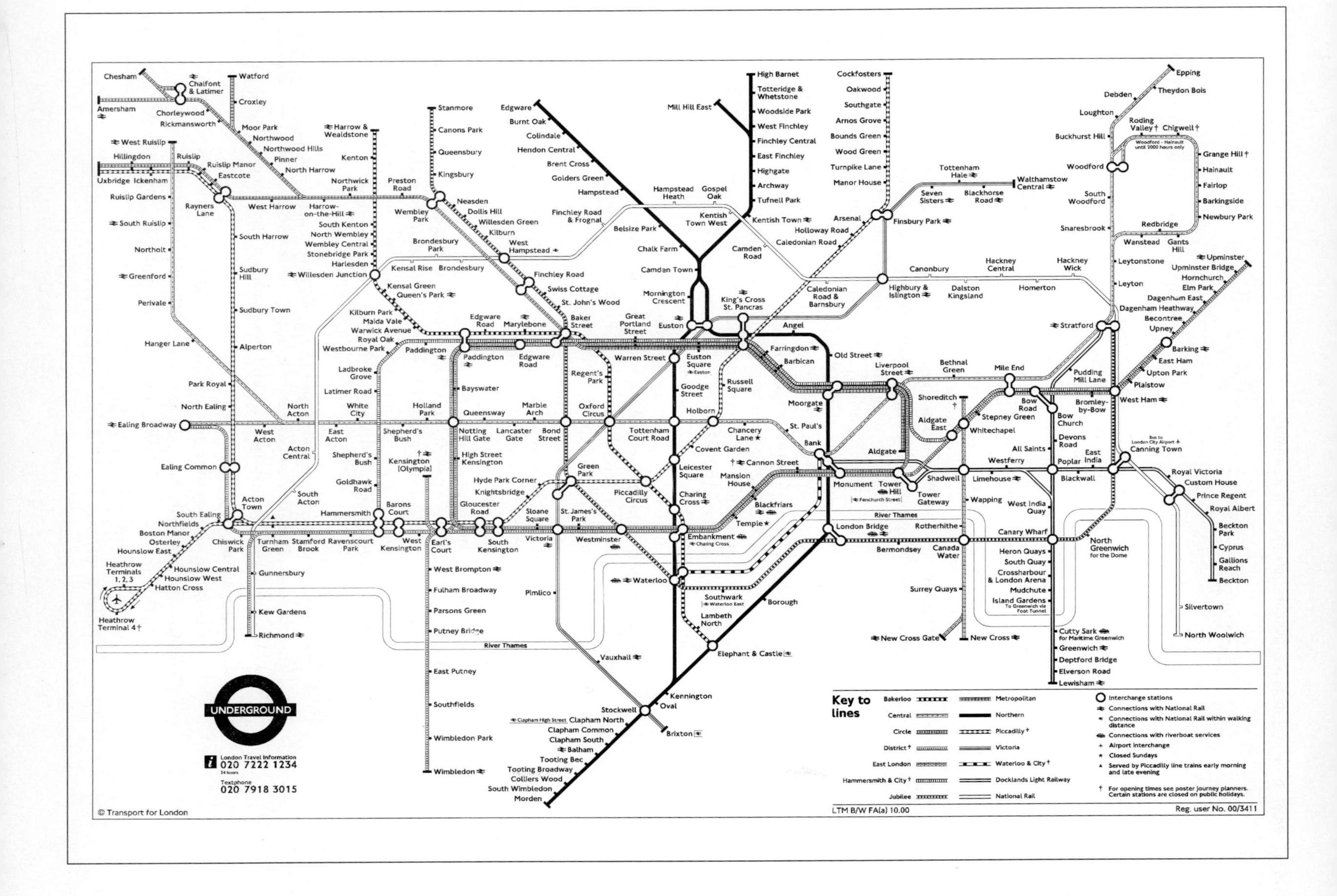

Key to lines
Bakerloo
Central
Circle
District †
East London
Hammersmith & City †
Jubilee
Metropolitan
Northern
Piccadilly †
Victoria
Waterloo & City †
Docklands Light Railway
National Rail
Interchange stations
Connections with National Rail
Connections with National Rail within walking distance
Connections with riverboat services
Airport interchange
Closed Sundays
Served by Piccadilly line trains early morning and late evening
† For opening times see poster journey planners. Certain stations are closed on public holidays.
River Thames
UNDERGROUND
London Travel Information 24 hours 020 7222 1234
Textphone 020 7918 3015
LTM B/W FA(a) 10.00
Reg. user No. 00/3411
© Transport for London

## COLOR

Like graphic symbols, *color* can become a dimension of each visual form. It plays an increasingly important role in recording and communicating technical information. Computer-generated maps of the seafloor, for example, use color to indicate depths and elevations, from deep blue for the greatest depth to red for the higher elevations. Shading adds an almost three-dimensional element. Scientists discussing the structure of cells often use color to code different filaments of interest. Color helps identify individual lines in a graph and key them to a legend that provides more explanation about the source and meaning of the line. It adds a dramatic element to tables and matrixes, which otherwise might seem dull.

Color can also be instructive. For example, electronics manufacturers often color-code the cables and sockets on their machinery. The colors ease the job of instructing customers in setting up the equipment. Rather than having to describe cable locations, the manual simply notes, "Plug the purple cable into the purple socket." Rustproofing paint applied to metals that must endure harsh weather, like deep-sea oil rigs, bears different pigments for different layers of application. Maintenance staff can discern easily how much wear the surface has endured by the color of the exposed paint layer. The buoys that identify ownership of lobster traps are color coded. The colors are registered on the lobstermen's licenses. In the water, lobstermen can easily identify their own and others' traps.

When you compose a visual in color, keep in mind, as with all other visual elements, how the reader will see your document or screen. The reader may well not see what you see. Find out if the medium appropriate to your communication can handle color. Build in devices in case your visual will be reproduced in black and white. The London Underground map, for example, uses color in the original to designate the various lines. But each is also identified by other graphic devices that allow proper interpretation without color.

Color is one of many visual strategies that help you record and structure your thinking. Such strategies are also powerful tools for communicating information to others. Avoid using them simply to dazzle or to make bad information look good, but do use them ethically and creatively to enhance your audience's ability to understand and to act.

### CHECKLIST: Composing Visuals

**1. Use visuals to record, inform, and persuade an audience**

**2. Compose in visuals rather than text to**

Compress a discussion that would be lengthy in text
Clarify otherwise complex information
Encourage speedy and effective decision making
Cross borders of language and culture
Present information that resists textual presentation

3. **Select the visual form appropriate to the context**
   Compare discrete units (numbers or words) in tables
   Show trends and relationships in graphs and diagrams
   Describe and instruct in drawings and photographs
   Use maps to show connections in a landscape or an argument
   Accommodate the reader's ability to read visuals
   Highlight significant information
   Adjust to the available technology for producing and displaying visuals
4. **Use graphic symbols as elements of a code that**
   Represents repeated information in other visual forms
   Describes processes, systems, and sites internationally
   Helps readers navigate documents and Web sites
   Represents functions of a computer program
5. **Use color to inform, enhance, and instruct**

## EXERCISES

1. Write a set of directions to accompany a letter to a speaker who will be addressing your organization. Tell the speaker how to get to the right location for the talk. Use both text and visuals.

2. At your instructor's request, write a memo that responds to one or more of the following tasks:

   **a.** Examine the visuals in the *Wall Street Journal, Scientific American,* and a technical journal in your field. How do they differ? How are they alike?

   **b.** *Architecture,* published by BPI Communications, Inc., uses both drawings and photographs highly effectively. Survey several issues of the journal to determine what material the editors express in photographs, what in drawings, and what in words. Note also the design of pages in the journal.

   **c.** Examine the visuals in five corporate annual reports. They may be five years of reports from the same company or the reports from five different companies. Do the visuals aim primarily to convey information or to serve as decoration? What do they say about the company's character, its culture, the reliability of its reports? How are they integrated with the text?

   **d.** Compare an aerial photograph, a topographic map, and a highway map of the same area. What differences occur in the three visuals? How might each contribute to a report?

   **e.** Find two articles on broadly the same subject, one in a technical journal and the other in a semitechnical or popular account (see Chapter 21). What visuals does each author use to appeal to the audience? How much background in mathematics or graphic interpretation does the audience need to understand the visual?

3. Explain the rules of some game, for example, soccer or baseball, to an audience unfamiliar with the game. Use only visuals. Look at an issue of a sports

magazine or watch such explanations on television to develop a strategy for your presentation.

4. Draw a flow chart of the tasks needed to prepare a formal report.

5. Arrange into a table the following information about a proposed course on the feeding habits of wildlife:

   The lecture schedule is based on the ten-week quarter. Lecture will be three hours a week. The material is divided according to species. The first two weeks will concentrate on deer. Weeks 3 and 4 will be devoted to rabbits. The next two weeks will cover the three squirrel species found in the northeastern United States. One week each will be assigned to the pheasant, quail, and ruffed grouse. The last week of lectures will cover the wild turkey. Diets of adults, immatures, and reproducing females will be studied for all species.

6. The following statistics represent the sales per hour of pizzas at three stores in a college town on a Friday night:

   8–9 P.M.: X Pizza, 36; Y Pizza, 47; Z Pizza, 12
   9–10 P.M.: X Pizza, 28; Y Pizza, 54; Z Pizza, 56
   10–11 P.M.: X Pizza, 10; Y Pizza, 62; Z Pizza, 61
   11–midnight: X Pizza, 10; Y Pizza, 58; Z Pizza, 121

   Illustrate these data in a line graph, a bar graph, and a table. Which version seems most effective to compare the effects of Z Pizza's late-night buy-one-get-one-free coupon? To compare sales at each shop? To show how sales change hourly?

7. From whatever information interests you, develop a visual. Reproduce it on an overhead that you show to the class. Discuss why you chose the visual form you did. You may develop your information from your observations of yourself or others, for example, the amount of time you sleep in a week; the amount of time each day you devote to such activities as studying, eating, and, of course, sleeping; the amount of time you think your professor assumes you spend in studying, chatting with friends, and the like. Or you may use information derived from your classes or from a research project.

## FOR COLLABORATION

1. On a team of three, develop a "visual essay" or a poster session (see Chapter 22) on the model of the *Historical Atlas of Canada* or the Wharton Life Interests Project. Select a topic that has high visual content or at least potential, and then explain its features mainly through visual devices. You can create a document or a hypertext or multimedia program. If your instructor directs, you may also use this exercise to create a set of instructions in visual form (see Chapter 17).
2. On the same team, or another, evaluate the following candidate symbol signs that represent the disposing of litter. The signs would be attached to

litter bins. Evaluate them on their semantic, syntactic, and pragmatic dimensions, as you read in *Crossing Cultures: International Signs and Symbols.* (Source: AIGA)

# COMPOSING TEXT

"THE IMPORTANCE OF CORRECT USE OF LANGUAGE LIES NOT ONLY IN BEING ABLE TO REPORT RESEARCH WELL; IT IS WITH LANGUAGE THAT WE DO MOST OF OUR THINKING"

W.I.B. BEVERIDGE

In addition to visuals, you'll use text to communicate with readers. This chapter discusses four units of text: paragraphs, sentences, lists, and headings. Sometimes, especially in a collaborative venture, you'll need to write only a few units. In an individual effort, you may need to write all the text in a document and thus are likely to use every kind of unit.

## PARAGRAPHS

The paragraph is the most basic and important unit of text (we don't *speak* in paragraphs). As you group ideas in paragraphs, you employ a major strategy of information design. To work, a paragraph has to do two apparently contradictory things at once. It must stand still and move. It stands still in that it is unified and adequately treats the essential aspects of one point; it moves in that it begins at one point and develops toward another. Within each paragraph

- Motivate the reader.
- Lead the reader from familiar to new information.
- Show hierarchy or parallelism.
- Use topic sentences and transitional devices.

CROSSING CULTURES

## Viewing Paragraphs

A good paragraph in English develops linearly; it takes a straight line from one thought through the examples or illustrations that support it. We talk about a "train of thought." That norm is hardly universal across cultures, as the figure shows. According to one researcher, Arabic paragraphs, for example, develop in a "complex series of parallel constructions, both positive and negative" (Kaplan 403). They display coordinate ideas unlike the subordination considered beneficial in English. Chinese and Korean writing uses an "approach by indirection. . . . The development of the paragraph may be said to be 'turning and turning in a widening gyre.' The circles or gyres turn around the subject and show it from a variety of tangential views, but the subject is never looked at directly. Things are developed in terms of what they are not rather than in terms of what they are" (Kaplan 406).

A communications consultant in Japan found that the concept of a paragraph itself was a new idea to the technical professionals he was working with, along with issues like varying levels of generality, unity of subject matter, and coherence (Stevenson 328). Instead of paragraphing, the Japanese use a technique called *kishoten-ketsu,* a pattern that differs from those common in both Arabic and English. It emphasizes beauty, surprise, indirection, emotion, and impressions. One linguistics professor remarks, "First, you have the subject, *ki,* then you raise it, *sho,* next roll it, *ten,* and then . . . you end it beautifully, *ketsu*" (Dennett 116). According to a Japanese physicist, writing a technical text in Japanese "takes three or four times the time and energy to convey the same idea (as English)" (Dennett 116). He also notes, "You cannot say your conclusion in the first state. If you translate from English to Japanese, the translated material must be in a sense vague . . . so you get good Japanese" (Dennett 116).

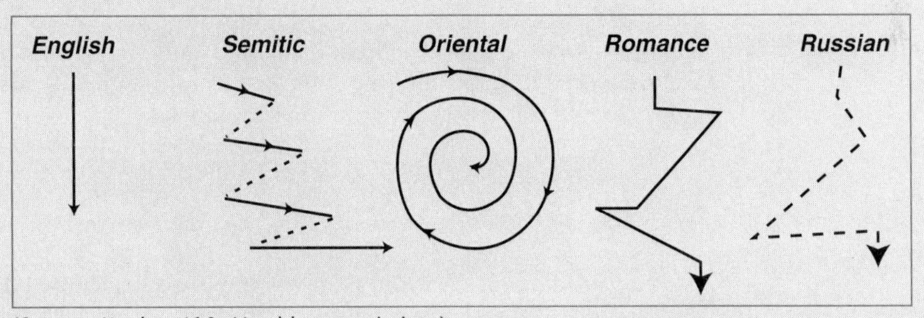


(Source: Kaplan 410. Used by permission.)

### Motivating the Reader

Composing your text in units of paragraphs motivates the reader to move through the discussion. Remember to *create* paragraphs, a guideline often forgotten by writers of e-mail messages. Indenting the first word of a paragraph or double-spacing between single-spaced blocks makes an attractive page or screen. In addition, adjust the length of paragraphs to reflect stages in the development of your discussion—easy material in longer paragraphs, more difficult in shorter ones, for example. Adjust the length, too, to allocate your reader's attention. Fend off monotony by avoiding a succession of short, choppy paragraphs or a dismal collection of long ones. Avoid especially using long paragraphs at the beginning and end of a discussion, segments that require special strategies to draw the reader's attention.

In measuring the length of your paragraphs, calculate size in final form, not in draft. Paragraphs that look like a good length on your screen may be too long when they appear on a Web page or are set in narrow double columns of type (like those used in a newsletter or technical journal). They may be too short on a printed page of single-column type. Visualize what the paragraph will look like to the reader.

## Leading the Reader from Familiar to New Information

Start a paragraph with material that is familiar to the reader and move to the new. Note that approach in the following two paragraphs, an original and a revised version of the opening of a report on methods for immobilizing cadmium in soil. The information in each is the same, but they differ in plan. The original immerses the reader immediately in new information—cadmium contamination—rather than beginning with the familiar and concrete—the sources of such contamination. The revision begins with the familiar.

> ***Original.*** Soils represent major sinks for metals like cadmium that are released into the environment. Soil does not have an infinite capacity to absorb metal contaminants and when this capacity is exhausted, environmental consequences are incurred. Contamination of soils by cadmium and other heavy metals has become a global concern in recent years because of the increasing demands of society for food production, waste disposal, and a healthier environment. The main causes of cadmium contamination in soils are amendment materials (e.g., municipal waste sludge) and fallout from nonferrous metal production and power plants.

> ***Revised.*** Such sources as mines, smelters, power plants, and municipal waste
> *new ————————————→ familiar*
> treatment facilities release metals into the environment. These heavy metals,
> *new → familiar*
> especially cadmium, then find their way into the soil. The soil does not have an
> *new ————————→ familiar*
> infinite capacity to absorb these metals. Instead, unabsorbed metals move
> *new*
> through the soil into the groundwater or are extracted by crops that take the contamination into the food chain.

Individual sentences in the paragraph you've just read are also connected through a familiar-to-new structure; a subject familiar to the reader appears at the beginning, and the end tells the reader something new. This structure keeps readers from falling through the cracks of an argument and provides writers with an easy opening to a next sentence—just rephrase the end of the one before, as in the following paragraphs (italics added).

> In high-wage countries like the United States, industries stay competitive in a global marketplace through *innovation. Innovation* can lead to better production processes and better performing products. (National Science Board, 168)

> Solar systems divide themselves into two categories: *active* and *passive*. An *active* solar system uses a mechanical pump or fan to transport heat from the collector to a *storage tank* for later use. While *such equipment* is more elaborate and costly than that for a passive system, it offers at least two distinct *advantages*. One *advantage* is its ability to *achieve hotter, more useful temperatures*. This *increased efficiency* leads to another advantage: the need for *less collector area*. The *collectors* fit more easily onto existing buildings than the bulkier passive equipment.

## Showing Hierarchy or Parallelism

To help readers understand and remember, show the hierarchy from main points to subordinate supporting points, and show the parallelism of equal items.

For more about hierarchy and parallelism, see Chapter 7

**Hierarchy.** The most-to-least important pattern in the following paragraph demonstrates the hierarchy in popular sources about science and technology:

> Television continues to be the most frequently used information source. Ninety-five percent of American respondents indicated that they watched at least an hour of television news almost every day. Nearly two-thirds reported listening to an hour or more of news on the radio almost every day. On the print side, 56 percent of adults reported that they read a newspaper almost every day, while 28 percent read a news magazine regularly. Conversely, only 9 percent of adults reported that they read a science magazine regularly. This array of results points to a high level of information consumption in both the broadcast and print media among American adults. (National Science Board 201)

This paragraph, from a segment of a report about science magazines, arranges statistical information derived in a survey. It moves from the most frequently used (and most familiar) source to the least common. The final sentence summarizes the paragraph ("This array of results") and points to the next paragraph, which discusses another source—public libraries. Repeated phrases ("almost every day") and words that signal a change of topic or emphasis ("On the print side," "Conversely") aid in connecting the sentences.

**Parallelism.** The following paragraph uses parallel structure to demonstrate the contrast between two approaches to writing computer code:

> The bricoleur resembles the painter who stands back between brushstrokes, looks at the canvas, and only after this contemplation, decides what to do

> next. For planners, mistakes are missteps; for bricoleurs they are the essence of a navigation by mid-course corrections. For planners, a program is an instrument for premeditated control; bricoleurs have goals, but set out to realize them in the spirit of a collaborative venture with the machine. For planners, getting a program to work is like "saying one's piece"; for bricoleurs it is more like a conversation than a monologue. In cooking, this would be the style of those who do not follow recipes and instead make a series of decisions according to taste. While hierarchy and abstraction are valued by the structured programmers' planner's aesthetic, bricoleur programmers prefer negotiation and rearrangement of their materials. (Turkle and Papert 136; from *Signs* 16:1 [1990] © 1990 by The University of Chicago)

The paragraph makes the approaches specific by identifying two individuals: the "bricoleur," or hobbyist, and the "planner." Each sentence adds a concrete description of behavior to complete the reader's picture. The sentences, too, are *parallel,* that is, expressed in similar form; parallelism of expression reinforces the parallelism in logic. Sentences 2, 3, and 4 all begin "For planners" and describe planners before bricoleurs. Parallel construction has the further advantage of presenting new information in a familiar form. In addition, analogies to processes the reader is expected to be familiar with (painting, conversation, and cooking) help the authors explain a new concept—programming style.

### Using Topic Sentences and Transitional Devices

You may preview in one paragraph the core idea of the next (as you've just seen). You may also keep that idea in mind without directly expressing it. But most of the time, you'll aid your reader by expressing the idea explicitly, up front, as the first sentence in a paragraph. Note this approach in the following paragraph, the second in a series (thus "another") on indicators concerning money spent on international research and development:

> International coauthorship of scientific articles represents another indication of enhanced collaboration. In 1991, 11 percent of the world's articles were internationally coauthored—this is twice the percentage from a decade earlier. This increase in international cooperation is evident in several fields, but especially in physics, mathematics, and earth and space sciences. Although U.S. researchers still collaborate most frequently with colleagues in the United Kingdom and Germany, there has been increased cooperation with France, Japan, and Italy. (NSF xviii)

Familiar-to-new order also connects the second and third sentences through the phrase "This increase."

Beginning every paragraph with a topic sentence may risk boring the reader. In most cases, take the risk: A strong topic sentence gives you a track to run on and your reader an aid to skimming and remembering the document. Use the sentence to connect paragraphs too: "Having considered the disadvantages of e-mail, we turn to its advantages"; "These statistics must be explained." In addition, insert directional signals as appropriate to show turns in the discussion: "on the contrary," "despite these statistics," "considering this fact." Devices such as these, if used carefully rather than automatically, guide the reader within a paragraph and from paragraph to paragraph. Repeating key terms may also help: "This circumstance," These devices," Such ideas," "But these conditions." Such phrases or summaries further reinforce the pattern of familiar to new. In a long discussion, you may devote a paragraph to reminding the reader of what has gone before and previewing what will come next. Such transitional paragraphs add no new information but mark the writer's kindness to readers:

> This review of our company's environmental performance last year leads us to consider the changes in our own policies and the regulatory environment that will affect our performance in the coming year.

## SENTENCES

A paragraph connects two or more sentences. Adjust sentence length to accommodate the reader, to segment information, and to show your personal taste. It's generally easier to go wrong in long sentences than short ones, so save long sentences for occasions when you know your reader is skillful in reading, and you feel in control of the writing. Don't fear short sentences. The period is one of the most underused—and desirable—marks of punctuation in scientific reports and articles. In general, avoid a succession of either long or short sentences; when you've written three long sentences, make the next one short. For more sources of advice about writing sentences, see *Electronic Edge: e-Grammar.*

Let's look briefly at how you can design information in a sentence to show hierarchy or parallelism. Before doing so, however, we should talk about the use of active and passive voice in technical texts, an often-debated issue.

### Active and Passive Voice

In English, verbs clarify what the subject of the sentence did or does. What happened? What's happening? This element is usually the sentence's main news. Verbs have many attributes, including *voice:* active and passive. "The report explains" is an example of active voice; "The explanation is included in the report" is passive. When you use active verbs, you show the subject doing something. A passive verb shows the subject being acted upon and often needs a phrase to show who did the action, as in the following:

> ***Active:*** The customer bought ten tons of steel.
>
> ***Passive:*** Ten tons of steel were bought by the customer.

Many passive sentences convert easily to active ones:

> ***Passive:*** The distribution along the x–y direction is markedly affected by temperature.
>
> ***Active:*** Temperature markedly affects the distribution along the x–y direction.

A CLOSER LOOK

## An International Style

When you write for an international audience that may read English as a second (or third or fourth) language, choose words with particular care.

**1.** *Control connotations.* When the sentence

"The spirit is willing but the flesh is weak."

was translated into Japanese, the result (as translated back to English) became

"The alcohol is strong but the meat is rotten."

The connotations of the first sentence were misread as representing something literally.

**2.** *Use words consistently.* The word *while,* for example, can have several meanings (Kirkman):

*Time:* He works while he eats lunch.

*Opposition:* Host-caused bone losses are not easily preventable while environmentally caused ones are.

*Concession:* While invasive surgery is appropriate in some situations, most patients would prefer a noninvasive approach.

Avoid words like *since* (meaning both cause and time) and *take* that can be ambiguous. At the least, use the terms consistently to evoke only one meaning in any one document.

**3.** *Avoid culture-bound phrases.* These phrases gain their meaning from a national or organizational context and are thus meaningless to those outside that context:

- Do a u-turn in research direction
- To blue-sky a solution to a problem
- Go the whole nine yards
- Turn from poacher to gamekeeper
- Ballpark figure
- The point man of our lab team

**4.** *Avoid colloquialisms.* Colloquialisms are a level of language sometimes called "street talk" or "slang." Such language is the opposite of formal expression. Be formal when you address international readers, at least until you know that they understand, and perhaps prefer, a more conversational style. American English in particular builds phrases with prepositions that can radically alter the meaning of a term, as in the following several uses of *hang:*

- To get the hang of something
- To hang out
- To hang in there
- To have a hang-up about
- Hang up the phone

Here are some other colloquial phrases that confuse the uninitiated:

- Give 110 percent
- Chill out
- We're making it up on the fly
- I've had it
- He swears by this technique

**5.** *Avoid telegraphic style.* Use articles (*a, an, the*) and full spellings. Instead of saying

"Meet, Brixton lab, 3:30, brg questionnaire & survey results for joint tabulation,"

use complete expression:

"Come to a meeting at the Brixton lab at 3:30 P.M. Bring the questionnaire and your results from the survey. We will tabulate them as a group during the meeting."

Although the active is often preferable, the passive may be necessary and effective in technical documents where there is no actor or agent (as in an automated process) or when the actor is unimportant:

> The sludge containing the calcium sulfate and calcium sulfite is pumped from the settling tanks to small ponds, where it is stabilized by the addition of fly ash and lime and then partially dried.
>
> The cancers are divided into three broad groups.

The passive also helps when you want to avoid accusing the person who is really at fault. A complaint manager might say, "The mower has not been lubricated according to warranty instructions," when someone complains about defects. It would be less tactful (if more concrete) to say, "You did not oil the machine when you should have." Moreover, the passive keeps prose impersonal. Scientists and engineers use it to deemphasize the actor and place the emphasis on the action. They also use it to avoid including the pronouns "I" or "we" in their writing because some journals do not allow such usage.

But the passive has downsides:

- Increases the potential for grammatical errors with modifiers drifting around an unexpressed subject.
- Starts trains of prepositional phrases.
- Fosters roundabout and vague expression.
- In long stretches, drains the vigor from the text.

These downsides have led some publications to welcome the use of "I" when

- Emphasis does indeed belong on the agent.
- Responsibility for an action needs to be defined.
- The author wishes to indicate results or conclusions without supporting data.

Active voice is more economical and direct. So don't just slide into the passive; use it only when appropriate.

## Hierarchy

Design your sentence to emphasize the hierarchy of information when one idea in it is more important than another.

> ***Flat:*** The condensed water then runs down the cover to a collection device, and the concentrated brines are disposed of.
>
> ***Hierarchical:*** The condensed water then runs down the cover to a collection device that stores the concentrated brines for later disposal.

Much technical and scientific information reflects a hierarchy of cause and effect:

> Because soils and rainfall vary, the soil moisture in individual fields must be monitored.

You may need to look beneath a string of modifiers to find the right connections or causal links:

*Unclear Connections:* In Japan, rice farmers use irrigation water contaminated with cadmium from mining operations, and the farmers show severe symptoms of cadmium poisoning.

*Causal Links:* In Japan, rice farmers who use irrigation water contaminated with cadmium from mining operations show severe symptoms of cadmium poisoning.

A modifying clause at the beginning of a sentence produces anticipation and prepares the reader for something significant:

If we could once isolate the cause of the disease, then we could end the epidemic.

**Dangling Modifiers.** When you design information in a hierarchy, however, make sure you connect the modifying element clearly to the word it modifies. If not, you have created a dangling modifier, a phrase or clause that floats in a sentence. It either modifies nothing or attaches itself, wrongly, to the nearest available noun. When you want to be clear, attach the modifier. The introductory phrase in the following sentence dangles:

*Dangling Modifier:* After reaching northern Alaska or the Arctic islands, breeding occurs in the lowlands.

To correct, supply the proper subject:

*Corrected:* After reaching northern Alaska or the Arctic islands, the whistling swans breed in the lowlands.

Or expand the participial phrase to a clause:

*Corrected:* After the whistling swans reach northern Alaska or the Arctic islands, they breed in the lowlands.

Or rephrase the participle as a noun in a prepositional phrase:

*Corrected:* On their arrival in northern Alaska or the Arctic islands, the whistling swans breed in the lowlands.

Avoid modifiers that look in two directions and thus cause ambiguity:

*Ambiguous:* I've been trying to place him under contract here for three years.

Does that mean you have been trying for three years or his contract runs for three years? To indicate a three-year contract:

*Clear:* I've been trying to place him under a three-year contract.

Is the warning in the following sentence for car drivers about the trains—or for the drivers of the trains about cars?

*Ambiguous:* The crossing signals do not provide a warning to drivers of approaching trains.

CLEAR: The crossing signals do not warn drivers that a train is approaching.

**Inverted Subordination.** Make sure the final structure of the sentence matches your intention. Problems occur when the subordination is inverted, that is, when main elements are expressed as modifiers:

*Inverted Subordination:* Damage of surfaces by fretting is initiated by mechanical wear produced by the vibrating of the one surface on the other.

It's hard to know what's doing what in the sentence. The sequence of cause and effect becomes clearer when participial phrases are sorted into clauses with strong verbs:

*Proper Subordination:* When one surface vibrates on another, this mechanical wear causes fretting, which damages the surfaces.

The following sentences also show how proper subordination helps clarify meaning:

*Inverted Subordination:* The cutthroat trout is a spring spawner over gravel pits in riffles of streams producing 3,000 to 6,000 eggs according to weight.

*Proper Subordination:* The cutthroat trout, which spawns in spring over gravel pits in riffles of streams, produces 3,000 to 6,000 eggs according to the trout's weight.

*Inverted Subordination:* Cracking is a serious problem causing reduction of strength.

*Proper Subordination:* A serious problem, cracking reduces a material's strength.

**Positive Phrasing.** To avoid a buildup of modifiers, use positive phrasing. This advice, of course, applies to the grammar of your sentence, not its content. Readers can process positive statements more easily than negative ones.

*Unclear Negative:* If the cooling system does not have sufficient coolant it will not be able to keep the engine running at a reasonable temperature.

*Clearer Positive:* A proper amount of coolant will help keep the engine running at a reasonable temperature.

*Unclear Negative:* When openness does not exist in an organization, many people are not kept informed of potential problems.

*Clearer Positive:* Openness in an organization helps ensure that people are kept informed of potential problems.

*Unclear Negative:* Do not send messages that cannot be made public.

*Clearer Positive:* Send only messages that can be public.

*Unclear Negative:* He did not neglect international issues.

*Clearer Positive:* He accounts for international issues.

## Parallelism

As in page and screen design, visuals, and paragraphs, parallelism in a sentence or sentences reinforces the logic of a series and builds rhythmic expression. An *enumerator* term sets up the logic; the grammatical form of the first item governs the form of all the others:

In the nineteenth century, proponents of domestic science encouraged proper training and education for its practitioners, codified standards of

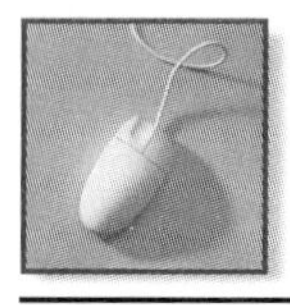

ELECTRONIC EDGE

## e-Grammar

You're writing at midnight. You know you have problems with sentence fragments, but you can't remember what a fragment is. Or a dangling modifier, another of those errors your instructor keeps finding in your text and noting in your margins. What to do?

Review your textbook. Or, for more details, visit the Web. Many excellent grammar advisers await you there. To find them,

- Enter a keyword (like "sentence fragment" or "grammar") in a search engine.
- Visit almost any college or university site; these usually have links to writing advice.
- Visit the Web site for your text: www.prenhall.com/andrews.

The grammar advisers you'll find in your search take several shapes. A major one is an online writing laboratory, or OWL. Purdue university has one of the best (and oldest) OWLs: http://owl.english.purdue.edu. Writing centers at other colleges and universities also provide guidelines for good writing that you can access 24 hours a day.

Second, many instructors of writing courses offer Web pages with advice for writing in their classes, advice that often transfers to other classes. These sites are usually linked to university home pages. The English department is a good place to start your search. Third, grammar hobbyists, like the fans of rock groups or movie stars, create Web sites to demonstrate their loyalty to the subject. These sites reflect the personal quirks and interests of their authors. They may be harder to access directly and can be less authoritative than university sites. They come and go as their authors' interests vary. But they may also be more fun.

Fourth, book, journal, and newsletter publishers offer advice on writing. These commercial sites include advertisements, which may be distracting, and they may tease you with only enough information to entice you to purchase their products. Ones devoted to the resume and cover letter may encourage you to buy a resume kit from them. But they can be useful.

Writing sites vary in their content and ease of use. Some focus on the nuts and bolts: frequently misspelled and misused words; verbs (tense, active and passive voice, agreement with their subjects); punctuation and capitalization; parallel structure; documenting sources. Others expand beyond these topics to models of successful essays and reports, tips on the writing process as well as overcoming writer's block, and discussions of voice and tone. Self-quizzes and other activities invite you to practice what they preach. Sophisticated sites turn these into interactive exercises. The best sites ease your search with tables of contents, alphabetical listings linked to pages that discuss that topic, and a search function.

Not all sites are authoritative, so you may want to check one against the other. Check, too, the *origin* of the site. On the Web, it's easy to enter a site devoted to British English, or the English of the Indian subcontinent, or other non-American sites. Such browsing enriches your understanding of the many Englishes, but be careful to follow the guidelines for the English your audience expects—which may indeed be British.

> good practice, and sought status for professional housewives equal to that of other professionals, mostly men, in law, medicine, education, and engineering.

The next example lists roles (the enumerator term, not expressed) that the subject (he) performed, each expressed as a noun:

> It was not as a surgeon, as the maker of his great museum, or even as a discoverer in science that he revealed his greatness.

Look for ways to straighten the parallelism of a rambling series:

> ***Unparallel:*** This overgrazing results in damage to the range, lower-quality livestock, and alters the numbers and distributions of other organisms, including small mammals.
>
> ***Parallel:*** This overgrazing damages the range, lowers the quality of the livestock, and alters the numbers and distributions of other organisms, including small mammals.

In the next pair of sentences, you see how parallel construction hits a logical snag if the enumerator term doesn't work:

> ***Unparallel:*** The valving improvements we seek will increase reliability, accessibility, and maintenance and allow application to all sizes of valves.
>
> ***Parallel:*** The valving improvements we seek will increase reliability and accessibility, decrease maintenance, and allow application to all sizes of valves.

The term *increase* is set up to control each of the following items. But it doesn't work for *maintenance*—presumably, the improvements should decrease maintenance. The second sentence corrects that problem in logic and brings the last item "allow" more directly into the series. When you compare two items, tell the reader both halves.

> ***Incomplete:*** The rotation span of an unmanaged forest is much longer than a managed one.
>
> ***Complete:*** An unmanaged forest has a much longer rotation span than does a managed one.
>
> ***Incomplete:*** Often a farmer is able to plant his crops earlier in the spring because the field is dry.
>
> ***Complete:*** If he adopts this tillage system rather than traditional techniques, a farmer is often able to plant crops earlier in the spring because the field is dry.

## LISTS

Within a sentence or paragraph, or as a unit in itself, you can design appropriate information in enumerated *list* form. The chief element in the design, of course, is parallelism. Lists

- Segment information into small, easily digested elements.
- Sort and rank details better than running text.

- Increase the white space on a page.
- Vary the look of units on the page, especially in text that is otherwise dense.
- Reinforce the parallelism of a series or sequence.

Your thoughts may naturally come to you as elements of a list, for example, steps in a process, items in an inventory, lines in a budget, criteria for a system, advantages or disadvantages of a proposed activity. Or in revising a draft, you may discover that running text in a paragraph contains a series of parallel elements that would become more accessible to the reader in list form.

Keep each element short, preferably one line. Order a sequence logically: by time, by priority, alphabetically by first letter, and the like. Introduce the element with a word (first, second) or number (1., 2.) that indicates ranking. Use an icon or bullet (•) for an unranked list. Omit punctuation at the end of each element unless it is a complete sentence.

Sometimes short paragraphs are used as list elements, especially in the conclusion or recommendation section of a report. If you use this structure, however, make sure a visual cue or highlighting (for example, boldface on a key term that starts each paragraph) allows the reader to scan the page to see the paragraphs as items in a list.

## HEADINGS

In technical documents, another unit of text—a heading—serves a critical role in showing the reader in advance the organization of the discussion. The keywords or phrases displayed as headings

For more about headings, see Chapter 7

- Ease skimming and foster remembering.
- Add an attractive design feature.
- Provide directional signals that help readers follow the turns of your presentation.

The best headings occupy no more than one line; usually a few words suffice. It's easy to write descriptive headings ("Introduction," "Discussion," "Conclusion"). These terms may move directly from the specifications for your document to the document itself. Genre conventions also often dictate appropriate terms for headings. In an IMRAD report, of course, you'll use the headings "Introduction," "Method" (sometimes "Materials" too), "Results," and "Discussion." A proposal will most likely have a section titled, "Statement of Work."

But within these general sections, or when your Web site or document is not constrained by set headings, choose *informative* headings, which indicate directly the content of the section. Note that all the headings of this chapter take an informative approach. Like many journal articles and technical reports, this chapter (and others in the book) avoids the heading "Introduction," assuming that the first unit of text is, of course, an introductory one.

To create headings, pull out the key terms from the unit of discussion the heading labels. Your outline is a good source, although you may think of richer terms as you write. Keep headings parallel in logic and expression. Group similar material under headings of similar rank. Avoid single subheadings (when you divide something, you come up with at least two). Adopt the style for headings specified by the person or organization you are addressing. Whatever the form, be consistent.

Another form of heading is the subject line for a memo or e-mail message. The line serves an important role in drawing attention and predicting an appropriate response, so it deserves careful composition. Don't simply leave in the default, often "Re: Your mail," message in an e-mail system.

For more about memos and e-mail, see Chapter 20

Headings provide important signals to readers about how to read and what to remember, so you'll want to practice writing headings well. That skill will carry over into the effective development of the other textual units you've read about in this chapter: lists, sentences, and whole paragraphs. You'll combine these with visuals to implement the design of an information product that will work for the reader—and for you.

## CHECKLIST: Composing Text

1. **Adjust text units to the context**
   The reader's capacity for paying attention
   Logical divisions of the information
   Expected structure of the information product
   Best size or type for the medium
2. **In paragraphs and sentences**
   Lead the reader from familiar to new information
   Show hierarchy or parallelism explicitly
   Choose the active over the passive voice of verbs if possible
3. **Use enumerated lists to**
   Reinforce the parallelism in a series
   Add white space on the page
   Break up an overly text-based page or screen
4. **Use headings to**
   Help the reader skim a print document to find items of interest
   Provide active links in a Web site
   Alert readers to the topic of a memo

## EXERCISES

1. Identify the main point in the following paragraph and reorganize the sentences around it to achieve a coherent unit. Hint: underline "Hauser Lake Dam" each time it appears and line up information around it.

   (1) Around 1900 an attempt was made to use steel as a major construction element in three large dams. (2) However, after the failure of the Hauser Lake Dam, the idea was generally abandoned. (3) Today steel dams are used only as temporary cofferdams needed for construction of permanent dams. (4) Basically a framework covered with riveted steel plate is used for the steel dam. (5) A masonry abutment is used to anchor the steel work into the

reservoir. (6) A typical design is shown in Figure 19. (7) The Hauser Lake Dam in Montana failed on April 14, 1908. (8) Steel dams decreased in popularity after this. (9) There were only three major steel dams made. (10) These were the aforementioned Hauser Lake Dam, the Red Ridge Dam in Michigan and the Ash Fork Dam in Arizona. (11) The basic design of these dams can be seen in Figure 20. (12) The few number of these dams that were built made stability design a matter of calculation rather than experience. (13) The Hauser Lake Dam failed due to overturning.

2. The following paragraph provides some confusing signals to readers about what they are expected to know and what material is expected to be new to them. It also announces topics it doesn't really treat in the way announced. Revise the paragraph to straighten its approach to the reader.

   *Foam applications* is the phrase used to describe the process which applies additives to a moving paper web as foam. This process has many advantages over conventional liquid applications. Two problems have hindered the progress of this new technology. The process must be established in the paper industry. The speed limits encountered in the current application technology must be increased in order to expand the number of paper machines which can potentially use foam. The development work run at the Placerville plant is attacking both of these problem areas and this report summarizes the achievements which have been accomplished there during the year.

3. Here are some redundant expressions common in technical prose. Discuss *why* the phrases are redundant, and add any similar phrases you find in your reading. Keep this list handy as you write so you avoid such expressions.

   - completely eliminate
   - collaborate together
   - few in number
   - estimated at about
   - component part
   - background experience
   - past history
   - oval in shape
   - hot in temperature
   - red in color
   - they are both alike
   - adequate enough
   - a funding level of $100,000 in magnitude
   - final end
   - cheaper in cost
   - true facts
   - close proximity

4. Technical writers often smother their verbs by transforming action into a noun or by using unnecessarily long verbal phrases. Such usage diminishes

accuracy and bores readers. The following pairs show how you can clarify verbs.

| Smothered | Clear |
|---|---|
| make contact with | call, write, visit |
| make a purchase, give approval to | buy, purchase, approve |
| have a deleterious effect on | harm |
| have a tendency to | tend to |
| have an influencing effect on | influence, affect |

Good verbs give energy to sentences. Note how transforming a noun back to a verb energizes the corrected sentence in each of these pairs:

***Transformed Verb:*** Explanation of the variables is included in the report.
***Strong Verb:*** The report explains the variables.

***Transformed Verb:*** Camouflaging the vehicle is the next step you should perform.
***Strong Verb:*** Next, camouflage the vehicle.

Unsmother the verbs in the following sentences:

**a.** Mating of grizzlies takes place in the month of June.

**b.** There was a variation in weight.

**c.** A review of our customers' needs for such compounds was undertaken.

**5.** Convert the following passive sentences into the active voice.

**a.** Control of distortion can be accomplished with the following machines.

**b.** The solution of this problem has been obtained by adding sufficient deoxidant.

**c.** As soon as the explorers reached the ice, camp was set up.

**d.** An improvement in quality has been made.

**e.** The values are found to be in agreement.

**6.** Correct errors in parallelism and incomplete comparisons in the following sentences.

**a.** Seawater may consist of whatever type of water the ship is operating on—fresh, brackish, or seawater.

**b.** Factors contributing to increased pollution included population increase, industrial expansion and their waste, pesticides poisoning wildlife, and people who used their trucks and automobiles more.

**c.** High nutrient levels in our waterways result from deforestation, mining, farming, and from the manufacturing segment of our economy.

**7.** Straighten the hierarchy in the following sentences so that main ideas stand out from their supporting evidence.

**a.** It is important to mention, when talking about losing weight, that changing a diet should not be the only way of losing weight, changing a diet with an appropriate daily exercise regimen.

**b.** There are currently two bills being considered by Congress that intend to enhance the vaccination program for infants in rural areas and inner cities.

8. Make a list of the errors cited by your instructor in each assignment you submit. Review the list with each assignment. Try to make sure that no item appears more than once—and that no additions are necessary by the end of the term.

9. Visit Web sites devoted to helping you write (see *Electronic Edge: e-Grammar*). The writing center at your school or university may be a place to start, then follow its links. Also try http://webster.commnet.edu/writing/writing.htm.

## FOR COLLABORATION

1. Assume your team of two or three has come up with the following items of information after a brainstorming session on the Delaware Estuary. You want to include these in an introductory section to your report on a proposed regional program that will institutionalize wise management and conservation of the estuary. Develop a central point for your description and a plan for ordering the information. At your instructor's request, write the introductory paragraph or concluding paragraph—or entire description.

   - over 90 percent of the area meets the "swimmable and fishable" criteria of the Clean Water Act
   - definition of an estuary (transitional zone where salt water from sea mixes with fresh from the land)
   - much cleaner today than ever in this century
   - major problem: the reach that flows between Philadelphia and Camden has lots of bacteria and a higher than acceptable level of dissolved oxygen
   - productive habitat
   - breeding, spawning, and feeding grounds for fish
   - migratory waterfowl, reptiles, mammals live there
   - fishery is down since 1900
   - buffer upland areas from flooding
   - overfishing has reduced levels of estuarine-dependent species
   - greenway trails being developed to connect historic sites and recreational facilities
   - habitat destruction and lowered water quality led to decline in fishery, so shad, Atlantic sturgeon, and striped bass are below historic levels, although some improvement
   - estuary naturally protects drinking water—filters pollutants and sediment
   - heavy use of water for industry and homes threatens long-term supply
   - some fish die when they get caught in power plant water intakes
   - show all the states included in the estuary program area (map?)
   - toxic substances are at elevated levels in sediment and in the organisms that live in the estuary so there have been fish consumption advisories
   - new public parks in the area
   - show the three zones (upper, transition, lower)
   - the fragmentation of the habitat, especially loss of freshwater wetlands, means that organisms which need specific types of habitats aren't thriving

- new forms of shipping have led to new commercial uses and reinvigorated urban waterfronts
- deterioration of the estuary's ability to buffer the impact of pollutants
- poor land management means that lots of agricultural land and natural habitat is eaten up with houses

2. Take a page from an assignment you wrote for this class or another class. Don't write your name on the page, but include a code that indicates authorship to the instructor. Hand in the original and three copies to the instructor. The instructor will divide the class into teams of three people each and distribute three sets of copies to each team. Using the guidelines for group review in Chapter 2 and those on style in this chapter, each team member will review all three assignments distributed to his or her team. As a group, the team will discuss the good points and weaknesses of each assignment and then rewrite as necessary to create three papers that all team members agree meet standards of good prose. These rewrites will then be handed in to the instructor who will return them to the authors.

# EXPLAINING

"MUCH OF WHAT YOU SEE DEPENDS ON HOW YOU LOOK."

***ALICE'S ADVENTURES IN WONDERLAND***

This chapter provides guidelines for designing visuals and text to fulfill one of the most important roles of technical communicators: explaining. In an explanation, you help readers understand some concept, object, action, place, or system. Your explanation may occupy an entire document, or it may be a segment of a memo, letter, report, proposal, or set of instructions. To compose an explanation, call on one or more of three strategies, each discussed in this chapter: defining, classifying, and describing.

## DEFINING

You can explain by defining the terms you use to represent your subject. What is *sustainable agriculture?* Define each term to help the reader understand the concept. If your reader is familiar with the idea of agriculture in general, but not with the idea of *sustainability,* focus on that new practice within the context of what the reader will find familiar. The extent of your definition and its location in the text depend on the reader's degree of familiarity with the concept or item and the purpose that you and the reader share.

### Length and Location

You may devote an entire report to defining a significant or controversial concept, for example, "electronic superhighway," "wellness," "strategic research," or "sensitive species." Or in an article or report that has other purposes, you may include a definition in the introduction that sets up your approach. To establish a new definition of "forestry," an article about the dean of a forestry school begins by contrasting the traditional and the new image of a forester:

> Forestry is not what it used to be. In myth, the forester was a solitary, leathery-skinned logger, ax on shoulder, and able to scale the tallest tree or dance on a log running down a wildwater stream. Nowadays, that image has been replaced by one that is a bit more, well, corporate. The modern forester, armed with laptop computer and cellular telephone, probably considers himself—or herself—a forest manager, and like managers of every persuasion, the guardians of the timber must somehow balance the conflicting demands that come with the wooded territory. There's profit and loss, of course, as well as dealing with habitat preservation, recreation, endangered species, and the like. Managing the woods, not to mention the larger environment, now requires infinitely more than lacing on hobnailed boots and wielding a crosscut saw. (Fellman 48)

This definition creates a picture that motivates the reader to continue reading. Less extensive definitions pop up from time to time to explain a term that's not in the reader's vocabulary. For example, when the reader is at least somewhat familiar with the concept underlying a term and the context is already clear, you can often use a simple synonym or phrase to clarify the meaning. Such a definition works well for a general audience whose purpose in reading doesn't require an elaborate technical understanding. Saying, for example, that *compression* makes things shorter and *tension* makes things longer may define those terms adequately for a popular audience, if not for engineers. Such informal definitions allow you to avoid interrupting a long explanation. Synonyms often appear after an expression like "that is" or in parentheses:

*tomalley* (liver) in lobsters
*meteor* (shooting star)
*boring hole* (deep, narrow penetration)
*kryopreservation,* that is, freezing embryos for future use

You can also place definitions in a sidebar, a text unit included in a margin, or box a definition within a column of type as an interest-getting device that interrupts the flow of text around it. A definition in an explanatory note at the bottom of a page keeps the text uncluttered by information only a few readers may need. To keep definitions even farther from the main text and to accommodate readers whose range of knowledge varies widely, use a glossary, a separate list of terms and their definitions, usually included in an appendix to the text. Such a glossary is an essential tool to establish fixed definitions for translation.

## Attitude

Depending on your reader and purpose, adjust the length and location of your definitions in the text. Also adjust how much of your attitude toward the concept you let show. Many technical definitions are matter-of-fact; others reflect your values and persuasive interests. The following two definitions of clearcutting, a forestry practice, show differences in attitude between paper companies and environmentalists opposed to the companies' practices:

**Paper Company**

True silvicultural clearcutting—the removal of an entire stand of trees in one cutting—is a useful forest management practice that can promote the growth of a high quality forest. Clearcutting has both short-term and long-term benefits as part of an overall forest management plan. (Scott 2)

**Opponent**

Clearcutting is: "a hemorrhaging of the forest," or "part of a mountaintop shaved down to the ground, with the nearby denuded mountainside almost certain to erode during the next rain." (Austin 3)

In defining a term, be aware of biases that may skew the meaning. In addition, watch for unintended side effects when you use metaphors (like "hemorrhaging"), that is, when you give something a name that belongs to something else. The term may embed a value system your readers would find incompatible if they really examined it. For example, the term *kingdom* in biology reflects the political system—kings—current when it was coined. *Food chain* implies a hierarchy leading up to humans (Coletta). Biologists might seek less politically charged metaphors if they were naming the classification system today. As with any writing, be aware of the connotations, the associated meaning of the terms you use, when you explain something in a definition. Be sure, too, you accommodate the reader by using terms that are more familiar and clear to the reader than the term being defined and by not repeating the term in the definition.

## Formal Definition

When a synonym or short phrase isn't enough to convey a term's meaning to the reader, you will need to compose the definition more formally. Name the item to be defined, fit it into a class of such items (its *genus*), and then note what distinguishes it from other items in the class (*differentia*). The principal verb of a formal definition is usually some form of the verb *to be.* Other verbs, however, are possible: *can be thought of as, stands for, represents, refers to,* and *occurs when.* Place the term in a category that is as narrow as possible, avoiding mere generalizations like *thing* or *factor.* Consider the following examples of formal definitions:

*Term* *Genus* *Differentia*
Flood stage is the height of a river above which damage starts, typically because the river overflows its banks.

*Term* *Genus* *Differentia*
Diet is the food and drink normally consumed by an individual or group.

*Term* *Genus* *Differentia*
Virtual reality is the computer representation in text and graphics of spaces and situations that engage participants as if they were real.

| *Term* | | *Genus* | *Differentia* |
|---|---|---|---|
| Cloning | is | a procedure | in which a cell from one embryo is used to grow another one that's genetically identical. |

| *Term* | | *Genus* |
|---|---|---|
| Human tissue | is defined as | groups of cells, organized and operating together, |

*Differentia*
that form identifiable structures of the body.

Even facetious definitions use this formal pattern, as illustrated by a sign over the entrance to a university faculty club:

> A committee is a group of the unwilling, appointed by the unfit to do the unnecessary.

Sometimes you might place the definition before the term to be defined so the reader has a familiar image with which to understand the new term:

> Any substance made of carbon and hydrogen, such as wood, coal, or petroleum, is called "organic."
>
> A reservoir of hot water mixed with methane gas and trapped underground at high pressure in sediments of impermeable shale is known as a "geopressurized system." (U.S. Committee for Energy Awareness 15)

## Stipulative Definition

In *Through the Looking Glass,* Humpty Dumpty tells Alice that *glory* is a "nice knockdown argument," explaining that the term means "just what I choose it to mean—neither more nor less." When you define terms, you may need to limit their meanings to serve a particular purpose in your document, although try to avoid being as arbitrary as Humpty Dumpty. Such limited or stipulative definitions often appear in an introduction so you and your reader will share the term's meaning in a special context:

> In this report, the term *competencies* refers to the knowledge or skills that an employee has or needs. We focus on three competencies: technical, business, and relational.

The following definition allows the author to limit the subject and make sure the audience doesn't think of water or soil pollution:

> The term *pollution* as used in this report refers to the presence of pollutants in the atmosphere brought about by the discharge of such noxious substances as lead, sulfur, and carbon compounds from gasoline-propelled vehicles.

## Expanded Definition

If the term you are defining forms the core of your discussion, you may expand your explanation to include the term's derivation, a comparison with a similar object or concept, examples, visuals, or a description of its operation.

**Show the Term's Derivation.** Explain the term to the reader by retracing its derivation or etymology. To find the origins of commonly used words, look in desk dictionaries. For more technical terms, check specialized dictionaries and thesauruses in the appropriate field. To define an acronym, start by clarifying what the letters stand for. A PERT chart, for example, derives from *P*rogram *E*valuation and *R*eview *T*echnique. A LAN is a *l*ocal *a*rea *n*etwork. Noting the history or attributing the source of a term also serves as a good method of definition. To define *hypertext,* for example, you might provide a narrative history of its development.

**Compare the Term to Something Else.** As with the definition of forestry you read earlier, which compares two images of foresters, define a concept or item by *comparison*. For example, you might explain the Internet (many people do) as a highway conducting messages around the globe. That explanation uses an analogy, a comparison that shows the similarities in two dissimilar things. Through *analogy,* you show how something new to the reader is like something the reader already knows. You might also use a *metaphor,* as long as you are careful about its connotations. Messages are *traffic* and an overload of messages creates a *traffic jam;* cost mechanisms are *toll plazas;* connections are *on ramps*. Those with a special interest in global usage sometimes refer to the network as the "*infobahn,*" after the autobahns of Germany. People adverse to computers look for the "information shunpike." The following definition of *hypertext* uses an analogy to print:

> *Hypertext* is a function of computers. But it is like footnotes in a print text that invite the reader to jump from the sentence she's reading and land on a note someplace else—at the bottom of the page or the end of the document.

A U.S. government document on reducing noise levels in a house begins its definition of "cross talk" with an analogy. Note that the definition precedes the term to be defined:

> Metal ducts act like speaking tubes on a ship. Noise from one room may be carried to another room through an unlined duct serving both rooms. For example, the problem of "Cross Talk" frequently occurs in a common return duct serving back-to-back bathroom installations. (U.S. EPA 61)

**Give an Example.** Help readers visualize the term in action by providing a "for instance" or "for example." A definition of *biomass* in a document on electricity generation begins with examples of what's included in the term:

> Biomass includes plants, manure, and ordinary garbage. Biomass feedstocks can be processed into liquid, gaseous, and solid fuels. One of the best known biomass fuels is ethanol, derived from corn and sugar crops. Ethanol can be mixed with gasoline to produce gasohol. (U.S. Committee for Energy Awareness 10–11)

The following definition of *bycatch* uses examples:

> *Bycatch* and *incidental catch* are terms that describe sea life fishermen throw back in the water. For example, fishermen throw back juveniles too small to

be kept legally and skate that has little market value. Often, such sea life is so badly injured that it can't survive. The terms also sometimes refer to fish netted unintentionally but kept because of their market value. For example, a cod fisherman in the Gulf of Maine might net some haddock that he'll then sell if it's of legal size.

The following definition of *gene* also conforms to the species-genus-differentia pattern, with examples in the differentia.

> A gene is a hereditary unit. The hereditary units—that is, the genes—determine whether a person will have blue or brown eyes, curly or straight hair, be tall or short, etc. (Novo Nordisk 7)

**Use a Picture.** To help readers see a concept, show them a picture. Figures 10.1 and 10.2 use this technique. Figure 10.2. also shows how an extended definition can be the linchpin of a technical or corporate program. It

For more about pictures, see Chapter 8

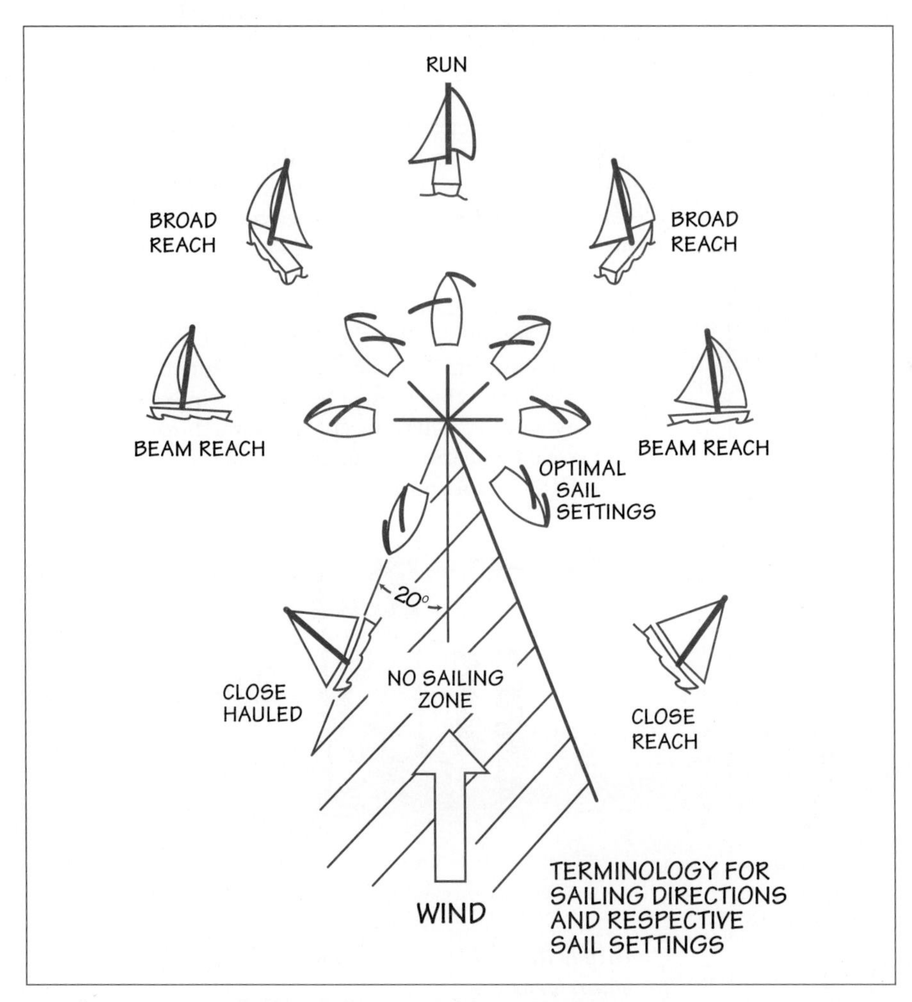


**FIGURE 10.1 A Visual Definition of Terms in Sailing**
(Courtesy of Joan M. Klair.)

# What Is Quality?

Sometimes it's easier to look at both sides of Quality to understand this concept. Most of us can relate stories about our experiences with poor quality in either the goods or services we have purchased. For instance, you may have purchased a pair of shoes only to have them fall apart. Or perhaps you went for a haircut and, after looking in the mirror, felt like you needed to go into hiding. Why did you consider these goods and services to be of poor quality? The answer is simple: You did not get what you wanted. Your requirements for shoes that would hold up under normal conditions were not met. You wanted only one inch to be cut off your hair, but the stylist took two inches.

In the latter case, while you think you received a poor quality haircut, someone else may think it's a good haircut. When we use subjective words, such as "goodness" and "excellence," to describe Quality, we miss the real reason why we are satisfied or dissatisfied with the product or service. **The key to defining Quality lies in fulfilling expectations or requirements.**

The first absolute of Quality Improvement tells us that we need to redefine Quality in a concrete way — as conforming to customer requirements and giving customers what they need, want, request and when they want it. This concept is summed up in a phrase you'll be hearing often: Do it right the first time! or DIRTFT for short.

*Quality is defined as conformance to requirements.*

FIGURE 10.2 **Defining Quality**
(Courtesy of the Sun Company.)

reproduces a page from a brochure that addresses a companywide readership at the Sun Company, which inaugurated an extensive program to instill a concept of "quality" in all its processes. "Quality," it notes, is "definable, measurable and manageable," and the brochure aims to prove this point.

**Describe Its Operation.** If the term pertains to a process or system, define it by describing that activity, as in Figure 10.6 on page 220, an expanded definition of the term *skeleton construction.*

## CLASSIFYING

You can also explain concepts and items by classifying them. You saw how classification helps you to group and thus define terms. More broadly, identifying common elements in natural and abstract phenomena helps you understand what things are and what makes things run. You put items into categories. For example, when a college biology department held a sale to get rid of equipment it no longer needed, it used classification to sort its discards onto different tables and describe the contents to potential buyers:

Table 1: glassware (beakers, petri dishes, pipettes, test tubes, volumetric flasks)

Table 2: machinery (stimulators, cameras, balances, shakers)

Table 3: cages (for rats and mice, some simple, some elaborate)

Table 4: books (textbooks, research documents, manuals)

Table 5: charts and models (including a mold of half a human head)

Classification helped the department determine what it had and helped the buyers find what they wanted. To classify items:

- Determine your purpose in sorting the items.
- Select a principle for placing items in a group (size, weight, advantages, materials of construction, use, and the like) that matches the purpose.
- Apply the principle to each item.
- Change the principle if any item doesn't fit, or discard the item.

Then describe the resulting groups in parallel terms, as in the department's use of nouns that named objects. Here are some other examples of classification:

**For more about parallelism, see Chapters 7 and 9**

- Anthropologists divide societies into two broad categories by their posture: those who sit and those who squat. Europeans and Americans, for example, tend to sit; many Asian and African societies find squatting more appropriate.
- The American Psychiatric Association's *Diagnostic and Statistical Manual of Mental Disorders* (DSM) describes every mental disorder by classification. The fourth edition, 886 pages, includes more than

300 disorders (up from 106 in the first edition) sorted into some 20 categories.

- An encyclopedia classifies items of general knowledge by alphabetizing keyterms; you know that information on *bridges* will come after information on *beer.* The Yellow Pages of a telephone book list individual services under general categories to ease accessibility. The categories are arranged alphabetically.

The most vivid form for showing classification is a table. Figure 10.3 shows a classification scheme that has become an icon, the Periodic Table of the Elements. It helps researchers understand and describe basic elements as well as search for new ones.

## DESCRIBING

You can *describe* as a way of explaining something to your audience. A good description establishes the essence of an item, system, concept, place, or process. This section provides guidelines for writing such descriptions. Because of the significance of a special form of description—giving instructions—an entire chapter (Chapter 17) covers those techniques.

Descriptions, like definitions, come in many sizes. You may devote an entire fact sheet, brochure, or report to description. Or your description may provide evidence for a more persuasive purpose, for example, recommending a new accounting procedure to your supervisor, selling a roofing

**Periodic Table of the Elements**

Metals
Nonmetals
Metalloids
Noble Gases

1 — Atomic Number
H — Symbol
1.00794 — Atomic Mass

| | | | | | | | | | | | | | | | | | |
|---|---|---|---|---|---|---|---|---|---|---|---|---|---|---|---|---|---|
| 1<br>**H**<br>1.00794 | | | | | | | | | | | | | | | | | 2<br>**He**<br>4.00260 |
| 3<br>**Li**<br>6.941 | 4<br>**Be**<br>9.01218 | | | | | | | | | | | 5<br>**B**<br>10.811 | 6<br>**C**<br>12.011 | 7<br>**N**<br>14.0067 | 8<br>**O**<br>15.9994 | 9<br>**F**<br>18.9984 | 10<br>**Ne**<br>20.1797 |
| 11<br>**Na**<br>22.9898 | 12<br>**Mg**<br>24.3050 | | | | | | | | | | | 13<br>**Al**<br>26.9815 | 14<br>**Si**<br>28.0855 | 15<br>**P**<br>30.9738 | 16<br>**S**<br>32.066 | 17<br>**Cl**<br>35.4527 | 18<br>**Ar**<br>39.948 |
| 19<br>**K**<br>39.0983 | 20<br>**Ca**<br>40.078 | 21<br>**Sc**<br>44.9559 | 22<br>**Ti**<br>47.88 | 23<br>**V**<br>50.9415 | 24<br>**Cr**<br>51.9961 | 25<br>**Mn**<br>54.9381 | 26<br>**Fe**<br>55.847 | 27<br>**Co**<br>58.9332 | 28<br>**Ni**<br>58.69 | 29<br>**Cu**<br>63.546 | 30<br>**Zn**<br>65.39 | 31<br>**Ga**<br>69.723 | 32<br>**Ge**<br>72.61 | 33<br>**As**<br>74.9216 | 34<br>**Se**<br>78.96 | 35<br>**Br**<br>79.904 | 36<br>**Kr**<br>83.80 |
| 37<br>**Rb**<br>85.4678 | 38<br>**Sr**<br>87.62 | 39<br>**Y**<br>88.9059 | 40<br>**Zr**<br>91.224 | 41<br>**Nb**<br>92.9064 | 42<br>**Mo**<br>95.94 | 43<br>**Tc**<br>(98) | 44<br>**Ru**<br>101.07 | 45<br>**Rh**<br>102.906 | 46<br>**Pd**<br>106.42 | 47<br>**Ag**<br>107.868 | 48<br>**Cd**<br>112.411 | 49<br>**In**<br>114.818 | 50<br>**Sn**<br>118.710 | 51<br>**Sb**<br>121.75 | 52<br>**Te**<br>127.60 | 53<br>**I**<br>126.904 | 54<br>**Xe**<br>131.29 |
| 55<br>**Cs**<br>132.905 | 56<br>**Ba**<br>137.327 | 57<br>***La**<br>138.906 | 72<br>**Hf**<br>178.49 | 73<br>**Ta**<br>180.948 | 74<br>**W**<br>183.85 | 75<br>**Re**<br>186.207 | 76<br>**Os**<br>190.23 | 77<br>**Ir**<br>192.22 | 78<br>**Pt**<br>195.08 | 79<br>**Au**<br>196.967 | 80<br>**Hg**<br>200.59 | 81<br>**Tl**<br>204.383 | 82<br>**Pb**<br>207.2 | 83<br>**Bi**<br>208.980 | 84<br>**Po**<br>(209) | 85<br>**At**<br>(210) | 86<br>**Rn**<br>(222) |
| 87<br>**Fr**<br>(223) | 88<br>**Ra**<br>226.025 | 89<br>**Ac**<br>227.028 | 104<br>**Rf**<br>(261) | 105<br>**Ha**<br>(262) | 106<br>**Sg**<br>(263) | 107<br>**Ns**<br>(262) | 108<br>**Hs**<br>(265) | 109<br>**Mt**<br>(266) | 110<br>(269) | 111<br>(272) | 112<br>(277) | | | | | | |

| | | | | | | | | | | | | | | |
|---|---|---|---|---|---|---|---|---|---|---|---|---|---|---|
| Lanthanide Series | 58<br>**Ce**<br>140.115 | 59<br>**Pr**<br>140.908 | 60<br>**Nd**<br>144.24 | 61<br>**Pm**<br>(145) | 62<br>**Sm**<br>150.36 | 63<br>**Eu**<br>151.965 | 64<br>**Gd**<br>157.25 | 65<br>**Tb**<br>158.925 | 66<br>**Dy**<br>162.50 | 67<br>**Ho**<br>164.930 | 68<br>**Er**<br>167.26 | 69<br>**Tm**<br>168.934 | 70<br>**Yb**<br>173.04 | 71<br>**Lu**<br>174.967 |
| Actinide Series | 90<br>**Th**<br>232.038 | 91<br>**Pa**<br>231.036 | 92<br>**U**<br>238.029 | 93<br>**Np**<br>237.048 | 94<br>**Pu**<br>(244) | 95<br>**Am**<br>(243) | 96<br>**Cm**<br>(247) | 97<br>**Bk**<br>(247) | 98<br>**Cf**<br>(251) | 99<br>**Es**<br>(252) | 100<br>**Fm**<br>(257) | 101<br>**Md**<br>(258) | 102<br>**No**<br>(259) | 103<br>**Lr**<br>(260) |

**FIGURE 10.3 The Periodic Table of the Elements**

system to a customer, or encouraging a developer to select the site you own for building a marina.

## Specifications

One genre devoted entirely to description is a *specification,* which details the configuration, components, and operation of a piece of equipment. It may also describe processes: standards for testing and measuring the quality of a product and methods for building, installing, or writing something. The American Society for Testing and Materials (ASTM), for example, establishes test standards that it publishes as specifications. Operations manuals establish the limits and best practices for using equipment. Purchasing agents develop specifications for items they want to buy and circulate these to vendors in a request for bid or request for proposal. Figure 10.4 shows a food label that specifies the content of the item. The label itself was also written in response to a specification: the 1994 Nutrition Labeling and Education Act that specifies standards for the content and format of the label, including type size and the use of boldface to highlight the most important

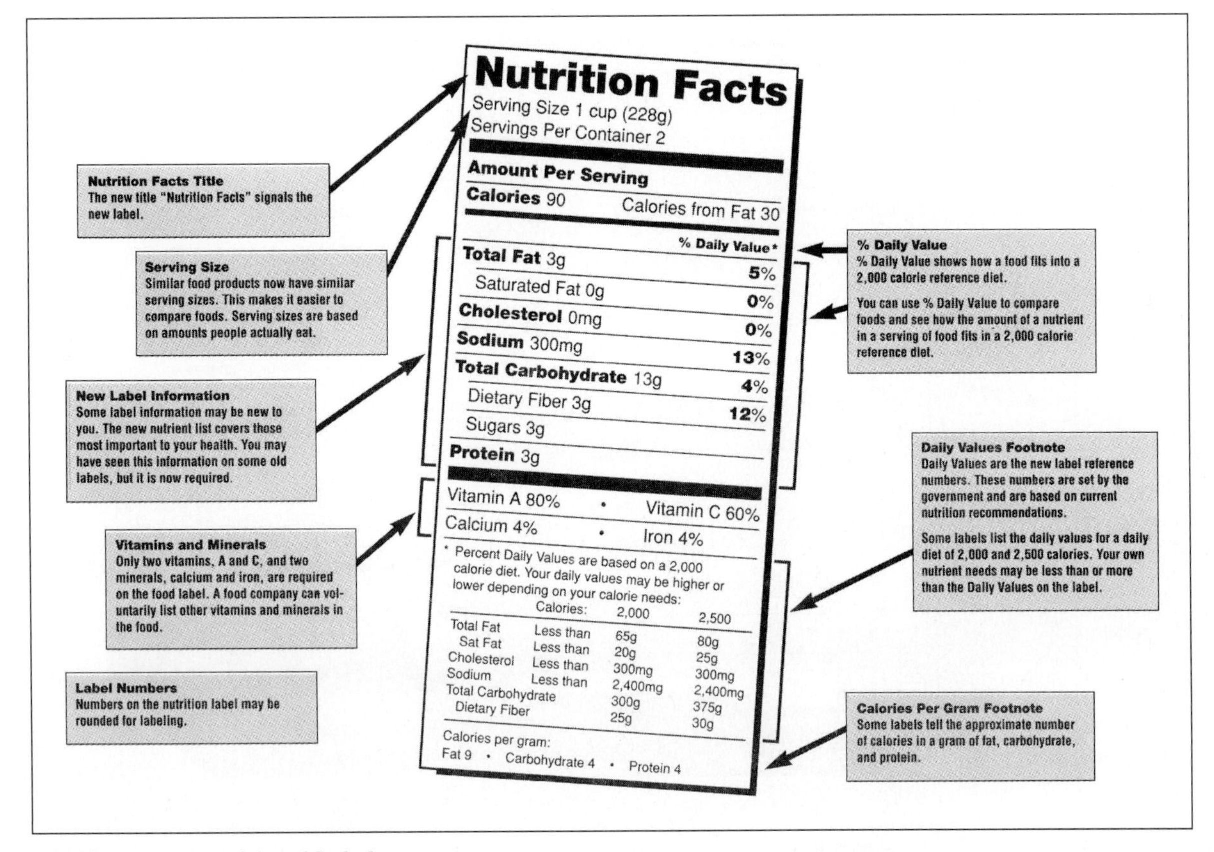


FIGURE 10.4 **Nutritional Label**
(Courtesy of the Mazer Corporation.)

information. Figure 10.5 shows how specifications (and a drawing) help describe a ship that responds to oil spills.

## Expanded Description: Mechanism, Organism, or System

You'll need to adjust the language and the amount of detail in your description to your audience. Define terms that may be unfamiliar to the reader and explain any facts the reader might not find self-evident. For example, the following paragraph expands the factual description of the oil spill response vessel to accommodate a general audience:

> Oil spill response equipment includes the stern-mounted Transrec 350, which is rated as capable of recovering up to 443,520 gallons of oil and oily water per day; assorted other skimmers; oil containment boom; and a 32' support boat to tow the boom in the water. In addition, the ship has a storage capacity of 168,000 gallons for recovered oil and oily water, and two 700 gpm oil/water separators to clean oily water.

If your purpose requires an extended discussion in, for example, a report or segment devoted to the item you are describing, organize your

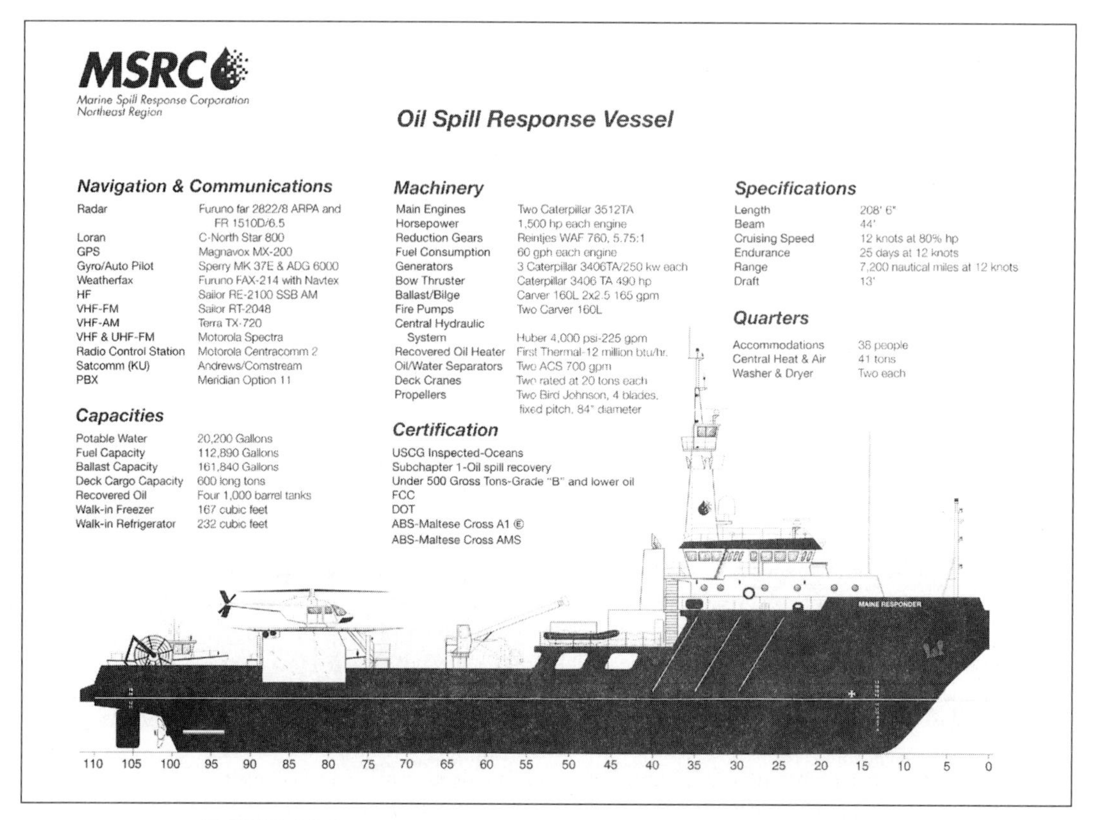


**FIGURE 10.5 Specifications for an Oil Spill Response Vessel**
(Reprinted with the permission of the Marine Spill Response Corporation.)

information to provide an overview first and then an analysis of significant features. Here is one effective pattern for describing a mechanism, organism, or system:

1. Introduction. An overview
   - Definition
   - Purpose/benefits
   - General principle of operation
   - General appearance (often a visual)
   - Division into components
2. Component-by-component analysis
   - Purpose of component
   - Appearance
   - Details: shape, size, relationship to other parts, material, color, and so on
3. Conclusion
   - Special points/advantages (or disadvantages)

Figure 10.6 provides an extended description in this pattern. Within the description, keep the subject consistently either singular or plural. If you begin by describing "the vessel," don't arbitrarily switch to "vessels" in the middle. Keep verb tense consistent; use the present tense for a general picture.

Use terms the reader understands, and define any new terms the reader needs to proceed with the description. Control the details to present just enough to the reader. Arrange those details after an overview, at least when you address U.S. audiences. You may find, however, that other readers prefer the details first. One expert notes that Japanese readers, for example, are frightened off by seeing the big picture right away. They prefer to be introduced to the parts one at a time before encountering the whole (Subbiah 16). As always, test your audience and purpose and deviate from the guidelines you've read here when you know your reader will welcome the deviation.

The following paragraph takes a different tack in describing the special features of an organism (a living thing): the daylily (*Liliaceae*), genus *Hemerocallis:*

> This is one of the great genera in all horticulture, for Daylilies are the plant that every gardener dreams of. They offer glorious colors in every shade except blue, a long season of bloom (though individual flowers last but a day), and a robust disposition that makes them easy to transplant, easy to maintain, quick to multiply and entirely free from pests. With a bit of care in selection, a garden of Daylilies can be built to bloom from June until September, and recent breeding has produced colors of truly astonishing beauty, creams and pinks and lavenders plus bicolors that bear no resemblance to the bright orange of the species. Planted together in large numbers, they quickly crowd out weeds and make a carefree and colorful ground cover that will be there when all of us are gone. Best of all, Daylilies are

**Definition—a dream plant**

**Special features and advantages**

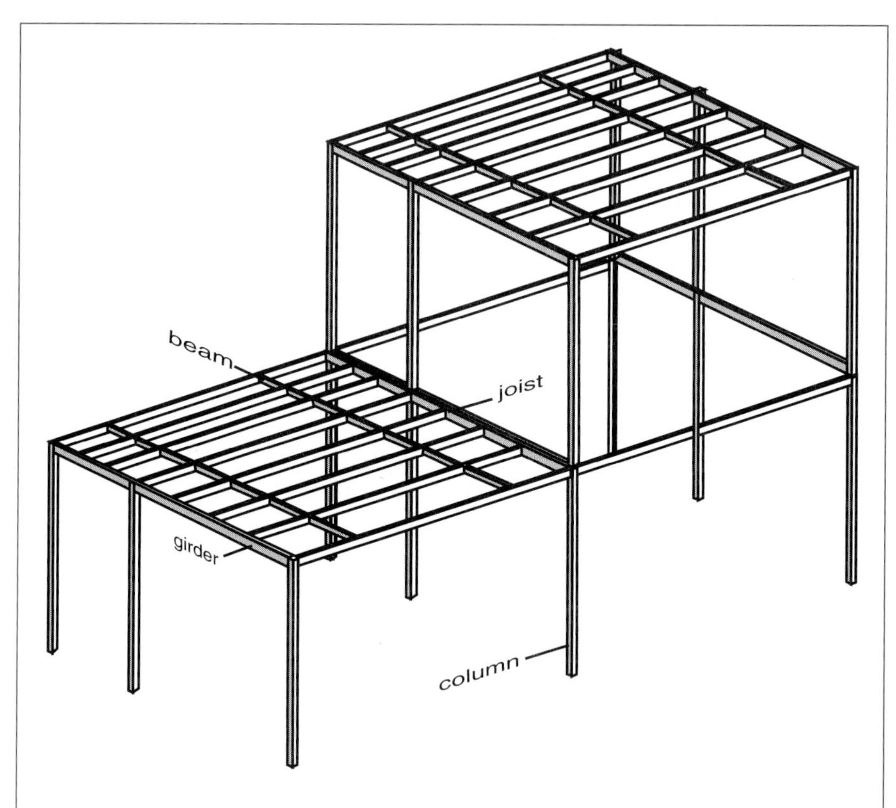


Formal definition
Purpose and components explained through an analogy
Principle of operation shown in a comparison
Component-by-component analysis
Definition

Skeleton construction is a system for framing a building. This system uses such structural elements as beams, columns, joists, and girders to support the loads of the building, just as bones are used to support the weight of the human body. Skeleton construction differs from wall-bearing construction, which uses walls instead of individual linear elements to support the loads.

The structural elements that create the skeleton of a building are referred to as members (see figure). The two basic types are beams and columns. A beam is a horizontal member that transfers loads laterally along its length to its support points. Beams are supported by columns, vertical members that transfer loads from the entire building to the foundation.

More detailed analysis of components and relationships

In addition to these basic elements, several specialized members are used in skeleton construction. Closely spaced beams called *joists* support the floor or the roof. Larger beams, called *girders,* support the joists. Vertical or diagonal bracing members connect beams and columns from one level of a building to beams and columns on a different level. These members reinforce the building against the horizontal forces of wind or earthquakes.

The final components of skeleton construction are the connections that transfer the load from one member to another. The connections are an important part of the frame. Several types are possible depending upon which members are being connected and whether the system is braced or unbraced. Bolts, rivets, or welding make the connections.

Special advantages and conclusion

Skeleton construction is a modular type of framing in which the structural elements form a 3-D grid. The inherent geometric order of this grid aids the designer in initiating and reinforcing a building's functional and spatial organization.

FIGURE 10.6 **Describing a System: Skeleton Construction**
(Source: Riggi et al. Graphic adapted by Erin Gantt-Harburcak.)

generally inexpensive (fancy new varieties are the exception), and some of the oldest varieties offer grace and beauty that is unexcelled. Among the miracles of modern Daylily breeding are forms called tetrapolids, which have double the normal chromosomes. The result is larger flowers with heavier texture and many more blooms per stem over a longer period. Many of these varieties are sensational performers that make a good thing even better, but some are, to our eye, clunky and awkward, and some big names have been excluded from our list for that reason. (White Flower Farm 71)

One special form of daylily

Accompanied by two color photographs that depict a group of daylilies and a tight shot of one lily (Figure 10.7 on page 221), this paragraph nicely describes the special features of the flower. As you read, however, you probably noticed that the voice in the daylily description is different from that in the passage on skeleton construction (Figure 10.6). The author is intrusive, the language more colorful: "every gardener dreams of," "glorious colors," "truly astonishing beauty," " be there when all of us are gone." You probably suspected, rightly, that it comes from a sales catalog. The writer conveys his positive attitude toward the flower as a device for persuading the reader.

FIGURE 10.7 **Daylilies**
(Courtesy of White Flower Farm, Litchfield, CT. Photo by Michael Dodge.)

Although appropriate in marketing, such subjectivity may be suspect in a technical report or article, where convention dictates a more neutral approach and requires that authors place a check on feelings and self-display. Concrete, verifiable features of the item being described take precedence over the impression that the item makes on the writer or the

A CLOSER LOOK

## Picturing Things

Egbert Woudstra, a professor at the University of Twente in The Netherlands, uses some of the following questions to help determine what information needs to be illustrated in popular articles about technical information and how to select the right form of illustration. Descriptions in such articles usually center on a *material object* (like some piece of apparatus) or a *process* (chemical, physical, or production). Following the questions are suggested standard forms of illustration to answer them. A "schematic" is a schematic diagram as opposed to a "drawing," which refers to a realistic illustration (Woudstra 93–95).

**Object Questions**

1. What is it/what does it look like in whole or in part? (photo, drawing, or schematic)
2. What attributes/characteristics (technical specifications) does it have? (table)
3. How is it built/what parts does it consist of? (photo, drawing)
4. What is (are) its function(s)? (drawing or schematic, perhaps with text labels)
5. What is (are) its application(s)? (table with overview, series of photos)
6. How does it work? (block diagram or flow chart, photo, schematic)
7. What theory is involved? (text)

**Process Questions**

Distinguish between continuing processes and discrete or batch processes with separate stages.

1. How is the research done? (photo of the experimental set up or of, say, a focus group in action)
2. How is an object produced? (photo of production line or installation)
3. What stages are there? (series of photos, drawings, strips)
4. What object is carrying out the process or a part of it?
5. What is (are) the object's functions? (drawing or schematic with labels)
6. What is (are) the product(s) or stages of the process?
7. What is (are) its applications? (table with overview, series of photos)
8. What data are known about (part of) the process? (table, graphs)

**Context Questions**

1. What organization/industry will use the process? (photo of building)
2. Who is (are) the researchers or producers? (photos)
3. Where is the work done? (photo of the office or lab)
4. What forms of cooperation are involved? (photo, organizational chart)
5. How is it financed? (table of sources, pie chart)

writer's attitude toward it. Here's one rewrite of the *Hemerocallis* description that meets those constraints:

> Daylily (*H. liliaceae*). A popular plant, daylilies appear in every color except blue. Colors include cream, pink, lavender, and bicolors. If carefully selected, a garden of daylilies can bloom from June until September (although individual flowers last only one day). Daylilies are generally inexpensive, hardy, and pest free; they transplant easily, multiply rapidly, and require little care.
>
> When planted together in large numbers, they crowd out weeds. Newly bred forms called "tetrapolids" have double the normal chromosomes. They thus have larger flowers with heavier texture and more blooms per stem over a longer period.

Overview

Specific details

Advantages to the gardner

All writing, of course, is a matter of selecting information, and to that extent any description is subjective. Technical marketing documents (like the catalog) and proposals may be more overtly subjective than reports. Be aware, too, that neutrality in description is not a global preference, especially in high-context cultures that measure the truth of a statement largely in terms of who said it. In addition, even in the United States, a preoccupation with objectivity is beginning to seem dull at times, even misguided or silly.

## Expanded Description: Site

You'll also need to consider the appropriate level of subjectivity when you describe a place or site. Such descriptions often form the core of engineering reports. In writing, be careful to avoid disclosing your attitude toward the place, if such disclosure is not appropriate. Do make clear, however, your perspective on the site, where you (and the reader) are looking from (Figure 10.8). Provide details that help the reader to visualize the territory from that perspective. Consider the following passage:

> Much of the annual precipitation occurs as snow, which increases with elevation. In the Muddy River drainage (2,000 feet at the New Muddy River and 2,100 feet at the Smith Creek Road snow courses), the snowpack averages from 2 to 4 feet in depth with 9 to 15 inches of snow water content. At the 3,200-foot elevation near June Lake, the spring snowpack averages 7 feet in depth with 37 inches of snow water content. At 4,400 feet elevation at the Plains of Abraham on the northeast slopes of Mount St. Helens, the spring snowpack averages 13.6 feet in depth with 77 inches of snow water content. (Gifford Pinchot National Forest 79)

Oveview: Precipitation increases with elevation

Details keyed to elevation

The passage moves from lower to higher elevation, and details concerning precipitation are organized by significant marks of elevation. The next passage moves from east to west:

> Most of Delaware is near coastal waters, and no part of the State is farther than eight miles from tidal water. Delaware's coastal waters include the Atlantic Ocean to the three-mile limit, Delaware's portion of the Delaware River and Bay, the three Little Bays of southeastern Sussex County (Rehoboth, Indian River, and Assawoman), and many tidal streams. Excluding

Delaware described in terms of relationship to water

**FIGURE 10.8 Drawing of a Utility Plant**
The plant is shown from the point of view of a low-flying aircraft, a perspective that is increasingly common, especially in computer-generated drawings and animation. (Courtesy of Atlantic Generation, Agent for Vineland Cogeneration Limited Partnership.)

> these streams, the State's coastal waters are 493.3 square miles in area and approximately twenty percent of Delaware's area. (*Proposed Coastal Management Program* 1)

## Expanded Description: Process

For more about processes, see Chapter 17

Finally, you may need to describe a process in order to explain something to the reader. The events in a sequence or steps in a set of instructions are processes: how an office or a snowmaking system runs or how the fishing stock in an area becomes depleted; how to use a tool, operate a computer program, apply for a loan, or get a car to work. Your purpose will dictate different writing approaches. Here are the three most common purposes for describing processes:

- To record: *How* I or we do (or did) something or how something happens.
- To inform: *Why* I or we do (or did) something or why something happens.
- To instruct: How *you* (the reader) can do something.

Let's assume, for example, that you'd like to expand the description of a daylily to show development and change in the lily over time. The original description presents a snapshot: the lily's shape, components, size, color, and the like. To show the process of growth, you might

- Record, through a series of daily photographs or notations on key dimensions, the growth of one lily over a summer.
- Inform readers about the reasons for the changes they would see as the lilies rise from their bulbs into full plants.
- Instruct gardeners in methods for planting and tending daylilies.

This section provides guidelines for recording and informing; Chapter 17 provides an extensive discussion of instructing.

**Recording.** In an office, a manufacturing plant, or a laboratory, and in the field, you record processes in notebooks, journals, or reports to preserve the details. That record may be merely your own notes for yourself, or it may address other readers:

- Colleagues who need to reproduce and thus verify the process
- Supervisors who check on your work
- Attorneys and patent officials if your procedure is something new for which you'll seek a patent
- Standard-setting organizations that evaluate whether your organization's procedures indeed match their norms for quality

Make sure your record is accurate and complete. For laboratory notes, write in longhand in a bound notebook with numbered pages to help convince readers that the notes have not been altered after the fact.

When you record a process or system, you let it tell its story. You can also write a narrative that tells your own story about something you have done or will do. Such a narrative approach is common, for example, in trip reports: when I left, what I did each day, when I returned. Similarly, you may report progress on a project as a narrative. Others may review narratives you write for the record, but the major reader is often only you or other members of your team.

**For more about progress reports, see Chapter 16**

**Informing.** When you write to inform, however, your major reader is always someone else—perhaps many other people. Although close colleagues may be able to read your notes and understand exactly what you mean, you often need to interpret and explain your information for other readers. Select from the record the details that will convince them without boring them, overwhelming them, or lying. Often, for example, trip reports and progress reports fail because they only record, without also explaining why that record matters. Similarly, computer documentation fails when it shows functions only from the system's perspective ("I can create two-column text!" "I can find and replace!") without telling readers what the function means for them. The narrative of method in a proposal fails unless you give reasons for the reader to choose *your* approach.

**For more about proposals, see Chapter 14**

In writing to inform, provide introductory and concluding segments to help guide your reader through the information and interpret the factual record. The following outline suggests one approach to writing:

1. Introduction
   - Definition of the process
   - Purpose/benefits
   - General principle of operation or history of the process
   - List of major steps
2. Step-by-step sequence
3. Conclusion
   - Special points/advantages or disadvantages
   - Relationship of this process to others

The following passage demonstrates this technique. It introduces an article on brewing, and, in particular, on the role enzymes play in the brewing process. Note that punctuation reflects British usage ("Brewing" 26).

Overview and definitions

> Traditionally, beer is produced by mixing crushed barley malt and hot water in a large circular vessel called a mash copper. This process is called 'mashing'.
>
> Besides malt, other starchy cereals such as maize, sorghum, rice and barley, or pure starch itself, are added to the mash. These are known as adjuncts.
>
> After mashing, the mash is filtered in a lauter tun. The liquid, known as 'sweet wort,' is then run off to the copper, where it is boiled with hops.
>
> The 'hopped wort' is cooled and transferred to the fermenting vessels where yeast is added. After the fermentation, the so-called 'green beer' is matured before the final filtration and bottling.

Lead-in to detailed explanation

> This is a much-simplified account of how beer is made. A closer look reveals the importance of enzymes in the brewing process.

You may decide that a visual is the best way to describe a process, with a caption or label that helps the reader know what to look for, as in Figure 10.9.

## COMBINING TECHNIQUES

Defining, classifying, and describing are useful techniques for explaining technical information. Each has a certain purpose and pattern. You may develop one segment, or even an entire report or article, through one technique. But you may also weave these techniques together. Such interweaving creates effective explanations in the examples reproduced in Figures 10.10 and 10.11. Figure 10.10 is a short article and Figure 10.11 is a segment from a technical report. Knowing these techniques provides a shortcut as you assemble materials in a way that both makes sense and holds the reader's attention.

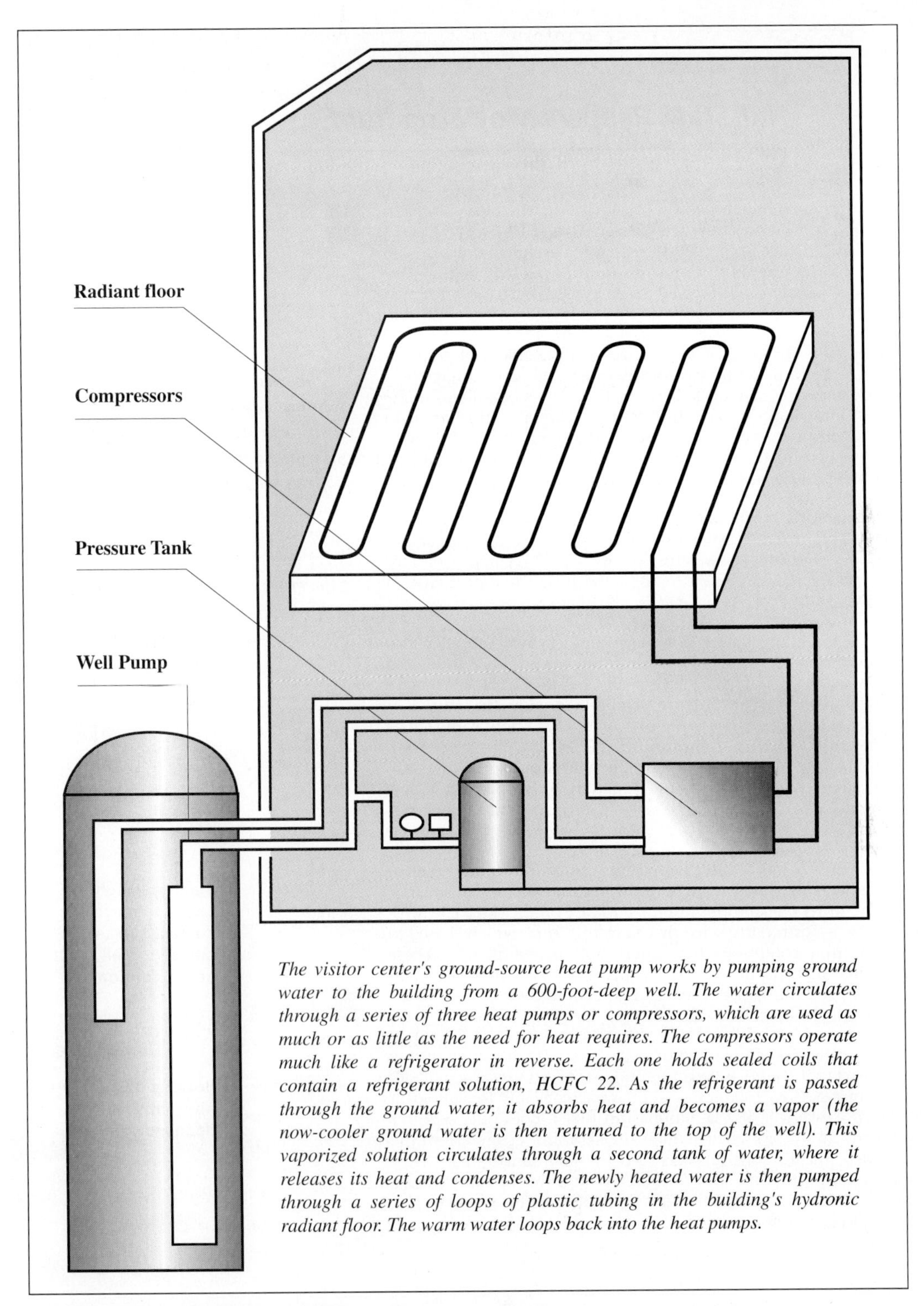


**FIGURE 10.9 How a Ground-Source Heat Pump Works**
(Courtesy of Larry Needle and the Maine Audubon Society.)

## *Hot Stuff: Peppers May Stop Pain*

On Christopher Columbus's second voyage to the New World in 1494, the expedition's physician, Diego Alvarez Chanca, opened a spicy new chapter in mankind's search for useful chemicals from plants. Chanca wrote the first detailed account of a plant in the West Indies—the chili pepper—that seemed to have medicinal value for skin problems and other woes.

**Introduction to the topic wrapped in a story and pun, "spicy"**

**Formal definition**

Five hundred years later, the active chemical ingredient in chilies is carving out a niche in modern medicine as a new and surprisingly effective pain reliever. The chemical is trans-8-methyl-N-vanillyl-6-nonenamide, better known as capsaicin (pronounced kap-sa'-i-sin). Capsaicin is the most potent of a group of compounds that give chilies their fiery bite. Chili peppers contain anywhere from 0.1 percent to 1.0 percent capsaicin, depending on the variety, soil, and climate.

*" the active chemical ingredient in chilies is carving out a niche in modern medicine as a new and surprisingly effective pain reliever."*

**Description of chilles and their benefits**

Chilies apparently synthesize the compounds as defensive chemical weapon to ward off predators. Man is the only mammal known to eat and savor these tongue-searing, eye-watering plants. Man has eaten chilies for thousands of years and has used them for centuries as ingredients in herbal medicines.

**New topic—research-and another pun, "firing"**

In addition to finding a place in modern pain relievers, capsaicin also is firing a wide variety of research. Hundreds of research reports involving capsaicin now are being published each year. Chemists, for instance, are trying to modify its molecular structure to produce analogs that work better. Medical scientists are using capsaicin to understand how pain signals are transmitted from one nerve cell to another. They have uncovered indications that capsaicin works by depleting the body's stockpiles of a substance involved in the genesis of inflammation. Inflammation is the process that causes pain and tissue damage in arthritis and many other diseases.

**Tie-in through capsaicin to narrower subject-topical medicines for pain relief and definition**

Capsaicin already has begun to appear in topical medicines sold for joint and muscle pain in the United States. Topical medicines are creams, ointments, and other preparations intended for use on the skin. The most familiar of these is Zostrix, marketed by GenDerm Corp., which is used to alleviate pain from conditions that affect millions of people. These include arthritis, diabetic neuropathy (a nerve damage that occurs in many diabetics), and shingles, that painful skin condition caused by reactivation of the chicken pox virus in adults.

**Process description: how capsaicin deadens pain**

Despite its centuries of use in crude pepper extracts, it is only during the last few years that scientists began to understand how capsaicin deadens pain. Thomas A. Gossel, professor of pharmacology and toxicology at Ohio Northern University in Ada. emphasizes that it does not work like many liniments and creams that redden the skin and create a burning sensation. These medications are classified as "counterirritants." They work by producing irritation, which causes a burning (or sometimes a cooling) sensation in the skin. This superficial pain confuses nerve cells, muting perception of deeper pain from joints and other inflamed tissue.

**FIGURE 10.10 Hot Stuff: Peppers May Stop Pain**
A short article that uses definition, classification, and description to explain how a chemical relieves pain. (Courtesy of the American Chemical Society.)

**More details about an element in the process: Substance P**

Application of capsaicin to the skin usually does cause a burning sensation and redness. But capsaicin relieves pain in a different way—by depleting the body's stockpiles of a peptide, a protein-like compound called substance P. Substance P is believed to be the main neurotransmitter for relaying pain signals to the brain. Neurotransmitters are chemicals that carry nerve signals across the gap, or synapse, separating one nerve cell from another.

*" capsaicin relieves pain...by depleting the body's stockpiles of a peptide...the main neurotransmitter for relaying pain signals to the brain."*

Substance P is produced and stored in certain nerve cells, or neurons, including "sensory" neurons that carry pain signals from the skin, joints, and muscles. If one of these cells senses a painful stimulus, it produces an electrical signal that travels along the length of the neuron. When the signal reaches the end of one neuron, it causes release of substance P. which flows across the gap to the next neuron. Substance P binds to receptors on the next neuron, causing it to generate another electrical signal that continues on to the brain, where it is perceived as pain.

When applied repeatedly to the skin, capsaicin disrupts this process. It causes sensory nerve fibers to use up their stocks of substance P and somehow prevents them from replenishing the depleted supplies. With less neurotransmitter available, neurons can't transmit pain signals effectively and patients feel less pain. The process is reversible. After people stop using capsaicin, substance P—and pain—return. Because capsaicin is specific for nerve cells involved in transmitting pain, other nerve function is unaffected.

Substance P also is believed to be involved in transmission of signals that the brain interprets as itching or pruritus. Capsaicin thus also is used for intense itching, such as that occurring in patients undergoing kidney dialysis.

**Summary of process and goal of new research**

New research also suggests that substance P may play a role in causing inflammation, which leads to tissue damage in arthritis and other conditions. By depleting substance P, capsaicin thus may treat arthritis itself and not just alleviate the pain.

Natural capsaicin is the most effective compound available for depleting substance P. according to Gossel. But researchers are trying to enhance its effects by modifying the structure of the capsaicin molecule. One major goal is to find a way of modifying capsaicin to reduce or eliminate its main side effect, a burning sensation on the skin. Although the burning usually disappears after a few days or weeks of using capsaicin, it can be severe enough in some sensitive patients to require use of anesthetic ointment.

Topical creams such as Zostrix contain only small quantities of capsaicin—0.025 percent or 0.975 percent—because small amounts of the compound are most effective. The creams are prepared with capsaicin that is extracted from red chili peppers. Needless to say, technicians take great care to avoid contact with the concentrated capsaicin powder. They usually wear full-body protective suits to avoid inhaling the powder or getting it into the eyes, nose, or mouth.

FIGURE 10.10 **Hot Stuff: Peppers May Stop Pain (Continued)**

1. Definition of "land management" and its environmental benefits

2. Another aspect—the complex politics of land management

3. Problems with a local rather than regional approach

4. Definition of what results from a lack of planning: sprawl

5. Negative effects of sprawl classified

The Delaware Estuary Program

## Land Management

The way we manage land—where we put office buildings, highways, parking lots, houses, stores, farms and parks—is an issue that is important to the Delaware Estuary Program for a variety of reasons. Some of these ① reasons relate to water—the development of land can have significant effects on the quality and quantity of water that flows over it (rainwater and melting snow), through it (streams and rivers) and under it (ground water). Others relate to wildlife and the loss of habitat for the many species of plants and animals that reside here.

② But land management, in addition to being an important environmental issue, is a complex political issue. Traditionally, local governments in New Jersey and Pennsylvania have determined how land is used through local planning and zoning. In Delaware, the traditions are similar, but the counties play a more significant role, primarily in unincorporated areas. In all three states, land management is a carefully guarded local prerogative.

There are throughout the Estuary watershed many examples of individual communities that have implemented successful planning and resource protection programs within their borders. But even in ③ these communities, development and growth proceed without consideration of their cumulative effect on the region's natural resources, including the Estuary. It is rare, in fact, for local decision-makers to consider the regional implications of the land use decisions they make because neither New Jersey, Pennsylvania, nor Delaware's land use statutes encourage them to do so. In addition, without substantial technical or financial assistance, it is unreasonable to expect local governments, many of which rely on volunteer planning boards and commissions, to spend time thinking about issues as they relate to places beyond their own borders. Natural systems, however, are not confined by, and rarely are contained in, arbitrary political boundaries.

Whatever its causes, there is a lack of a regional land use perspective in the Estuary region. This is significant because over the past several decades growth in all three Estuary states has been typified by low density ④ housing and highway-oriented strip commercial development, all spread over large areas. This pattern of growth and development is referred to as suburban sprawl, and it has a variety of negative cumulative effects on the ecological health of the region, defined here as the Estuary study area.

Among other negative effects, sprawl:

- Consumes vast quantities of land, including prime agricultural land;
- Fragments natural habitat, creating small, less ecologically valuable spaces;
- Makes it necessary for people to drive everywhere they go, causing highway congestion and air pollution—and air pollutants ⑤ eventually end up in the Estuary, falling with rain;
- Exacerbates nonpoint source pollution problems as more and more area is covered with impervious material;
- Costs local governments huge sums of money for new sewer systems, streets and highways, referred to as "infrastructure," and for the delivery of public services over wide areas.

FIGURE 10.11 **Land Management**
A segment from a technical report that combines strategies to explain a concept. (Courtesy of the Delaware Estuary Program.)

## CHECKLIST: Explaining

1. **Define the term for a concept, object, or process you don't expect the reader to know**

   Adjust the definition's length and location to your emphasis and the reader's need

   Control the display of your attitude

   Use familiar terms in the definition

   Expand the definition of an important term

2. **Classify a group of items to help readers understand**

   Select a simple sorting principle

   Apply the principle to all items

3. **Describe a mechanism, organism, site, or process**

   Select the language and details to match the reader's prior knowledge and your joint purpose

   Keep your perspective consistent

## EXERCISES

1. Write an extended definition of one of the following terms to submit to your instructor. Note the audience for whom you are writing at the top of your definition:

   transportation
   pollution
   diet
   hypertext
   information
   energy
   design

2. If you are at work on a report, prepare a formal definition of a major term you'll include in the report. That definition might then appear in your report's introduction.

3. Note any elements of subjectivity in the following description. What advice would you give to the writer for revising the passage in a more objective style?

   I'll never forget the first time I saw *Hemerocallis liliaceae,* although it was 15 years ago. I had been allowed a glimpse of a garden in Connecticut opened to the public only one day a year. The ruling presence there was its owner, a woman, as they say, *d'un certain age,* who courted both flowers and people with skill.

   The lily had a trumpet-like appearance, fully extended in orange extravagance. Really, the lily wasn't one flower but many, a cascade of color that rioted across the landscape. It—or they—demanded attention. A close-up look revealed finely spun petals that curled outward like the fingers of dancers from Thailand. Light shone through each blossom, stopping only at the green, earth-centering stems.

4. Develop a visual and a textual description of mechanism, organism, site, or system to submit to your instructor. Some suggested topics:

   a highway interchange
   a local area network
   the site for a new building
   equipment necessary to use energy from the sun
   a mouse
   soybeans
   skis
   a soccer field

   Note your audience at the top of the page or include that information as part of your description. Clarify whether you are describing the item in general or one particular item and establish your point of view.

5. Change the audience for the document you created in response to Exercise 4. Write the description for the new audience. Attach a memo to your instructor commenting on how a difference in audience caused you to write (and draw) differently.

6. The following paragraph compares the calories in a pint of shrimp fried rice from two different restaurants, called RA and RB. Classify this information better to bring out the contrast.

   RA's pint of shrimp fried rice contains fewer calories than RB's. RA uses a special brown Chinese rice; this brand of rice has 300 calories per pint. RB uses white rice, which is approximately 400 calories per pint. RA and RB use different amounts of shrimp: RA includes 10 small shrimp and each has 20 calories, which amounts to 200 calories per pint; RB adds 15 small shrimp, each shrimp containing 20 calories, which amounts to 300 calories per pint. Both restaurants add soy sauce for flavor: RA adds 2 tablespoons of soy sauce, which amounts to 10 calories, whereas RB adds 4 tablespoons, which amounts to 20 calories. The restaurants also use scallions for flavor: RA uses 1/4 cup, which is 30 calories and RB uses 1/3 cup, which is 45 calories. Finally, oil is used to prevent the rice from sticking while it is being cooked: RA uses 2 tablespoons of oil, which is 200 calories, and RB uses 4 tablespoons of oil, which amounts to 400 calories. The total caloric intake of a pint of shrimp fried rice from RA is only 740 calories compared to RB's, which is 1165 calories.

7. Analogies and metaphors are useful ways to understand and explain a process. Note the role of analogy in the following passage:

   Piglets like to play. Confined often in small spaces, they tend to nip at each other, and sometimes those nips are more like bites. The bites often lead to infections. In the United States, the breeders commonly treat those infections with antibiotics that are often expensive and increase the cost of raising pigs.

A practically minded farmer in Estonia, however, looking for a less expensive alternative, chose to redirect the pigs' attention. He provided toys—odd bits of pvc pipe and other small, harmless objects. If the piglets like to play, he reasoned, give them something other than each others' ears and tails to play with. The approach worked: the pigs avoided the bites that lead to infections and thus the farmer avoided the need for antibiotics.

As you read textbooks and other literature in your discipline, look for *analogies* and *metaphors* used to explain a process. At your instructor's direction, photocopy one passage that uses such an approach, provide a complete citation for the passage, and write a brief covering memo that evaluates the effectiveness of the analogy or metaphor.

**8.** Draw, in a standard form (like a flow chart), a diagram of some process. Then write a brief description to explain the process to a specific audience. Name the audience at the top of your description. Your goal is not to teach the audience how to perform the process but to enhance the audience's understanding of how the process is performed.

**9.** The following description of a process demonstrates problems in point of view and emphasis. Even if you're not familiar with the chemical terms, outline the paragraph and suggest a method for restructuring it. For your restructuring, determine an appropriate purpose and audience. At your instructor's direction, rewrite the passage.

To develop a shampoo requires particular chemical components to be considered. Cleansing is the primary function. Since the basic chemical ingredient is a detergent, a cleansing agent of one or more anionic surfactants needs to be selected. However, this may be assisted by nonionic or amphoteric surfactants, especially in a mild shampoo, to lower irritation of the anionic surfactant. Secondary, a foam enhancer from the fatty acid alkanolmides or amine oxides groups is used to build density, reduce stripping effect of detergent, and impart minor conditioning. The next step in the procedure addresses its contribution to styling, masking of mechanical defects ("repair" to "damaged" hair), luster, ease of combing and fragrance, as these are performance characteristics sought after by customers. The customized recipe has many options to utilize various chemical properties which result as: deep conditioner, viscosity modifier, fragrance, clarifying agent, preservative, color, opacifying agent, and pH control. Finally, an effectively formulated shampoo will remove surface oils, dirt and debris from the hair shaft and scalp, leaving the consumer thoroughly satisfied with the finished product.

## FOR COLLABORATION

**1.** Describe in words a simple mechanism or an uncomplicated object. Do not identify it. Exchange your description with another student and ask her or him to draw a sketch to accompany the description.

**2.** Provide a sketch of a simple mechanism for another student, who will then write the accompanying text.

3. In a team of three students, build a simple device (a paper airplane or a mechanism composed of Legos or other toy building blocks). Exchange devices with another team and write a description of that team's device as well as the process you think the team went through in building the device. Retrieve your device in exchange for the description, which the builders of the device will then test for accuracy and clarity, as you test their description of your device.

CHAPTER 11

# PERSUADING AND PROVING



"THE AUDIENCE TO WHICH SCIENTIFIC PUBLICATIONS ARE ADDRESSED IS NOT PASSIVE; BY ITS CHEERING OR BOOING, ITS BOUQUETS OR BRICKBATS, IT ACTIVELY CONTROLS THE SUBSTANCE OF THE COMMUNICATIONS THAT IT RECEIVES."

JOHN ZIMAN

When you explain something to a reader, you help them understand that term, object, process, or concept. That explanation may also play a role in achieving another purpose: *persuading* your reader to act in the way you intend. This chapter provides guidelines for designing information to persuade. It is the most customer-oriented kind of design.

## PERSUADING

To sense the differences between explaining and persuading, take the sentence "The cat is lying on the table." This is a statement of fact. You could expand the description: The cat is black and white, eight pounds, mixed breed, two feet long, rather low slung, sweet face, named "Rover." That expansion adds some subjective qualifiers, like "rather low slung" and "sweet face," but it still aims at helping an audience picture the cat. Suppose, however, that you are trying to determine the source of cat hair in your food or back up your opinion that the cat (or more likely the cat's owner) lacks discipline. Then the statement about the cat's position on the table takes on a new role as evidence. You use it as backing for your case about cat hair. In addition, in a household where cats and people don't normally dine together, you could use that statement persuasively to mean, "Get the cat off the table."

To persuade an audience, you say something (or, sometimes, you don't say something) that causes them to

- Change their mind.
- Agree that something you think is significant is indeed significant.
- Make the decision you recommend.
- Approve the course of action you'd like to undertake.
- Hire you.
- Buy your product.

Many years ago, the Greek philosopher Aristotle identified three strategies for persuasion: *ethos,* the appeal of the character of the speaker or writer; *pathos,* the emotional appeal; and *logos,* the logical appeal. In Europe and America today, these strategies still dominate. Emotional appeals, for example, often underlie advertising and marketing. For scientific and technical presentations, logical argument usually takes priority, with attention paid, too, to the authority of the speaker. Emotions tend to be dismissed. Elsewhere in the world, Aristotle's strategies don't always hold water, as *Crossing Cultures: Strategies of Persuasion in Arabic and Japanese* suggests. In the next section, you'll read some advice about persuasion in that international context. The rest of the chapter returns to the Aristotelian tradition as it has been updated by technical professionals writing in English. You'll read guidelines for persuading through proving, that is, using logic to create a sound argument that convinces your audience. Persuading and proving are complex processes, but these guidelines provide a starting point.

When you persuade, you adjust your ideas to your audience and adjust your audience to those ideas. To accomplish that adjustment, you have to understand your audience's values so you can create common ground, establish your authority, and participate in their process for making decisions and getting others to comply with them. As you do so, be prepared for change. All generalizations about culture are risky, and the global marketplace is putting many traditional values to the test. Young people in particular in Japan, the United States, and most countries in Europe are becoming more alike in their education, income, lifestyles, professional experience, and consumer habits. In addition, the growth of multinational companies fosters similarities in organizational culture across the globe. These similarities ease technical discussions, so long as you are aware that deeper differences may also present themselves.

## Creating Common Ground

**For more about the values that underlie persuasion, see Chapter 3**

First, identify the values that govern your audience's thinking in the situation at hand. You have values and norms for action you accept as true, and so does your audience. When people are like-minded, a simple appeal to a shared value (like keeping cats off dinner tables) can be enough to encourage agreement and action. Good teams, for example, can "read each other's minds." In the low-context situations frequently found in today's global economy, however, you often have to create common ground between you and your audience through some of the following strategies.

**Look at What People Do for Clues About What They Believe.** Take, for example, shopping. Although Americans are generally comfortable using credit cards for purchases, many Germans prefer to pay as they go. This value, however, inhibits purchases online, which depend mainly on credit. Enter a start-up company that serves as a clearinghouse between Internet merchants who are eager to sell to German customers and those customers who are increasingly connected to the Internet. Because many Germans and other Europeans *do* have bank accounts that allow transactions similar to debit-card ones in the U.S., the start-up serves as a clearinghouse between online merchants and the banks, verifying and processing bank payments. The start-up developed its business by knowing a core value of its customers' customers. Recent changes in other German shopping habits reflect in particular the arrival of U.S.-based superstores with a pattern of long opening hours. For years, shop hours in Germany have been severely restricted. Most retail stores have closed daily from 1 to 2:30 P.M. and nightly at 6:30 P.M., Saturdays at 2 P.M., and Sundays all day. Why? Labor unions and the owners of small stores who fear competition from superstores give some practical reasons, but most commentators point instead to fear that longer hours would destroy the German family, whose Sunday walks and evening meals would be upset as people drew apart to work and shop. These restrictions are gradually moderating.

Let's look at a few other examples that show how habits reflect values. In Japan, professionals put in long days of formal office work and long nights of socializing; that socializing plays an important role in establishing and maintaining the relationships essential to good business. The emphasis on relationships differs sharply from the task orientation frequently seen in U.S. organizations.

Many cultures are less addicted to speed than Americans, who often think that something fast, like e-mail, is bound to be better than a slower message and that short-term (and thus faster to see) results deserve more attention than long-term ones. Americans tend to measure things in minutes, whereas many other people think in centuries.

The personal computer, a perfect fit with an individualistic culture, is less welcome where people think more collectively. Where people mistrust intangibles, electronic accounting is suspect; in Japan, for example, it's not uncommon for shopkeepers to check up on a computer with an abacus or to carry large amounts of cash and avoid electronic bank transfers.

**Be Patient.** Listen for clues about what people are thinking. Understand that some cultures frown on saying "no," and thus you should understand a comment like "that would be very difficult" to mean "no" and not press. Take time from talking about the issues to talk about how you and others in the conversation feel the conversation is going and to confirm agreements.

**Accept Difference.** When you persuade someone, you have to build common ground while still recognizing differences. Incorporate those differences in your approach. For example, understand the reluctance of people whose view of life is fatalistic ("What will be, will be") to forgo their pleasures in smoking or drinking to excess even if you show them their increased risk of cancer or other diseases. Your discussion of long-term consequences won't be persuasive with people who don't pay much attention to long-term risks. Similarly, recognize other situations in which people's values compete. In determining strategies for managing a forest, environmentalists may emphasize the need to preserve endangered species of wildlife, whereas farmers and loggers may see the land as an economic resource.

Recruiting practices also differ. Although many European and Asian companies look like U.S. organizations, they seek different attributes in a recruit. Japanese firms, for example, emphasize character over accomplishments, and recruiters weigh heavily a candidate's integrity, trustworthiness, team behavior, and compatibility with the organization. U.S. recruiters tend to emphasize an applicant's credentials, summarized in a resume, to see what he or she has accomplished. Japanese interviewers like to ask candidates about their religion, home life, upbringing, and other personal information and have found U.S. legal constraints against such probing in employment evaluations difficult to deal with. As an applicant for a job in Japan, be prepared to respond to such questions.

### Establishing Your Authority

**For more about cultural context, see Chapter 1**

Second, in addition to creating common ground, find out what it takes to establish your authority or credibility as someone worth listening to. In a high-context setting, your authority, and thus your trustworthiness, derives from your membership in the group. It's less simple when you need to establish that authority. You establish credibility by how you behave and what you say. You create a persona, online, in print, or in person, that persuades the reader to take you at your word. Here are some suggestions for achieving a persuasive persona.

**For more about persona, see Chapters 7 and 21**

**Demonstrate the Personal Qualities Your Audience Values.** Determine those qualities and balance them against qualities valued in your own culture. If you pride yourself on individual assertiveness and direct speech, you may have to take a quieter approach to be heard by people who find such assertiveness offensive.

**Show Your Knowledge of Authorities the Audience Respects.** Citing other authorities helps validate your own work, justify your stance, and show that you recognize the collaborative nature of all scientific and technical pursuits. In cultures where relationships are highly valued, demonstrating your position in a network of the right people may be enough to prove your credibility.

**Acknowledge Your Perspective.** Although objectivity is valued in Europe and North America, you still must recognize that even in that tradition, and especially in other settings, people present information from a particular point of view. "All scientific inquiry requires a perspective," notes a leading scientist, "for research is not only investigation but also interpretation.

Scientists' perspectives condition what they perceive as important for the advancement of science, as well as the design of research and the weight given to conclusions" (Frankel B2). Another expert comments,

> Every age sees the world through the lens of its own obsessions, using its latest inventions as metaphors. For René Descartes to see animals as machines, there first had to be machines. It takes nothing away from Charles Darwin's observation that species compete and adapt to note that Darwin himself came from an industrializing society whose members were being forced to compete and adapt. (Berreby E5)

To encourage an audience to believe you, recognize you have a point of view and they may see things differently.

**Show That You Are Well Informed.** The nature of information varies across cultures. You enhance your credibility when you recognize where information is needed and produce the right information in a form your audience can understand.

### Identifying the Process of Persuasion

Third, identify the process of persuasion that your audience finds natural. That process, too, embeds cultural values. Aristotelian argument is not only acceptable but desirable, for example, in the United States and the United Kingdom. People debate issues openly in courts of law, statehouses, and meeting rooms. That's how individuals distinguish themselves and establish their authority. The goal is winning, and the process is often described in military terms: *tactics, opponents, a war of words, battle lines,* the *global arena.* An investigation of an opponent centers more on finding weaknesses and vulnerabilities than on understanding them. In addition, the argument can take on a life of its own separate from the arguers.

But other cultures have different approaches. The Japanese, for example, tend to avoid open discord and make decisions on intuition, on prevailing moods, and on a need to keep harmony. The message resides in the situation and in the people, not as an objective abstraction. The Japanese language, too, is less precise and pointed than English, more tolerant of ambiguity. Verbs and nouns are not specified in person or number, and open-ended or obscure statements provide a kind of "buffer zone" that avoids personal attack. Such indirection helps save face for everyone. In an example of how some people misunderstand these differences in persuasive process, one consulting firm that trains professionals in international negotiation notes it can help Europeans "get behind the inscrutable mask" of their Japanese colleagues. Ambiguity is not a mask, and thinking that way only reinforces the wrong approach.

In addition, Japanese organizations traditionally arrive at decisions and settle issues collaboratively. Proposals for actions or other decisions are circulated throughout the organization to gain consent in advance—to build a context in which the decision then becomes inevitable. In traditional U.S. companies, an executive often imposes a decision made individually. The U.S. system tends to be "fast-slow-slow-slow": a quick decision

CROSSING CULTURES

## Strategies of Persuasion in Arabic and Japanese

There is an Arabic proverb: "Enough repetition will convince even a donkey." This proverb suggests one technique for convincing and persuading people: repetition. Researchers note that a typical Arabic approach is "repeating, rephrasing, clothing and reclothing [a] request or claim in changing cadences of words" (Koch 48). The goal is not to convince the reader through reasons but to bring the reader into the writer's point of view. "Persuasion is a result . . . of the sheer number of times an idea is stated and the balanced, elaborate ways in which it is stated. . . . Ideas flow horizontally into one another" in parallel constructions (52). Underlying this approach is the assumption that persuasion is about truth. In traditional Arabic society, which is hierarchical and religious, truth is established, agreed on, and close to the surface. It is not a matter for individual debate and not in doubt—unlike the debate and doubt in many democracies. For Arabs, attempting to *prove* the truth through reasoning is counterproductive. Moreover, their very language, as in the Koran, is sacred. Idea and language are one. Truth is revealed and validated through the elegance of its presentation (Koch 47–60).

Japanese proverbs show a different strategy for persuasion: silence. One proverb goes, "Sickness enters through the mouth; misfortune comes out of the same place." Another: "When you speak your lips get cold." Similarly, a Chinese expression (found in a fortune cookie) says: "From listening comes wisdom; from speaking, repentance" (Gall). Contrast those maxims with a common U.S. expression, "The squeaky wheel gets the grease." As participants in a high-context culture, the Japanese are comfortable with silence. That comfort often means that outsiders see them as enigmatic and inscrutable. Americans tend to reward fluency, to speak out assertively on issues of interest, and to be uncomfortable with silence—attributes that sometimes causes them to give in too easily when they negotiate. Faced with silence, they see opposition. They then lower their bargaining position in their eagerness to talk. Their counterpart's silence is an effective persuader.

that may take some time to implement. The Japanese is "slow-slow-slow-fast" because consent is achieved in the process so the final implementation is speedy. In an international setting, then, don't just assume your approach to persuasion is always right. Instead, recognize that persuasion is demanding and complicated. Be patient. Show goodwill. Be polite. Take time to understand how your audience thinks. Use that understanding as a basis for further discussion, and be willing to change your mind.

## PROVING

In scientific and technical circles today, the major form of persuasion is argumentation, that is, making statements and backing them up. The arguments appear in the pages of professional journals, in presentations at meetings, and online in newsgroup discussions, e-mail messages, and Web sites devoted to topics of interest. They also anchor more overtly persuasive documents like progress reports and, especially, proposals. To argue effectively as a technical professional, you have to prove the truth of what you are saying. Readers are likely to say, "show me." You need to give them good reasons for believing you.

**For more about proposals and articles, see Chapters 14 and 21**

## Asserting

Statements you make that are either true or false are called *assertions*. Your first task in reasoning is determining the likelihood that your audience will agree with your assertion. You can probably assume ready agreement if the statement is easily verified or certain, that is, if observation or experience could prove it. Let's look at some examples:

*The cat is lying on the table.* If the audience, too, can see the cat, or trusts you know cats, they are likely to accept your statement as true.

*A meter is made up of 100 centimeters. There are 24 hours in a day.* People in many places on the globe would say "of course" to each of these statements. Neither raises a question because each is true by definition. A meter, by definition, is made up of 100 centimeters. A day, by definition, is divided into 24 hours. If your audience accepts this vocabulary, they have to agree your statements are true.

*Seattle received 5.3 inches of rain in March.* This statement is less obviously true. If you carefully observed the weather gauge in your Seattle backyard, or if the statement were reported by a reliable source, like the National Weather Service, your reader would probably accept this as true. By contrast, however, consider a measurement of snow accumulation provided by a ski area whose marketing interests might encourage the largest possible number.

*More cars collided with moose in Maine this year than last year.* An audience would probably accept this assertion as long as certain terms were defined (like what "colliding" means and whether "year" is the calendar year or the hunting season). An authoritative source, like the Maine State Police, would also bolster credibility.

## Providing Good Reasons

When you introduce uncertainty in your statements and are more subjective, that is, when you move from fact to interpretation, you increase the likelihood that some audiences will disagree with you. Simply saying something, even with conviction, even in capital letters, would not prove your point to a doubting or unwelcoming reader. You need to provide good reasons to believe you. Let's look at some examples of assertions that need proof.

*Asian students have a keener knack for science than Americans.* To prove this, you and your audience need to agree on the major terms, for example, what group of people Asian refers to and how someone demonstrates a knack for science and how you have tested it. *Knack* could mean performance—hands-on science—as opposed to testing ability. In addition, you

A CLOSER LOOK

## Persuasive Design

A simple visual can focus political and technical argument, as the development of the Eating Right Pyramid (opposite page) shows. The traditional method for representing the basic four food groups was a pie chart (logically enough) with equal pieces. The pie suggested the need for variety. But it didn't address two other needs: moderation and proportionality, that is, eating different amounts from each group to maintain good health. So the U.S. Department of Agriculture launched a campaign for a new graphic metaphor to explain that message. They hired consultants to test several images with a wide range of audiences from children to employees of major food manufacturing corporations. The consultants presented many shapes: shopping carts, tabletops, a bowl, and the pyramid. The bowl was a leading contender because it addressed well the need for variety. Moreover, the meat and dairy industries and makers of oils and sweets preferred the bowl because it suggested the nutritional equality of all food and deemphasized the advice to use their products "sparingly." The pyramid, they felt, had a "good-food, bad-food" connotation. Bread and pasta producers, however, liked being "basic." A health group objected to putting sweets and fats at the top because the positioning might encourage a misreading that such foods should be eaten first. But most observers agreed that the Eating Right Pyramid conveys the USDA's major message: Eat less fat (Burros C1, C4).

A more controversial image is the "Mediterranean Diet Pyramid." Proposed by nutritionists and epidemiologists from the World Health Organization and Harvard's School of Public Health, this graphic depicts elements of a common diet in the Mediterranean region where the rates of chronic diseases are low and life expectancy is high. This pyramid adds wine

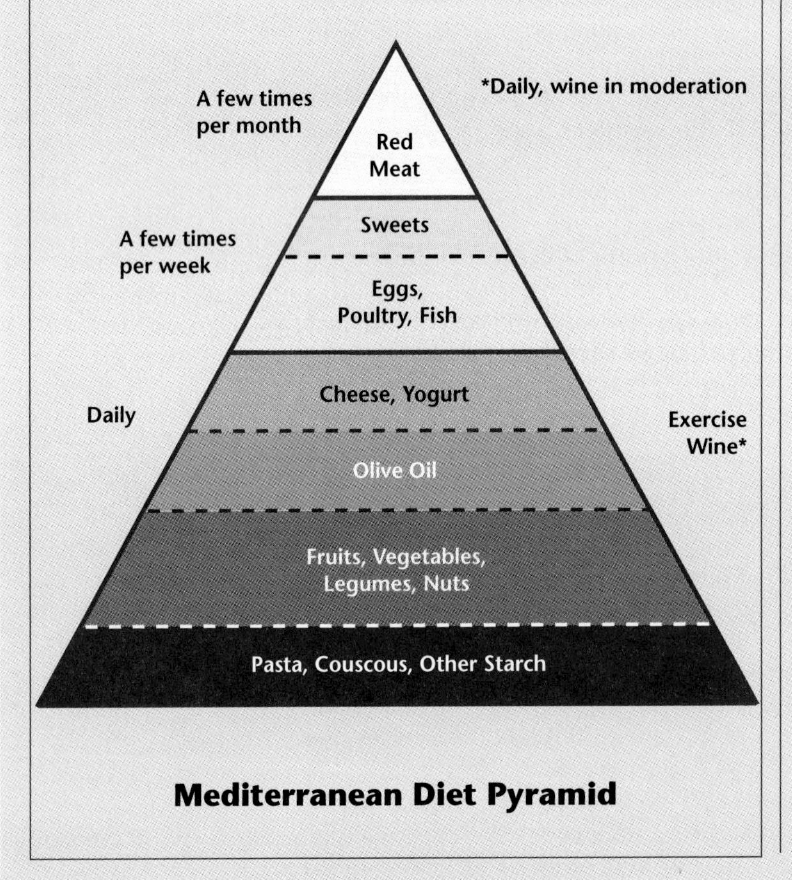


**Mediterranean Diet Pyramid**

and exercise to the mix, but critics charge that it provides unreliable advice because the research that led to it was confined to a population only of men between 40 and 59 years old. The potential benefits of alcohol are also hotly debated. For years, the U.S. government maintained that advocating its use was ineffective public policy, regardless of whatever small benefits it may have. The "Dietary Guidelines for Americans" of 1996, however, noted that "Alcoholic beverages have been used to enhance the enjoyment of meals in many societies throughout human history," an indication that moderate alcohol consumption can be acceptable (Miller B1).

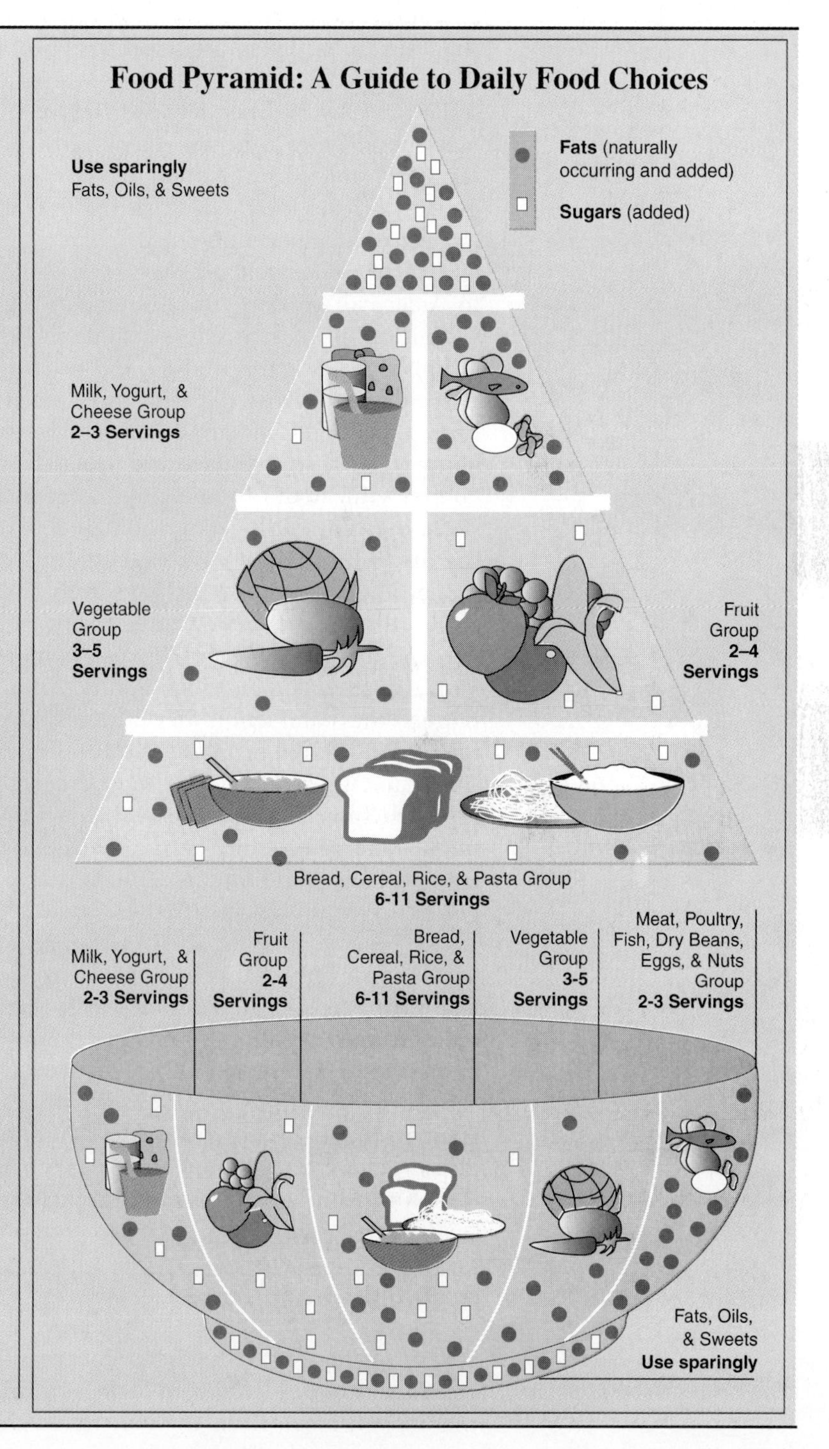

need to control the sample of people being discussed. You could, for example, measure performance on standardized advanced placement tests. For the sample, you could take all the twelfth graders in one city, or a random sample of twelfth graders who take the tests in any given year. But different attitudes toward testing or facility with English, for example, could skew the test results. There are additional problems: Tests may not be the best indicator of skills, one city may be exceptional in some way, and science instruction may be unequal in different high schools; that is, some schools may spend more time on science instruction or may offer better science teachers. And so on. Although the statement may look like a fact at first glance, a closer look shows that the tests for proving it are not so simple and its terms introduce ambiguity.

*More money spent on education means better educated students.* Particularly to American readers, this statement might seem like something "true by definition" because it's almost a maxim of public policy. Yet proving the relationship between spending and learning may be illusive. Expressed as a percentage of gross domestic product, the expenditure on education of both Japan and Germany is less than in the United States. Yet by many measures, including standardized tests, Japanese and German students are "better educated." In addition, someone may point out that educational spending is not a linear phenomenon, with each additional dollar producing an additional benefit. A school district may spend a large amount of money and see no result, when simply adding a few more dollars might produce a disproportionately large result.

*Basic research is more important than research in engineering.* One problem with this statement is its ambiguity. If it means, "I like to do basic research better," the statement is an opinion that you can't really prove. If the statement appears in a proposal to a funding source, it may be intended persuasively and thus mean that the reader should give more money to basic research than engineering research. An engineer reading this statement might take offense at the implied opposition between "basic" and "engineering" and point out an error in reasoning because some research in engineering could be classified as "basic."

These statements, then, are uncertain, and to prove them you need good reasons. You support a new or previously uncertain idea with information that is certain, well known, easily verifiable, or agreed on. To be persuasive in your reasoning, you also make sure to select information your audience will find compelling, and you structure that information in a way that gains both their attention and their consent or compliance. In the following sections, you'll read a few guidelines for developing good reasons. These guidelines fall into two broad patterns commonly used to connect ideas and their backing in the European tradition: deduction and induction.

## DEDUCTION

**For more about classification, see Chapter 10**

When you reason and present your reasoning *deductively,* you move from a statement about *all* items in a class of things to statements about *one* item; that is, you move from the general to the specific. As a technical professional, use deductive reasoning to develop a new conclusion from two statements, one about classes of items or people and the other about their characteristics. Deduction, then, is reasoning from *statements*.

### Reasoning Through Deduction

The classic form of a deductive argument is a three-part device called a *syllogism*. The three syllogisms that follow are all valid, if simple, arguments. If you agree to the backing assertions (labeled 1 and 2), you must agree that the conclusion (statement 3) is true too.

**A.** 1. All spruce trees are conifers.
2. All the trees in this stand are spruces.
3. All the trees in this stand are conifers.

Syllogism A arrives at its conclusion by fitting one class (spruces) into another class (conifers).

**B.** 1. All the computers in the multimedia laboratory have CD-ROM drives.
2. This computer is in the laboratory.
3. This computer must have a CD-ROM drive.

In Syllogism B, statement 1 asserts that all members of the class of computers share a characteristic (CD-ROM drives). The second statement fits one computer into the class; thus it displays the characteristic of the class (it has a CD-ROM drive).

**C.** 1. No marshlands should be paved for parking.
2. This property is a marshland.
3. This property should not be paved for parking.

Syllogism C presents another typical form: a negative statement 1, which leads, then, to a negative conclusion.

These examples greatly simplify a process with extensive rules and elaborate possibilities. For practical purposes, the three-part scheme is often reduced to just two parts, a conclusion and one assertion, with the other implied ("Because this property is a marshland, it should not be paved over.").

### Persuading Through Deduction

Thinking deductively also helps you on a larger scale to arrange the segments of an entire persuasive document. Based on deduction, classical scholars identified a five-part arrangement for presenting an argument. This arrangement is a good starting point for many documents today, including especially proposals and discussions of policy:

Introduction
Statement of facts
Confirmation

Refutation

Conclusion

In the *introduction* you present your major conclusion, followed by the *facts* that set up the background, often in a narrative. Then you *confirm* your major conclusion by providing your supporting assertions, your good reasons. In the *refutation,* you meet any objections head on. You point out any ways that arguments opposed to your own are wrong, because they depend on either false reasoning or false information. In the *conclusion,* you reiterate your major assertion as something especially certain because you have proved it. Figure 11.1 argues its point effectively in this pattern.

## INDUCTION

When you reason by induction, you move in a different direction from the general-to-particular approach of deduction. You start with particulars, with observation and experience, the results of an empirical investigation. From this evidence, you find similarities that allow you to formulate a statement about all things in a category, or you develop a way to connect seemingly miscellaneous events or items. Induction is the basic tool of science in the European tradition. Conclusions derived through induction are often then presented deductively, as in Figure 11.1.

**For more about empirical research, see Chapter 4**

### Reasoning Through Induction

Inductive reasoning is a process of developing and testing hypotheses, statements about similarities or connections. To arrive at a hypothesis, you leap from observations about some things across a field of things you have not observed to a statement about all things. The more instances you collect, the less risky your leap. Your hypothesis helps you determine what experiments to conduct or what to look for in a situation.

A good hypothesis is generally simple. One that is forced or has too many exceptions is probably not useful. On the other hand, the hypothesis should not exaggerate or oversimplify. When testing a hypothesis, you should

- Consider the facts widely.
- Confront contradictory evidence with open eyes.
- Work with a numerous and representative sample of instances.
- Prove that alternative explanations are unworkable.

A good hypothesis can also explain more than just the evidence on which it is based and can make accurate predictions. When your hypothesis has been sufficiently tested and other experts agree to it, it is considered true, at least for the present. Two major strategies for developing a hypothesis are thinking by analogy and analyzing causes and effects.

**Analogies.** Through analogy you try to understand how elements of a new phenomenon relate to one another in a way similar to how elements of a

The need to reduce the amount of construction and demolition waste by saving reusable items is greater than ever before.

**Major assertion: need to save reusable items**

Although comprehensive figures on the percent of building materials in the waste stream are not available, some data on waste wood are available. The Environmental Protection Agency (EPA) estimates that wood and lumber waste alone comprises 3.7 percent of the waste stream (1).

**Statement of facts: statistics**

What does the figure 3.7 percent mean? For illustration purposes, let's look at an average-sized home with 1,600 square feet. This home may have 20 windows, 15 doors, 14 light fixtures, 300 pieces of hardware (includes hinges, knobs, pulls) and 1,000 board feet of flooring. A commercial building has even more. In 1991, 545 demolition permits were issued in Hennepin County (one of seven counties in the Minneapolis—St. Paul metro area). So, in one year, that could mean 10,900 windows, 8,175 doors, 7,630 light fixtures, 163,500 pieces of hardware, and 545,000 board feet of flooring were sent to a landfill.

**More backing: an example**

The conclusions of a feasibility study of salvaging and recycling building materials (2) found that *reuse of materials (as opposed to recycling) is the best method for decreasing the amount of construction and demolition waste in the waste stream.* The reasons for reuse include: requires less energy, less processing and less transportation; has lower adverse environmental impact; is more viable for local/small-scale operations; requires less investment capital; and involves fewer regulatory guidelines.

**Confirmation: Reason 1: Reuse is better**

While there is a large supply of construction and demolition materials, the feasibility study found *an interest (one-third of home-owner survey respondents and half of contractor survey respondents) in using second-hand building materials.*

**Confirmation: Reason 2: People want second-hand items**

According to the study, *many salvage yard customers consider hand-crafted woodwork to be superior to new. A used solid-wood paneled door is better built, more attractive and usually less expensive than its modern counterpart, the hollow-core door. Also, many traditional styles aren't readily available elsewhere except as expensive reproductions.*

**Confirmation: Reason 3: Second-hand items can be better**

From a practical standpoint, saving and reusing quality items will not end the continuous construction and demolition waste stream, but it will reduce the volume of material that enters the waste stream. From an aesthetic standpoint, reusing architectural elements will add unmatched old-world charm, beauty, and value to home improvement projects.

**Refutation: Reusing will not end construction waste**

**But it will reduce volume and add charm**

1. Hennepin County Comprehensive Recycling Study, July 1985.
2. Salvaging and Recycling Building Materials: Feasibility Study, David B. Mason, 1990.

**FIGURE 11.1** **Reusing Architectural Elements: The Impact on Our Environment**
(Courtesy of Architectural Antiques, 801 Washington Avenue North, Minneapolis, MN 55401.)

For more about analogies, see Chapter 10

known phenomenon relate. A key analogy underlies much scientific research, namely, that what researchers demonstrate in the laboratory will be replicated in a similar population or situation in the world at large. Medical researchers, for example, test therapeutic drugs on animals to determine their potential efficacy in humans. Engineers build pilot plants to anticipate problems in the design of structures and processes before full-scale production.

Analogy also helps technical professionals develop new concepts. For example, an analogy to how humans might respond to earthquakes is helping engineers design buildings that resist such shocks. Traditional systems were passive; reinforcing supports and isolation mechanisms allowed the building to absorb the quake's energy. New systems respond actively like humans, yielding instantly but then pushing back. Such active systems include devices for automatically producing counterforces.

**Cause and Effect.** A second common form of inductive reasoning is analyzing the cause of some event or phenomena or determining the effect that will come from some cause. It can be tricky as you sort through necessary causes that must be present for the effect to occur and contributory causes that may also be in the picture. For example, observers may conclude that low wages are a necessary cause of low morale in an organization. But the wages may be only a contributory cause, along with other causes, for example, lack of safety precautions, inadequate health protection, or the absence of benefits. Your perspective influences what you see as a cause. Notes one scientific researcher, "The cause of an outbreak of plague may be regarded by the bacteriologist as the microbe he finds in the blood of the victims, by the entomologist as the microbe-carrying fleas that spread the disease, by the epidemiologist as the rats that escaped from the ship and brought the infection into the port" (Beveridge 126). Here are some other examples of cause and effect reasoning:

- The harvest of fish off the coast of New England declined 62 percent between 1982 and 1993. The fishery collapsed. Why? One hypothesis is that the U.S. federal government caused the collapse. In the 1970s, the government provided loans and tax credits that motivated investors to outfit a large fleet of big boats with high-tech gear; that fleet fished the waters beyond their ability to regenerate. Many fishermen (especially those who own small boats) would argue that the buildup of big boats was indeed the necessary cause of the fishery collapse. But contributory causes including pollution, the destruction of some fish habitat, and a rise in the number (and appetite) of seals, which prey on fish, may have been more significant. A few scientists also dispute the notion that the fishery did collapse. They see the decline in the mid-1990s as part of a long-term natural cycle.
- One condition of modern life is sleepiness. You may observe this condition among other students in your class. College students, for example, are among the most sleep deprived. From age 17 to 25, humans have a greater need for sleep than at any other time except early childhood (Brody C14). Young adults need ten hours of sleep, but

most get only six. What causes sleepiness? One answer is light bulbs. People should go to bed at nightfall and rise at dawn, a schedule that at the equator would mean 9 hours of sleep. But artificial light extends the day. In addition, work patterns now set many people on 24-hour clocks. In sleep labs, where there are no lights, clocks, or alarms, people demonstrate the need for about 8 or 8½ hours of sleep (Brody C14).

Sleepiness can be the *effect* of tensions, artificial light, television, and perhaps other causes. It can also be a cause. Sleepiness can compromise "alertness, ability to remember, judgment, reaction time, and mood" (Brody C14). Sleep-deprived drivers have caused many accidents. Sleepiness also enhances alcohol's negative effect; one beer to the sleep deprived is as impairing as a six-pack to the well rested.

- Assessing public health risks also requires cause and effect reasoning, for example, in reviewing the dangers posed by electromagnetic fields that radiate from power lines. In 1995 the American Physical Society said public fears about the fields were "groundless" and diverted billions of dollars that might have been spent on more serious environmental problems. Physicists were skeptical of a link between power lines and cancer. Although the electricity flowing through such lines does produce magnetic fields that can penetrate the human body, the strength of those fields is low. A power line can radiate fields of 5 to 40 milligauss; by comparison, an electric can opener radiates about 280 milligauss at a distance of about 1 foot—and all fields drop sharply with distance. Moreover, people are constantly exposed to the earth's natural magnetic field of about 500 milligauss. Magnetic fields of 3 to 4 milligauss have been linked to childhood leukemia, but many of the studies reporting such linkages have been criticized as "ambiguous or flawed." Nonetheless, activist groups have forced utilities to move or shield power lines at an annual cost estimated at $1 to $3 billion that some scientists believe is misdirected (Broad 19).

## Persuading Through Induction

The power line example shows some of the challenges in inductive reasoning, especially when it is applied to matters of public policy. To be *persuasive,* you must select and arrange your argument effectively *in the reader's terms* and *from the reader's point of view.* Here are some guidelines.

**Analogies.** You can never *prove* anything with an analogy, but you can make your audience see a situation vividly and concretely. You can make something unfamiliar seem familiar, as in the comparison you just read between power lines and can openers as producers of radiation.

- To support her argument about the difficulty of measuring estrogen by-products that appear in tiny amounts in blood and urine, for example, one researcher noted that it is like finding a grain of sugar in a swimming pool or a particular drop in more than 13 million gallons of milk (Carnevale B1).

- To argue that humans are altering the landscape at a much faster pace than traditional agents like rivers and wind, an environmentalist notes that if all the rock and soil moved in such activities as construction and mining in the United States each year were dumped into the Grand Canyon, the Canyon would be filled in 400 years.

Analogies are effective when your audience finds the similarities relevant and significant and can't point out any dissimilarity that is even more important.

**Examples.** Analogies can help you persuade an audience to see how things that seem different are really similar, and thus they will accept your new idea because it is like a familiar one they already agree with. When you want to represent an entire sample, that is, to show the similarity in similar things, you use an example, as in the example of an "average-sized" home in Figure 11.1. Like analogies, examples may demonstrate your point although they can't prove it. A good example should be impressive and not easily invalidated by a counterexample.

Researchers have used examples to persuade government agencies that a good way to prevent crimes is to think of crime as a disease (an analogy) that can yield to techniques demonstrated as effective in treating diseases. Popular talk about crime as an "epidemic" only underlines the fact that epidemiological techniques are appropriate. Central to such techniques is the concept of a "tipping point." Epidemics are not linear phenomenon, with each additional effort producing a corresponding improvement. Instead, interventions have disproportionate consequences depending on when and how they occur. The following example helped persuade decision makers about the appropriateness of the tipping point theory. A Stanford University psychologist, Philip Zimbardo, parked a car on a street in Palo Alto. It remained untouched for a week. He had an identical car parked at the same time in a similar area of the Bronx, New York, but with its license plates removed and its hood up. The car was stripped within a day. Zimbardo then smashed a window on the car parked in Palo Alto. It, too, was stripped within a few hours. Zimbardo concluded that "disorder invites even more disorder—that a small deviation from the norm can set into motion a cascade of vandalism and criminality" (Gladwell 38). The tipping point was the broken window.

**Cause and Effect.** A pattern of cause and effect provides an effective way to organize a persuasive discussion. The following paragraph, for example, uses such a pattern to help support a transportation department's appeal to use an alternative to salt in winter road care:

Effect: potholes

Cause: salt; familiar-to-new sentence pattern

> Sodium chloride causes potholes in highways. Because concrete is porous, the salt penetrates the concrete until it reaches the reinforcing steel beneath. When the salt reaches the steel, the steel rusts, and the rust occupies a larger volume of space than the steel from which it came. The expanding rust then causes the concrete to crack and form potholes.

Figure 11.2 provides another example of persuasion through cause and effect reasoning.

**Inductive Organization.** Although you sometimes turn an inductive argument around when you present it to an audience, that is, you begin with the conclusion, one form of technical document does maintain a largely inductive order. That is the reporting format known by its initials, IMRAD, common in reports and articles on empirical research. In using this approach, you begin with an *introduction* that details the problem initiating your research; you then describe your *materials and methods,* that is, your procedure; next, you discuss what your procedure achieved or demonstrated, your *results,* usually in quantitative or visual terms. At the end, you provide your *discussion,* in which you use analogies, examples, and cause and effect reasoning to defend your hypothesis and persuade the audience to accept it as true.

**For more about IMRAD reports and articles, see Chapters 15 and 21**

## UNCOVERING ERRORS IN REASONING

When you use deduction or induction to prove or persuade, watch out for some common traps in such reasoning. These traps are *fallacies,* which are either false statements or false arguments. False statements refer to the subject of the argument; that is, they are untrue. False arguments are missteps as you assemble a line of reasoning in support of your point. Broadly speaking, all false arguments are *non sequiturs,* that is, generalizations or conclusions that "don't follow." Such lapses in a chain of reasoning reduce your

After nearly a week of wet weather, the snowpack is saturated with moisture. Temperatures dropped unexpectedly fast this morning, producing a crusty surface, particularly at the upper elevations.

**Effect: saturated snowpack**
**Another bad effect: crusty surface**

On the lower elevations, this crust is thin enough that it should break up with skier use. However, to ensure the best possible conditions, we are delaying opening this morning for 1/2 hour to permit grooming operations to start. We'll also be closing certain trails as the day progresses to permit grooming. You may expect these trails to re-open once they've been groomed.

**Balance of long-term good ("best possible conditions") with short-term inconvenience ("delaying opening")**

Because surface conditions this morning are less than what we'd like to offer, we invite you to clip today's ticket after skiing today, and bring it back any time prior to December 25. This ticket will provide you with a $10 discount on your next day of skiing.

**Goodwill compensation**

**FIGURE 11.2 Using Cause and Effect Reasoning to Persuade**
This notice was attached to the windshields of automobiles in a ski area parking lot. (Courtesy of the Sunday River Skiway Corporation.)

ELECTRONIC EDGE

## A Calculated Risk

Statistics play a powerful role in proving a point and persuading an audience. The expression "The numbers speak for themselves" attests to their power. Recent advances in information technology have greatly expanded our capacity for deriving and manipulating statistical evidence. One area of study that has benefited significantly from this capacity is the assessment and management of *risks,* an interdisciplinary field in which psychologists, engineers, public policy analysts, epidemiologists, and statisticians study the probability that something harmful will occur and the costs of measures to prevent or mitigate such occurrences. Studying risks depends to a large extent on computers and new quantitative techniques to record and analyze vast databases (Ross). Such studies allow people to assign numbers to possibilities that previously were only vaguely felt. Those numbers, in turn, "can embolden us to feel that we are masters of the universe. Or they can make us feel despondent under the burden of making correct decisions" (Ross 43–44).

The numbers help people make decisions about personal behavior. Several studies provide hard evidence of everyday risks. Based on a computer database of information on traffic fatalities, scientists can closely define the risks of driving a car. For example, the driver of a small car is 17 times more likely to die in a head-on crash than one in a larger car. If you wear a seat belt as you drive, you reduce your likelihood of dying in an accident by 42 percent. Similarly, large-scale public health studies prove correlations between smoking cigarettes and identifiable adverse health effects. The U.S. National Cancer Institute has calculated that one serious sunburn increases one's risk of cancer by as much as 50 percent. Over a lifetime, one in seven people in the United States could develop skin cancer from overexposure to the sun (Ross 47).

*Understanding* probabilities expressed in numbers can cause problems for nonexperts. So to make the risks more concrete, policy advocates may translate them from a ratio (for example, a risk of 1:1000) to a comparison with other risks. The risk of dying from exposure to asbestos, for example, is lower than the rate of death from swallowing a toothpick (Ross 44). To translate the probability of dying from an airline accident (1:7 million), one expert notes that a flier who boards a jet each day can expect to die every 19,000 years (Passell F3). Smoking one cigarette cuts 5 minutes off the smoker's life (Ross 46).

Even given the numbers, however, nonexperts often take more subjective and perhaps

credibility and help justify your audience's rejection of your ideas. If you can point out a fallacy in someone else's reasoning, however, that will help you defuse that argument. Here are some common ways that statements fail to "follow" one another.

### Faulty Generalization

You generalize falsely when you leap to a conclusion without adequate evidence. Such generalizations are a common problem in induction. Your

emotional approaches when they are asked to rate the risks they face. They tend to rate higher the risks they can't control (like natural disasters and nuclear power) than voluntary ones (like smoking). They also overestimate spectacular deaths (like airplane accidents) and underestimate common ones (heart disease, stroke). They underestimate "natural risks," like sunburn, which are easy to protect against. In addition, they build houses in areas which are prone to earthquakes, high tides, or flooding.

Assessing risks, then, is a matter of art as well as science. It frequently requires trade-offs among equally compelling numbers supporting different decisions, and information technology only increases both the *quantity* and the *quality* of those calculations. Beyond personal behavior, the numbers also serve a persuasive purpose in debates on public policy. Otherwise abstract questions like "How much is a life worth?" are yielding to precise—and highly debated—figures. "If seat belts cost, say, $50 per car," notes one source, "and equipping a million cars with seat belts will save 1,000 lives, the regulators must be assuming that lives are worth at least $50,000 a piece" (Passell F3). Considerations of cost, however, usually don't appear in safety regulations, so some policies are cost effective and some are extremely expensive. Required child immunization programs have been shown to be highly cost effective, less costly than treating the disease. On the other hand, many researchers suggest that expensive government programs to clean up toxic wastes may actually contribute to a shortening of life for many people when money for cleanups is taken from health care programs for the poor, like prenatal care and nutritional supplements. Requiring cars to be more fuel efficient to reduce air pollution causes manufacturers to build smaller and lighter cars that put people at greater risk of death in an accident.

New quantitative techniques also make some regulations obsolete. For example, new instruments can measure the presence of various chemicals in foodstuffs down to parts per trillion. Laws from the 1950s forbade any chemical that caused tumors in lab animals from being used in processed food. Scientists can now find traces of chemicals that would not have shown up with 1950s technology. One expert notes that coffee contains 1,000 natural chemicals of which only 26 have been tested and 19 of those produce cancer in laboratory animals; if it were a synthetic, coffee would be banned under current U.S. law (Ross 48).

observations may be irrelevant, unrepresentative, or not numerous enough. You may oversimplify or you may fail to incorporate cultural differences in your generalization.

- "E-mail makes life easy." Many people find it easy to compose and send e-mail messages, but much current research also points to the stress that an abundance of such messages places on readers, the not-so-subtle pressure to respond to each message and to respond rapidly, the lack of civility that such rapid responses often demonstrate, and the like. The generalization doesn't stand up to scrutiny.

- "When in Rome, do as the Romans do." Perhaps. But this generalization, interpreted to mean that you conform to the customs prevailing where you work, could mean seriously compromising ethical standards if, for example, the local culture endorses child labor and your company does not.

### Implied Assumption

You often take things for granted in a discussion, for example, that being healthy is better than being sick, and thus people would want to behave in ways that promote health. But sometimes what you take for granted is false, dubious, or unverified, and that false assumption misleads the reader. Consider the following:

- "Because this is a modern, sealed, energy-efficient building, it is the best environment in which to work." The implied assumption here is that energy conservation creates good working conditions. A counterargument, however, might point to evidence that tight buildings with inadequate ventilation trap noxious fumes that may cause workers to experience nausea, fatigue, and dizziness, a condition now known as "sick building syndrome."
- "The neat, orderly world of the past is gone, and traditional leadership approaches are inadequate." The implied assumption is that the past was "neat and orderly," a grand generalization, of course, and something that many readers would not see as true. As a writer, you would have to first establish the changed conditions that now require new "leadership approaches."
- A picture in the newsletter of a large chemical company showed an executive unrolling before him a 20-foot-long chart detailing the steps required for one plant to meet the standards of just one federal agency. The implied assumption: If regulations take so much paper, they must be excessive. Someone taking a different perspective might point out, reasonably, that the length of the flow chart of the regulatory process is not necessarily proportionate or even related to its appropriateness.

### Begging the Question

Another form of implied assumption is common enough to have its own name: begging the question. In doing so, you ignore a specific question that is at issue in the discussion and thus accept as fact something that needs proof. For example, if you ask, "Do you read your e-mail more frequently during the week or on the weekend?" you beg the question of reading e-mail: The listener may not normally read e-mail at all.

- "During the seven-day trapping period, nine Microtia were captured, one in a live trap and eight in the snap traps. With a ratio of eight to one, it is concluded that snap traps are more effective and easier to use than live traps in the capture of small mammals." The evidence helps support "more effective" but begs the question concerning ease of use.

- "I worked 15 hours on this report, did five revisions, and got a C. She tossed in her first draft and got an A. Your grading isn't fair." The conclusion begs the question of quality in the final product. The grade represents what the work accomplished, not how long the work took.

### Either/Or

Your conclusion also "doesn't follow" if you see in a situation only one alternative when there are none—or more than two, as in the following examples:

- "Books, not bricks." Angered by plans for a series of brick walkways they saw as demonstrating an overemphasis on the physical dimensions of the campus, faculty argued that the money budgeted for more walkways should be spent instead on books for the library. The money, however, wasn't available for books because it had been earmarked for capital improvements and helped keep local contractors busy during an economic slump.
- "To solve our need to conserve energy, we must have innovation and imagination from architects, not legislation from Washington." This statement sets two alternatives against one another. In practice, energy conservation requires a combination of imagination and enabling legislation.
- "If you're not at your desk, you're not working." This statement first implies that sitting at a desk equals working, a false assumption (some people daydream at desks). In addition, not every job requires sitting at a desk.
- "If you don't care what you look like, you don't care what work you do. This lack of good dress shows a careless approach to everything." If you are scruffily dressed as you read this book—and you are still reading it attentively and profitably—you've invalidated this argument.

### One-Valued Thinking

When your thinking is based on only one set of values, you fail to incorporate how people from different cultures or different value systems might see things differently. For example:

- *Customs.* To archaeologists, Native American artifacts found buried in the southwestern United States are items for study, collection, and display in museums. To the Native Americans, the artifacts are sacred connections to their ancestors that should remain where they are.
- *Observations.* As observers of nature, native populations often see things that nonnative scientists miss on their brief visits, for example, peculiarities in the mating cycles or life cycles of marine mammals in the Bering Sea.

### Faulty Causality

When you try to prove that something caused something else, be aware of several potential problems. For example, you may be tempted to say that something that precedes something else causes it, when the relationship is only one in time. The fallacy is so common that it has a name (derived from the Latin): *post hoc ergo propter hoc*—"after this, therefore because of this." You stayed out late Tuesday night; you failed Wednesday's exam. The late night might have caused the failure—or the real cause might have been a failure to study or to keep current on readings. Economists have to guard carefully against this trap when they measure how monetary and other policies affect economic conditions. Here are some other examples:

- You may feel sleepy after you eat a big lunch and blame the lunch for your lull. But experiments with people who didn't eat lunch, and still experienced a lull, show the relationship to be simply chronological.
- You take some medicine. Your symptoms disappear. It's easy, then, to think the medicine caused the relief. Under controlled circumstances, however, it may be shown that the medicine is not the cause of the relief but some by-product or other activity is—like lying down after taking the medicine.

Another trap to avoid in determining causes is mistaking a cause for an effect. Several scientists, for example, have correlated heavy smoking and drinking with heart disease. A recent study, however, suggests that both the smoking and drinking and the heart disease may be effects of yet another cause: early alienation of the subject from his or her family. The resulting tensions lead both to drinking and smoking and to hormonal changes that reduce the effectiveness of the body's immune system.

### Faulty Comparison

You've seen that analogies, examples, and other forms of comparison can help create an effective argument, but you mislead the reader when your comparison hides more significant differences, as in the following example:

- "More people would rather go fishing than get married. A 1999 survey revealed that the American public bought 90 million fishing licenses but only 9 million marriage licenses. Fishing is thus ten times more popular than marriage."

Differences in the licensing procedures, the valid duration of the respective licenses, and the nature and purpose of the two populations (those who fish and those who marry) mean this conclusion is false, if still amusing.

### Misuse of Numbers

Statistics lend themselves to manipulation in support of almost any position. As a technical communicator you have a special obligation to be honest with numbers as well as words (Huff and Geis). Make sure, for example, that your numbers add up. As one expert notes, "If you add up last year's reported imports and exports for all of the countries in the world, world imports exceeded world exports by more than $100 billion." The conclusion (made with tongue in cheek): "We are running a huge global deficit in our interplanetary trade. Space aliens are stealing American jobs" (Krugman).

Also look for hidden (and more significant) extremes in evidence presented as an average:

- "The average passenger originating at the Portland Jetport takes four trips a year." This statement suggests that Portland, Maine, is populated by travelers. More significant, however, are the extremes: It could be that 80 percent of the travelers take only one trip a year and 20 percent take eight or more.

The numbers that present the results of empirical research often express estimates or extrapolations researchers make while leaping from a sample of observations to a conclusion about all instances. When those numbers also become part of a debate about public policy and spending money, then advocacy groups sometimes exaggerate to gain attention and present as fact something that is really a guess. Health statistics can be particularly alarming, especially in the popular press, which may report spectacular findings (a journalistic maxim is "If it bleeds, it leads") without appropriate qualifications about sample size, method of taking the sample, and the like. Some projections are wrong because they are too linear when the phenomena they are based on are not linear. Some, derived from self-reporting, don't correct for people's reluctance to talk about sensitive topics like drug use. Some fail to account for such social changes as better testing for a disease that can explain what seems to be a new epidemic. Some derive from outdated statistics. In addition, because life expectancy in the United States this century has gone from 48 to 73 years for men and from 51 to 80 years for women, people are dying from diseases they would not have lived to develop earlier. One projection that 4 million Americans have Alzheimer's disease, which seems frightening, was based on a study of only one community, too small a sample. In addition, practitioners differ in their definition of Alzheimer's disease, it isn't easy to diagnose in routine exams, and it progresses gradually (Crossen B1).

The risks of disease, too, can often be seen as less bleak, even negligible, when compared to other risks considered normal (see *Electronic Edge: A Calculated Risk*). For example,

> The radioactive emissions from nuclear power plants online in 2050 would increase the number of cancer deaths by 8.7 per year.

As stated, this picture seems bleak. But compared with other risks considered normal, the hazard turns out to be negligible. A scientist has concluded that, on an individual basis, equivalent risks include being a fraction of an ounce overweight, smoking 0.03 cigarette per year, and driving 1 mile a year. To deal with numbers honestly, you must present the broad picture.

## THE ART OF PERSUASION

Although deduction and induction help you structure your writing so you can establish your authority, prove your point, and persuade your reader, we should round out this discussion by noting that logic and good character alone are often not enough to convince an audience. When scientists talk about the "elegance" of a theory, they are describing a quality that is more than the mere accumulation of evidence. The aesthetics of the argument may be as persuasive as its logic. When several researchers simultaneously arrive at the same explanation of an occurrence, the one who reports it most effectively often is credited with the discovery. So you need to think well and then express that thinking in appropriately attractive expression. The guidelines in this chapter will help you create an effective presentation of your ideas, and the beauty of that written argument may be its most convincing feature.

For more about explaining, see Chapter 10

### CHECKLIST: Persuading and Proving

1. **Determine what it will take to persuade your audience**
   - Identify the values that underlie the situation
   - Build common ground if your values and your audience's differ
   - Establish your authority
   - Acknowledge international differences in the process of persuasion
2. **Use the Aristotelian appeal to logic as a strategy for persuading by proving**
   - Know when your audience will require proof of your assertions
   - Use deduction to derive good reasons through a series of statements
   - Use induction to argue from observations
   - Arrange your argument to meet your audience's expectations, either general to particular or particular to general
3. **Avoid errors in your own reasoning**
4. **Use your knowledge of such errors to identify weaknesses in other people's reasoning**

### EXERCISES

1. Scientific journals as well as more popular accounts provide discussions of many controversial issues. Such discussions often have persuasive purposes. They may aim to convince readers that something is true, to gain the readers' compliance with a particular point of view, or to encourage readers to take a particular decision or action. Examine at least three articles on one of the following controversial subjects (or another subject of your choosing). Summarize the arguments in a review of literature (see Chapter 21):
   - Potential health risks from genetically engineered fruits and vegetables
   - Global warming
   - Infant formula use in developing countries

- Ethical issues in human genetic engineering and cloning
- Competing demands of environmentalists, commercial fishermen, and sport fishery in waters being depleted of fish
- Priorities for development in areas with endangered species of wildlife
- Privacy on the Internet

2. As the result of a survey, researchers concluded that college-educated people in the United States earn 57 percent more over their lifetimes than those without a college degree. In a class discussion, develop some causes for that effect. For example, a global economy means that low-skilled workers in the United States are competing with a worldwide work force willing to work for less. New technology, too, like robots and automatic teller machines, reduces the need for less skilled workers and thus depresses wages. Similarly, develop reasons to explain the results of a national driving center survey, which shows that a person with a master's degree is 28 times more likely to fall asleep while driving than someone with only a high school education.

3. Collect examples of the errors in reasoning described in the chapter. Hint: Advertisements for technical products often provide a good source. Look for visual errors as well as verbal ones.

4. What assumptions are implied in the following statements?

   a. People who like their work ought not to get good salaries.
   b. *Environment* is a broad term that embraces problems of air and water pollution, birth control, urban transportation, and everything that affects or threatens human survival.
   c. We can no longer afford the luxury of designing for conditions that do not exist.

5. Internet discussion groups and chat rooms often provide forums for argumentation. If you belong to such a discussion, observe closely how one "thread" or theme is developed. How do the various discussants establish their credibility? What values are taken for granted in the group? What are the key issues? What assertions need to be proven? What kinds of statements or observations provide the proof? Write a brief e-mail message to your instructor answering these questions and citing particular instances from the electronic debate.

6. Collect syllabi from at least three classes in which you have been enrolled. In addition to describing the course, a syllabus has a persuasive aim—to get you to act in the way the instructor feels is right. A syllabus establishes the authority and character of the teacher as it demonstrates logical reasoning. Write a memo to your instructor in which you analyze the persuasive and argumentative strategies in the three syllabi. One approach: Briefly describe your criteria for a good syllabus and measure your collection against them. Are the syllabi useful? Complete? Misleading in any way? Another approach: Determine how the instructor asserts his or her authority and characterize the instructor's persona (see Chapter 7). Does the syllabus show an instructor who is friendly? Intimidating? Describe how those attributes are demonstrated: word choice, choice of details, length of the syllabus, format and design, typeface, and the like.

## FOR COLLABORATION

In a team of three, find statistical data on some topic of interest. *The Statistical Abstract of the United States,* published by the Bureau of the Census, is one good source. Photocopy the statistics; then create a visual, on an overhead, that develops a comparison among the statistics or shows trends (see Chapter 8). Show the visual to the class and discuss why you decided on that particular visual form. Note any limits or discrepancies in the original data. Be careful to avoid the misuses of numbers you read about in this chapter.

# CHAPTER 12

# REVISING



"GOOD WRITING SHOULD PROCEED REGULARLY FROM THINGS KNOWN TO THINGS UNKNOWN, DISTINCTLY AND CLEARLY WITHOUT CONFUSION. THE WORDS USED SHOULD BE THE MOST EXPRESSIVE THAT THE LANGUAGE AFFORDS, PROVIDED THAT THEY ARE THE MOST GENERALLY UNDERSTOOD. NOTHING SHOULD BE EXPRESSED IN TWO WORDS THAT CAN BE AS WELL EXPRESSED IN ONE; THAT IS, NO SYNONYMS SHOULD BE USED, OR VERY RARELY, BUT THE WHOLE SHOULD BE AS SHORT AS POSSIBLE, CONSISTENT WITH CLEARNESS; THE WORDS SHOULD BE SO PLACED AS TO BE AGREEABLE TO THE EAR IN READING; SUMMARILY IT SHOULD BE SMOOTH, CLEAR, AND SHORT, FOR THE CONTRARY QUALITIES ARE DISPLEASING."

**BENJAMIN FRANKLIN, 2 AUGUST 1733**

As you manage and design your information, and especially when you're ready to release a document or invite readers into your Web site, check to make sure you've created the best product for the context. Is your document or site customer oriented? This chapter reviews some strategies for looking back on your design with the reader in mind, a process called "revising" or "editing" or "testing."

## REVISING A DOCUMENT

When you look back at a document, do so in stages, checking on big issues before little ones.

### Starting with the Big Picture

First, take care of the big picture:

- What is the main point?
- Is it the *right* point?
- Is it obvious and clear?
- Have you met the criteria of the assignment?
- Is the information accurate and complete?

Check the content. Your document won't work if it misses the point, fails to be honest, addresses the wrong reader, or addresses the right reader at the wrong time. In checking content:

- Place major statements in prominent positions, in headings or the lead sentences of paragraphs.
- Check that restatements in the abstract, introduction, and conclusion match. Avoid contradicting yourself.
- Provide explanatory background if your main point will cause difficulty for the reader because, for example, it's complex or unexpected.
- Identify places where you need to extend the discussion of main points with subpoints and definitions.
- Look for ways to shorten the text by deleting unnecessary specifics.
- Eliminate statements that raise unnecessary questions.
- If you anticipate resistance to your main point, show the reader explicitly how the subject at hand affects her or him.

Verify the accuracy and completeness of supporting information or note explicitly any gaps or problems. In addition to being inaccurate, your information is wrong if it is

- Information the reader didn't want.
- Not current if the reader needs current information.
- Scattered without emphasis or divisions.

### Reviewing the Design

For more on design as well as composing, see Chapters 7, 8, and 9

The big picture check helps assure the document's *effectiveness:* It does the right thing. A second check centers on *efficiency:* doing things right. An efficient document is one that is well designed; it helps the reader find, understand, and remember your message. It clearly establishes both the hierarchy of main and sub-points and the parallelism of equal items. To review the design, first check the headings:

- Does all information under one heading really belong there?
  - Is the relative emphasis indicated by the headings appropriate?
  - Are there enough headings to be meaningful?
  - Does an excess of headings dissolve the emphasis and continuity?
- If you have more than three levels of heading, how can you reorganize to reduce the levels?
- Are all headings parallel?
- Are there any single subheads? When you divide a unit, you must create at least two subunits.

Then, review the other units of discussion:

- Sort familiar from new information and begin with the familiar.
- Follow the format of the appropriate genre (for example, proposal or memo or letter).
- Summarize periodically in a long report and forecast the next section.
- Divide long paragraphs into smaller ones.
- Review the layout and typeface for ease of reading.
- Check that pages are numbered consecutively.

## Revising Visuals

Figure 12.1 provides guidelines for editing visuals within a technical or scientific document. In addition, check that visuals are well placed to break up units of text, and reread your text to see if any information presented there might be better presented in visual form.

## Revising Text

Edit your text to make sure your expression is correct and your voice as it comes off the page is a welcoming one. Figure 12.2 shows how the guidelines you have just read can help turn a weak draft into effective text. Correct any errors and inconsistencies in word choice, spelling, sentence structure, and punctuation.

How much revising should you do? The answer reflects your own writing process, the importance of the document you are writing, and the demands of the reader. If you knew your point and planned well in advance, your draft may require little editing. If you used your draft as a way to discover your ideas, you may need to spend more time changing around your expression and visuals. In addition, not all documents require the same polish. A simple e-mail message you send to a colleague may contain grammatical problems and spelling errors but still do its job. Such lapses in a

- Place a visual near where you discuss it.
- If you discuss a visual in several places, then include it in an appendix, perhaps as a foldout.
- Avoid unnecessary redundancy between a figure and a table or between a photograph and a schematic diagram.
- Make the data in a table or figure agree exactly with numbers in the text. If the cost of a part changed between when you wrote the draft and when you created the figure, rectify the change in both.
- Use terms consistently between figures and text.
- Give every visual a number (or letter) and label.
- Mention every visual in the text—don't let any enter unannounced.
- Keep a separate list of visuals as you draft, and check and renumber as necessary.
- Make sure labels read in the same direction (generally horizontal to the page).
- Connect labels and the parts they identify with arrows or lines that are clear but not intrusive.
- Indicate clearly with footnotes or references the source of all information you did not observe or derive yourself.
- Set up all tables or figures that are part of a series in the same form. The rules of parallelism that apply to language apply also to visuals.
- Check the arithmetic. Make sure, for example, that columns in a percentage table add up to 100 percent if they ought to, and explain if they don't.
- Clearly differentiate between actual numbers and percentages.

FIGURE 12.1 **Guidelines for Editing Visuals**

For more about typeface and design, see Chapter 7

proposal to a major client, however, or in a job application letter, could eliminate you from the competition.

## REVISITING A WEB SITE

The term *revisiting* suggests some additional checks on a Web site beyond those appropriate to any document.

1. Is it *fast?* Users expect speed, even users who might be more patient with print. With one hand or finger on a mouse, users are ready to click on something else if your site doesn't give them something to engage in immediately. Give them something, perhaps text while graphics load, or hints about the graphics. A site that takes too long to load won't appeal.
2. Is it *cool?* Hardworking sites that provide essential scientific and technical information may escape the cool test, but most others won't.

**Before Revision**

The Comprehensive Environmental Response Compensation and Liability Act Information System (CERCLIS) is the United States Environmental Protection Agency's (EPA) inventory of potential hazardous substance release sites. The CERCLIS list for Cumberland County maintained by the EPA, Region 6, updated February 12, 2001, indicates that the subject site has not been investigated as a potential uncontrolled hazardous substance release site.

**Begins with new, rather than familiar information**

**Familiar information—the subject site—buried, along with the main point about it**

Based upon our review of CERCLIS information and our conversation with Paul Jones from the state Department of Natural Resources and Environmental Control four CERCLIS sites appear to occur within about a 5-mile radius of the site. The name, street address and status of these four CERCLIS sites are as follows:

**Comma needed after long introductory phrase**

| Site | Address | Status |
|---|---|---|
| Dower's farm | 12 Pine St. | Preliminary assessment performed; no further remedial action planned |
| Bridge St. | 20-40 Bridge St. | Preliminary assessment performed; no further remedial action planned |
| Mill dump | Spencer St. | Preliminary assessment performed; no further remedial action planned |
| Olan Rd. | Olan and Canal Sts. | Preliminary assessment performed;<br>no further remedial action |

**Status column in table is redundant and wastes reader's time and document space**

None of these four sites have been listed on the National Priority List (NPL) of Superfund sites.

***None* takes a singular verb**

**After Revision**

The 15-20 Flower Ave. site has not been investigated as a potential uncontrolled hazardous substance release site. The standard inventory of such sites is the Comprehensive Environmental Response Compensation and Liability Act Information System (CERCLIS) of the United States Environmental Protection Agency (EPA). The CERCLIS list for Cumberland County (maintained by the EPA, Region 6, and updated February 12, 2001) does not include

**Begins with familiar information, fully identified**

**New information woven in where reader is ready for it**

**Nice given/new order to connect sentences (CERCLIS . . . CERCLIS)**

FIGURE 12.2 **Applying the Guidelines for Revision**
The revised text is clearer and easier to read.

the Flower Ave. site. A review of CERCLIS information and a conversation with Paul Jones of the state department of environmental resources, however, indicate that four such sites may occur within about a 5 mile radius of the site:

Brief informal list with addresses

| Site name | Address |
|---|---|
| Dower's farm | 12 Pine St. |
| Bridge St. | 20-40 Bridge St. |
| Mill dump | Spencer St. |
| Olan Rd | Olan and Canal Sts. |

Summarizes status of other sites for emphasis and ease of reading

After performing a preliminary assessment of these four sites, the state plans no further remedial action at any one of them. None is listed on the National Priority List (NPL) of Superfund sites.

FIGURE 12.2 **Applying the Guidelines for Revision (continued)**

Commercial sites have set a high level of design, a design that mixes entertainment with information. In that context, if you want to attract attention, you need to imitate some of this commercial approach, especially in your use of visuals and graphics. Use an appropriate contrast between text and background. Use a readable typeface accessible to any browser.

For more about branding, see Chapter 7

3. Is it *branded?* Is the site owner's persona clear and consistent through every page?
4. Is it *interactive?* Readers create their own text from your text, even more than with a print document. They may come to your site via a link from another site and thus lack the context of your home page. Help them find their way around with navigational tools. Help them get back to your home page—from any page.
5. Is it *richly linked?* Are links easy to identify (by repeated highlighting color, for example)? Are there enough links? Of the appropriate kind? With appropriate one-sentence tag line descriptions? Are the links accurate and up to date? Are there places needing more links? Or, conversely, have you sent too many visitors away to links that won't bring them back to your page? And is that a problem?
6. Does it have *appropriate titles and keywords?* These enable search engines to bring interested users to your site.
7. Is it *divided into modules?* Viewers scan more than they read, and they generally don't like to scroll. Divide the site into small units of text and visuals, preferably one screen for each unit. If you go beyond one screen, show the visitor clearly how to get back home.

ELECTRONIC EDGE

## Electronic Style Guides

A highly specialized form of electronic style guide is a *structured document processor* (SDP). Built into a word processing system, it prompts you to insert the appropriate information at the appropriate location in such rule-bound documents as catalogs, price books, procedures, or regulations. In doing so, it

- Reduces the decisions you need to make about structure, style, and format.
- Prevents you from placing certain information wrongly.
- Automatically formats the document.
- Stores reviewers' annotations online.

The system can also generate routine documents automatically by, for example, searching through a database. The SDP can be networked so that multiple writers can assemble one unified document (Norman and Grider).

SDPs are essentially electronic strategies for creating traditional documents, and they work best within one company or organization. Other strategies are being developed to foster the creation of new electronic message forms and formats. The core strategy is an international language for encoding electronic text. The language moves across various forms of hardware and software, that is, across various "platforms." One such language is Standard Generalized Markup Language (SGML). When you use SGML, you create a message that is open ended. Readers can print it, store it, and manipulate it in their own system. The Association of American Publishers, for example, created an SGML standard for books and serials that several publishers in the United States, the Netherlands, and Germany have adopted. The Society of Automotive Engineers has also adopted SGML as its standard for the delivery of emissions data worldwide as well as shop manuals. The U.S. Securities and Exchange Commission uses SGML in its Electronic Data Gathering, Analysis, and Retrieval (EDGAR) project that accommodates the electronic filing of financial information (Gilmore).

An enhanced markup language that is even more flexible and accommodates communication across the Internet is Hypertext Markup Language (HTML), the standard for the Web. HTML does limit your options in text display. You may, for example, like to use letters or Roman numerals in an ordered list, but the HTML standard for "ordered list" is Arabic numerals, and that's how your text will appear. The more important advantage of HTML, however, is its ability to go beyond single documents. With HTML, you can break out of the concept of a document to create hypertext links to information available at your own site and throughout the Web. Those links depend on a system of "anchors," known as Uniform (or Universal) Resource Locators (URLs), that embed the access method and address of the computer where the information being linked to is located.

8. Does it *prioritize?* Does the physical placement of items on the home page and on each screen reflect the priority you place on the information displayed?
9. *Can readers reach you?* Provide an e-mail link that allows visitors to reach you with comments or requests.

## MEETING STANDARDS

As both a student and a technical professional, you'll have some important help in revising. That help is standards that apply to scientific and technical documents. Whenever you write a document, find out from the audience—for example, your professor—which guidelines apply, and follow those dictates exactly. Technical professionals often write *to spec,* that is, within clear-cut specifications for content, format, and style.

Project teams, corporations, professional associations, publishers, professors—all establish guidelines for writing. Such guidelines circulate in both internal documents ("style sheets" and "house standards," as in Figure 12.3) and published compilations. Figure 12.4 lists some frequently

*Legal issues.* The use of company name and trademarks as well as citing other companies' trademarks

*Publishing protocol.* Guidelines for several types of publications, including technical papers, journal/magazine articles, brochures, pamphlets, press releases, advertisements, and other documents

*Company logo.* Proper use of the logo in various media

*Document design.* Packaging, page size, margins, white space, typeface, type size, column width, leading, justification, heads and subheads, color, photos and artwork, "final touches"

*Document organization.* Front matter, body, and back matter

*Writing and style.* Grammar, spelling, word choice and usage, punctuation, capitalization, abbreviations, numbers, mathematics, lists

*35 mm slides.* Planning the design, preparing the text, using graphic elements, applying color, numbering

*Overhead transparencies.* Planning the design, preparing the text, sample transparency with graphic, sample transparency with text

*Appendices.* Service marks, preferred spellings, units of measurement abbreviations, symbols, subscripts, and the like

FIGURE 12.3 **Sections of an In-House Communications Handbook**
When you write within an organization, you'll need to check what you say against corporate standards. This list suggests the extent of those standards.

(Courtesy of Barbara Harrison. Reprinted by permission of Barbara Harrison and The Wordmaster, Inc.)

American Chemical Society. *www.acs.org. The ACS Style Guide.* 2nd ed. 1997. New York: Oxford University Press. Also see guidelines for each ACS journal at the Web site.

American Institute of Physics. *www.aip.org. AIP Style Manual.* 4th ed. New York: The Institute, 1990.

American Mathematical Society. *A Manual for Authors of Mathematical Papers.* Rev. ed. Providence: The Society, 1990.

American Medical Association. *www.ama-assn.org. AMA Manual of Style.* 9th ed.

American National Standards Institute. *Electronic Manuscript Preparation and Markup: American National Standards for Electronic Manuscript Preparation and Markup.*

American Psychological Association. *www.apa.org. Publication Manual of the American Psychological Association.* 4th ed. Washington, D.C.: The Association, 1994. See also Harold Gelfand and Charles J. Walker, *Mastering APA Style: Student's Workbook and Training Guide.* 1990. The Association.

Butcher, Judith. *Copy-Editing: The Cambridge Handbook for Editors, Authors and Publishers.* 3rd ed., 1992. Cambridge: Cambridge UP.

*The Chicago Manual of Style.* 14th ed. Chicago: U of Chicago P, 1993.

Council of Biology Editors. *Scientific Style and Format: The CBE Manual for Authors, Editors, and Publishers.* 6th ed. Cambridge, Cambridge UP, 1994. In 2000 the council changed its name to the Council of Science Editors. *www.councilscienceeditors.org.*

Institute of Electrical and Electronics Engineers. *Information for Authors.* New York: The Institute, 1966.

International Committee of Medical Journal Editors. *Uniform Requirements for Manuscripts Submitted to Biomedical Journals.* Philadelphia: American College of Physicians, 1993.

Li, Xia and Nancy B. Crane. *Electronic Styles: A Handbook for Citing Electronic Information.* Medford: Information Today, 1996, *http://www.uvm.edu/~ncrane/estyles.*

Modern Language Association. *www.mla.org.* Joseph Gibaldi, *MLA Handbook for Writers of Research Papers.* 5th ed. See also Joseph Gibaldi, *MLA Style Manual and Guide to Scholarly Publishing.* 2nd ed. 1998.

Microsoft. *The Microsoft Manual of Style for Technical Publications.* 2nd ed. 1998 (in print and on CD-Rom).

United States Geological Survey. *Suggestions to Authors of the Reports of the United States Geological Survey.* 7th ed. Washington, D.C.: U.S. Government Printing Office, 1991.

U.S. Government Printing Office. *Style Manual.* Washington, D.C.: U.S. Government Printing Office, 1973.

Walker, Janice R. and Todd Taylor, *The Columbia Guide to Online Style.* New York: Columbia UP, 1998. *www.columbia.edu/cu/cup/cgos/index.html.*

Xerox Corporation. *Xerox Publishing Standards. A Manual of Style and Design.* New York: Watson-Guptill, 1988.

FIGURE 12.4 **Selected Style Guides for Writing in Scientific and Technical Fields**

used style guides produced by a range of different organizations and disciplines. When you write, and when you revise, check for conformance with organizational policy concerning copyright and permissions, release of proprietary information, advocacy of particular positions, endorsement of particular products, and citation of sources used in preparing your work. One problem with standards and recipes is the constraint they may place on your creativity. If the specifications are heavy-handed, they may flatten the style of any documents that conform to them and keep the texts from reflecting changing conditions. But these are risks in any form of communication. Following the recipe means you have less to think about when you prepare routine documents.

## PREPARING A DOCUMENT FOR TRANSLATION

Most documents you write will require some degree of revising. In addition, you'll find that some documents will also need to pass other tests. One such test is translation, either before the document reaches readers or by the readers themselves. A relatively simple grammar and easy acceptance of new terms favored the adoption of English as an international language of science and technology. But when you use English for an international audience, be aware that non-native speakers may not read or hear the same meanings you assign to terms. Be aware, too, that people in different cultures adhere to different conventions in the way they express dates, numbers, quantities, and symbols, as *Crossing Cultures: Meeting the Conventions* suggests. They also use different approaches to punctuation and style—even within English. International standards, especially those being developed in the European Union, are encouraging greater cross-cultural uniformity. But some differences remain, and some confusion. The confusion can have disastrous results; investigators point to a mixup between English and metric units in navigational data as the main reason the *Mars Climate Orbiter* went off course in 1999 and probably destroyed itself.

To avoid confusion in language, start by making sure your English is excellent. In addition, familiarize yourself with both the language and the cultural context of your readers. Find out, for example, the level of literacy and the style of teaching and learning common in the country. Review how they count, measure, and express numbers; how they sort lists (the 26-letter alphabet from A to Z is not universal); in what direction they read; how they address people and write postal addresses and phone numbers; and the like. Accommodate the technological sophistication of the country. Electronic delivery is inappropriate, for example, where electrical service is spotty, phone lines few or exceedingly expensive, and computers rare. For that context, plan on paper text. In addition, you often have to adjust the image area of documents created in North America so they can be printed on the size of paper common elsewhere.

Consider the cultural predispositions of your reader. For example, the amount of detail you should supply varies from culture to culture. One French technical writer notes, "Consider the reader is intelligent. Give fine details only when necessary (the 'step by step' syndrome is boring for the reader; here in France, American documentation is sometimes considered laughable because of the incredible details)" (Borghi). Given the context of litigation and liability in the United States, however, you may be required to include details.

These considerations are especially important when your document will be translated (Bell). To ensure an accurate translation

- Prepare a glossary of technical terms, their specialized meanings, and equivalents in the target language.
- Write out any acronym or abbreviation at its first use.
- Watch pronouns. If what they stand for in the original English is unclear, that ambiguity will probably lead to mistranslation.
- Use short sentences—the written equivalent of speaking slowly.
- Use active verbs and straightforward subject/verb/object order.
- Avoid dependent clauses and compressed constructions like strings of modifiers before nouns.
- Reduce the amount of text, perhaps by shifting some discussion to visuals.
- Structure documents in small modules to give the non-native speaker or translator breathing room.
- Avoid casual side comments.
- Be formal in your level of language and approach to the reader.

In addition, "Think like a stateless person" (Borghi):

- Avoid culture-bound references to sports teams, food items, political debates, and other current events.
- Avoid humor, especially wordplay, which often causes problems in translation. For example, a London employment agency's logo is a stylized bull's eye and the words "Aim Hire." Understanding that design requires you to understand more than those two words; you need to know about guns (or maybe darts—or arrows) and shooting ranges; about the phrase "ready, aim, fire"; about the sound of "hire," which could be heard as "higher"; and about the concept of "aiming high" to reach some goal. Translating that concept would be difficult.
- Allow room for text to expand in translation. English words are often shorter than their equivalents in other languages and they can often express ideas more concisely, so design your documents to allow at least 20 to 30 percent more room for a translated text.
- Provide enough white space on each page so you can keep page design and page numbers consistent across several translated versions of one document.

CROSSING CULTURES

## Meeting the Conventions

People in different cultures adhere to different conventions in the way they express dates, numbers, quantities, and symbols. They also use different approaches to punctuation and style—even within English. International standards, especially those being developed in the European Union, are encouraging greater cross-cultural uniformity. But some differences (and a bit of confusion) remain. Here's a brief review (Miller 96).

**Time**

| | |
|---|---|
| United States | 9:22 P.M |
| United Kingdom | 21.22 |
| Switzerland | 21,22 |
| Germany | 21.22 |
| Computer systems | 21:22:00 |

Most computers also indicate relationship to Greenwich Mean Time (GMT); for example, Germany is +001 and the eastern United States is −005.

**Date**

| | |
|---|---|
| United States | October 11, 1997 or 10/11/97 |
| United Kingdom | 11th October 1997 or 11/10/97 |
| On the continent | 11.10.97 |
| International standard | 1997-10-11 |

**Calendar**

| | |
|---|---|
| United States and Europe | Gregorian |
| Israel | Gregorian and Hebrew |
| Arabic countries | Gregorian and Islamic |
| Japan | Gregorian or by Japanese imperial era |

**Numbers**

| | |
|---|---|
| United States | 4,000.20 |
| United Kingdom | 4,000.20 in general usage; 4000.20 in scientific |
| France | 4 000,20 |
| Germany | 4.000,20 |

**Currency Symbols**

| | |
|---|---|
| United States | $1,000.30 |
| United Kingdom | £1,000.30 |
| Germany | 1000,30 DM |
| Euro | €1000,30 |

Most international standards place a three-letter code for the currency unit before the number, as in USD 1000.30 (U.S. dollar); GBP 1000.30 (Great Britain pound).

Translating text is expensive, and thus many companies emphasize that you should write in an English that can be widely understood. Consider using visuals to convey essential information, especially visuals that do not include words. Another technique is to adopt a layered approach, particularly for large online documentation systems. SAP AG, a German-based international software company that produces a popular integrated business system, uses such an approach (Elliott). Based on a concept of "information layers," the approach aims at minimal translation of only the top

**Systems of Measurement**

| | |
|---|---|
| Everywhere but the United States | Metric distances in kilometers and centimeters (except the United Kingdom, which uses miles)<br>Temperatures in degrees Celsius |

**Paper**

| | |
|---|---|
| United States | Letter (8½″ × 11″) and legal (8½″ × 13″) |
| Europe | A1 (1 square meter); the most common is A4 (210 mm × 297 mm); also, B sizes |
| Japan | JIS-B4 and JIS-B5 |

**Addresses**

| | |
|---|---|
| United States | House or office number comes before the street (15 Broadway) |
| Germany | Street name is usually first (Bergstrasse 12). |
| France | House number is followed by a comma (49, rue de St. Paul) |
| Switzerland | Office or house number (sometimes with an indication of its level—"bis," for example, for "lower") comes before the street, and the postal code before the city:<br>2 bis, rue Ad-Fontanel<br>CH-1227 Carouge<br>Switzerland |
| China | City name is the first line, then the street, then the individual |

**Punctuation (a few examples)**

| | |
|---|---|
| United Kingdom | No periods with US, UK, Dr, Ms, eg, ie |
| United States | Quotation marks are single marks inside double |
| United Kingdom | Quotation marks are double marks inside single |

**Sorting Order**

| | |
|---|---|
| United States | A to Z |
| Denmark | includes letters after Z |
| Latin America | *Ch* is a single character placed after *c* and before *d* |

layer, the one users most frequently access. If all goes well, that layer should be all the user needs. Fewer users will delve into deeper levels. How many levels the company translates first depends on the market, that is, the number and importance of current and potential customers who speak that language. In addition, the company considers the availability of translators and the likelihood that customers will accept instructions in a foreign language. A layered approach can cause problems, however, if a user suddenly encounters a foreign language after clicking on a hypertext link or pursuing some topic into a deep layer. Improved software for translation is helping companies overcome some of these problems and deliver messages to customers in their own language.

## TESTING FOR USABILITY

When you write a procedure or set of instructions, that document will need to pass other tests, generally called validation and verification.

- *Validation* is a check for accuracy. For example, an instructional manual may not be entirely valid if it lacks information concerning a last-minute enhancement in a new product or if it misrepresents the configuration of a device's components. Many manuals are prepared before a product is finalized, so a final test of the manual is to check it against the last version of the product it documents.
- *Verification* is a test of usefulness and readability. Ambiguous language, insufficient or overly abundant detail, confusing format, confusing visuals—these may cause a set of instructions to fail.

For more about instructions, see Chapter 17

Because of the importance and prevalence of procedures, communications researchers have developed specialized techniques for verification. One technique is a usability test. In such an approach, you gather a sample of potential users of the document and then ask a series of questions as they work with it. Here, briefly, are some guidelines for conducting such a test.

*Before* you conduct the test,

- Develop a list of issues that you think may cause problems in your instructions.
- Define exactly what you are testing.
- Select a sample of people who represent well the intended users of your product.
- Write a profile of those users and tie their characteristics to the issues you think might cause problems (see *A Closer Look: A Usability Test*). Rate them on a scale from novice to expert.
- Develop a specific set of tasks for the test.

*During* the test,

- Watch users and ask them to comment aloud as they work through your written instructions.
- Or ask them to respond to a questionnaire that evaluates their ability to understand or follow the instructions or to remember the steps.

*After* the test, compile your results and edit the instructions to correct any weaknesses.

For more about proposals, see Chapter 14

Revising and testing help ensure the quality of your documents. This chapter has reviewed methods for looking back at what you have written, on paper or online, with the reader's needs in mind so you connect well with the reader across differences in information, in style, and in language.

A CLOSER LOOK

## A Usability Test

A technical writer posted the following summary of a usability test to a newsgroup discussion. He tested the effectiveness of a tutorial and a task-oriented reference guide that aimed to help users draw chemical structures with a computer program (Courtesy of Trevor Grayling).

> The goal of the documents, as defined in the document plan, was to get the average user "up and running" and able to draw the chemical structures they were interested in with only occasional reference to the reference manual. . . . Given the goal, we devised the following usability test:
>
> **1. SELECTING A REPRESENTATIVE AUDIENCE.** We found seven users in house who represented the skill sets of our end users; some had no experience at all with previous versions of our products, some had limited experience, and some were experts with earlier versions. They also had varied chemistry backgrounds.
>
> **2. DEFINING TEST CRITERIA.** Since the goal was to be able to "draw structures with only occasional reference to the reference manual . . . ," we created four "average" chemical structures, three of which did need reference to the manual for some specific functions. A successful test would be one where ALL the users would be able to EXACTLY duplicate the structures given them on paper. The test criteria have to be defined before the test (!) and very specific. There was no time limit because time was not part of the goal of the documentation set. How they used the documentation was also not part of the test criteria.
>
> **3. CONDUCTING THE TEST.** We set up seven computers in a conference room. We provided each user with the tutorial, the reference, and four chemical structures on paper. We told them: (a) that this was a test of the documentation and not of the testers, to reduce anxiety; (b) that they were to exactly duplicate the paper structures on screen in terms of the exact bond length, bond angles, atom symbols, and so on; (c) that they could use any of the documentation they saw fit, but that they were not obliged to use it if they didn't want to; (d) that there was no time limit; (e) that we would be looking over their shoulders and taking notes; (f) that only if they were well and truly stuck and unable to proceed could they ask for assistance.
>
> **TEST RESULTS.** All users did complete all four structures. Time for the test varied from 40 minutes to 1½ hours. Some dutifully worked through the tutorial first, some only referred to it when they got stuck, and some skimmed through it looking for the "gist" of the product. Two users got temporarily stuck but we refused to help them, telling them to rely on the material provided (we could see that they weren't really stuck).We obtained tons of useful information by observing the users, even though the test was successful. Although we had ourselves tested the documentation thoroughly, we noted paragraphs of text and instructions that were correct but ambiguous. If something could be read in two different ways, it would be. We also noted the need to put the "getting out of trouble" section much nearer the front of the tutorial.

## CHECKLIST: Revising

1. **Revise to ensure that the big picture is clear**
   Focus on one main point—the right one
   Make that point obvious
   Confirm the accuracy and completeness of information
   Ensure compliance with the assignment
2. **Review the design of the information and of the information product**
   Help the reader find, understand, and remember your message
   Review headings and other devices that make the structure stand out
   Make sure the text sorts new from familiar information and moves from the familiar to the new
3. **Revise expression, as needed, to be correct and appropriately polished**
4. **Review any visuals**
   Well labeled and clear
   Consistent with the text but not redundant
5. **Revisit a Web site to make sure it is**
   Fast loading and interactive
   Attractive and indicative of its owner's persona or brand
   In modules that are clearly marked and prioritized
   Richly linked
6. **Meet any standards that govern the document**
7. **Write especially well for translation**
   Prepare for a cultural translation as well as a language one
   Keep text and visuals simple and unambiguous
   Allow for expansion
8. **Apply a usability test to any instructions**

## EXERCISES

1. The following description responds to an assignment to "describe the growth of something." The author chose soybeans, whose growth pattern he knows well. But he moves through a series of false starts in getting to the assignment. Outline the topics presented and then revise the description to support a core idea.

   > The study of plants had its start when the land-grant colleges were first being formed. Before that time there was very little science to the process of plant growth. Agronomists or plant scientists observed that a plant did actually grow better when a fish was placed with the seed at planting time.
   >
   > I think most of us now know that the reason for this was the nutrients provided by the decaying fish. But do we understand the process that makes a plant grow? How about a soybean, for instance? A soybean is an agricultural plant that is being used not only for farm feeds, but for the making of synthetic fibers, plastics, and meatless food.

The seed is made of protein, oil, and carbohydrates, all of which are very important organic products. These materials are used for energy when the seed starts to grow. Inside each seed is a small part called the germ, which is actually alive. If kept away from moisture, the seed will still be capable of growing after many years.

If you place a soybean between a couple layers of damp paper towels in a warm spot, the growth process will start. The seed will absorb water, swell, and finally split open.

The soybean is a dicotyledon, or two-part seed. In between the two halves is the germ that grows. The germ obtains its energy from the cotyledons, which contain the proteins and carbohydrates. The root comes out and, directed by gravity, heads downward. When the root has found a place to support itself, the plant starts rising toward the light and unfolds. By this time, very small leaves have formed between the two cotyledons, and with the addition of sunlight, water, and nutrients from the roots, will make its own food. The cotyledons have given up all their food and fall off. In a few weeks, the world has another plant.

2. Revise the following passage to remove the ambiguities, inconsistencies in terms, and culture-bound references that would make it difficult for a non-native speaker of American English to understand—or even translate:

   His take on the problem is that the new system is a kludge and the manufacturer's literature on it is a total whitewash. At the end of the day, their representatives wouldn't be there for us. We'd be out on a limb with a half-baked system that would take us hundreds of man-hours to get on-stream. Take my word for it, we'd better fold our cards and head home. I think the decision against buying the system is a no-brainer.

3. Obtain a copy of the style guide that applies to writing in your field and review its instructions for preparing a report. Summarize those instructions in a brief memo to your instructor. If you are working on a report, attach to your memo a draft of your bibliography for the report written in accordance with the style guide's instructions.

4. Write a letter requesting information (see Chapter 18). Then ask someone who is fluent in another language to translate it into that language. As that person translates, or when the translation is finished, interview the translator about how such a message would be conveyed in that culture and about any difficulties encountered in translating. Then turn in to your instructor your original letter, the translated letter, and a brief report on your interview. Discuss these translations in class.

5. The following passages come from a report evaluating whether a site is potentially a wetland. The "before" is a draft version; the "after" is edited text. Analyze the differences in the passages.

   **Wetlands and Soils**

   **Before.** A review of the National Wetlands Inventory (N.W.I.) map for the tristate 7½ Minute Quadrangle prepared by the United States Department of the Interior, Fish and Wildlife Service, indicates that the subject site does not appear to contain mapped wetland areas. An omission of wetlands from the N.W.I. map does not connote a lack of wetlands on-site. Therefore, the apparent lack of mapped wetlands should not be construed to mean that no wetlands are present on the subject site.

**After.** This preliminary assessment indicates that the site probably does not include regulated wetlands. No wetlands appear on the site as shown on the National Wetlands Inventory (NWI) map for the Tristate 7½ Minute Quadrangle. The NWI map, however, which is prepared by the U.S. Department of the Interior, Fish and Wildlife Service, sometimes fails to show wetlands that then become apparent on a field inspection.

**Flood Levels**

**Before.** A review of the Federal Emergency Management Agency (FEMA), Flood Insurance Rate Maps for unincorporated areas north and south of the city indicate that the subject site may exist within the 100-year flood plain. The 100-year flood elevation for the Rogers River is set at 10 feet by FEMA. Flood levels along the lower tidal portion of Misaly Creek may be controlled by the Rogers River flood levels. As such the subject property, which is mapped by the U.S.G.S. to be at an elevation below 10 feet, may be within the 100-year flood plain of Misaly Creek.

**After.** The site may be within the 100-year flood plain of Misaly Creek. Flood levels along the lower tidal portion of that creek are controlled by Rogers River flood levels, which the Federal Emergency Management Agency (FEMA) sets at 10 feet. The FEMA Flood Insurance Rate Maps and the USGS map show the site to be at an elevation below 10 feet and thus potentially within the 100-year flood plain.

## FOR COLLABORATION

In a team of three, develop a short document, for example, a set of procedures (Chapter 17). Then test it according to the usability guidelines in this chapter. Determine what standards or legal codes apply to such a document. Edit it for accuracy, clarity, and suitability for translation. Verify its approach with a sample of readers to make sure it works in practice. Write up your testing procedure and results in a memo to your instructor that you attach to your original (and perhaps revised) document.

PART FOUR

# Developing Technical and Scientific Documents

CHAPTER 13

# ABSTRACTS AND EXECUTIVE SUMMARIES



"LESS IS MORE."

MIES VAN DER ROHE

An *abstract* is a form of summary. An abstract accompanies many documents in both academic and research settings: reports, proposals, articles, dissertations, or papers to be presented orally. It may be a unit (the first) of a parent document, or it may appear separately on a Web page or in a collection of abstracts, like *Dissertation Abstracts* or *Chemical Abstracts.* Another form of summary, an *executive summary,* is common in business settings for corporate or government reports and proposals. Because writing an abstract or executive summary helps you discover the real essence of your work, the process has high value for you. An abstract or other summary is also highly valuable for your readers, who are more likely to read that than any other part of a report or article. A major product of research is reports and publications; proper abstracting helps ensure that reports and articles will be used. In this chapter you will learn how to write an abstract as well as an executive summary.

For more about indexes and abstracts, see Chapter 5

## PURPOSE

All the audiences of a document with an abstract will at least skim the abstract:

- Managers who might not read the technical discussion in a report
- Accountants interested in a general overview of a project
- Cataloging librarians who just need to know where to file a document

- Researchers perusing a collection of abstracts to determine if they should consult a parent document; sometimes the abstract contains all the information the researcher needs.
- Attendees at conferences reviewing abstracts in a preliminary program to decide which events to attend and to formulate the questions they wish to ask of the speakers
- Program chairs of conferences to determine which proposed papers to accept for presentation

In addition, many more abstracts than articles find their way into translation, so an abstract can help when the original document is in a language the reader doesn't understand. Sometimes, too, an abstract is the only information source available; many physicians in developing countries, for example, rely on abstracts alone to help them treat patients (Evans 5).

Information retrieval organizations like Chemical Abstracts Service sometimes hire professionals to write abstracts. But, like many similar organizations, they prefer to have the author of the original document write the abstract. That approach speeds the process of disseminating information and ensures accuracy because the abstract comes from the person who did the work. As a student and as a professional, you need to know how to write an abstract. Depending on the audience's purpose in reading the abstract, take one of two approaches: descriptive or informative.

### Descriptive Abstract

A *descriptive* abstract (sometimes called a *topical* abstract) briefly indicates the topics covered in the parent document. It provides no conclusions or supporting evidence and rarely exceeds one or two sentences regardless of the length of the original. Many begin "This [report] [article] discusses," as in the following descriptive abstract:

> This article discusses a Singapore-based research project that analyzed how cultural background affects an applicant's performance in a job interview.

Although useful to catalogers, the descriptive form is less useful to researchers who look for substance rather than mere categories. The statement "The cross section was measured" (descriptive) is less valuable to researchers than "The cross section was found to be 6.25 cm" (informative).

For more about abstracting literature, see Chapter 5

### Informative Abstract

As the name implies, an *informative* abstract tells more, for readers who need more information. It doesn't talk about the report or article but about the investigation. It provides the content of a document in a nutshell, as in this informative abstract for the article on job applicants, titled "Chinese Cultural Values and Performance at Job Interviews: A Singapore Perspective" (Wong and Lai 9):

> In a country like Singapore, which is rated high in power distance and low in individualism (using Hofstede's dimensions of national cultures), interviews for entry-level positions in multinational corporations (MNCs) may reveal subtle clashes in culture. To test this hypothesis, we analyzed transcripts of job interviews involving nine English-speaking applicants from Chinese backgrounds and two experienced interviewers from Anglo-American MNCs in Singapore. Our assumption was that a person's cultural background and upbringing influence his or her performance at job interviews. The findings reveal that Chinese applicants tend to defer to the interviewer (i.e., superior) and focus on the group or family, besides being averse to self-assertion. Hence, applicants from a Chinese background may be disadvantaged when being interviewed for jobs with MNCs which are heavily influenced by Anglo-American culture.

The abstract must be intelligible and complete in itself; it should not be necessary to read the original report or article in order to understand the abstract. Readers of the job interview article, for example, would be familiar with Hofstede's dimensions of national cultures. The "guide for authors page" of a journal usually provides specific instructions about the composition of abstracts for that journal. Figure 13.1 shows other sample descriptive and informative abstracts. Brief informative abstracts that provide a main conclusion sometimes appear on the table of contents page of a magazine or journal. Figure 13.2 shows such an approach in the British journal *Nature.*

## Hybrid Abstract

In practice, the neat distinction between informative and descriptive abstracts is sometimes blurred, although the two points of view represent real differences, and purists object strongly to a hybrid approach. In general, maintain one point of view, but at times you may find an organization or publisher asks for a hybrid. The following abstract of a journal article presents the background from an informative point of view and then turns to a descriptive overview of the article:

> Despite the increasing interdependence of global economies, there is surprisingly little cooperative product development done by related international divisions of multinational industrial corporations. Cooperative development could offer multinational corporations the means to pool expertise and other resources available in various regions to produce globally competitive products. The two main obstacles to joint international development are the failure of management to recognize the extent to which sociocultural aspects affect their own decision-making processes and those of their international counterparts; and even having recognized that sociocultural aspects have a significant impact, more managers are ill prepared to deal with them. A case study is presented of a large Swiss electronic-equipment manufacturer with a U.S. subsidiary in the same field. Qualitative and descriptive data from interviews with managers and engineers are reported and related to the sociocultural backdrop in which they make their decisions. Recommendations and areas of further study are also presented. (Grinbergs and Rubenstein 22 © 1993 IEEE, used with permission.)

**Thesis: lack of cooperative development**

**Reason: two obstacles**

**Research method: a case study**

**Description of article's content**

**The Use of Models: Nineteenth-Century Church Architecture in Québec**

Defines the exhibit's topic: models

*Descriptive:* This exhibit consists of four models used by architects in the mid-nineteenth century to describe their projects to clients.

Explains why models were used

*Informative:* By around 1850, new stylistic and typological diversity in architecture and growing professionalism among architects had led to drawings more detailed than ever before—and laypeople now had difficulty interpreting them. Canadian architects began to make greater use of models, partly because they wished to make architectural projects more accessible to clients.

Provides details and the rationale for the exhibit

The four models exhibited—for St. Andrew's Presbyterian Church (ca. 1849), Chalmers Free Church (before 1850), Église Sainte-Hélène in Kamourasaka (ca. 1847), and Église Sainte-Anne-de-la-Pérade (ca. 1855)—will, along with archival materials (including documentation of lost models), illuminate a little-known architectural practice and contribute to the general architectural history of North America.

**Technology Toys and Sex Roles in America, 1920–1940**

Overviews the thesis concerning women in science

*Descriptive:* The author examines the role of toys—especially "technology toys"—in socializing American children during the decades of the 1920s and 1930s and draws a conclusion concerning the disproportionately small representation of women in the professional fields of science and technology.

Relates toys to career decisions

Sets up comparison

*Informative:* During the decades of the 1920s and 1930s, American toys socialized children into appropriate sex roles. This was especially true of "technology" toys—those thought to embody modern examples of useful science and technology. For girls, these were usually small electrical appliances which prepared girls for careers of cooking, cleaning, and laundering. For boys, they fell into one of four categories: tools, vehicles, construction sets, and science outfits. Unlike girls, boys were encouraged by their technology toys to be bold, inventive, constructive, and curious about the principles underlying modern science and technology. Although seldom acknowledged, the result was to teach girls "their place." Boys, on the other hand, were deliberately encouraged to develop their personal capacities to the fullest extent possible. It seems likely that this distinction accounts, in part, for the disproportionately small representation of women today in the professional fields of science and technology.

FIGURE 13.1 **Descriptive and Informative Abstracts**

("Models" courtesy of the Centre Canadien d'Architecture and "Technology Toys" courtesy of Carroll Pursell. Reprinted by permission.)

**CONTENTS**

**2 Jan. 1997 Vol. 385 Issue no. 6611**

**ON THE COVER**



*Nature* on the Web — http://www.america.nature.com

Nature America Inc.

◀ Optical resonators are components of lasers and other optical devices. In a symmetric resonator, light trapped by total internal reflection circulates around the perimeter. The properties of a new type of asymmetric resonant cavity are described on page 45. In these cavities, chaotic motion of the light rays leads to a highly anisotropic pattern of emission. Cover shows the onset of directional emission in such a device.

**THIS WEEK...**

**RNA on the move**
Elongation factor G (EF-G) is a GTPase involved in the translocation of bacterial ribosomes along the messenger RNA during protein biosynthesis. The 'textbook' view of the mechanism of translation assumes that EF-G/GTP binds to the ribosome and induces a conformational change allowing translocation; only after translocation is the GTP hydrolysed as EF-G leaves the ribosome. But the Article on page 37 suggests that, in fact, the GTP hydrolysis step precedes and greatly accelerates the ribosomal rearrangement that leads to translocation. News and Views, page 18.

**Straying asteroids**
The Trojan swarm is a large population of asteroids — possibly even larger than the asteroid belt itself — orbiting the Sun at about the same distance as Jupiter. A new study of the orbits of the Trojan swarm suggests that the orbits are not stable indefinitely and that there are more than 200 'escaped' Trojans at least 1 km in diameter currently roaming the Solar System, a few of them on Earth-crossing orbits. Such Solar System 'strays' are of interest not least because they sometimes make themselves felt on Earth, for instance in the form of the Cretaceous–Tertiary impactor. Page 42.

**Targeted attack on tumours**
A new concept in antitumour agents is descri[illegible] the use of engineered cells to deliver a cytoto[illegible] tumour itself. The oncoprotein HER-2 is over[illegible] surface of breast, ovarian and other tumou[illegible] class of tumour-specific killers are lympho[illegible] modified to express an antibody specific to H[illegible] a *Pseudomonas* exotoxin, so that the anti[illegible] cell to infiltrate the tumour, at the same ti[illegible] toxin. Extensions of this technique could be [illegible] other tumours, and also in autoimmune dis[illegible] infection.

**Time to face the Muzak**
Events recommended for commemoration in this year's Anniversary Commentary include the discovery of the electron, the invention of the sewing machine, the establishment of the American Association for the Advancement of Science, the publication of Johann Hevelius's lunar map, the first description of t[illegible] and the origins of piped music. Page 13.

Negative result: a first gl[illegible]



**TROJAN asteroids, which may outnumber the asteroids in the asteroid belt, are objects that orbit the Sun with the same mean semi-major axis as Jupiter, but lead or trail the position of Jupiter in its orbit by ~60°. One very interesting aspect of the Trojan swarms is that a significant number of asteroids are on orbits that analytic theory suggests should be unstable[1]. Here we present the results of long-term dynamical integrations of the Trojan asteroids that enable us to investigate the stability of the swarm population. We find that the orbits of the swarm asteroids are not stable indefinitely–the gravitational effects of the giant planets have reduced the swarms' outer boundaries over time. We estimate that there are over 200 escaped Trojan asteroids with diameters >1 km currently roaming the Solar System, a few of which may be on Earth-crossing orbits.**

*Nature*® (ISSN 0028-0836) is published weekly on Thursday, except the last week in December, by Macmillan Magazines Ltd (Porters South, 4-6 Crinan Street, London N1 9XW). Annual subscription for USA and Canada US$495 (institutional/corporate), US$145 (individual making personal payment). Canada residents please add 7% GST (No. 140911595). US and Canadian orders to: *Nature*, Subscription Dept, P. O. Box 5054, Brentwood, TN 37024-5054, USA. Other orders to *Nature*, Brunel Road, Basingstoke, Hants RG21 2XS, UK. Periodicals postage paid at New York, NY 10010-1707, and additional mailing offices. Authorization to photocopy material for internal or personal use, or internal or personal use of specific clients, is granted by *Nature* to libraries and others registered with the Copyright Clearance Center (CCC) Transactional Reporting Service, provided the base fee of $9.00 an article (or $1.00 a page) is paid direct to CCC, 222 Rosewood Drive, Danvers, MA 01923, USA. Identification code for *Nature*: 0028-0836/97 $9.00+$1.00. US Postmaster send address changes to: Nature, 345 Park Avenue South, New York NY 10010-1707, US. Published in Japan by Nature Japan K.K., Shin-Mitsuke Bldg, 36 Ichigaya Tamachi, Shinjuku-ku, Tokyo 162, Japan. © 1997 Macmillan Magazines Ltd.

v

**FIGURE 13.2 Table of Contents from *Nature***
Note the brief abstracts under THIS WEEK. These address a broad audience of readers. The insert shows the more detailed—and technical—abstract that begins the article itself, "Dynamical evolution of Jupiter's Trojan asteroids" by Harold F. Levison, Eugene M. Shoemaker, and Carolyn S. Shoemaker.

### Executive Summary

Abstracts are common in academic or research settings, and another form of summary, called an *executive summary*, is common in business settings for corporate or government reports or proposals. Abstracts rarely exceed 200 to 300 words; executive summaries vary considerably depending on the length of the parent document, with some summaries long enough to be bound separately. Most, however, fall between 200 and 1,000 words.

An executive summary follows the informative point of view but is geared to decision makers and thus skews the emphasis to highlight issues of importance to management. The executive summary of a proposal, for example, may be the only section read by all people evaluating the proposal. When they've read that overview, each then pursues the section demanding her or his special expertise, for example, the budget or the technical approach. Figure 13.3 shows a segment from the executive summary of a report.

**For more about proposals, see Chapter 14**

## WRITING AN ABSTRACT OR EXECUTIVE SUMMARY

You can apply many of the techniques for summarizing someone else's document that you read about in Chapter 5 when you write an abstract or summary of your own work. This section provides a few additional guidelines.

### When Should You Write?

Most writers assume the abstract is the last unit in the composition of a report. Writing it last makes some sense because you've worked out the main points and merely need to collect them from your text. In addition, writing the abstract after drafting the report or article

- May reveal ambiguities, contradictions, or redundancy in the parent document.
- May suggest how expression in the original might also be condensed.

Consider, however, writing the abstract *before* you write the longer document:

- Helps you clarify the story you need to tell.
- Helps direct the efforts of collaborators on a team report.
- Provides a text for the intended audience to review before you spend extensive amounts of time composing.

### What Information Should You Include?

Writing a descriptive abstract is easy. Just note the type of document ("This report," "This article,") and list the topics it centers on. Writing an informative abstract, however, is more challenging. If the publication you are

Other implications of recent economic studies:

• Actions to reduce greenhouse gas emissions today will impose costs long before benefits (if any) emerge.

• Costs rise at an increasing rate as the target levels of greenhouse gas emissions are decreased.

• Delaying a targeted emissions goal by a decade or so would have a small impact on cumulative emissions but could entail significant cost savings.

Greater resources focused on uncertain future climate issues will divert international attention from more certain and immediate problems of poor countries, and, in the U.S., efforts to improve competitiveness, health care and education reform, and the already full and costly environmental agenda.

***Policy Issues***

Policy making is complicated by scientific uncertainty, uncertainty about the economic impacts of potential climate change and policies to reduce emissions, the global nature of the issue, and the very long time scales involved.

The highest priority should be to improve scientific understanding of climate. Delaying more severe actions until a credible scientific basis can be established would avoid misguided, nonproductive, and costly steps. The potential for low-cost technological breakthroughs over the next half century could make a nation regret taking premature and expensive actions in the near term even if research later determines that carbon dioxide emissions should be constrained.

Certain actions are worthy of immediate support:

• Accelerate the pace of research into basic climate science and impact assessment.

• Identify and pursue measures that will reduce the threat of climate change, yet also make sense in their own right.

• Establish sustained research and development programs that improve the ability to economically produce and utilize energy with less potential for the accumulation of greenhouse gases.

• Expand efforts to understand and communicate the economic, social, and political consequences of both climate change and proposed policy responses.

FIGURE 13.3 **Executive Summary**
Concluding segment from a 3-page executive summary of a report, "Issues & Options: Potential Global Climate Change," produced by the Global Climate Coalition, Washington, D.C. 1994. Used by permission.

writing for dictates a specific structure for the abstract, then follow that structure, as in Figure 13.4, which shows the structure for abstracts of papers describing experimental or clinical research in the *European Journal of Surgery*. All of this information must be included in fewer than 400 words.

If no structure is imposed, think of your abstract as a prose form of your outline or table of contents (Figure 13.5). When you write

- Emphasize what the document says that is new, original, and significant.
- Stress the scope and objectives of the work (the problem and purpose).
- Report the chief results, conclusions, and recommendations, if any.
- Elaborate on the title and answers any questions posed by it.
- Describe procedures only briefly, with emphasis on any departures from the customary.
- Maintain the relative emphasis and organization of the original.
- Use abbreviations only if they are standard in the discipline.

Omit supporting examples and details so the main points come through. Avoid becoming bogged down, as in this example:

> A 72-gram brown Rhode Island Red country-fresh candled egg was secured and washed free of feathers, etc. Held between thumb and index finger, about 3 feet more or less from an electric fan (General Electric Model

*Objective:* Purpose of the study.

*Design:* Type of study, for example, whether double blind, retrospective, open, multisite.

*Setting:* Where the study was conducted: private practice, university department, district hospital, community.

*Subjects or material:* Number of patients, subjects, or animals, along with an explanation of how they were selected to give the reader an idea about the generalizability of the results.

*Interventions:* Any interventions, for example, the technique of an operation or duration and dosage of a drug regimen.

*Main outcome measures:* Methods for assessing the outcome of subjects or success of experiments.

*Results:* Main results, with an assessment of their statistical reliability and a note about any failures or withdrawals.

*Conclusions:* Main conclusions, with suggestions about any clinical applications.

**FIGURE 13.4 A Structured Approach for Abstracts in the *European Journal of Surgery***
(Source: Adapted from Evans 5.)

| Table of Contents | Abstract |
|---|---|
| The Need for Genetic Counseling<br>Definition of Genetic Counseling<br>Statistics on Genetic Defects | Genetic counseling is a service for people with a history of hereditary disease. One in seventeen newborns has some defect; one-fourth of the patients in hospitals are victims of genetic diseases (including diabetes, mental retardation, and anemia). Of every 200 children born in the United States, one has chromosome abnormalities. |
| Purpose of Genetic Counseling<br><br>The Counseling Process<br>Evaluating Needs<br>Taking a Family History<br>Estimating the Risks<br><br>Counseling the Family | Genetic counseling offers parents an alternative to genetically diseased children and assistance for those with children already afflicted. The first step in counseling is to evaluate the needs of the parents. A family history is prepared and the risk of future children's being afflicted is evaluated. The life expectancy and possible methods of treatment of any afflicted child can also be determined. Alternatives are presented. |
| Determination of a Genetic Disorder<br>Amniocentesis<br>Karyotyping<br>Fluorescent Banding<br>Staining | Four prenatal tests are used to determine if a genetic disorder is present: amniocentesis, karyotyping, fluorescent banding, and staining. |
| Advantages of Genetic Screening<br>Lower Cost<br>Increased Availability | The development of these four relatively simple methods has lowered the cost of genetic counseling and increased its availability. |

FIGURE 13.5 **Turning an Outline into an Abstract on Genetic Counseling**

No. MC-2404 Serial Number JC 23023, non-oscillating, rotating on "high" speed at approximately 1052.23 +/− 0.02 rpm), the egg was suspended on a string (pendulum) so it arrived at the fan with essentially zero velocity normal to the fan rotation plane. The product adhered strongly to the walls and ceiling and was difficult to recover; however, with the use of putty knives a total of 13 grams was obtained and put in a skillet with 11.2 grams of hickory smoked old-style bacon and heated over a low Bunsen flame for 7 minutes and 32 seconds. What there was of it was excellent scrambled eggs.

Here's a more streamlined version:

> Throwing an egg into an electric fan produced very good scrambling. The product was difficult to recover from the walls and ceiling, but the small amount recovered made an excellent omelet. A shrouded fan was designed to improve yield in preparation for additional experiments.

Work from an outline—either one you prepare before writing the draft document or one you construct when you have finished the draft. In the first sentence, clarify the objectives of the work, then note its scope (with terms like *brief, comprehensive, exhaustive*). Check the abstract against the parent document. Make sure the abstract matches what you say there and includes no information not in it.

A CLOSER LOOK

## Keywords

Keywords perform an essential role in bringing together authors and readers. As the term implies, such words help readers unlock the content of a document. Opening the door into information sometimes requires an intermediary between author and reader, someone who indexes the information, and that person, too, depends on keywords to compile the index.

Generally, nouns serve best as keywords, often with modifiers that show relationships among the terms. The more precise the noun phrase, the more powerful the key. The keyword phrase *technical communication* might specify one form of communication among others (for example, *business* communication, *science* communication). But in the context of other articles about technical communication, one would need more concrete indicators of the coverage in the article at hand: *e-mail collaboration among scientists, technical genres, environmental writing.* The terms *zinc copper blood cow* indicate the broad *topics* of an article, but they could identify very different approaches:

Zinc effect upon copper in blood of cow

Copper effect upon zinc in blood of cow

Zinc's mutual interaction with copper in blood of cow

Existence of zinc and copper in blood of cow

Effect of zinc supplements

Effect of zinc and copper on blood of cow

To help the reader know which approach the article takes, and thus how relevant it is to the reader's research, the author should specify the relationship (Borko and Bernier 88).

In addition, keywords should reflect the way the discipline shared by author and reader talks about the information. Some researchers go so far as to say that a discipline is defined by its terms, which reflect its view of reality. Those terms matter. Authors should invoke them as a critical shorthand for their readers. Terms change with new discoveries in a field, so authors should keep up to

### What Voice Should You Use?

An abstract is not a telegram. Use complete, grammatical sentences that include verbs, conjunctions, and articles. Telegraphic style may be appropriate for your notes: "Soy ink 80 percent biodegradable; classed as hazardous waste for pigment, heavy metal." But that style will not work in the abstract. Instead, write, "Although 80 percent biodegradable, soy ink is still considered a hazardous waste because the pigment is a heavy metal." In addition,

- Use subordination to condense information, but avoid overlong sentences.

date not only with how colleagues see things but also with how they identify in words what they see.

The terms used in the abstract and in the parent document must also be consistent. Because authors may prepare these documents at different times, or different authors may prepare each, a check on keywords provides a good check on more global consistency in the work. Especially in scientific and technical documents, synonyms are suspect; the same idea, event, mechanism, or organism should be referred to in the same way throughout to avoid confusion.

Finally, keywords must increasingly be appropriate for a worldwide audience. What people in the United States refer to as a *moose,* for example, is called an *elk* in Finland. One way around such differences is to use scientific terminology only, although even that approach is somewhat flawed because scientific names are not necessarily consistent around the globe. More difficult are references to political entities and to human beings, tribes, ethnic groups, and the like. Terms may carry emotional baggage, like the distinction between *Inuit* and *Eskimo.* Following the guideline that one should refer to people as they refer to themselves, one should use *Magyars* for Hungarians and distinguish between *pakeha* and *Maori* rather than referring to *New Zealanders.* Using highly specialized names, however, may reduce rather than expand the number of readers accessing the literature. But at the same time, it begins the process of introducing the new term (Meyer-Rochow 57).

Keywords have traditionally aided readers as an indexing tool, and now they are serving a new function in hypertext documents. Highlighted terms bring readers directly to linked sources in a CD-ROM and on the Web. The importance of keywords only increases in this new setting, and thus authors reap significant rewards when they develop these terms well.

- Use active verbs wherever possible; passives may be necessary if the publication you are writing for frowns on the use of "I."
- Report what was done in the past tense; report the implications of that activity in the present. "We tested the prosthetic devices in a sample of 20 sheep. The insertion procedure consists of four major steps."
- Scout for words that are not carrying their weight. You can almost always eliminate expletives, such as *it is, there are,* and *there is.*
- Ask a friend or colleague to check the abstract for completeness, proper emphasis, and objectivity.
- As you did before writing, review the instructions of the journal or organization to make sure you have fulfilled its requirements.

## KEYWORDS

In addition to an abstract, you may also need to provide a few *keywords* (known also as *index terms* or *descriptors*) that indicate where the document should be categorized in an information retrieval system (see *A Closer Look: Keywords*). The words may come from the title, from the abstract itself, or from the thesaurus that accompanies the leading journals in the particular discipline being discussed. The terms, often included at the bottom of the abstract, should match the expectations of readers searching for the document. For example, keywords for the abstract on technology toys (Figure 13.1) might be *technology, toys, sex roles, women professionals in science.*

## VALUE

Abstracting requires special language skills and discipline. It forces you to confront your own ideas and examine the relationships among them, their sequencing, their relative importance, and their relationship to the document's purpose and to the investigation's problem. Avoid restatement (although that's permissible in the original document). Good abstracts help people manage the overload of information about science and technology by providing readers with the essence of a longer document.

### CHECKLIST: Abstracts and Executive Summaries

1. **Write a summary to convey the parent document's central message in a nutshell**
2. **Use a *descriptive* abstract to alert readers, especially catalogers, to the topic of the document**
3. **Use an *informative* abstract to talk about the main news of the investigation for researchers**
4. **Use an *executive summary* to provide evidence for decision makers in a business setting**

**5. Write a summary before writing the document to clarify the main point, structure, and evidence**

**6. Depending on the context, include**

Purpose of the work
Procedure (if unusual)
Scope and conditions
Main results
Main conclusions and implications

**7. Avoid a telegraphic style; use complete sentences and a comfortable voice**

## EXERCISES

1. Determine the keywords that would apply to each of the abstracts given in this chapter.
2. Condense the following abstracts from student reports and make them more coherent.

### ADVANTAGES OF ULTRAFILTRATION FOR PROCESSING WHEY

The report was done for the Che-dar Company, which asked for information on whey-processing methods. A recommendation of one of these methods was also requested. Spray drying of whole whey, electrodialysis, and ultrafiltration were the processes discussed. After a review of the advantages and disadvantages of the three methods, it was recommended that ultrafiltration be used. This decision was based on the lower energy requirement, the low initial investment, and the greater profitability of ultrafiltration.

A description of the ultrafiltration process and cleaning procedures is given in the final section of the report. The whey is concentrated by passing it through a semipermeable membrane that allows water, lactose, and nutrients to pass through. Protein molecules are too large to pass through the membrane and remain in the concentrate. The permeate and concentrate are spray-dried to obtain lactose and protein-rich powders.

Cleaning and sanitation must be done on a regular basis because operating temperatures and whey components are favorable for microbial growth. The procedure consists of cleaning with detergents containing proteolytic enzymes, sanitizing, and rinsing with water.

### SILVICULTURAL PLAN FOR THE HORSTER FOREST

Accessibility to markets and the high prices being paid for timber have been responsible for the increased demand for development of silvicultural plans of management. The silvicultural plan of management presented in this report was developed for a fifty-acre hardwood forest in northwest Ohio. Five variables were taken into consideration in developing this particular plan of management. Those variables that were investigated are discussed in the following order: (1) soil types and pH values; (2) accessibility to the forest; (3) stocking levels present; (4) the value of the tree species present; and (5) federal funding programs. After the preceding variables were assessed, it was found that the limiting factor in the forest appeared to be the high level of stocking. Therefore the recommendations given at the end of the report are based heavily on control of stocking levels.

3. Write an abstract of one lecture in a technical class or of a guest lecturer's presentation on campus.

## FOR COLLABORATION

Write an abstract of your team's final report—in advance of writing the report—to see how the story comes out. Then use the abstract to coordinate the group's activities. To write the final report, expand on each of the points you established in the abstract.

# PROPOSALS



**"YOU CAN BUY IN YOUR OWN LANGUAGE, BUT YOU HAVE TO SELL IN THE LANGUAGE OF THE CUSTOMER."**

A proposal is an offer to provide a product or service or to do some kind of work to solve a problem. It can be a simple one-page document written by one person for one reader. It can be a multivolume document composed by a team of 300 people and read by 20 or 30 evaluators. And it can be something in between. Broadly conceived, proposals document ways to make the world a better place. They are major tools in business, in research, and in the classroom.

## PURPOSE

The purpose of a proposal is to show how, in a particular context, the world can be made better. In writing, you need to adjust your approach to each context. The contexts, and thus the proposals that respond to them, fall into three major categories, discussed in this section.

### Bid

In the simplest context, you provide a routine response to a well identified problem. The product or service you're offering comes off the shelf, and the major differentiating item is usually the price. Your proposal, often called a *bid,* may contain a few lines of text and columns of numbers indicating costs. Once the bid is accepted, the document becomes a binding contract, for example

- Distributors of cleaning supplies write bids to provide their products to corporations, restaurants, and other organizations with buildings to clean.
- Furniture companies write bids to sell desks and chairs to universities.

Less routine bids tailor a known procedure, product, or service, like business process software or sophisticated navigational systems, to a particular customer. In these circumstances, the bid may contain extensive discussion about how the item will be made to work for the customer:

- Defense contractors write proposals to governments worldwide to build or recondition ships and planes.
- Geotechnical engineers write proposals that describe the method they would use to determine soil conditions on a building site.
- Consultants write proposals that show how they would prepare an environmental impact statement for a project.

### Implementation Proposal

Beyond a simple bid, you may find yourself proposing to design and implement a new system or process. No off-the-shelf approach will do. Your document must account for the problem, as you see it, and the way to improve the situation. Figure 14.1 reproduces an implementation proposal. Here are some other contexts for such proposals:

- A nurse proposes the implementation of an improved triage system in an emergency room.
- An engineer proposes an oil filtering system to recycle old oil and thus save money in both purchasing and disposal.
- A technician proposes a new system for recording the food intake and weights of laboratory animals.

### Research Proposal

A third major context for proposals is *research*. In a research proposal, you define a problem, one that may not have been well recognized or defined. Then you describe a method for finding a solution, and you convince the reader that the method is likely to succeed. The audience may be a supervisor, a client, a sponsor, or a professor. Figure 14.2 reproduces a student research proposal. Major government funding agencies, like the National Science Foundation and the National Institutes of Health, and major private not-for-profit organizations, like the Robert Wood Johnson Foundation and the Rockefeller Foundation, award contracts and grants for research in response to proposals that compete for such money. The funding agency usually establishes a theme or priority for the kind of research it seeks; identifying the causes and treatment of diseases like AIDS and cancer, developing ways to prevent crime, or coming up with innovative techniques for handling waste are some representative priorities.

**MEMORANDUM**

Date: 30 October 2001
To: Chuck Blackman, Manager
From: Peter Sanchez
Re: New Method of Processing Inventory Shipments

This is a proposal to implement a new method for processing inventory shipments. The proposal derives from my own observations and from conversations with several other employees.

States the proposal

**Problems with the Current System**

Currently, shipments are not being opened and stored promptly. This delay in the availability of inventory has meant that employees find it difficult to meet customer requests. Inventory that remains in boxes or is not stored properly cannot be located or sold. In addition, employees also do not know the current status of merchandise. Third, stockpiles build up: when one shipment is not stored as the next one is received, it is twice as hard to catch up on the work. Inventory fills the aisles of the stockroom and makes it hard to move around and do one's job. These problems are leading to an even more serious one: a decline in employee morale and customer service.

Identifies four problems with current processing method

**Objective of the Proposed System**

The main objective of the proposed system is the timely processing of merchandise shipments. Such processing depends on

- Separating the process into discrete tasks: receiving, opening, verifying, and storing
- Delegating authority to one individual who is responsible for the process
- Designating certain employees as specialists in the process

Explains the objective of a new system

**Implementation of the Proposed System**

The proposed system reflects comments from current employees as well as my own observations of the process as it is now performed.

Except in rare situations, shipments arrive on Monday and Friday. For those two nights:

1. Schedule a larger work force. Three additional employees should work as processing specialists on evenings when shipments arrive.
2. Assign the same individuals to work those evenings so that the process becomes routine for them.
3. Separate shipment processing into discrete tasks. These tasks are receiving, opening, verifying, and storing. All three people can help to bring the shipment into the stockroom. They then separate the boxes on the basis of their content: shoes, apparel, and hard goods. Each person takes one of the three types of items and verifies the shipment against the order. When the shipment is verified, each person can properly store the merchandise.

Further explains steps to meet the objectives

FIGURE 14.1 Implementation Proposal

4. Delegate authority to one of the three specialists. In that way, one person can resolve any disputes without undue questioning.
5. Designate processing as a responsibility. Provide a specific timetable for each of the processing specialists.

**Benefits of the New System**

Acknowledges the cost while justifying the need

The cost of the extra employee time on shipment evenings is more than outweighed by the benefits of the new system. These include

- A more organized stockroom
- More easily accessible inventory
- Better customer service
- Enhanced sales
- Enhanced employee morale

For all these reasons, I'd recommend that we implement the proposed system as soon as possible.

FIGURE 14.1 **Implementation Proposal (continued)**

## SOLICITED AND UNSOLICITED PROPOSALS

Within each of these contexts, a proposal may be solicited or unsolicited, depending on who recognized the need or problem first. If it's the client, customer, sponsor, or professor who sees the problem and requests a solution, the result is a *solicited proposal.* When writers think of the problem on their own, they create an *unsolicited proposal.* In practice, however, as you'll see, this neat distinction breaks down. When you identify a problem that needs to be solved, you are likely to be more successful in proposing a solution if you can talk with the potential audience so they then *solicit* the proposal.

### Solicited Proposal

A solicited proposal responds to a direct request. That request may come from your professor in an academic setting. Outside the classroom, government agencies, businesses, and individual clients solicit responses to solve a need or problem. That solicitation may circulate in conversations or more formally in a classified advertisement or document called a Request for Proposal (RFP), Request for Bid (RFB), or Request for Qualifications (RFQ). RFPs may be as short as a paragraph or as long as a volume. The U.S. Department of Commerce publishes RFPs in *Commerce Business Daily* and describes its initiatives on its Web site. Other federal and state agencies also circulate RFPs, which often reflect political agendas, as in a federal program to purchase paper with a high content of recycled fibers. Here is the opening paragraph of an RFP, called a "Proposal Information Packet," circulated by e-mail and on the Web to defense contractors who might supply

Date: 12 April 2001
To: Professor Xien
From: Julie A. Fine, Tim Kelly, and Karen Romanelli
Re: Proposal for Our Technical Writing Research Project

For the final project in the course we would like to research the best habitat for conserving the Delmarva fox squirrel (DMFS). This historically abundant member of Delaware's mixed-forest fauna has recently undergone a serious decline in population. The audience for our report will be the Fish and Wildlife Division of the Delaware Department of Natural Resources and Environmental Control (DNREC). We will assume that they have asked us to analyze three parcels of land in Sussex County in order to recommend which would be the best site for conserving the DMFS.

Introduction ties the proposal to the assignment and establishes the context

**Statement of Problem**

Heading, like the others, is *descriptive* and standard for a proposal

The Delmarva fox squirrel, a native to this region, is an important subspecies as a habitat health indicator for mixed forest stands and as a valuable addition to Delaware's biodiversity (Bendel & Therres, 1994). Recently, however, DMFS populations have been falling rapidly. In 1967, the U.S. Department of the Interior placed this once-common creature on the endangered species list (32 FR 4001) with remnant populations surviving in only four Maryland counties (Deuser et al., 1990). In essence, the DMFS has been eliminated in Delaware, with several adverse consequences.

Brief review of literature helps identify the problem and establish its importance

First, a significant piece of Delaware's heritage has been lost. Second, the DMFS is native only to the Delmarva peninsula and occurs nowhere else in the world. Hence, if these trends continue, this subspecies may be lost completely. Third, this loss is a sign that healthy, suitable ecosystems are also declining. That means that other species are sure to disappear as well. To stop such a domino effect, we need to heed its warning and conserve the habitat.

Details about "adverse consequences" further explain the problem

Lead-in to next segment: objective

**Objective**

The objective of our research is to analyze three parcels of land in Sussex County that are candidate sites for conserving the mixed-forest habitat required by the DMFS and for reestablishing a viable population. We will then recommend the best site to the Division. We will base our recommendation on answers to the following questions:

1. What specific habitat requirements must be met for the DMFS population to survive?
2. How large is each parcel?
3. How much of each parcel is categorized as mixed forests and what other types of land cover occur nearby?
4. How available are the DMFS's required resources such as water (streams etc.), shelter, and food in each parcel?
5. How fragmented is each parcel and its composite mixed forest sites in terms of roads, highways, and railroads?
6. How has each area changed in the last 10–20 years due to succession and development, and how is it likely to change in the future?

Questions tie the objective to the problem and forecast necessary technical tasks to find answers

List form eases skimming

FIGURE 14.2 **A Collaborative Student Proposal to Conduct Research**
(Courtesy of Julie A. Fine, Tim Kelly, and Karen Romanelli.)

2

**Work is divided into phases and the tasks are reviewed**

**Uses the jargon of the field for an expert reader who expects it**

**Specific details add to the team's credibility**

**Schedule is reproduced on page 29**

**Reiterates the importance of the project**

**Describes team credentials to persuade reader to accept the proposal**

### Method

The proposed habitat assessment project consists of two phases. In phase 1, we will develop a preliminary recommendation based on interviews, library research, and computer mapping. We will base the recommendation on the appropriate Habitat Suitability Index (Allen, 1982). In phase 2, we will conduct an on-site investigation of the parcels to groundtruth our recommendation and test the soil and water. The second phase report will be an addendum to that of the first phase. This proposal pertains only to phase 1.

1. *Interviews.* We currently plan to interview David Hannab of the Division and Dr. John Mackenzie of the Food and Resource Economics Department at the University. We may add other interviews.
2. *Library research.* We will use the resources of the main library to read about the DMFS and its habitat. We will also search for information on the Internet.
3. *Digital data.* We will obtain and analyze digital data for the three parcels. We have available to us at the university TIGER line files and other land use files that we will import into ATLAS GIS and will then integrate into a base map. Once the base map is created, we will split land use information into layers to establish a database for analysis.
4. *Recommendation.* Based on our research, we will develop a recommendation concerning which parcel contains the most suitable habitat and will submit a progress report on April 27 and a final report on May 16.

The attached schedule details our tasks, their proposed completion date, and the allocation of tasks among team members. It also notes the schedule for our team meetings.

### Justification

The plight of the Delmarva fox squirrel is a pressing ecological problem. Unless corrective action is taken at once, we may lose this valuable part of nature forever. As a team, we represent three different kinds of expertise that can be focused on this problem. Julie is a natural resources major who has completed an independent study project on wildlife habitats in Sussex County. Karen is a civil engineering major with a specialty in transportation and landform. Both have extensive experience with global information systems. Tim is an English major concentrating in technical writing and is especially interested in graphics. All of us are strongly committed to enhancing the environment. With your approval, we will begin phase 1 immediately.

**FIGURE 14.2** **A Collaborative Student Proposal (continued)**

3

**References**

Allen, A. W. (1982). *Habitat suitability index models: Fox squirrel.* U.S. Dept. of Interior, Fish and Wildlife Service FWS/OBS-82 (pp. 10–18).

Bendel, P. R., & Therres, G. D. (1994). Movements, site fidelity, and survival of Delmarva fox squirrels following translocation. *American Midland Naturalist 132*(2), 227–233.

Deuser, R. D., Dooley, J. L., Jr., & Taylor, G. J. (1990). *Habitat structure, forest composition, and landscape dimensions as components of habitat suitability for the Delmarva fox squirrel.* U.S. Dept. of Interior, Fish and Wildlife Service general technical report, pp. 52–108.

**References are in American Psychological Association (APA) Style**

FIGURE 14.2 **A Collaborative Student Proposal (continued)**

the appropriate technology and thus understand the highly encoded description (note the acronyms):

> The Advanced Research Projects Agency (ARPA) and the Rome Laboratory (RL) are jointly soliciting proposals in the area of information science and technology in support of ARPA-Rome Laboratory Planning Initiative (ARPI). ARPI delivers enabling technology to a wide array of next-generation concept demonstrations including the ARPA Advanced Technology Demonstration (ATD) in Joint Task Force Command and Control, associated other ARPA ATDs that will benefit from command and control like functionality (e.g., Agile Manufacturing), RL ATD's USTRANSCOM Planning Tools and Air Campaign Planning Technology Insertion Program (proposed), and the Joint Directors of Laboratory's demonstration initiatives. It is strongly desired that technology developed by ARPI also impact the proposed national challenge problems that are envisioned to be part of National Information Infrastructure (NHI) programs.

### Unsolicited Proposal

An unsolicited proposal results from the writer's own perception of a problem. You have to convince the reader there is a problem before you can begin to write about solving it. This situation is particularly risky within organizations because you may point out a problem to the very person who created it. You also risk your own time in pursuit of a problem no one may see as such. But such a proposal may be just what you need, as in Figure 14.1.

## STRUCTURING THE PROPOSAL

Although proposals vary greatly depending on the context that gave rise to them, they are likely to contain most of the segments described in Figure 14.3. Draft your proposal with these in mind. In addition, attend closely to any specifications your reader gives you, as in Figure 14.4. These may be exceedingly detailed, down to the point size of type for the final document, method of electronic delivery, method for calculating the budget, and the like. Any small variation from these specifications may cause the proposal to fail. Take the guidelines seriously. *A Closer Look: Storyboards* shows one technique often used by proposal teams to structure their information. The following sections discuss each segment of a proposal.

### Introduction

Because many proposal evaluators decide after reading only the first paragraph or two whether to read further, pay particular attention to your introduction. The introduction sets the stage. Spend more time on the problem under review when the proposal begins at your own initiative then when you are asked to solve a problem the reader already knows about, perhaps all too well.

The following paragraphs open a proposal from an administrator at a large law firm that aims to correct a problem she has identified:

Identifies the problem: complaints

Suggests two causes of the complaints: diversity of pay scales and invoice periods

States a third cause: more cases without new support

States the proposal and high likelihood of success

> We've received a growing number of complaints from the insurance companies we work with concerning our billing practices. As you know, each company dictates rates of payment for the professional services of law offices. Each company has a distinct scale for senior and junior partners, senior and junior associates, clerks, and paralegals. Moreover, each company has a different invoice period. The number of insurance-related cases has grown steadily in our firm. But that growth has not been accompanied by a growth in the administrative systems to serve them. In the last year we've received too many complaints concerning rates that were charged either incorrectly or inconsistently and invoices that arrived either late or early. I propose to research software systems that will help us centralize accounting procedures in the office and correct these improper billing practices.

In the paragraph that follows, Johnson Engineers sells their approach to a solution:

Summarizes the RFP to show that Johnson understands the problem

Summarizes qualifications

Briefly states the proposal

> The water supply in Atwater County is steadily decreasing. Without a plan to alleviate the supply deficit, the county risks extreme water shortages and an inability to expand the existing system as the community grows. For this reason the Metropolitan District Commission (MDC) has requested an evaluation of the feasibility of a reservoir storage facility at one of three potential sites in the county. At Johnson Engineers, we offer over 23 years experience in conducting such studies. We propose to investigate the ecology, geology, and water availability of each of the three sites. On the basis of this investigation, we will recommend to MDC the best location for the proposed reservoir.

For more about memos, see Chapter 20

If you write the proposal as a memo, the heading provides a preview for the reader, who then comes to the memo itself with its topic in mind (Figures 14.1 and 14.2). The *implementation* proposal (Figure 14.1) shows in the introduction how the writer turned his own observation of a problem into a situation in which his supervisor requested the proposal. The

**Executive Summary**

**Introduction**

Summarizes your proposed activity by stressing benefits to the reader. Provides the context by briefly restating the problem (if the proposal is solicited) or establishing the problem's background and urgency (if unsolicited). Overviews the content and plan of the proposal for a wide readership.

**Statement of Problem**

Identifies the problem necessitating the proposed work, with enough detail to make the problem clear to the reader. Provides a review of relevant literature if that is needed to establish the significance or dimensions of the problem.

**Objective(s)**

Lists the specific, measurable outcomes you plan to accomplish. Explains the solution you will propose. In a research proposal, shows that your work will contribute to the theme established by the funding agency or the assignment made in class.

**Method or Activities**

Explains either how you will implement your solution (if the solution is known) or how you will conduct research to support your hypothesis about the solution (if it is unknown). Ties the activities directly to your objectives. Convinces the reader that your approach is reasonable, suited to your resources in people and facilities, likely to succeed, and better than the competitor's. May include a review of literature to show how the proposed approach derives from but improves on that of other workers, and is innovative and distinct. Notes compliance with any federal, state, or local laws in undertaking the work. Indicates your procedures for ensuring quality control.

**Management**

In a collaborative proposal effort, shows how the team will be coordinated, scheduled, and monitored. Profiles the key staff along with the extent of their participation in the proposed project (extended biographies may be included in an appendix).

**Schedule**

Places your implementation or research activities on a time line. Convinces the audience that the time line is realistic. Serves as the proposal at a glance.

**Justification**

Answers the question "Why you?" Provides your track record of relevant accomplishments. Assures the reader that adequate staff and facilities are available to carry out the project as outlined. Describes laboratory or field sites, specialized equipment, and computer and other information systems matched to the tasks. Convinces the audience that the project is worth doing.

**Budget**

Assigns dollar (or British pound or Deutschmark) values to all activities or resources mentioned in the proposal. Often read first as, in effect, the quantitative abstract of the proposal. Divides the total budget into categories.

FIGURE 14.3 **The Proposal Genre: Structure and Common Elements**

All proposals must be in the following format; nonconforming proposals will be rejected without review. A proposal must be in a single volume and include the following sections, each starting on a new page (where a "page" shall be no greater than 8½ inches by 11 inches in type not smaller than 12 pitch). The total page count for all sections below, except sections O, P, and Q, must not exceed 55 pages; over length proposals will be rejected without review.

A. A cover page including BAA number (BAA #94-38), proposal title, plus technical and administrative points of contact along with their telephone numbers and electronic mail address (if any) followed by an official letter.

B. A one page summary of the innovative claims for the proposed research, including highlights of new ideas and projected impact of the proposed research on the goals of the Planning Initiative.

C. A one page summary of the deliverables associated with the proposed research.

D. A one page summary of the schedule and milestones for the proposed research, including estimates of cost in each year of the effort and total cost.

E. A one page summary of any proprietary claims to results, prototypes, or systems supporting and/or necessary for the use of the research, results and/or prototype. If there are no proprietary claims, this must be stated.

F. A clear Statement of Work (SOW) of not more than 3 pages, detailing the scope, background, objective and approach of the proposed effort and citing content and timing of specific tasks to be performed as well as specific contractor requirements. If not included in section D, include here a detailed listing of the technical tasks/subtasks organized by contract year. Note: the SOW should contain a generic task to support a TIE or (IFD) and an optional generic task to support technology transition.

G. A description, no more than 3 pages long, of the results, products, and transferable technology expected from a prospective user's point of view. This section should contain a clear description of how results will be made sharable throughout ARPI and what use these results might be to other groups. In addition, this section should address specific innovative approaches the offeror will take to facilitate technology transition (e.g., send graduate students to a partnered company for a "summer with industry," conduct a summer school, etc.).

**FIGURE 14.4**
**Format Specifications from an RFP**
This segment shows the detailed instructions about format common especially in RFPs from the U.S. Department of Defense.

See Chapter 15 for the final report of this project

opening is short, appropriate for a business document. The *research* proposal (Figure 14.2) addresses the students' professor, who set several conditions for the final project to be detailed in the proposal:

- Identify and address an audience (not the professor).
- Choose a problem related to the environment.
- Use both empirical and document-based research methods.
- Use computer mapping.

The introduction immediately responds to the first three items and sets up later detailed discussion of all four.

A CLOSER LOOK

# Storyboards

*Storyboards* are a widely used, graphics-based technique for productive collaborative proposal writing. The approach derives from the motion picture industry, where a story line is broken out into a series of pictures mounted in sequence on a wall. A storyboard proposal is, in a sense, a graphic outline that defines each module of the proposal. Each board has at least one graphic, a heading, and brief text summarizing a section of the document (see the model below, courtesy of Chris Miller). These boards are mounted on the walls of a proposal preparation room. Team members walk past the boards, noting the approach and sequencing, any problems in emphasis, and any missing information. They write their comments on the boards. Sometimes they vote if a decision is needed between two or more plans. The team develops consensus and creates, in effect, a draft and visuals that need only some modest further work to be a successful proposal. Recently, the storyboard approach has been translated into software that allows for similar consensus building without the need for boards and walls.

Proposal Storyboard

Topic ____________________

Name ____________ e-mail ____________

Department ____________________

Argument ____________________

Write a topic sentence for each paragraph:

1) ____________________

2) ____________________

3) ____________________

4) ____________________

5) ____________________

Figure No ______ Caption ____________________

Approved ____________ Date ____________

519 DCA 12M

### Statement of Problem

The process of preparing a proposal starts with the identification and analysis of a problem, a situation ripe for change. If you have been *assigned* a problem, or if you are responding to an RFP, comb the assignment or the RFP for every aspect of the problem to be addressed. On that basis, decide if you indeed have what it takes to write the proposal. You can't turn down a classroom assignment, but you have that option in other contexts. In the classroom, too, select problems that are workable. Do you have the skills to study the situation or the solution to match the problem? Can you comply with the specifications? Do you have time to prepare the proposal?

As you gather information, make sure you are dealing with only one major problem per proposal. To sort through the problem, read articles in pertinent journals or magazines or search documents available online or in Web sites applicable to the topic. Such reading will aid you in understanding the dimensions of the problem and how others have tried to solve it. Get the facts straight. The Delmarva fox squirrel team includes a brief review of literature that helps identify the problem and establish its importance. The inventory proposal delineates four causes for the delay in processing shipments at the sporting goods store.

The following problem statement anchored another student proposal for research into a better system for crop rotation.

> Along the Atlantic coastal plain, the most commonly grown crops are corn and soybeans. Both crops are highly marketable and provide good yields. These crops are either grown continuously in the same field from year to year or in rotation, generally with one another. Rotating the crops has several benefits including decreased pressure from pests and disease, decreased soil compaction, and decreased need for fertilizer. Researchers have been investigating another crop to add to the rotation, preferably one that can be sold in a less competitive market than that for corn and soybeans. (Hamilton)

### Objective(s)

How do you propose to deal with the problem? Every proposal must have a core idea, its major sales pitch. This becomes most evident in the objectives segment. The objectives are outcomes that can be measured to tell you the problem is solved; to ease that measurement, objectives are usually stated in list form. They set the scope of the project. In an implementation proposal, it's the selling feature—like the "timely processing" of inventory. In a proposal for research, the core idea is your research questions or your hypothesis about what you'll find when your research is completed. In the Delmarva fox squirrel proposal, the students have one major objective (to analyze three parcels). They then make that analysis concrete through setting six research questions, each tied to the problem and each forecasting the necessary technical tasks to find the answers.

For more about hypotheses, see Chapter 11

From one hypothesis you can derive several different objectives. Here's an hypothesis concerning crop rotation:

- Kenaf (*Hibiscus cannabinus*) looks promising as a crop to use in rotation with corn and soybeans because the market value of its products is increasing and its management practices are similar to those for crops currently produced in this region. (adapted from Hamilton)

This hypothesis embeds at least three different objectives that could provide different avenues for research on kenaf:

- The objective of my research is to review the literature published in the last five years concerning the cultivation of kenaf in North America.
- The objective of my research is to determine the marketability of kenaf products in Virginia.
- The objective of my research is to assess the acceptability of kenaf as a rotation crop for farmers in Virginia.

### Method or Activities

How will you meet these objectives? In an implementation proposal, the methods segment shows the steps you'll take, as in the segment headed "Implementation of the Proposed System" in Figure 14.1. In a research proposal, you describe your method for testing the hypothesis or reaching the objectives. Your hypothesis may be your proposal's major selling point, but you may also find that your method is unique, and thus it holds the major appeal. Set the tasks you need to perform, perhaps divided into phases, as in the Delmarva fox squirrel proposal.

The more detailed and specific your description, the more convincing you'll be. For example, if an objective is to determine the likelihood that farmers will plant kenaf in Albermarle County, Virginia, as a rotation crop, then a task might be

- Survey farmers in Albermarle County.

This task may need further specification in subtasks:

- Obtain a list of farmers from the county extension agent.
- Write a brief questionnaire about rotation practices and preferences and review it with the county agent.
- Determine which farmers have Internet access.
- Send the questionnaire electronically to farmers who have e-mail access and by postal delivery to the others.

Readers evaluating your proposal will measure your activities carefully to determine if the scale or scope is appropriate (Should you, for example, limit yourself to one county? Survey the entire state? Question every farmer? Randomly select one out of every four farmers?) and if the plan is realistic and likely to succeed. Figure 14.5 represents visually the proposed tasks to be performed by a design team.

### Management and Schedule

How will you manage these tasks, and when? You need to show readers that you have neither underestimated nor overestimated the difficulty of the work or the time it will take to complete it. To that end, you provide a

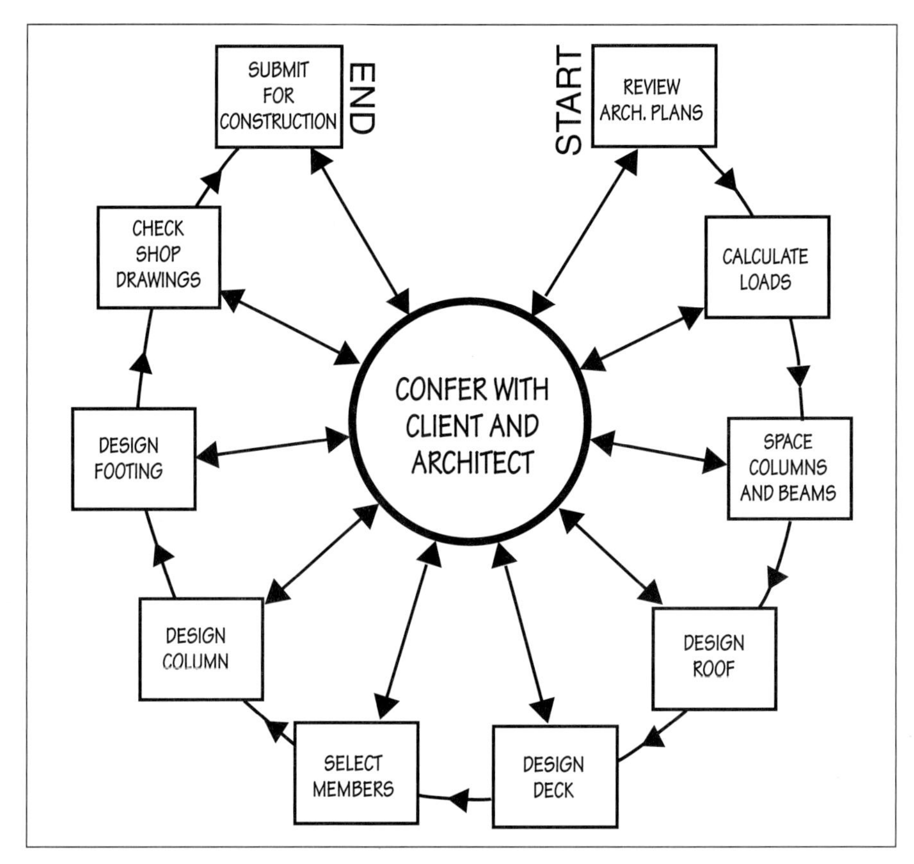


FIGURE 14.5 **Construction Development Tasks**
A persuasive visual that indicates tasks to be performed by the design team of structural engineers and emphasizes close communication with both the client (a developer) and the architect.
(Courtesy of Brian Tarantino.)

schedule of tasks and their completion dates, usually in visual form (see Figure 14.6). In a formal proposal for a collaborative project, you also provide information about team member credentials. Depending on the length and complexity of this information, you may include it as part of the methods or justification section (as in the Delmarva fox squirrel proposal) or in a separate segment of its own.

## Justification/Benefits

Why, in the end, should the reader accept this proposal? Your major selling point may be your core idea, your method—or you. Particularly in high-context settings, your reader's trust in your professional reliability may be enough to convince. As your personal resume supports what you can do on the basis of what you have done, so a proposal persuades on the basis of the proposer's track record and character. Avoid simply boasting "We make

| Tasks | Who | Apr | | | | | | | | | | May | | | | | | | | | | | | | | | | | | | | | | |
|---|---|---|---|---|---|---|---|---|---|---|---|---|---|---|---|---|---|---|---|---|---|---|---|---|---|---|---|---|---|---|---|---|---|---|
| | | F | S | S | M | T | W | T | F | S | S | M | T | W | T | F | S | S | M | T | W | T | F | S | S | M | T | W | T | F | S | S | M | T |
| | | 21 | 22 | 23 | 24 | 25 | 26 | 27 | 28 | 29 | 30 | 1 | 2 | 3 | 4 | 5 | 6 | 7 | 8 | 9 | 10 | 11 | 12 | 13 | 14 | 15 | 16 | 17 | 18 | 19 | 20 | 21 | 22 | 23 |
| Outline Proposal | Team | △ | | | ▽ | | | | | | | | | | | | | | | | | | | | | | | | | | | | | |
| Draft Storyboards | Team | | | | △ | | | | ▽ | | | | | | | | | | | | | | | | | | | | | | | | | |
| Meet w/System Integrators | Jim/ Kevin | | | | | | △ | | ▽ | | | | | | | | | | | | | | | | | | | | | | | | | |
| Meet w/Vendors | Jim/Kevin | | | | | | | △ | ▽ | | | | | | | | | | | | | | | | | | | | | | | | | |
| Review Storyboards | Jim/Kevin/ Harley | | | | | | | | △ | | | | | | | | | | | | | | | | | | | | | | | | | |
| Discuss Investment | Team | | | | | | | | | | | △ | | | | ▽ | | | | | | | | | | | | | | | | | | |
| Draft Exec Summary | Jim/Harley | | | | | | | | | | | △ | | | | | | | | | | ▽ | | | | | | | | | | | | |
| Draft Tech Proposal | Team | | | | | | | | | | | △ | | | | | | | | | | ▽ | | | | | | | | | | | | |
| Draft Initial SOW | Kevin | | | | | | | | | | | | △ | | | | | | | | | | | | | | | | | | | | | |
| Draft Cost Proposal | Dennis | | | | | | | | | | | | | | | | | | △ | | | | | | | | | | | ▽ | | | | |
| Draft Final SOW | Kevin | | | | | | | | | | | | | | | | | | △ | | | | | | | | | | | | | | | |
| Red Team | | | | | | | | | | | | | | | | | | | | | | | ◆ | | | | | | | | | | | |
| Deliver Graphics to CS | Team | | | | | | | | | | | | | | | | | | △ | | | | | | | | ▽ | | | | | | | |
| Create Graphics | CS | | | | | | | | | | | | | | | | | | △ | | | | | | | | | | | ▽ | | | | |
| Revise Proposal | Team | | | | | | | | | | | | | | | | | | | | | | | △ | | | | | | | | ▽ | | |
| Revise Exec Summary | Jim/Harley | | | | | | | | | | | | | | | | | | | | | | | △ | | | | | | | | ▽ | | |
| Edit/Proof | Harley | | | | | | | | | | | | | | | | | | | | | | | | | | | | △ | | | | ▽ | |
| Originator Review | Jim/Kevin | | | | | | | | | | | | | | | | | | | | | | | | | | | | | △ | | ▽ | | |
| Layout | Gloria | | | | | | | | | | | | | | | | | | | | | | | | | | | | | | △ | | ▽ | |
| Cost Review | Dennis | | | | | | | | | | | | | | | | | | | | | | | | | | | | | △ | | | | |
| Copy/Bind/Deliver | Gloria | | | | | | | | | | | | | | | | | | | | | | | | | | | | | | | | △ | |
| Due ARPA | | | | | | | | | | | | | | | | | | | | | | | | | | | | | | | | | | ◆ |
| | | 21 | 22 | 23 | 24 | 25 | 26 | 27 | 28 | 29 | 30 | 1 | 2 | 3 | 4 | 5 | 6 | 7 | 8 | 9 | 10 | 11 | 12 | 13 | 14 | 15 | 16 | 17 | 18 | 19 | 20 | 21 | 22 | 23 |

**FIGURE 14.6 Defense Industry Proposal Schedule**
The executive summary is outlined and drafted as a control device throughout the planning of the proposal.
(Courtesy of Lockheed Martin Advanced Technology Laboratories.)

the finest widgets in the world." Instead, be concrete: "Our widgets have operated for more than 25,000 hours without a single failure." Give the reader clear and compelling reasons to choose you (Stein). Detail your track record on similar problems. Also provide the organization's credentials and the available support system—in people and facilities. These resources must be adequate to complete the tasks successfully. If you lack appropriate expertise or resources, you may arrange a collaborative effort with another organization that provides those resources.

### Budget

How much will all this cost? In the budget, you identify the resources needed to carry out what the document proposes. Professors may not expect budgets in student proposals written as classroom exercises, but evaluators in the workplace often start their reading with the budget. They may see the budget as the distillation of the entire proposed project and a test of that project's realism and worth. As elsewhere in the proposal, closely follow any direct guidelines in preparing the budget. Calculate salaries or other payment to personnel as requested: by hour, by day, by month, or by some other unit.

## MANAGING A COLLABORATIVE PROPOSAL EFFORT

While Peter Sanchez wrote on his own, the Delmarva project represents a collaborative effort. Such collaborations are common in proposal writing, especially with a complex or large-scale problem or solution (or both). You may find it helpful to join forces with people who share your expertise, or represent other kinds of expertise, to develop an effective proposal. For example, a typical proposal from a defense contractor may require a large team. In one recent proposal effort, the team consisted of 40 people: 4 design engineers to work on an entire system, 12 engineers for components, 2 program managers, 2 cost estimators, 3 proposal specialists, 1 marketing person, 3 people from quality control, 2 from manufacturing, 2 from safety, 1 photographer, 2 illustrators, 3 word processors, and 3 editors. Another drew work from 300 engineers (Stein). The cost of preparing such proposals averages $500 to $700 a page.

### Agreeing on the Core Concept

Whether your proposal team is large or small, its first responsibility is to agree on the proposal's core concept. As in any collaborative effort, proposal teams can run into trouble when members approach the project with varying ideas about the proper outcome, varying levels of enthusiasm, and varying commitments to a task that is often not their main responsibility. A clear sense of purpose and good leadership help the team overcome these problems. They also prevent team members from duplicating each other's

CROSSING CULTURES

## Multinational Proposal Teams

Collaborating on a multinational proposal team requires extensive upfront planning as well as patience. U.S. style is often competitive, aggressive, detailed, and urgent. But many Europeans chafe at the "persuasiveness that must be written into U.S. proposals," they are not familiar with deciphering Requests for Proposals or negotiating detailed statements of work, and they take deadlines less seriously than Americans (Wiese MG-85). Americans are used to the immediacy of electronic communication, they resent delays, and thus they sometimes simply write over the contributions of their European teammates, creating an "understandable mistrust" (Wiese MG-85).

In addition to technical concerns, the team also must address such issues as currency exchange rate differences, export and import licenses, percentage of work to be conducted in each country, and patents and intellectual property rights. If a joint venture is to be established, it may require an international business plan, legal incorporation, its own logo or signature, and a physical headquarters somewhere. Translating the executive summary of a proposal, or an entire proposal, can add other difficulties, because many terms that are standard in the United States can mislead or offend in another context, because equivalent terms are not always available or may be debatable in another language, because the fast pace of writing may not be matched by similar speed in translating, and because any British members of the team may object if American English prevails.

To offset these difficulties, the American members of a multinational proposal team need to spend more time talking with their teammates and set a pace that allows for that process. As one expert notes, "If your company can do something in a day, it will take five companies two weeks" (Wiese MG-86). All members of the team, or at least representatives of each national partner, should sit in on discussions of draft components of the document. Such open discussion reveals variations in the way different companies do the same thing along with "national issues and sensitivities" and helps to resolve differences in emphasis, terminology, and values (Wiese MG-87). It also helps to develop a glossary and style guide so that the resulting proposal displays consistent usage and speaks with one international voice.

work or working at cross purposes, one member, for example, preparing extensive demographic data on a region that other members decide isn't relevant to a proposed survey. Sorting through core ideas and tasks early in proposal preparation and debugging potential solutions mean fewer rewrites and less overtime fixing a poor proposal.

### Assigning Tasks

As they carry out the tasks required to fulfill the proposal's purpose, team members usually form into three subgroups, each group performing a major proposal function: marketing, technology, and communication. Often, the groups perform their functions in a linear sequence. The marketing group, for example, begins the process by initiating or maintaining contact with the client or customer and then identifying the problem to be addressed. Marketing next turns the problem over to technical experts who determine whether the company is the right one to solve it and, if so, how to solve it. Finally, the communications specialists write the proposal. Several proposal experts, however, say the process works more effectively when representatives of all three functions collaborate throughout the processes (Stein). They thus can raise and resolve conflicts in advance of the competition external to the company.

In this coordinated approach, everyone contributes to the discussion and everyone writes. The experts at each function, however, take primary responsibility for developing certain kinds of information. The marketing experts, for example, analyze the context. They determine the customer's values and needs as well as the "critical success factors," that is, how the customer will measure the success of the project. To do this, they monitor potential opportunities, analyze the competition, and develop a strategy for tracking the customer (a "pursuit plan"). The technical experts provide the major content of the document. Depending on the proposal, that group may include scientists, engineers, lawyers, accountants, or other professionals. Some establish the detailed statement of the problem; the core idea, product, or service that will solve the problem; and the method for implementing that idea or researching how to prove it. That method may require elaboration in a step-by-step description of a procedure. Others, experts at financial matters, create the budget. If the proposal requires extensive legal commentary, then lawyers may prepare that information. Communications experts interweave the marketing and technical efforts to articulate the clear, persuasive message each proposal must convey. They review the entire proposal to make sure it is well organized, easy to read, and well suited to the customer. They also supervise the design and production of the proposal.

### Monitoring the Team

Team members meet regularly to make sure the work is proceeding appropriately. They may also keep in touch through frequent e-mail or phone contacts. As a group, they monitor their work. In addition, the leader keeps close tabs on progress. Three specific tools for monitoring are a schedule, a critical review, and an executive summary.

**Schedule.** Proposals are often written under near-panic conditions. To reduce the sense of panic and foster good thinking, even when deadlines are tight, the team needs to plan its activities. That plan is made concrete in a schedule that establishes each task and names the person responsible for that task. Begin with the final deadline and work backward, dividing the time available by the tasks to be performed. Note, by the way, that most final deadlines are firm; if your proposal isn't received by that date, it isn't considered. Figure 14.6 shows the schedule for preparing a complex proposal in the defense industry.

The schedule helps coordinate the team effort by reminding everyone of what has been done and needs to be done and providing the leader with information about any need to redirect the work or add to or reduce the team. Several schedule forms are common, and many of them are available in computer programs that automatically plot and update the sequence.

**Critical Review.** In some intensely competitive situations, the proposal team may also monitor the quality of its work through a critical review. For the review, similar to those that architecture students, for example, participate in at the end of a term, a panel of experts questions the team and evaluates the proposal. A common form in the defense industry is called a "Red Team Review." The Red Team consists of company experts who take the role of the customer or client and look for any weakness in content or presentation. Red Team comments then help the team revise their draft.

**Executive Summary.** A third monitoring device is the executive summary, a statement of the proposal in miniature. Although it is sometimes written after the fact, the summary can serve an important role in providing a blueprint for the team if it is written before the subteams develop their detailed information. It helps ensure a coherent and persuasive message as the team identifies the problem, explains its approach, and justifies its approach as the best one. A critical review often focuses heavily on the executive summary (Figure 14.7).

**For more about executive summaries, see Chapter 13**

## DELIVERING THE PROPOSAL

How your proposal *reads* and *looks* conveys an important message about how your proposal will *work*. Review your draft for style and design.

### Style

Assess carefully your audience's values and way of seeing things and gear your expression accordingly. Nigel Greenwood, a technical writer in France, for example, notes that French proposal writers consider it only respectful to assume that readers can work out the details on their own. They thus emphasize imagination and elegance of language over explicit, factual content and cultivate ambiguity rather than clarity. In addition, the French tend to prefer abstract ideas and to rely on authority, while people in Anglo Germanic cultures like that in the United States focus on more prosaic issues like who will do the work and how the solution will be implemented. So French proposal writers may use long, sophisticated sentences that digress with abandon and follow several thoughts at the same time. An evaluator in the United States would expect a more rigorously linear structure.

In addition, be careful to use the audience's specialized language, and, at the same time, avoid your own code words that the audience might not understand. Check the RFP, for example, for key terminology and make

## Innovative Claims A

### *The information associate will enable users to act on critical battlefield data in minutes rather than hours.*

Problem: Distributed information spaces similar to the Internet are emerging as critical elements of military operations. The Army's Tactical Internet will provide numerous information sources that could improve commanders' ability to make critical decisions. Information sources on this network might allow an analyst to identify critical targets, such as mobile tactical missile launchers, in time for the commander to neutralize their effect on the battle. However, to exploit this potential will require that users become more and more expert in information access; even then, crucial data may be overlooked in the mass of information flooding the network.

Proposed Contribution: We propose to solve this problem by developing a scalable *information associate* that will use its own knowledge of the infosphere to satisfy user information goals, and free users to focus on their primary tasks.

To illustrate this concept and its benefit, the graphic shows how an information associate will identify critical data throughout the tactical network, and then locate critical target indicators in a distributed information space on a land battlefield.

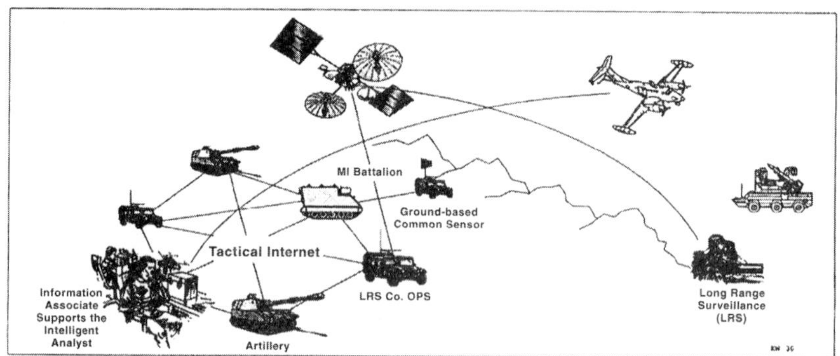


*The information associate will enable users to locate, identify, correlate, and act on critical, battle-deciding data in minutes rather than hours.*

Enabling Technical Developments: We propose a new approach to the development of an information associate. The associate will consist of a collection of networked intelligent agents directed and managed by the associate. As the network expands, this agent-based associate will expand with it, scaling up and making the network itself the associate.

IIA-1

Use or disclosure of proposal data is subject to the restriction on the title page of this proposal.

**FIGURE 14.7**
**Executive Summary**
The draft of this summary helped coordinate the proposal team.
(Courtesy of Lockheed Martin Advanced Technology Laboratories.)

sure you use those terms in your response. Avoid synonyms (calling something a "transmitter" in one place and a "beacon" in another) because they may confuse readers. Spell keywords the way the audience spells them, even if that's different from your organization's style.

To match the needs of most U.S. evaluators, be positive and bold with verbs, avoiding passives ("what is to be done") that cause ambiguities. Instead, note that specific people—*I* or *we* or a particular person whose name appears as the subject—will take specific actions. Differentiate among such hard-worked words as *analyze, determine, fabricate, study, define,* and *characterize.* Use the simple present tense for general descriptions and the future tense for specific statements of future activity:

- This proposal outlines a strategy for analyzing three parcels of land in Sussex County.
- We will conduct an on-site investigation.

### Design

No one would approve a proposal on looks alone, but a good-looking proposal predisposes the reader toward reading and toward a good opinion of you. So pay attention to design as you draft the proposal and especially as

**For more about design, see Chapter 7**

**ELECTRONIC EDGE**

## B2B

Businesses are increasingly buying from each other electronically. Digital connections improve management of the supply chain and help save on inventory costs. Electronic order forms save costs in printing, duplicating, and storing; ease accounting and database management in general; and speed delivery and response. Thus many routine bids are sent and evaluated automatically online. The growth of the Web has only heightened the appeal of e-business, and many more companies are getting into the act. Business-to-business electronic commerce—B2B in the shorthand of the Internet—has ratcheted up in scale (Oppel BU3).

The simplest B2B transaction occurs one to one when a buyer visits a seller's Web site, finds a product or service that fits a need, and places an order (Mihm 28). Many businesses buy their computers in this way. "Horizontal" or "functional" sites offer a range of products that apply to almost every business. Such products are referred to collectively as MRO (maintenance, repair, and operation) supplies. They include floor wax, stationery, industrial solvents, and the like. Some auction used equipment.

"Vertical" sites or markets, unlike horizontal ones, are industry specific. They offer everything from raw materials to finished products. ChemConnect, for example, offers a range of products from raw chemicals to finished plastics and resins. They also offer expertise at selecting the right product.

When many suppliers need to submit bids to one or two major buyers, those buyers tend to build their own exchange. General Motors, for example, built such a hub and has been joined by the other big automakers.

The greatest opportunity for start-up and other software companies to make money on B2B comes in markets that are fragmented, like agriculture, with many buyers and sellers joining and competing to conduct transactions. Participants will pay the intermediaries subscription or other fees for maintaining the exchange or small commissions on each transaction.

FIGURE 14.8 **Preliminary Segments of a Well Designed Proposal**
(Courtesy of Lockheed Martin Advanced Technology Laboratories.)

you develop the package you'll send to your client, sponsor, supervisor, or professor. As with a personal resume, good design enhances both readability and credibility. You can be creative for some proposals: One company that sells underwater drilling components, for example, uses an animation system that shows how a yet-to-be designed piece of equipment might operate. The demonstration aids the manufacturer in convincing oil companies to purchase the item. In Figure 14.8, you see segments of a well designed proposal to the U.S. Department of Defense. Its simplicity reinforces a message of cost effectiveness to an audience that might be suspicious of glossy treatment. Use lists and headings within the text to aid people who skim proposals and to reiterate your core sales message.

## PASSING THE TEST

A proposal has one central message: I (or we) can solve your problem or improve your situation. To convey that message

- Identify the problem or situation.
- Explain your solution or improvement or detail your method for finding a solution.
- Justify why the client or customer should choose you.

Conveying that information convincingly is the role of the proposal.

Your proposal will meet a critical audience, and many proposals fail that review. One evaluator for a government agency said she rejected 85 percent of the proposals sent to her on the basis of unconvincing problem statements. Other reasons for rejecting a proposal include the following:

- The solution doesn't match the problem.
- The research idea is trivial or overcooked.
- The methodology is inappropriate, not well thought through.
- The amount of work proposed is unrealistic for the time frame.
- The proposal doesn't comply with the RFP's stated specifications for content, format, or schedule.
- The proposed budget is too high.
- The proposal includes special provisions that conflict with stated policy of the client or customer.

When a proposal is rejected, it's a good idea to find out why. Government agencies, for example, usually arrange such postreview, or debriefing, conferences. The conference can help you shape future proposals.

When the proposal is accepted or approved, you undertake the work. The proposal turns into a binding contract specifying the amount and conditions of the work. As part of that contract, you may write progress reports at predetermined intervals. If the work requires research or development, its end may be marked by a final report and perhaps an article for colleagues.

**For more about progress reports, see Chapter 16**

A well written proposal provides a good framework for writing the final report. But the two are not the same. Although it includes information about method, the final report emphasizes results. Where the proposal looks forward, the final report looks back, and thus you approach the two documents differently.

For more about reports, see Chapter 15

Whether your proposal is a one-page memo or a multivolume collaborative venture, remember it is a sales document. As such, it should be accurate and promise only what you can deliver. It should meet the customer's needs. And it should be a showroom for your good ideas.

## CHECKLIST: Proposals

1. **Write a proposal to offer a product or service or to do work to solve a problem**

   Write a *bid* to offer a routine response to a known problem

   Write an *implementation* proposal to design and implement a new system or process

   Write a *research* proposal to investigate a difficult problem by a new method

2. **Read solicitations (Requests for Proposals) or look around for problems not previously identified**

3. **Structure the proposal to match the request or your profile of the problem**

   Statement of problem and benefits

   Objective

   Method or activities: your core idea

   Management

   Schedule

   Justification

   Budget

4. **Manage a collaborative effort**

   Generate and agree on the core idea, the selling point

   Assign specific tasks and schedule them

   Monitor progress to ensure quality, responsiveness to the sponsor, and timeliness

5. **Deliver the proposal in an attractive package, in the right medium**

6. **Follow through, both when your proposal is accepted and when it is not**

## EXERCISES

1. Justify your request for funds to carry out a research project that might make you eligible for a scholarship or an honors program. First, obtain a copy of the requirements for the project; then write a response that conforms to those requirements.

2. In a memo to the chair of your major department, propose a revised set of requirements for graduation. Be sure to indicate what the present requirements are. Then describe the criteria you think the requirements should meet and note why the present course requirements fall short of or exceed that standard. Finally, describe your proposed changes.

3. Write a memo to your college librarian proposing that the library subscribe to three (or more) journals not currently in their collection (see Chapters 5 and 21.). Explain their value to students, indicate the audience each appeals to, and give the publication data and price.

4. Write a proposal to the person in charge of campus operations to suggest a change in a university policy concerning, for example, student hiring, access to free recreational programs, campus safety, recycling, or computer use. Assume that your proposal is unsolicited. Review any current policy guidelines and detail the problem before you describe your proposed solution.

5. The following brief proposal suffers from several problems in both concept and expression. It is, for example, more a statement about what the author wants than about what the reader might need. Such egotism is a common proposal fault. The proposal is also overly general. Is the plan of attack convincing? What *problem* is the author describing? What is the structure? How can this proposal be improved? By "contours," the author means large mounds of dirt that serve as design features in housing developments and natural buffers against street noise and unpleasant views.

   The use of contours in the landscape is a too seldom used device for astetic enhancement. Contours can and should be utilized in the home landscape for these reasons.

   As a tool to be used, contours can be the backdrop for a focus point of flowers. Have you ever seen a flat rock garden?

   As you drive through the suburbans, have you noticed the boring overall effect from yard to yard? By adding a new dimension by the use of contours a new life is established in the lawn.

   The use of contours have gained much attention in the area of hiding unwanted areas from sight or sound.

   Ideas of contours can be found in the landscape books, but they remain vague. A publication is needed in the use of contours for aesthetic and functional purposes.

   I am a recent graduate in landscape architecture. Using money from odd jobs in that study I have travelled across America and Europe to further my own education in what is lacking. I've found that the area that is growing in attention is the use of contours. Students and professionals alike are looking for an informative study.

   Working freelance I can study where I want, with whom I want, and at the best time. I would compile the best information available from books, interview professors and professionals in landscape architecture across the country, and do field studies to enhance the written. By working in such a manner, a minimal budget will be required, and an excellent publication will be produced.

A progress report will be available on April 27. The final publication will be received by May 25. This will be followed on June 7 with an article that can be used to publicize the work, suitable for use in any number of popular magazines.

A demand is present for an informative publication on the use of contours in the landscape. Working by myself, with a small budget such a publication will exist to your companies credit. I ask for your permission to start research immediately. With little risk, you have so much to gain.

6. Highly competitive written proposals are common in the United States but are less common in other countries. Prepare a review of literature on the cultural dimensions of influencing decision makers or on the meaning of the term *proposal* in different languages, and write up your review in a brief memo addressed to your class (see Chapters 20 and 21).

7. Guidelines for writing proposals are widely available. Find articles, brochures, or Web sites that offer such guidelines, and summarize the advice from several sources about one particular kind of proposal. For example, the Web site of the National Science Foundation provides guidelines for its grants (www.nsf.gov). For advice on writing to nongovernmental organizations, visit the Web site of the Foundation Center (http://fdncenter.org/onlib/ufg/index/htem). The Earthwatch Institute provides information on environmental grant making (www.earthwatch.org).

## FOR COLLABORATION

At the beginning of a team reporting project, submit a proposal to your instructor that outlines the problem you will address, the audience for the report, your procedure (including the tasks you'll undertake and how you'll manage team assignments), your schedule, and the intended outcome. The proposal will help you coordinate teamwork and allow the instructor to evaluate your approach and suggest any corrections to it if necessary.

CHAPTER 15

# REPORTS



"IF IT ISN'T WRITTEN DOWN, IT DIDN'T HAPPEN."

A report furnishes a permanent record of some work and its outcome. Many research institutes make a point of saying, "Reports are our only product." Reports also help individuals and organizations find more efficient methods of production, raise profits, develop new products or markets, and meet social and environmental responsibilities. Like proposals, reports range from short, informal memos to multivolume documents. And like long proposals, long reports often represent collaborative efforts. In this chapter you will learn how to write a report.

## PURPOSES

First, determine your purpose in writing the report. That purpose also reflects the reader's reason for reading the report. Two major purposes for writing are to inform and to persuade; similarly, two major purposes for reading are to understand and to decide or act. Although informing and persuading can overlap, it's helpful to think of one major purpose as you select and structure information. If you aim to enhance your reader's understanding of a project, an event, a site, a concept, and the like, write an *information report,* which is largely descriptive. If you aim to help the reader make a decision or act in a specific way, write a *decision-making report,* sometimes called an *action report* or *recommendation report.* Your approach in that context is more overtly persuasive and argumentative.

### Information Report

An information report may include the results of personal observations, experiments, interviews, surveys, and meetings. It may provide background on a situation or a position, as in *A Closer Look: White Papers.* You may also write an information report to

- Define a concept.
- Describe a situation or event.
- Present findings from empirical research.
- Review the literature on a topic.
- Show the financial status of your organization.
- Show how your organization complies with environmental or other standards.
- Outline a marketing approach.
- Define a policy.
- Summarize what you found out on a trip.

### Decision-Making Report

In a decision-making report, you deploy your evidence in a well structured argument that aims to help the reader solve a problem, make a decision, or act in the way you think is right. For example, you may write a decision-making report to

- Examine the feasibility of an action.
- Justify an action by providing good reasons.
- Define the options and limitations in a situation.
- Provide recommendations.

An information report may lead to a decision-making report. For example, you might write an information report to summarize research comparing the reproduction rates of bald eagles in different areas of North America. Or you might describe your own empirical investigation into why bald eagles along the Maine coast are the slowest to reproduce. If you find that the birds display high levels of contamination by PCBs and mercury, and those toxins are inhibiting reproduction, you may then write a decision-making report that recommends action to reduce pollution and increase the bald eagle's reproduction rates.

## CONTENT

Second, gather and select information. Their purposes differ, but the two kinds of reports can draw on the same warehouse of information:

| Information (Descriptive Report) | Decision-Making (Problem-Solving Report) |
| --- | --- |
| • Review of methods for drying corn | • Comparison of corn-drying methods to determine which method is most effective for farmlands in Iowa |
| • Geotechnical evaluation of the Northwest coast site to measure it against U.S. criteria for a designated wetland | • Recommendation to a developer against purchasing a Northwest coast site |
| • Results of empirical research to determine the effects of exercise on the mental health of college students | • Guidelines for college students on how to reduce stress through exercise |

Determining which kind of report to write, and on what topic, is usually not too difficult on the job; the intended reader often makes the assignment. In your career, you may write reports on a wide variety of topics. For example, one increasingly common topic for such reporting is the environment, as *Crossing Cultures: Monitoring the Global Environment* suggests.

As a student learning about the construction of reports, however, you often must begin by choosing a subject to collect information about. Use your understanding of the purpose of reports to focus your choice. Let's assume, for example, that you'd like to write a report on solar energy. How can you narrow that topic? To prepare an *information report,* classify or divide the topic into appropriate components or categories, as in the following list of general categories applied to solar energy:

**For more about choosing a topic, see Chapters 4 and 5**

| | |
| --- | --- |
| source | wind power, tides, solar gain |
| type | active or passive |
| location | solar installations in a particular city or area |
| building | solar churches, solar residences |
| material | new materials for constructing photovoltaic cells |

To prepare a *decision-making* report, identify audiences who might be interested in problems or issues associated with solar energy and then gather and select information that will help solve their problem, as in these examples:

### Home Owner

- Examine a house to determine the feasibility of converting its present heating system to use solar energy.
- Design a solar system for a new house.
- Retrofit a solar system onto an existing house.

A CLOSER LOOK

## White Papers

The term *white paper* refers to a special report that a company issues to state a position. The term derives from British diplomatic usage: In England, the term *blue book* refers to an extended treatise on government policy (bound in blue covers) and a *white paper* refers to a shorter report not suitable for binding. That term was adopted in the United States to cover an official statement of policy on a special political, economic, or social issue. Technical professionals employed by corporations, government agencies, and consulting companies prepare white papers that contain

- Details of a proprietary approach
- Background for a product or service
- Product positioning and competitor analysis
- Rationale for one alternative strategy over another
- Defense of an action or product that went wrong

To enhance at least the appearance of objectivity in white papers, some companies hire outside consultants to prepare and sign the document. Although such documents often serve as marketing tools that promote a company's product or point of view, they tend toward "neutral, conservative formatting, a complete lack of hype and slickness in production, and a convincingly objective tone." Notes one writer of such documents, "The more obnoxious, unpalatable, and self-serving your position is, the more toned-down your White Paper should be. I would recommend 12-point Courier [typeface that looks like the product of a typewriter] for heavy-duty industrial-strength White Papers" (Plamondon).

State Energy Agency

- Recommend that the agency develop a program to increase awareness of solar energy among schoolchildren.

## DESIGN

Third, design information to achieve your purpose. Like other genres of technical communication, reports often follow conventional approaches. Learning those approaches will help you meet your obligations to your reader and to your information. This section first provides guidelines for structuring report introductions, often the hardest task in writing reports, and then shows how to follow up on the introduction's promises in three different genres of reports: an explanatory report, an IMRAD report, and a recommendation report. The first two types are information reports and the third is a decision-making report.

## Introduction

Whether you are writing an information report or a decision-making report, you need to perform two critical tasks in the introduction: draw the readers' attention and orient them to the content and the structure of the report. To draw attention, consider some of the following strategies used in these model paragraphs.

### State Your Purpose Directly

The purpose of this report is to present the results of a survey concerning the use of campus computer sites during the current academic year. All 11 sites participated in the survey. Academic Computing Services questioned both site monitors and student users about their level of satisfaction with the management of the sites, with currently available systems and connections to the Internet, and with logistics: hours of operation, accessibility of computers, and training programs. After a brief description of the survey methods, the report summarizes comments in each of these three categories: management, resources, and logistics.

### Establish the Significance of Your Topic

The first campus computer sites mainly attracted students in technical programs like engineering, science, and accounting. Recently, however, computer applications underlie instruction in every course. Many students arrive at the university with excellent general computing skills, especially skills at e-mailing and surfing the Web. But they often lack even a basic understanding of classroom computing, for example, using sophisticated statistical software, developing PowerPoint slides, or finding and evaluating research sources on the Web. Students are simply expected to demonstrate those abilities.

Student users thus place increased demands on site attendants to provide instruction in these tools. Time for that instruction, however, only adds to the tasks attendants must perform and causes inefficiencies in site usage. Moreover, the students are often frustrated in trying to accomplish their required course work. That picture, at least, emerged from meetings with several site attendants. To determine the level of satisfaction—or dissatisfaction—among current users of the sites as well as attendants, Academic Computing Services conducted the survey whose results are summarized in this report.

### Correct a Misunderstanding

As a group of site attendants representing each of the university's 11 computer sites, we have compiled this report to present our concerns about the current status of the sites. The recent survey conducted by ACS covered some of the pertinent issues, but we feel that several other topics need to be addressed before the university will have an accurate assessment of satisfaction levels at the sites. In particular, we think the ACS questionnaire did not provide enough emphasis on who is responsible for training students to use computers and on which applications the sites should support. This report thus focuses on these two issues while also providing some additional comments on each of the three categories of information requested by ACS: management, resources, and logistics.

CROSSING CULTURES

## Monitoring the Global Environment

One topic that leads to many reports is the environment, and both governments and corporations are recognizing that environmental issues cross national borders. To deal with such issues, organizations like the United Nations have convened conferences on the environment. These include the first "Earth Summit" in Rio de Janeiro in 1992, which was followed by the Kyoto World Summit in 1997, which focused on global warming. The goal of such meetings is to develop guidelines for environmentally responsible conduct.

Enlightened self-interest, public pressure, and government regulations are forcing organizations both to recognize the environmental consequences of their activities—like pollution and depletion of natural resources—and to account for them in financial statements and other public disclosures. For example, responding in part to issues raised by the *Valdez* oil spill, oil companies document their environmental responsibility in reports addressed to various audiences—the government, shareholders, neighbors, and the concerned public. Other companies who are members of the International Chamber of Commerce (ICC) write reports that show their compliance with its "Business Charter for Sustainable Development."

Companies are thus developing a more "green" outlook and documenting their environmental performance through reports that respond to specific guidelines as well as emerging business requirements and changing public expectations. Furthermore, as one major Danish multinational company who issues such a report, Novo Nordisk, notes, "behind the scenes, the accountancy profession increasingly recognizes that industry's environmental liabilities are likely to mushroom—and is beginning to think about how companies should reflect such liabilities in their annual reports and accounts" ("Novo Nordisk Comes Clean" 9).

Deciding what to include in such reports is a "free for all," according to one authority; at this point, only about 110 of the world's 37,000 multinational companies now produce them (Boulton 14). KPMG, an international management consultancy, recommends as good practice reporting bad news as well as good, detailing "site-specific environmental performance" measured against environmental targets, soliciting outside verification, and accounting for environmental costs and liabilities.

Some reports simply boast about the company's goals—without noting how they are being met. Others, like one produced by a U.K. home improvement retailer, go into detail about a corporate search for environmentally sustainable supplies of timber. In the United States, companies "face strong incentives" from the Securities and Exchange Commission to disclose their environmental liabilities in part because other government agencies keep tabs on environmental violations. Thus KPMG found that U.K. companies with U.S. listings provided more information than those listed only in the United Kingdom. In the United Kingdom, an Advisory Committee on Business and the Environment made up of corporate executives has been working with the government to establish voluntary standards for reporting (Boulton 14).

### Characterize Your Intended Audience

This report summarizes the results of a survey ACS conducted over the past three weeks to determine levels of satisfaction among attendants and student users at our computer sites. Because the report addresses an internal university audience, we have not detailed the locations and characteristics of the sites. The report also assumes familiarity with the software applications supported by ACS and with course structures and academic calendars at the university.

### Refer to a Direct Request

On 15 May, David Folsom, Executive Vice President, asked Academic Computing Services to prepare a report assessing how satisfied students are with the current operation of our computer sites. To answer his question, ACS conducted a survey of student users. In addition, to enhance the validity of our findings, we also surveyed site attendants. This report presents the results of both surveys.

### Bring the Readers Up to Date

The last five years have seen significant changes in the computing picture on campus. The major change is a vast expansion in the population of students using computers, both in their rooms and at campus sites. While five years ago the population of such users was restricted largely to students in technical fields, today students are required to use computers in all their courses. These new users have intensified the demand on our current resources. They have, for example, increased the need for site attendants to provide instruction on pertinent applications and hardware. Such instruction is essential, in part because professors rarely take time to tell students how to accomplish such tasks as using sophisticated statistical software, developing PowerPoint slides, or finding and evaluating research sources on the Web. Students are simply expected to demonstrate those abilities. When site attendants have to provide such instruction, that creates inefficiencies in the use of site resources and heightens frustration levels. Even more significant, the burgeoning of new applications has also caused pressures on ACS to expand its expenditures as the sites try to keep up to date.

These model paragraphs show how you might open a report. In addition, you might note any limitations in the scope or approach of your report. If, for example, your reader expects you to cover the costs of equipping and monitoring computer sites, but you have decided instead to focus only on the operation of different sites, mention the reasons for that limited scope in your introduction. For most reports, it's a good idea to end the introduction with a statement about the report's plan, for example, "The following report is divided into three segments, each detailing results on one of the survey's main categories: management, resources, and logistics." That final statement then leads the reader smoothly into the report's discussion.

## Explanatory Report

Each paragraph concerning campus computer sites might begin a different explanatory information report about the topic. An explanatory report, as the name implies, helps the reader understand some concept, event, situation, or

For more on description, see Chapter 10

object. Particularly if the report centers on a mechanism, organism, system, or process, it may become an extended definition or description.

**Compliance Plan.** The audience may dictate the report's plan when it demonstrates compliance with a request or legal requirement. For example, consider Figure 15.1, which shows the required structure of one common form of explanatory report, an Environmental Impact Statement (EIS). This structure is dictated by the U.S. federal government, which requires agencies or other organization to file reports when they propose some action, such as commercial developments, explorations for natural resources, and new parks or recreation areas, that might have potentially harmful impacts on the environment. The report begins with a description of the proposed action and then details the current environmental conditions and how they might change.

- A detailed description of the proposed action and need
  - Location of the project (in text, maps, and aerial photographs) and boundaries of the affected area
  - Objectives of the project and developmental phases, including the impacts of each phase if more than one
- A description of the existing environmental conditions, including especially schools, hospitals, and communities that might be sensitive to the proposed changes as well as socioeconomic conditions (taxes, employment levels, and the like)
- A discussion of the environmental impacts that will probably result from such action
  - Short-term and long-term impacts
  - "Avoidable adverse impacts" and "unavoidable adverse impacts," including discussion of any measures that could mitigate these impacts
  - Impacts on human health and aesthetics (for example, an increase in the level of noise or air pollution; a decrease in scenery or removal of historic natural or man-made artifacts)
  - Any irreversible commitments of natural resources necessary for the action
- A description of any alternative actions that may prevent some or all of the adverse effects. This description includes the costs and environmental impacts of each of those actions, which fall into three categories:
  - Projects with similar objectives
  - Projects with different objectives
  - The "no project" alternative

**FIGURE 15.1** **Required Structure for an Environmental Impact Statement, One Genre of Government Report**

(Adapted from Warden and Dagodag.)

**Classification Plan.** Another plan is to organize information through classification, as you can see in the following outline of a report that classifies alternative solutions to the problems of a regional transportation system into six categories (CRGOC):

1. Freeway alternatives
   a. Capacity expansion and operational improvements
   b. HOV (High Occupancy Vehicle) lanes
   c. Incident management
   d. Ramp metering
2. Arterial alternatives
   a. Capacity expansion
   b. Operational improvements
   c. Computer signal systems
   d. Access management
   e. Incident management
3. Transit alternatives
   a. Rail or busways
   b. Bus service
4. Demand management alternatives
   a. Parking policies
   b. Employee commute plans
5. Land use alternatives
6. IVHS (Intelligent Vehicle-Highway System) alternative

As the last segment of such a report, you may simply present the last topic promised in the introduction, after you remind readers briefly about that promise. The transit report, for example, ends with a discussion of intelligent vehicle-highway systems. If you said you would detail the extent of flooding in three areas, readers would be alerted to the end if you simply devote the last section to the third area. If your information is complex or extensive, however, you may end with a summary. The summary adds no new information but instead pulls together the report's major points.

## IMRAD Report

If your information report describes empirical research, such as a laboratory report or a clinical report, your audience probably expects a common structure for such reports, called IMRAD, whose name is an acronym for the report's components: introduction, materials and methods, results, and discussion. Figure 15.2 shows the traditional IMRAD structure.

**For more about IMRAD, see Chapter 11**

**Introduction.** To draw attention and orient the reader to an IMRAD report, use the introduction to create, as one group of experts defines it, a research space (Swales, quoted by Harmon and Gross). Do this in three moves:

*Move 1:* Define a research territory.

*Move 2:* Stake out a niche within that territory.

*Move 3:* Occupy or defend that niche.

For more about reviews of literature, see Chapter 21

Begin by setting the context—the static picture. Note prior research (that's the "review of literature") to establish the scope and importance of

**Introduction**

1. Who asked me to look into this?
2. Why did they ask *me* or our team?
3. Why did they see this as a problem?
4. Who else knows about this and has written about it (review of literature)?
5. Has anything happened like this in the company or as recorded in the literature before?
6. If so, what was the outcome then?
7. What limits did I have in solving this (budget, time, overlap with someone else's responsibility)?
8. What priority does this have in the company?
9. What was I specifically asked to do?
10. Are there any hidden agendas?

**Materials and Methods**

1. Where did I look for information?
2. What chief areas did I study?
3. Whom did I talk with?
4. What did I read?
5. What surveys or observations did I make?
6. What tools or machines did I use?

**Results**

1. What results did I obtain from my work? What did I find out?
2. Are these results accurate? How do I know?
3. Are these results valid? How do I know?

**Discussion**

1. Do the results show any trends? short term? long term?
2. What do they add up to?
3. How do they relate to other findings?
4. Do the conclusions match my assignment?
5. Have I overlooked anything I was asked to do?
6. So what? In the reader's terms, and the organization's terms, what does all this matter? Future work?

FIGURE 15.2 **Organizing an IMRAD Report**

the problem. Then cite what disrupts that static condition—a gap, an inconsistency, a piece that doesn't fit the puzzle—and explain why that disruption is undesirable. Finally, note how or perhaps why your research resolves that problem—and lead the reader into the report to find out.

**Materials and Methods.** To assess the validity of your findings, your reader will expect a detailed discussion of how you conducted your work. The following paragraph begins such a discussion in a report on the design of a windmill:

> First, I obtained wind data from the weather station. The data were used to determine wind velocities and the distribution of hours of useful wind throughout the year. I then studied the efficiency of three types of wind wheels—multivane, turbine, and propeller—in extracting energy from the wind. In particular, I looked at the airfoil cross sections of a number of propeller blades. These data provided information for the design.

**Results.** You are likely to present your results in quantitative or visual form, often a series of tables or charts. You may also use diagrams and photographs. If appropriate, compare your results with expectations based on theory or published research. Note any limitations on your findings, including statistical reliability, any bias in the sample, or inconsistencies in data.

**Discussion.** The discussion section shows what the results add up to, their implications and significance. You "occupy the niche"; that is, you show how your findings close the gap, resolve the inconsistency, add a piece of the puzzle, or solve the problem. You provide your conclusions about the evidence. Often, you trace a pattern of cause and effect, as in these segments from the discussion section in a report on a major coastal storm:

> The severity of the problem at Dewey Beach is in part due to the lack of a dune system in front of the buildings, the progressive loss of beach sand due to beach migration inland, the lack of a significant natural source of sand in the surf zone and just offshore to naturally replenish the beach, such as is found at Rehoboth Beach, and the impact of a migrating coastline on structures not designed to be affected by waves and surf. (Ramsey et al. 22)

**Because this is familiar information to the reader, effect and cause are briefly described**

> As stated in the report of January 4, 1992 (DGS Open File Report No. 36), the worst case scenario for a coastal storm impacting Delaware's coast that would produce severe and life-threatening conditions is as follows:
>
> 1. A slow-moving storm with tropical-to hurricane-force winds.
> 2. Landfall over the southern Delmarva Peninsula that places the Delaware coast in the storm's northeast quadrant.
> 3. Continuation of the storm over several tidal cycles.
> 4. Landfall during high tide or an astronomical high tide. (Ramsey et al. 23)

**List form makes each criterion stand out separately**

List your conclusions or present them in paragraphs. Listing makes the value of each conclusion seem equal. Paragraph form allows you more ease in discussing relative emphasis and priority.

### Recommendation Report

The report on computing sites explains levels of satisfaction, the report on regional transportation classifies alternatives, and the report on coastal storms describes a worst case scenario based on empirical research. If the readers of such reports posed a different question to the writers, or the writers had a different purpose, then each might become a decision-making report: recommending new procedures for campus computer sites, or advocating the best transportation alternative, or recommending a technique for mitigating coastal storms. To persuade the reader to accept your recommendation, you need to create an effective argument. Let's assume, for example, you think demand management is the best transportation approach. If you anticipate that the reader will welcome, or at least not oppose, that recommendation, then use this organization:

- Introduction: brief overview of the problem and statement of the recommendation: demand management
- Description of the criteria for making your recommendation
- Brief review of each rejected alternative, along with reasons for rejecting it, in order from the least likely to the most likely
- Longer discussion of demand management (the most likely)
    - definition of terms, if needed
    - detailed analysis of why each method for managing demand will work
    - parking policies
    - employee commute plan
- Final statement of the recommendation: initiate demand management practices

If you anticipate disagreement, consider withholding your final recommendation until you have detailed all the alternatives and convinced the reader to agree that only the last one you present—demand management—will work.

**Introduction.** The introduction to a decision-making report, then, draws attention either directly or indirectly, depending on your expectation about how the reader will respond. Here, for example, is a direct introduction to a recommendation report:

**The writer's technical assignment**
**Brief description of method**
**Solution to assigned problem**

John X. Fancher requested that Enviro Consultants Ltd. prepare a remodeling plan to conserve heat in his family's 50-year-old farmhouse. This report describes the plan, which is based on a thorough examination of the house. Recommended improvements should retain heat for longer periods, lower the fuel bills, and increase his family's comfort.

**Limited scope of the report**

The plan indicates recommended actions, but it does not discuss in detail what methods or materials to use. For these reasons, it does not include specific cost estimates.

**Overview of report plan**

The report is divided into two parts: the causes of heat loss and recommended solutions to alleviate these problems. Briefly, the causes of heat loss are heat transmission through the building materials and air infiltration (passage of air through openings in the structure). Transmission of heat is simple to understand, so most of Part 1 centers on air infiltration.

Brief summary of contents

Part 2 details the recommended solutions in seven categories:

| | |
|---|---|
| weather stripping | storm windows and doors |
| caulking | vapor barrier |
| windbreak | entrance foyer |
| insulation | |

The student decision-making report in Figure 15.3 uses a more indirect approach. Its indirection responds less to anticipated resistance from the reader, however, than to an expectation that the reader, trained as a scientist, would prefer an approach similar to that in the IMRAD report, adapted for a decision-making purpose. In addition, the specific recommendation appears prominently in the executive summary, which the reader would see before delving into the discussion.

**Criteria.** In Figure 15.3, the writers retrace their research process as backing for their final recommendation. They state their criteria for the selection in the executive summary and through the questions they include as the key to their research method. In any decision-making report, you need to include such criteria. Do so in the introduction, if the list is not long, or in a separate section if their extent warrants. In Figure 15.3, the writers answer the questions for each habitat parcel being examined.

**Conclusions and Recommendations.** Your examination of the alternatives leads to conclusions on which you base your recommendations, as in Figure 15.3. A reader can often predict your recommendations based on your conclusions, but they are not the same. Conclusions look to the past; recommendations look to the future. Recommendations are the life blood of decision-making reports. Here is one segment from the recommendations section of a report on "Growth in a Global Environment":

> The Global Climate Coalition recommends a policy including the following options:
>
> - Accelerate the pace of research into basic climate science and impact assessment.
> - Identify and pursue measures that will reduce the threat of climate change, yet also make sense in their own right.
> - Establish sustained research and development programs that improve the ability to economically produce and utilize energy with less potential for the accumulation of greenhouse gases.
> - Expand efforts to understand and communicate the economic, social, and political consequences of both climate change and proposed policy responses. (Global Climate Coalition 44)

Parallel imperative verbs in list form emphasize action

## FINAL TEST

After you have drafted your discussion, make sure it presents your information—and you—in the best light and meets your readers' needs. Figure 15.4 provides a list of ways in which reports go wrong. After describing each

JKT Consultants, Inc.

Environmental Specialists
University of Delaware
Newark DE 19717

Telephone (302) 837-8969

# Recommended Habitat For Conserving the Delmarva Fox Squirrel in Sussex County DE

Final Report: Phase 1

**Prepared for**

Division of Fish and Wildlife
Delaware Department of Natural Resources
and Environmental Control
Dover DE 19901

**Prepared by**

JKT Consultants, Inc.
University of Delaware
Newark DE 19717

16 May 2001

**FIGURE 15.3** **A Collaborative Recommendation Report**

(Courtesy of Julie A. Fine, Tim Kelly, and Karen Romanelli. Maps reproduced courtesy of the Delaware Department of Agriculture.)

## Executive Summary

The Delmarva fox squirrel (DMFS) has been a severely declining subspecies on the Delmarva peninsula for many years. To reverse this trend, the State of Delaware has offered three parcels to the Fish and Wildlife Division for the conservation of DMFS habitat and the reestablishment of a viable population of this native squirrel. The Division asked JKT Consultants to assess each parcel and recommend the most suitable one for these purposes. We used three criteria in our determination:

1. The presence of large, concentrated mixed or deciduous forest regions
2. The absence of fragmenting features such as highways, railroads, and secondary roads
3. The presence of streams in the forested regions

The land cover surrounding each parcel was also considered in the evaluation. We found the Hugwell/Johnson Parcel to be the best choice of the three. It contains the highest concentration of mixed and deciduous forests and is not fragmented by highways, railroads, or secondary roads. In addition, three streams are present within the forested area. Finally, the parcel is largely surrounded by forest cover which serves effectively to increase the available habitat. Overall, we feel that this parcel offers the best habitat and would yield the most successful conservation efforts.

**Brief statement of the problem**

**The assignment and authorization for the report**

**The conclusion stated directly**

**Justification**

**Recommendation**

## Introduction and Background

The Delmarva fox squirrel (DMFS) is the largest North American tree squirrel. It is a native and historically abundant member of Delaware's deciduous and mixed forest fauna (Bendel & Therres, 1994). This squirrel has always been an important subspecies as a habitat health indicator and as a valuable addition to Delaware's biodiversity.

Recently, however, DMFS populations have been falling rapidly. In 1967, the U.S. Department of the Interior placed this once common creature onto the endangered species list (32 FR 4001) with remnant populations surviving in only four Maryland counties (Deuser et al., 1990). In essence, the DMFS habitat has been largely destroyed throughout the state of Delaware, and the DMFS is now a locally extinct subspecies.

This decline has several adverse consequences. First, a significant piece of Delaware's heritage has been lost. Second, the DMFS is native *only* to the Delmarva peninsula and occurs nowhere else in the world. Hence, if these trends continue, this subspecies may be lost completely. The third and perhaps most frightening consequence of the DMFS's decline as a subspecies is the similar fate it projects for Delaware's other native species. The disappearance of the DMFS from Delaware is a sign that healthy, suitable ecosystems are also declining. As this happens, other species are sure to disappear along with the DMFS.

This condition has alarmed many members of both the public and the government. Therefore, in an attempt to reverse recent trends, the State of Delaware has offered three parcels of land located in Sussex County, DE, to the Fish and Wildlife Division (hereafter referred to as the Division) of the Department of

**Establishes the significance of the topic**

**Notes a current problem**

**Detailed description of the consequences of the problem**

**Need for solution and thus the request**

**FIGURE 15.3 A Collaborative Recommendation Report (continued)**

Natural Resources and Environmental Control (DNREC). One of these parcels will be used to reestablish a viable population of the DMFS and preserve the now vanishing habitat that this and many other species rely upon. It thus falls to the Division to select one of these parcels as the best candidate for these purposes and implement a conservation plan in that area.

To meet this objective, the Division asked JKT Consultants to analyze the three parcels and to recommend the most suitable site for the conservation of the DMFS. The recommendation also reflects the appropriate Habitat Suitability Index. To meet this goal, we addressed five questions in our research:

Criteria stated as questions that framed research tasks

1. What specific habitat requirements must be met for the survival of a viable population of DMFS?
2. How large is each candidate parcel?
3. How much of each parcel is categorized as mixed forests, and what other types of land cover are found nearby?
4. What is the availability of necessary resources for the DMFS such as water (e.g., streams), shelter, and food in each parcel?
5. How fragmented is each parcel and its composite mixed forest sites in terms of roads, highways, and railroads?

Limitation on scope

This report contains the findings of phase I of our inquiry. Unfortunately, we were unable to locate historical data on the three parcels and therefore could not complete that segment of the proposed research. However, information on the regions surrounding the individual parcels provided us with ample insight into the recent development trends in these areas.

## The DMFS Habitat

Answer to the first question

The basic land cover preferred by the DMFS is one in which a large portion of the vegetation can be classified as deciduous upland trees. The ideal forested land cover for this species has a sufficient canopy closure (40–60 percent) and minimal understory (<30 percent). Therefore, mature stands of upland deciduous or mixed forest provide the ideal habitat for the conservation of this species (Bendel & Therres, 1994).

Description organized by classification. Note that DMFS is consistently singular

In addition to land cover, the DMFS must have suitable food available. The staple diet of the DMFS includes mast (fruits produced by deciduous trees such as oak, walnut, beech, and sweetgum), insects, roots, and bird eggs (Allen, 1982). The DMFS also eats agricultural crops such as corn and soybeans when available (Brown & Yeager, 1945). This squirrel obtains water mainly by consuming succulent foods (Allen, 1943). Although water is generally not a limiting factor to the squirrel for this reason, it readily uses streams and other water sources when present (U.S. Forest Service, 1971).

The home range of the DMFS in the eastern United States averages 5 to 10 acres and increases as the availability of suitable habitat increases (Adams, 1976). Females are the less mobile of the two sexes and are thus more susceptible to adverse habitat changes. Therefore, it is in the DMFS's best interest to inhabit the largest expanse of land possible (Nixon et al., 1980).

FIGURE 15.3 **A Collaborative Recommendation Report (continued)**

3

This habitat description is listed in the Habitat Suitability Index Model for the DMFS (Allen, 1982) and validated by several current researchers (D. Hannab, personal communication, 5 April 199-; J. Mackenzie, personal communication, 11 April 199-; R. Northrop, personal communication, 17 April 199-). Overall, land containing large mature stands of upland deciduous or mixed forests with available water supplies and minimal fragmentation will maximize the Division's chances of establishing a viable population of DMFS.

### Parcel-by-Parcel Description

This section of the report describes each parcel in terms of questions 2 through 5 in our research plan (Figure 1).

**Advance organizer helps reader prepare for next sections and reinforces the logic of the plan, which moves from least to most appropriate parcel**

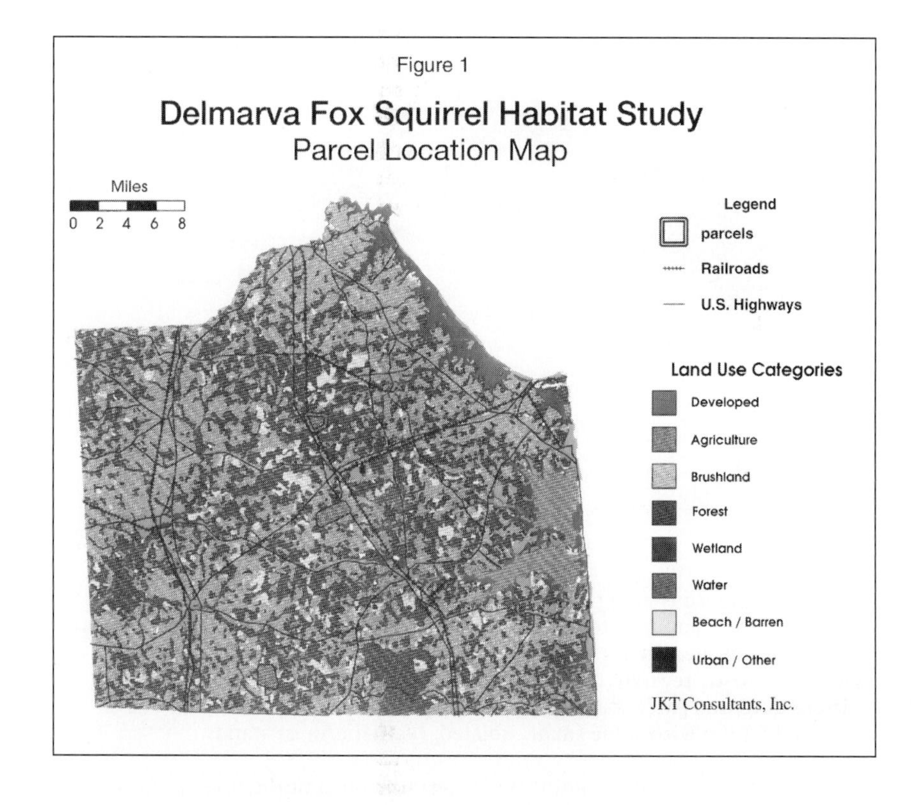


**FIGURE 15.3** A Collaborative Recommendation Report (continued)

4

## Branson Trust Parcel

Parcel descriptions are parallel in plan and organized to answer the five questions. Size (question 2) and land cover (question 3)

The Branson Trust Parcel encompasses 645 acres (perimeter = 4.97 mi) of land in central Sussex County (Figure 2). It is bordered to the east by U.S. Route 113 and is located about two miles south of Georgetown. The majority of this parcel is composed of mixed and deciduous vegetation (e.g., maple, oak, poplar). Also, smaller regions of coniferous (e.g., pine, cedar) vegetation occur in this region. Agricultural plots make up the remainder of this parcel's acreage.

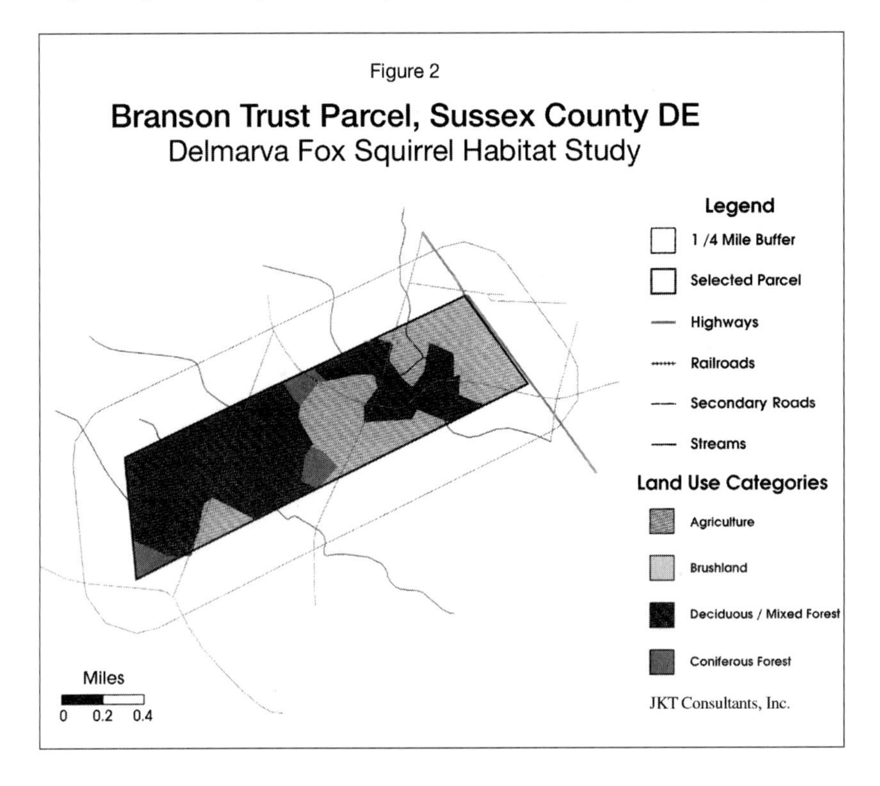


Availability of resources (question 4) and fragmentation (question 5)

U.S. Route 113 runs along the edge of a large agricultural field and therefore does not bisect any forested habitat. Four streams are included in this region and are all located in the mixed/deciduous forest habitats. However, several secondary roads severely fragment the forested regions. In addition, the area surrounding this parcel has been heavily deforested. The predominant cover is now agriculture with some small, isolated, remnant forest stands.

Conclusion on Branson Trust

Although Branson Trust contains the largest total area of the three parcels and a high percentage of mixed/deciduous forest, the fragmentation and local deforestation inhibit the survival of many species.

FIGURE 15.3 **A Collaborative Recommendation Report (continued)**

5

**Hudson Estates Parcel**

Answers to questions 2–5

Hudson Estates is a 519 acre (perimeter = 3.73 mi) parcel located in southwestern Sussex County about 6 miles east of U.S. Route 301 (Figure 3). Eighty-three percent of this parcel's land is classified as mixed or deciduous forest. The other 17 percent is classified as agriculture, brushland, and coniferous forest cover.

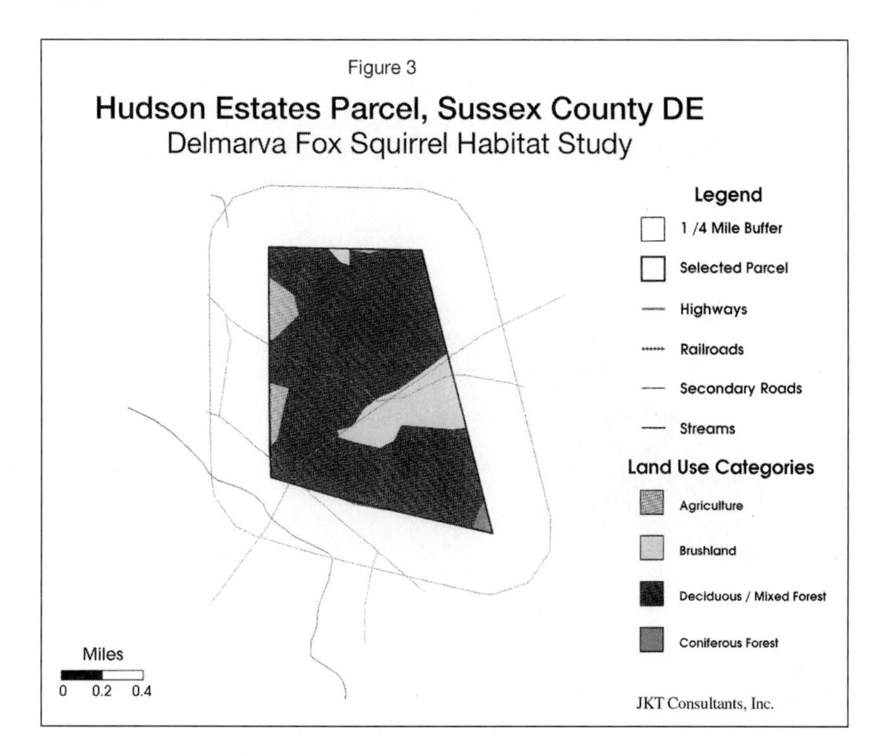


Highways and railroads are notably absent from this site. However, three secondary roads are present, all of which fragment the forested regions. Moreover, no streams flow through any region of this parcel, leaving all resident species with no ready supply of water. The area surrounding Hudson Estates has been partly deforested. The deforested areas have been replaced largely with agriculture and brushland.

Conclusion on Hudson Estates

Even though the Hudson Estates Parcel contains the largest acreage of mixed/deciduous forest, these regions are severely fragmented. In addition, the absence of streams and surrounding forest cover further detracts from the suitability of this parcel as wildlife habitat.

FIGURE 15.3 **A Collaborative Recommendation Report (continued)**

6

## Hugwell/Johnson Parcel

Answers to questions 2–5

The Hugwell/Johnson parcel is located in north central Sussex County along the Maryland and Delaware Railroad about 4 miles north of Georgetown (Figure 4). The railroad makes up the western border of the site. Over half of this parcel's 440 acres (perimeter = 3.47 mi) is mixed/deciduous forest. The remaining land in this parcel is classified as coniferous forest, agricultural plots, or brushland. Weaving through these regions are segments of streams and secondary roads.

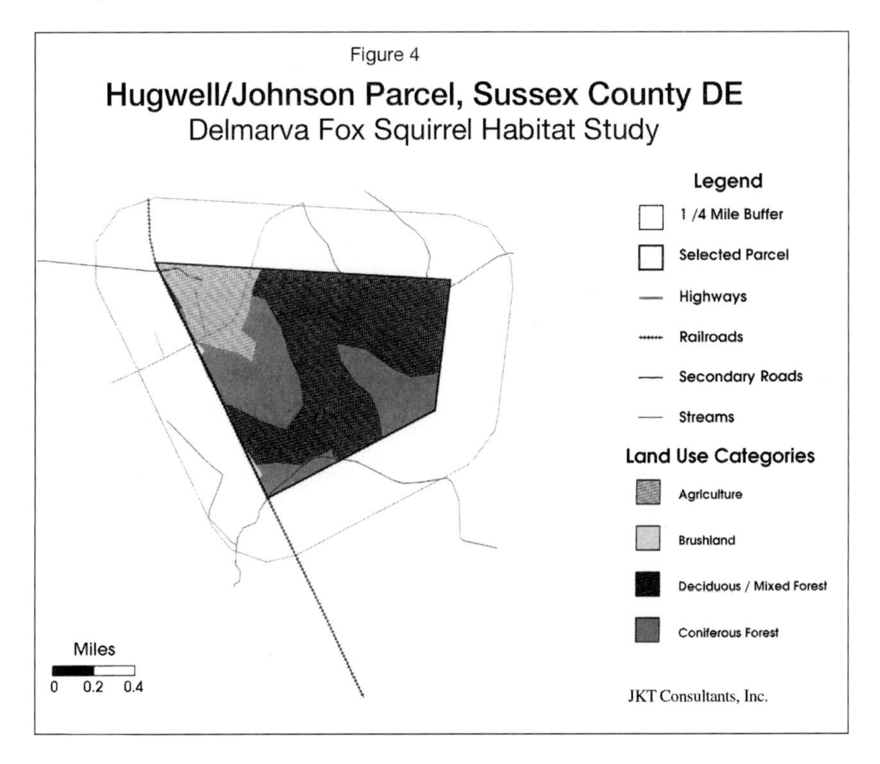


Conclusion on Hugwell/Johnson

The Maryland and Delaware Railroad, which marks the western border, does not intersect any interior portion of the parcel, and therefore does not fragment the constituent habitats. In addition, this parcel contains only one secondary road which bisects only agriculture and brushland. Again, no forested habitats are fragmented. Several streams flow through this parcel. The bulk of these streams occur in forested regions. This parcel is located in one of the most forested regions of Sussex County. Small patches of agriculture and brushland are also interspersed throughout this region.

Overall, the Hugwell/Johnson parcel encompasses significant tracts of healthy forested ecosystems. The low degree of fragmentation and the presence of several streams qualify this parcel as suitable habitat for a wide variety of species.

FIGURE 15.3 **A Collaborative Recommendation Report (continued)**

7

## Conclusions

Although the Branson Trust Parcel contains the largest regions of mixed/deciduous forest, the high degree of fragmentation and the lack of surrounding habitat detract from its overall suitability for DMFS habitat (Figure 2). While the presence of streams in the forested regions is attractive, it does not compensate for these negative qualities. Hudson Estates also contains a high percentage of mixed/deciduous forest. Again, however, the fragmentation by secondary roads and the relative lack of forests in the surrounding area negate this advantage. The lack of streams further reduces the attractiveness of this parcel as the prime DMFS habitat.

**The authors summarize their argument as backing for the recommendation to come**

Despite the relatively small size of the Hugwell/Johnson Parcel and its forested habitats, many attractive qualities make this parcel the best candidate. First, no railroads, highways, or secondary roads fragment any regions of the forested area. The DMFS and other resident species can thus move freely throughout their ranges. Second, the large percentage of forested areas around the parcel effectively increases the amount of habitat available to the squirrel. Finally, the presence of streams in the forested areas enhances the suitability of the parcel. Although not necessary for survival of the DMFS, the streams provide emergency reserves of water to the entire population.

## Recommendation

We thus recommend that the Division select the Hugwell/Johnson Parcel for reintroducing the DMFS and conserving its habitat. We wish the Division every success in this conservation effort.

**The recommendation, brief and direct**

8

## References

**Citations in APA format**


Adams, C. E. (1976). Measurements and characteristics of fox squirrel, *Sciurus niger rufiventer,* home ranges. *American Midland Naturalist 95, (1),* 211–215.

Allen, A. W. (1982). *Habitat suitability index models: fox squirrel.* U.S. Dept. of Interior, Fish and Wildlife Service FWS/OBS-82.

Allen, D. L. (1943). *Michigan fox squirrel management.* Michigan Dept. of Conservation, Game Division Publication 100.

Bendel, P. R., & Therres, G. D. (1994). Movements, site fidelity, and survival of Delmarva fox squirrels following translocation. *American Midland Naturalist, 132 (2),* 227–233.

Brown, L. G., & Yeager, L. E. (1945). Fox and gray squirrels in Illinois. *Illinois Natural History Survey Bulletin, 23 (5),* 449–532.

Deuser, R. D., Dooley, J. L., Jr., & Taylor, G. J. (1990). *Habitat structure, forest composition, and landscape dimensions as components of habitat suitability for the Delmarva fox squirrel.* U.S. Dept. of Interior, Fish and Wildlife Service general technical report 52–108.

Nixon, C. M., Havera, S. P., & Hansen, L. P. (1980). Initial response of squirrels to forest changes associated with selection cutting. *Wilderness Society Bulletin, 8 (4),* 298–306.

U.S. Forest Service. (1971) *Wildlife habitat management handbook: Southern region.* USDA Forest Service FSH 2609.


**FIGURE 15.3 A Collaborative Recommendation Report (continued)**

1. Unclear purpose:

   *"I don't know why I'm reading the report."*
2. Lack of conclusions:

   *"The emphasis is invariably upon what the person did and how it was done; I want to know what all of this means for me and what I do. Often, conclusions are missing or at best only hinted at. Readers shouldn't have to be mind readers."*
3. Insufficient contextual background:

   *"Nothing is worse than being plunged into a report without having been given a clear idea of the problem, situation, or need that motivated the report or the work in the first place. Writers assume too much. Even when it's something I'm supposedly familiar with, I need some brief reminder at the beginning of the report."*
4. Muddled sentences, garbled expression:

   *"At the very least, I expect writing that is clear. I don't want to have to struggle to decipher meaning. I don't have the time or the patience."*
5. Too much or not enough detail:

   *"On the one hand are writers who swamp you with unnecessary details and data until you're ready to cry for mercy—'Get to the point!' On the other hand are those writers who fill their reports with one glowing generalization after another so that all you're left with are questions and a craving for particulars. The first group of writers unnerves me; the second group makes me suspicious."*
6. Unclear relationship among ideas:

   *"Trying to figure out how idea A is connected to ideas B, C, and D is frustrating. Sometimes writers expect readers to read between the lines and provide the logical connections. Most readers, however, aren't going to expend the effort, or, if they do, they just might make the wrong connections."*
7. Unexplained jargon:

   *"While it is true that every field or discipline has its own specialized terms, to inflict these upon the unknowing and unsuspecting reader is not only unfair, but also rude."*

FIGURE 15.4 **Major Weaknesses of Technical Reports**

(Source: Christine Barabas, *Technical Writing in a Corporate Culture.* Copyright © 1990. Reproduced with permission of Greenwood Publishing Group, Inc., Westport, CT.)

problem, the researcher who compiled this list quotes a comment from a report reader about the problem. Write your reports well, so your readers don't have occasion to make such remarks.

Use headings and other design devices to speed your reader through your report, and adapt your expression, as in other documents, to the readers' expectations and level of expertise (Figure 15.3).

## THE REPORT PACKAGE

To fulfill their obligations to the reader, some reports need to incorporate additional elements beyond the discussion itself. The discussion becomes part of a larger package that includes a letter presenting the report to the reader, a title page, an abstract or executive summary, and references. Because of their significance, guidelines for preparing abstracts and executive summaries occupy an entire chapter of this textbook (Chapter 13). In this section, you will learn about preparing the other elements you saw in Figure 15.3, along with additional elements: a foreword or preface, glossary, and appendix. These elements accompany formal reports, often those written from one organization to another. Internal reports usually circulate in memo form and rarely require such extensive treatment.

### Letter or Memo of Transmittal

The letter or memo of transmittal is a permanent record of your delivery of the report to the reader (Figure 15.5). Sometimes a standard form is appropriate, with individualizing information filled in the blank spaces. But you can also use the occasion to converse with the reader. In your letter (or memo, if the report is circulating inside the organization) you announce the subject or title of the report, the authorization, and the date of authorization. In addition, you may

**For more about letters, see Chapter 18**

- Briefly discuss the content of the report.
- Note for this particular reader the most pertinent sections (perhaps those the reader assisted with).
- Offer to answer questions about the project.
- Remind the reader about any changes in scope or approach since the project was approved.
- Acknowledge the assistance of people who helped you prepare the report.

Clip the letter to the report's cover or include it as the first page inside the cover. The length of the letter depends on the extent of explanation or courtesy needed. The tone reflects your relationship with the reader, from informal to formal. It's usually best to take a formal approach if you are unsure.

### Title Page

The title page presents the title, of course, along with the names of the authors and the audience, the date, and any other necessary information about the report's origin and use. Your reader notices the title first, although you may consider it almost an afterthought. Take time to create a good title. Be brief, clear, and comprehensive. Use terms your intended

**JKT Consultants, Inc.**

Environmental Specialists
University of Delaware
Newark DE 19717

16 May 2001

Tom Moran
Fish and Wildlife Division
DNREC
Dover DE 19901

Dear Mr. Moran:

**Notes the purpose of the letter in a formal tone**

In response to your request of 3 April, we have analyzed three parcels of land in Sussex County to determine their appropriateness as habitat for the Delmarva fox squirrel. We are pleased to present the enclosed report, which concludes phase 1 of the project; it summarizes our analysis and describes each of the parcels.

**Provides the answer: the recommendation**

On the basis of our research, we recommend that the Fish and Wildlife Division select the Hugwell/Johnson Parcel. The report documents the reasons for our recommendation.

**Acknowledges assistance**

We would like to thank several people who helped us bring this project to a successful close. In particular, we are in debt to Dr. John Mackenzie for his assistance with ATLAS GIS and to Dr. Robert Northrop who helped us set the environmental criteria and access the appropriate Habitat Suitability Index.

**Ends on a pleasant note and looks to the future**

JKT Consultants has also thoroughly enjoyed working with you on this project. We recognize the importance of habitat studies and would be happy to assist your Division in other such projects should the need arise.

If you have any questions regarding this report or would like to discuss future plans, please feel free to write or call.

Sincerely,

*Julie A. Fine*

Julie A. Fine, Chief Analyst

Enc. report

FIGURE 15.5 **Letter of Transmittal**

readers would use to search for your report in an information retrieval system—and place those terms up front. Place terms like *final report* in the background. Here are some effective titles:

> *Bayside Tract Management Plan* (the proper name of the location precedes the more generic term *management plan*)
>
> *Supporting Research and Development in the NHS: A Report to the Minister for Health*
>
> *The Health of the UK's Elderly People*

Avoid being cute or punning (save such titles for popular articles):

For more about popular science articles, see Chapter 21

> Report: *Recommended Practices to Prevent Traffic Accidents on Freeways*
>
> Article: *Mass Murder on the Freeways*
>
> Report: *Recommended Removal of Interior Wall*
>
> Article: *Wall of Conflict*

A title that depends on jargon or terminology explained in the report is not sufficiently informative. *An Explanation of Pips and Poops in the HRT* may at first glance be amusing, even intriguing, but it is meaningful to only a very few specialists. More appropriate is *An Explanation of Sudden Power Surges and Drops in a Homogeneous Reactor Test.*

### Preface or Foreword

A *preface* or *foreword* may accompany an extensive or a controversial report, especially one that is bound and published. Conventionally, a preface is written by the report's author, while a foreword reflects the views of another authority who comments on (and usually praises) the report. A preface often explains what the author intended to accomplish in the report. It may call attention to some unusual aspect or indicate particular problems in dealing with the investigation, and it sometimes acknowledges help. The foreword discusses the qualifications of the author and the effectiveness of the report.

### Glossary

A *glossary* acquaints the reader with the meaning of technical terminology, abbreviations, and symbols (if any) used in the report. Some authorities discourage the use of glossaries because they require the reader to flip back and forth through the report. If you decide to include a glossary in your report, you can place it right after the table of contents so all readers see it, or you can put the glossary in an appendix so it is available to those who are not familiar with the technical terms used but not in the specialist's way. Glossaries help you accommodate a wide range of readers with one document, but they should not substitute for a careful approach to using and defining terms in the text. Glossaries are particularly helpful in documents that will be translated.

### References and Appendix

For more about citing sources, see Chapter 5

Two other elements you may include in a final report are a list of references (or bibliography) and an appendix. If you cite the literature in your report, properly credit the sources from which you borrowed information, as in Figure 15.3. An *appendix* furnishes supplementary material not essential to the development of the report that would, if included in the discussion, interrupt continuity. A properly labeled appendix is an appropriate place for

- Complete tabulations of data
- Sample calculations
- Detailed quotations
- Copies of letters that add to the validity of the evidence

**ELECTRONIC EDGE**

## Online Reports

Many companies and government agencies are turning to online information management systems to create and deliver their reports internally. Moreover, they may use the Web to deliver reports to customers and clients internationally. Such reports reach readers more rapidly than print documents while at the same time reducing the costs in production, storage, and consumption of a natural resource associated with paper.

Writers face new challenges in preparing reports for a wired world. A simple and imperfect solution is to "dump" the paper document into the system, ignoring most if not all of the differences readers encounter between paper text and text on a screen. But such ignorance is not bliss for the reader. Writers of paper documents highlight information and ease readability through page design and text elements like tables of contents, indexes, and abstracts. It's easy for readers to skim a paper report to get its gist and find what they need.

Writers of online text have to compensate for the small size of the screen and difficulties in skimming by creating devices that help readers navigate through a series of screens and images. Online presentation does offer some compensating features, however, including powerful search and retrieval strategies, hypertext links, help files, and the like. These let readers pull information not just from one report but from a database that includes a series of reports to create the one compendium of information needed to support a particular decision. For example, one application provides reports, updated monthly, on international business prospects sorted into five categories: company, country, sector, legislation, and news. According to its creator, "90% plus of all business questions can be filtered down to those particular topics . . . 70% of the queries are related to the first topic: company" ("Marketplace" 6).

Properly used, online delivery makes it easier to customize a report as you segment information in units that different readers can use in different ways.

# CASE
## REPORTING AN ACCIDENT

Last evening, there was a minor accident in the lab where you work as a technician.

Although unauthorized personnel should not be in the lab, the partners and children of staff do often come in to pick them up after work.

Last night, your young son came in to collect you, and in the short time he was at your bench, he managed to get something in his eye. The substance caused no damage, but it was sufficiently irritating to cause the child great distress at the time.

Your first instinct had been to rush the child to the nearest eye-wash station, which you discovered did not work. Fortunately, another one was not too far away, and despite a panic, you were able to irrigate the eye sufficiently promptly and adequately.

This situation, fortunately, had a happy ending. But it raises the need for *reporting*. Based on this discussion, draft two reports:

1. An *information report*. Describe the accident for the record.
2. A *recommendation report*. Recommend that the eye-wash station be restored to working order and perhaps, more broadly, that the lab's safety equipment be checked regularly and maintained effectively.

In addition, you might create a memo (see Chapter 20) to remind the staff that the protocol regarding unauthorized entry to the lab exists for a good reason and to encourage them to comply with it.

(Courtesy of Kaaren Blom and Laurie Hallam.)

- Copies of questionnaires
- Case histories

For example, the report on global change (see page 333) has three appendixes: a discussion of pertinent United Nations activities, an overview of U.S. policies, and copies of other national and regional proposals. Create one umbrella appendix or sort such materials into several separate ones. Use appendixes with discrimination, and avoid thinking of them as an attic or a method for bulking up an undernourished report.

Some organizations encourage writers to relegate all the details and supporting evidence to appendixes, a practice that shortens the report but can lead to problems in readability as people toggle between generalizations on one page and data dispersed someplace else.

Reporting plays a large role in the classroom and in professional life. Sometimes, you write a report for yourself. The act of composing information in a report helps you interpret facts and ideas and locate gaps you need to fill. More often, you use reports to carry messages to other people, to account for your activities, and to encourage understanding and action. If you keep in mind the different purposes for reports, in particular, the distinction between information reports and decision-making reports, you will be able to sort through seemingly complex reporting situations to focus your approach and create just the right document.

**For more about collaboration, see Chapters 2 and 14**

## CHECKLIST: Reports

1. **Determine your purpose for writing the report**

   Write an *information report* to help your reader understand a concept, item, or situation

   Write a *decision-making report* to help your reader solve a problem, make a decision, or act in the right way

2. **Select information that will support your purpose and your reader's need**

   Respond to any specific request from the reader

   Think of questions that will require answers

3. **Structure your report to conform to the genre appropriate to the context**

   In the *introduction,* draw the reader's attention and orient the reader

   To describe a concept, situation, event, system, or object, use an *explanatory information report*

   To describe empirical research, use an *IMRAD report*

   To lead the reader to a decision, use a *recommendation report*

4. **Test your report's approach, structure, and expression**

   Conform to the norms of the appropriate genre

   Meet the audience's needs

5. **Include supplementary elements as needed**

## EXERCISES

1. Mark with an "I" those subjects that seem appropriate for an information report and with a "D" those that suggest a decision-making report. After doing this, choose several subjects and provide titles that show how the same data could be used in both information and decision-making reports. For instance, the first subject below suggests a decision-making report and would be marked "D." Data could, however, be used for both types of reports, that is, *types* (information report) and *control* (decision-making report) of lawn weeds.

   Control of lawn weeds
   Fire warning systems for the home
   Pesticide use in Turkey
   Motorcycle maintenance
   Improving the merchandise displays in the dairy department of Pick 'n' Pay supermarkets
   Prevention of fraud in charitable organizations
   Development of a portable detector for plastic pipe and other underground objects
   Fire codes for dormitories
   Effects of entrained air on concrete
   Architectural glass for sun control
   Methods for treating waste at sea
   International standards for formatting documents

Oil reserves in Siberia
Trade agreements
The effects of accounting procedures on financial statements
Population control in the Baltic states
Aircraft safety in the United States and Italy
Recommended sprays for apple trees
Effects of wine drinking on health
Guidelines for forming a farm partnership
Entrepreneurial production of CD-ROMs for home use
Detecting food spoilage in sealed containers
Measuring the intensity of earthquakes
Earthquake-resistant design in Japan
Identifying types of soil and rock formation
Advantages and disadvantages of solar heating
An argument for (or against) gun control
Humane traps for animals

Some of these subjects may suggest a topic for your final report.

2. Present a numbered list of conclusions on any subject of your choice. Do not forget to include an introductory sentence. Rewrite the conclusions in paragraph form.

3. Construct a purpose statement for a decision-making report and a thesis statement on the same subject for an information report.

4. In the following example, distinguish between the conclusions and the recommendations for a report on "Bears: A Danger to Tourists in Our National Parks":

   A great number of accidents occur in our national parks because tourists insist on feeding bears. Bears if unmolested pose no danger to tourists, but if hand fed they often become aggressive, destructive, and dangerous. The most intractable animals should be removed from the park and transported to a less populated environment. Bears are wild animals; therefore, their behavior is unpredictable, especially when tourists dump garbage near their campsites or insist on getting out of their cars to feed them. Tourists who disregard warning signs should be shot. Some tourists take chances with bears in order to get "cute" photographs. Apparently, tourists do not realize the strength of these creatures, which can with one blow of a paw smash a heavy refrigerator case. All hopelessly vicious bears roaming the park should also be shot.

## FOR COLLABORATION

In a team of two or three, write a formal report, either informational or decision making, on a topic of your choice that is approved by your instructor. Review Chapter 4 concerning categories of problems and then pick a topic that is interesting enough to sustain the team through several weeks of research and writing. Ask questions. Establish the boundaries of your report by setting down the subject, problem, purpose, procedure, scope, and audience. You might include such information in a formal proposal (see Chapter 14). Here is an example:

*Question:* Will maintaining two rights-of-way connecting the same cities be economically feasible?

*Subject:* Conrail's request filed with the Interstate Commerce Commission pertaining to the abandonment of trackage in eastern Ohio.

*Problem:* Conrail now operates two parallel branches from Youngstown, Ohio, to Ashtabula, Ohio; both lines are in need of extensive repair because of lack of maintenance and the lack of funds.

*Purpose of Investigation:* The purpose is to determine whether maintaining two rights-of-way connecting the same cities is economically feasible and whether the operation of a single track will free money for maintenance that is now tied up in the operation of a second line.

*Procedure and Content:* The report will interpret the results of a survey concerning total daily train traffic on each of the two lines. These results will be used in an economic analysis emphasizing that the same total revenue can be realized by the use of one track daily instead of two. At the same time, daily maintenance costs can be cut as much as 48 percent. The success of a computerized train-scheduling method will be described to further justify the request. The method maximizes safety in the use of a one-track right-of-way.

*Audience:* The Interstate Commerce Commission (primary audience), Conrail officials (secondary audience).

CHAPTER 16

# Progress Reports



"EVERY PROJECT LOOKS LIKE A FAILURE IN THE MIDDLE."

ROSBETH MOSS KANTOR

Often, when you are in the middle of a project, particularly one that is lengthy, complex, or collaborative, it's a good idea to stop and ask, "How is it going?" Answering that question may require only a brief conversation with colleagues or a supervisor. Or it may lead to a written report, delivered online or in print. Such reports are called *progress reports,* or, sometimes, *interim* or *status reports.* They range from simple fill-in-the-blank approaches to lengthy discussions, but all progress reports meet the same broad purposes. In this chapter, you'll learn guidelines for meeting these purposes.

## PURPOSES

Most companies, government agencies, clients—and all managers—consider progress reports vital, even though many employees view them as interruptive, annoying, and boring. If the work is rolling well, you don't want to stop to write about it. If the work has been a series of problems, you don't feel you have any progress to write about. Whatever the state of your work, however, composing the report helps you

1. Check that you're keeping on schedule.
2. Record work accomplished over a specified period.
3. Convince a client, sponsor, or adviser that your work is on track and on schedule.
4. Assure readers that there will be no surprises at the project's end.

The first two purposes serve you directly. The interruption each report offers lets you survey and appraise the project as a whole and evaluate one phase. The report deadline may also motivate you to finish the work so you avoid embarrassment before your supervisor or your peers. In addition, writing during the project helps you remember details and, if the project works, approach the task of writing the final report with some text already at hand.

Purposes 3 and 4 pertain more to management. Supervisors and administrators consider progress reports essential managerial links. Such reports foster intelligent decisions on matters of money, time, equipment, and materials. They help managers and administrators establish and evaluate trends that may show whether a project should be continued, reoriented, or abandoned. Submitted to a sponsor, they often inspire renewed confidence in the feasibility of the work, help capture any necessary additional resources, and convey an optimistic image of you and your organization.

## FORM REPORTS

To reinforce the routine nature of progress reports, many professional, commercial, and academic organizations use standard approaches or forms that help writers supply the information decision makers need. Figure 16.1 shows a memo that covers, on one page, a standard approach to reporting progress:

- Work completed
- Future work
- Problems (if any)
- Correction (if there's a problem)
- Schedule status
- Budget status

Figures 16.2 and 16.3 show even more simplified fill-the-the-blank forms. These are particularly appropriate for information that is largely quantitative.

## ELEMENTS AND STRUCTURE OF DISCURSIVE REPORTS

A fill-in-the-blank approach, however, will not work in every situation. More complex projects may require more discussion. The core categories remain the same, but you need to provide explanations that serve your persuasive purposes. You may also have to amplify the core discussion with other segments, like an introduction, a conclusion, or recommendations. Figure 16.4, for example, is a discursive report, in complete sentences and paragraphs, that opens with an introductory statement summarizing the project. In this section, you will read guidelines for writing such reports, which may be the only tangible product of months of research.

**Blickle Grain Dryer Co.**

To: G. I. Smith
From: M. O. Eddy

Date: 11 August 2001
Reference: CD 500B

*Subject:* Progress report on energy conservation devices for grain dryers

Subject line alone establishes context

1. **Work completed (in %)**

| *Device* | *No. 1* | *No. 2* | *No. 3* |
|---|---|---|---|
| Design | 100 | 100 | 100 |
| Detail | 100 | 100 | 100 |
| Prototype/models | 100 | 100 | 100 |
| Construction | 100 | 80 | 60 |

Component-by-component presentation of the status of the entire project as of week's end

2. **Work to be completed (in %)**

| *Device* | *No. 1* | *No. 2* | *No. 3* |
|---|---|---|---|
| Construction | 0 | 20 | 40 |
| Test | 100 | 100 | 100 |
| Post-test changes | 100 | 100 | 100 |

Future work on each device

3. **Work performed this week**

We transported device no. 1 to the test site. It will be ready on schedule for preliminary tests on the layer-dryer when the corn harvest starts in late September. The construction of device no. 2 for batch dryers is progressing smoothly, and no difficulty in meeting the 15 September deadline is anticipated.

Details of this week's work

4. **Variance from schedule**

The construction of device no. 3 for continuous flow dryers is about five days behind schedule. A foundry error made it necessary to scrap one of the aluminum castings used in the hot air recycling mechanism.

Explanation of a problem and how it will be solved

5. **Correction of variance**

A new casting has been made and we have arranged with our machine shop to process it immediately upon delivery. Our prospects of meeting the 25 September schedule for it look good right now.

6. **Budget status**

| | |
|---|---|
| Amount spent this week | $ 2,676 |
| Amount authorized | 35,000 |
| Amount spent to date | 15,170 |
| Current balance | 19,830 |

Budget information in tabular form

FIGURE 16.1 **A Memo Reporting Progress in a Routine Structure**

**DAILY REPORT**

CUSTOMER ______________________ DATE ____________

SITE ______________________ ORDER # ____________

| ORDER NO. | FS | ES | START | END | JOB DESCRIPTION | INIT |
|---|---|---|---|---|---|---|
| | | | | | | |
| | | | | | | |
| | | | | | | |

WORK PERFORMED (Specify Item#, subcontractor, location, and description)

______________________________________________

______________________________________________

______________________________________________

| LABOR | | | | | EQUIPMENT | | | | |
|---|---|---|---|---|---|---|---|---|---|
| TYPE | MAIN | SUB | SUB | SUB | TYPE | MAIN | SUB | SUB | SUB |
| Foreman | | | | | | | | | |
| Operators | | | | | | | | | |
| Laborers | | | | | | | | | |
| Teamsters | | | | | | | | | |
| | | | | | | | | | |
| | | | | | | | | | |

dca082195 5M

Initialed by ____________

FIGURE 16.2 **Simple Form for Reporting Daily Progress on a Construction Site**

## Introduction

In a discursive report, introduce the core information with an opening statement that provides the context for the report. State that the document is a progress report, note the period covered, and summarize briefly, sometimes in quantitative terms, the status of the project. You may also, especially in a lengthy project, review the problem and objectives. Here's how a team of engineering students introduced their report to the faculty adviser:

> This progress report covers CS Environmental's activities between 1 April and 5 May on our senior design project. The goal of the project is to evaluate geological and soil conditions at the site of the proposed office campus and determine the best foundation type for Buildings A and B and the steel tank. As of 5 May, the project is 50 percent complete, as scheduled.

Another student team began its report on the redesign of work stations at a telemarketing firm with a summary of past and future work:

> The following report details our progress on the redesign of your work stations. We have completed the first phase outlined in our proposal of 12 October. In

Date:
Team members:
Project title:
Advisor:

Work completed in last two weeks:

________________________________________

________________________________________

Work planned for in coming two weeks:

________________________________________

________________________________________

Record of time spent:

| *Name* | *Hours Worked* | *Tasks Worked On* | *Tasks Planned* | *Initials* |
|---|---|---|---|---|
| | | | | |
| | | | | |
| | | | | |

Remarks:

________________________________________

________________________________________

Signed: ________________________ (advisor)

**FIGURE 16.3 Simple Form for the Biweekly Progress Report on a Senior Design Project**
The "initials" column reminds team members to sign—and thus agree with—the report, and the signature line indicates that the form functions in an approval route.

completing that phase, we have determined the problems in the current work stations that cause employee discomfort and contribute to your rising health care costs. We will spend the remainder of the project concentrating on these areas. Our next steps are to survey equipment currently available on the market that can address these problems and determine if installing this equipment would be cost-effective. We will also develop guidelines for reeducating employees to protect themselves from injury and recognize warning signs of developing problems. Where possible, we will suggest such reeducation as an alternative to new equipment. The rest of this report briefly summarizes our findings to date. (Courtesy of Mary Angerer and Rebecca Tinsman.)

## Work Accomplished

In the middle of the report, you assess progress toward your goal. The following discussion covers the most common approaches to presenting this information.

## JKT Consultants, Inc.

Environmental Specialists
University of Delaware
Newark DE 19717

Telephone (302) 837-8969

27 April 2001

Tom Moran
Division of Fish and Wildlife
DNREC
Dover DE 19901

Re: Progress on the Delmarva Fox Squirrel Habitat Study

Brief review of the context for the report

Dear Mr. Moran:

Quantitative summary of the status and assurance of compliance with schedule

As we agreed earlier this month, JKT Consultants is submitting this status report on our analysis of three parcels of land in Sussex County offered by the State as potential areas for reestablishing the Delmarva fox squirrel (DMFS). The analysis is part of a larger effort to preserve this severely declining species. We have completed half the work outlined for Phase 1 in our proposal of 12 April (see attached schedule). The project is thus on target and we anticipate that the remaining tasks will also be completed by our 16 May deadline.

Plan of the report follows the logical division of tasks announced in the proposal—by sources of information

**WORK COMPLETED**

JKT Consultants has collected information about habitat suitability from three sources: literature review, interviews, and digital maps.

Summary of information from the first source—the literature. Parenthetical citations in APA format

**Literature Review**

We first consulted the Habitat Suitability Index Model for the DMFS (Allen, 1982) and other documents to determine habitat requirements. In sum, the DMFS prefers a habitat of deciduous upland trees with sufficient canopy closure (40–60%) and minimal understory
(< 30%), so mature stands of upland deciduous or mixed forest would be best. The DMFS also requires a staple diet of mast (fruits produced by such trees as oak, walnut, beech, and sweetgum) (Allen, 1943; Bendel & Therres, 1994; Deuser et al., 1990) or agricultural crops like corn and soybeans (Brown & Yeager, 1945). The squirrel obtains water through consuming succulent foods, so water is not generally a limiting factor, although streams are used when present (U.S. Forest Service, 1971). The squirrel's home range averages 5 to 10 acres and increases as the availability of suitable habitat increases (Adams, 1976). Females



FIGURE 16.4 A Collaborative Student Progress Report

are the less mobile of the two sexes and are thus more susceptible to adverse habitat changes. Therefore DMFSs thrive when they have a large expanse of land (Nixon et al., 1980).

**Interviews**

To corroborate our readings, several of which date to the 1940s, we interviewed three experts: J. Mackenzie and R. Northrop of the University of Delaware and D. Hannab of your Division. We'll present the details in our final report, but in general, they agreed that land containing large mature stands of upland deciduous or mixed forests with available water supplies and minimal fragmentation will maximize the Division's chances of establishing a viable DMFS population.

**Connection to last paragraph ("to corroborate") moves into summary of information from second source: interviews**

**Digital maps**

With the help of J. Mackenzie, we located a digitized map that shows the land cover for Sussex County and a file that includes highways, railroads, secondary roads, and streams for the same area. We merged these files and created a base map of the entire county. We then input the coordinates of the three parcels into the map and located the parcels within the county in a way that allowed us to define their borders and identify prominent features. In brief, all are good candidates and encompass to some degree the required upland deciduous or mixed forest.

**Third source, as announced: digital maps**

**CURRENT WORK**

We are currently calculating the sizes of each region along with their concentration ratios, degrees of road and railroad invasion into desired habitat, and the availability of necessary resources. A "cookie cutter" operation has allowed us to aggregate data from various computer files to create rich pictures of each parcel. Each has both advantages and disadvantages, which we are investigating. The Hugwell/Johnson parcel has several streams in the forested regions, but those regions are relatively small; Hudson Estates is somewhat larger, but it has no streams and the forested regions are slightly isolated; the Branson Trust parcel has large regions of good habitat and many streams, but it is fragmented by secondary roads.

**Overview of current tasks**

More precise information will be needed to make an accurate recommendation. We are presently constructing a database for this purpose. We have uncovered a problem in locating a digital file with historical information on the three sites. We are now searching for a land use file with information from the 1970s. If we can obtain such a file, we will be able to judge both the maturity of the selected forests and the current trends in land use changes throughout the region. We have several leads concerning the location of such a file and are optimistic about our possibilities for obtaining the information.

**Notes a problem and possible resolutions**

**FIGURE 16.4 A Collaborative Student Progress Report (continued)**

3

Notes the next task: comparing the sites and recommending the best one

**FUTURE WORK**

Once we have compiled a consolidated database of the three parcels, we will compare characteristics of each site. We will begin the process by creating buffers around all relevant features. The buffers will allow a 0.1 mile zone around all highways and railroads (as undesirable habitat) and a 0.05 mile zone around all secondary roads. In addition to looking at the kinds of forests existing within these buffer regions, we will also consider the other desired habitat features, including concentration ratios and fragmentation studies.

Brief, positive ending

Finally, we will submit a report to the Division on 16 May. It will include our recommendation as well as an extensive analysis of our reasoning.

Sincerely,

*Julie A. Fine*

Julie A. Fine
Chief Analyst

4

**REFERENCES**


Adams, C.E. (1976). Measurements and characteristics of fox squirrel, *Sciurus riger rufiventer,* home ranges. *American Midland Naturalist 95,* (1), 211–215.

Allen, A. W. (1982). *Habitat suitability index models: fox squirrel.* U.S. Dept. of Interior, Fish and Wildlife Service FWS/OBS-82.

Allen, D. L. (1943). *Michigan fox squirrel management.* Michigan Dept. of Conservation, Game Division Publication 100.

Bendel, P. R., & Therres, G. D. (1994). Movements, site fidelity, and survival of Delmarva fox squirrels following translocation. *American Midland Naturalist, 132 (2),* 227–233.

Brown, L. G., Yeager, L. E. (1945). Fox and gray squirrels in Illinois. *Illinois Natural History Survey Bulletin, 23 (5),* 449–532.

Deuser, R. D., Dooley, J. L., Jr., & Taylor, G. J. (1990). *Habitat structure, forest composition, and landscape dimensions as components of habitat suitability for the Delmara fox squirrel.* U.S. Dept. of Interior, Fish and Wildlife Service general technical report 52–108.

Nixon, C. M., Havera, S. P., & Hansen, L. P. (1980). Initial response of squirrels to forest changes associated with selection cutting. *Wilderness Society Bulletin, 8 (4),* 298–306.

U.S. Forest Service. (1971). *Wildlife habitat management handbook: Southern region.* USDA Forest Service FSH 2609.




**FIGURE 16.4 A Collaborative Student Progress Report (continued)**

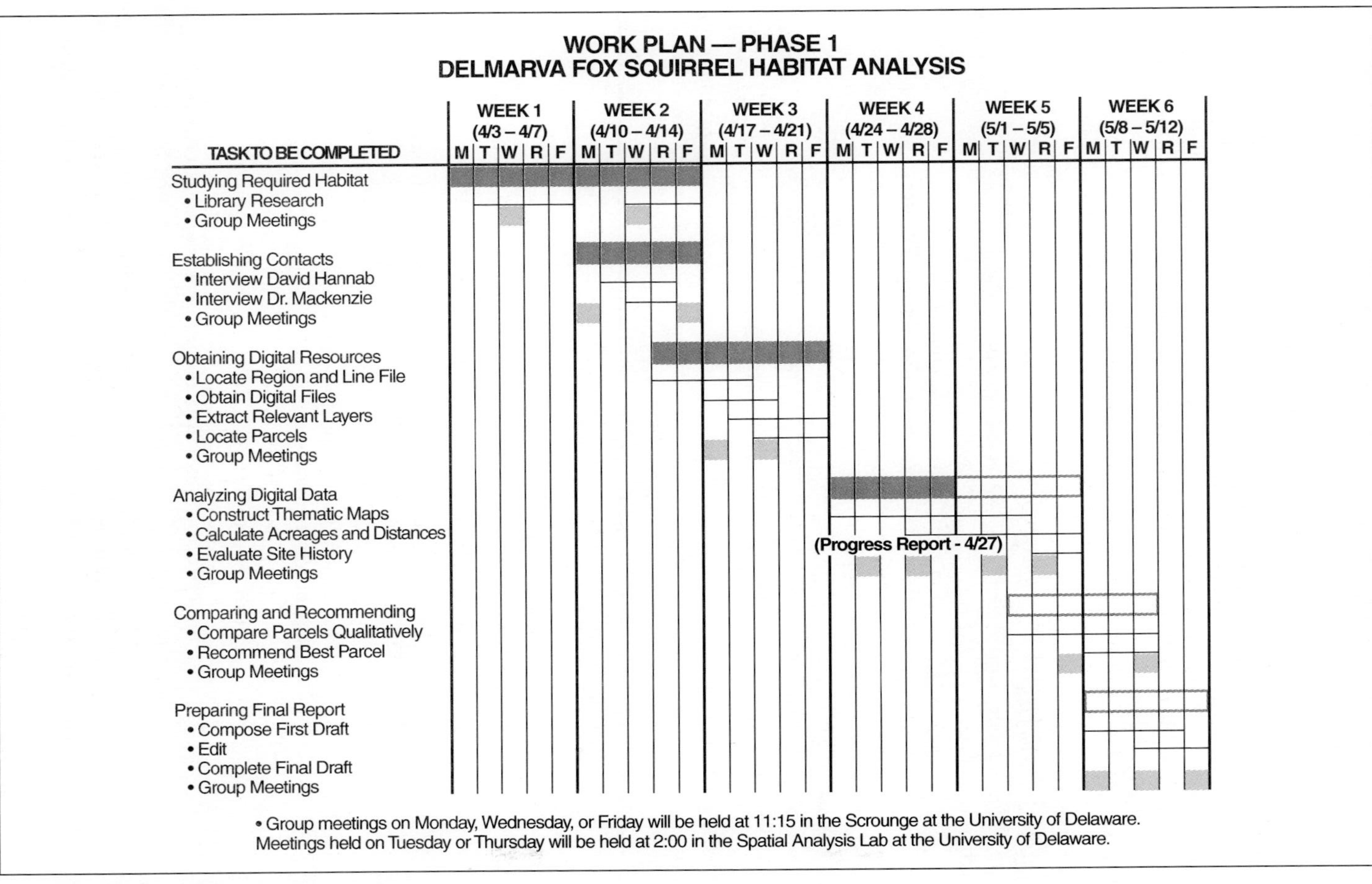


**FIGURE 16.4 A Collaborative Student Progress Report (continued)**

To demonstrate progress, the writers reproduced their proposed schedule (see p. 29), filling in the bars that represent work completed. Open bars indicate work to be done.

A CLOSER LOOK

## The Ethics of Progress Reports

The following "glossary" shows humorously how researchers may say one thing (in italics) . . . and mean something else (Graham).

**Introduction**

*It has long been known that* . . . I haven't bothered to look up the original reference.

. . . *of great theoretical and practical importance* . . . interesting to me.

*While it has not been possible to provide definite answers to these questions* . . . The experiments didn't work out, but I figured I could at least get a publication out of it.

**Experimental Procedure**

*The W-PB system was chosen as especially suitable to show the predicted behavior* . . . The fellow in the next lab had some already made up.

*High-purity* . . . *Very high purity* . . . *Extremely high purity* . . . *Superpurity* . . . *Spectroscopically pure* . . . Composition unknown except for the exaggerated claims of the supplier.

*A fiducial reference line* . . . A scratch.

*Three of the samples were chosen for detailed study* . . . The results on the others don't make sense and were ignored.

. . . *accidentally strained during mounting* . . . dropped on the floor.

. . . *handled with extreme care throughout the experiments* . . . not dropped on the floor.

**Results**

*Typical results are shown* . . . The best results are shown.

*Although some detail has been lost in reproduction, it is clear from the original micrograph that* . . . It is impossible to tell from the micrograph.

**Chronological.** The easiest way to present your work is chronologically: what you did from Day 1 until now, as in this example:

1. On December 1, I mailed 450 questionnaires to recreation directors in 20 Texas cities.
2. On December 6, I interviewed the mayor and the Director of Austin Recreation. Both approved of the program, although they questioned the feasibility of financing.
3. On December 8, I obtained a copy of Houston's program and found it similar to that in Austin.

Such a chronological approach, however, is usually not the best one. It makes the reader work too hard and thus may not be convincing. It lacks emphasis and may relegate to a subordinate position some key event—because it occurred, for example, on a Tuesday five weeks into the project—that should be highlighted to aid in a decision on the future course of the project.

**Task by Task.** Instead, you may plan the report to reflect the tasks outlined in the proposal. Detail the status of work on each task, as in Figure 16.4.

**Component by Component.** To report progress on a mechanism or system, proceed from one component to another component. The student report on the redesign of work stations divides its description of the work accomplished into the components of their redesign: video display terminal (VDT), keyboard, and chair. For each component, the team first analyzed the symptoms that appeared as a result of poor design and the features of the current stations that contribute to these problems. They then recorded the results of that analysis as a baseline for the redesign. Here is the record of their analysis of the keyboard:

> **Keyboard**
>
> *Symptoms:* Carpal tunnel syndrome, tendinitis, shoulder/elbow pain
>
> *Design Features:* The keyboards are situated at the edge of the desk in front of the screen. Many employees find their wrists "drooping" to rest on the desk while they type. Also, the keyboard is straight, forcing the employees' hands to come together in front of their bodies. Both the location and the shape cause wrists to bend out of a natural position, irritating the tendons and putting pressure on the nerve passing through the carpal tunnel. In addition, the height of the keyboard forces employees to bend their elbows past 90 degrees or to keep their shoulders hunched. Finally, the position of the mouse makes employees reach out and then rotate their shoulder in a way that causes strain over time. (Courtesy of Mary Anger and Rebecca Tinsman.)

**Decision Support.** Organizing information by time, task, or component helps you describe your work. Sometimes, however, you need a structure that's more overtly persuasive. You build an argument that aids the reader in making a decision about the project and assures the client or sponsor that all is under control. If, for example, you have encountered a problem, show how you plan to overcome it:

> Although we had anticipated including 22 buildings on the tour, we discovered after a preliminary run through the route that we could not complete the tour in the time allotted. Traffic delays—the normal rush hour chaos in Center City—ate up one-half hour. It took us an average of 8 minutes to herd people on and off the bus, another time factor we had not calculated in our preliminary design. The mechanics of simply getting from place to place reduced the impact and enjoyment of the places we did visit. We have thus pared the list of sites to 10, a number that allows us to sample the breadth of architecture in the city without sacrificing the ability for participants to savor what we present.

The following discusses an expansion in the tasks needed to accomplish the work:

> The majority of the work is proceeding on schedule; however, the Engineering Department has had to undertake an additional task, that is, installing two more monitoring wells downstream of the landform area. The wells presently being used to monitor the groundwater runoff do not comply with RCRA specifications. The Engineering Department has assured me that the new wells complying with the RCRA specifications will be on line by November 1.

If you don't know how to overcome the problem, use the progress report as a device for soliciting suggestions, as in this e-mail progress report from a student to a professor:

> Help!
>
> The University of London library isn't coming up with the sources I need on the automation of British steel production. I have British Steel's annual reports for the last 5 years, but these don't discuss their technology in detail (the focus is on finance and nice pictures). I have found some references to marketing difficulties because its products were defined in imperial and not metric measurements and to their "downsizing" (from 250,000 workers at one point to only 41,000 today). But I can't find sources on specific production techniques. I also can't find much in online sources, although perhaps that's just because my skills at such searches are pretty rudimentary.
>
> I could write to them, but I'm not sure about a specific office or individual to address there. I might also be able to visit their Scunthorpe/Teeside site, although they may not allow visitors. Do you have any suggestions?

## Future Work

At the end of the report, you generally provide a look ahead to work in the next phase. State the prognosis for the project in a tone that avoids either unbridled optimism or dark despair. If the project is not working—a condition you should make clear from the outset of the report—then the closing section recommends a change in attack or scope, maybe even abandonment if the problem looks unsolvable. But when the evidence warrants, the closing provides assurance of timely and successful completion. End on a positive, confident note. Here is a brief closing in list form:

> The tasks remaining include
>
> 1. Finding information about construction time for precast concrete structures.
> 2. Deciding on a framing system based on that information.
> 3. Consolidating individual team member reports and editing them into a final report that provides our recommendation.
>
> This project has gone smoothly with few complications. The team has worked together effectively, and we are confident that you will be pleased with the final report.

## Schedule and Budget Status

In addition to these discussions of work done and to be done, most progress reports also include two other segments, usually in visual form: a schedule and a budget. In these segments you show how you've spent the allotted time and money and how that expenditure compares to what you proposed at the beginning of the project. Some progress reports focus almost exclusively on these two segments. The visual form helps you emphasize the comparison between projected and actual status (Figures 16.5 and 16.6). When

| Task | Week 1 | Week 2 | Week 3 | Week 4 | Week 5 |
|---|---|---|---|---|---|
| Decide approach | xxxxxx | | | | |
| Visit site (all) | xxxxxx | | | | |
| Interview 3 employees (RR) | | xxxxxx | | | |
| Interview 3 employees (PX) | | xxxxxx | | | |
| Interview 3 employees (JJ) | | xxxxxx | | | |
| Review medical records (RR) | | | xxxxxx | | |
| Review work station literature (PX) | | | xxxxxx | | |
| Write progress report (JJ) | | | xxxxxx | | |
| Determine design options (all) | | | | — | |
| Discuss options with BB (all) | | | | — | |
| Select best design or method for dealing with current work stations (all) | | | | — | |
| Write final report (JJ) | | | | | — |
| Edit report (all) | | | | | — |

FIGURE 16.5 **Simple Schedule That Compares Proposed Work with the Current Status**
All items marked XXXX have been completed; those marked—remain for the next project phase (weeks 4 and 5). Initials indicate the team member responsible for the task.

| Item | Budgeted | Actual |
|---|---|---|
| Airfare to Philadelphia | 400 | 350 |
| Hotel (2 nights) | 250 | 275 |
| Per diem | 100 | 100 |
| Conference registration | 350 | 350 |
| Photocopying | 100 | 50 |
| Online searches | 325 | |
| Telephone | 125 | 75 |

FIGURE 16.6 **Segment of a Budget from a Progress Report**
The amounts in the "actual expenditure" column would be totaled (you see only a segment of the budget here) and that amount subtracted from the project budget to provide the current balance.

that comparison shows significant deviance, that is, unbudgeted expenses or a delay in the schedule, explain those problems in the text.

## PERSONA

For more about persona, see Chapters 7 and 21

In explaining what has gone right and wrong in a project, you convey as much through your *voice* in the text as through the information you present. The persona you establish in a progress report must be authoritative and convincing. As in a proposal, you are often talking about something you are not sure will work. In such circumstances, some writers put the best face on the situation. Perhaps acknowledging present difficulties, they still assure the reader that all will be well. Others prefer to dwell on the problems. Both approaches may lead to a distorted view of the project that deceives the reader. As you write, try for a balanced view, honestly assessing the situation and negotiating the steps needed to make the project successful. You waste both your time and your reader's if you fail to draw reasonable attention to problems.

Still, when you run into trouble in an investigation, you may find it hard to talk about that trouble. It requires special courage to own up to slips in the schedule, cost overruns, negative results, mistakes in the field or lab, or lack of success with a procedure. It's easy to become comfortable with ambiguous language that hides the problems. As one veteran reviewer of progress reports remarked, "When the contractor says he's 'virtually on schedule,' how far behind schedule is he? It's usually about 6 months." You do want to sound as positive as possible in a progress report, and thus some bending of the truth may be appropriate, especially because you may well be legitimately uncertain about outcomes. But you risk behaving unethically if you pad a report with meaningless phrases that can be interpreted in a number of different ways. *A Closer Look: The Ethics of Progress Reports* provides humorous examples of such padding.

When you write a discursive report, avoid causing more problems than you solve. Don't drown the reader in details. Avoid dwelling more on the past than on the future, more on accomplishments than on recommendations. Do provide what's needed for decision making. And do *write.* If you postpone writing, you create a damaging lag in information transfer. By the time the reader receives it, the report bears old news. So start with a simple approach, and add explanations as they seem warranted.

The first report you write on the job is likely to be a progress report because such reports are so common and because many people would rather have someone else—the new person in the department—write them. Different organizations employ different methods to manage projects. Follow the audience's guidelines in submitting a standard form or writing a discursive report, and be honest in your assessment of where the project stands.

ELECTRONIC EDGE

## Measuring Progress Online

Technical people rely increasingly on software to help them over the hurdles of writing a progress report. Once pertinent questions are selected, an "information manager" resident in a computer application prompts you to provide the right answers, perhaps even supplying a range of likely answers. With a few keystrokes or clicks of a mouse, you've filed your report. Here is a screen from a program that helps realtors continually update their tracking of house sales. (Courtesy of Chris Miller.)

Information systems can automatically generate reports at regular intervals concerning sales, inventory, worker hours and pay, and other routine information from databases that record such transactions.

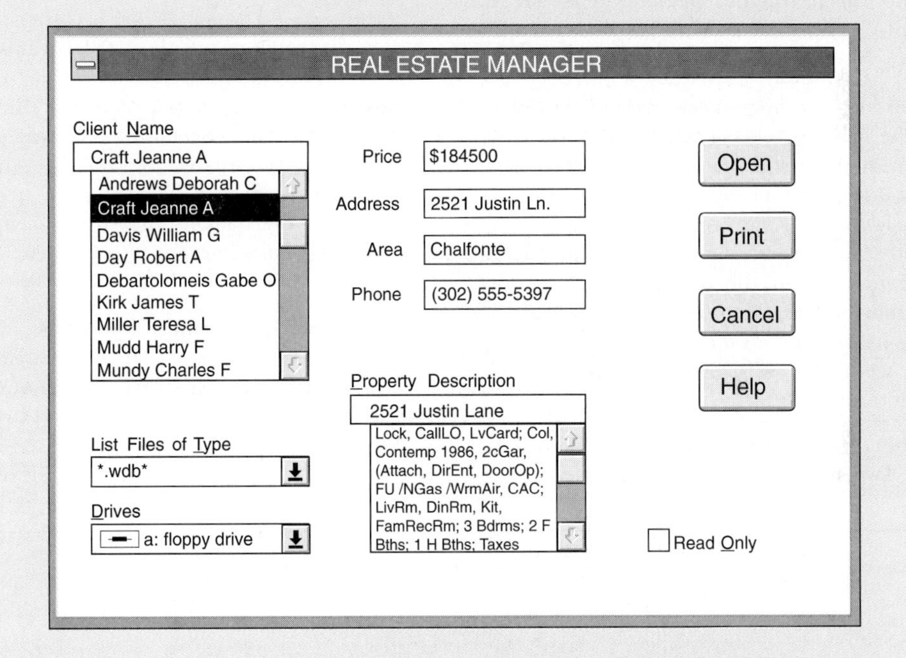


## CHECKLIST: Progress Reports

1. **Take time to assess projects in midcourse**
2. **Use that assessment to monitor your own work and record preliminary results**
3. **Meet the needs of management in monitoring projects and budgets**
4. **For many routine occasions, fill in a form report, online if appropriate**

# CASE
## FINDING YOUR FORM

Jim Smith is a technical writer for a small company headquartered in Philadelphia that produces a software package to track the sale of derivatives, complicated financial instruments. The program addresses a small, wealthy, intense, demanding international clientele, and the company maintains a staff in major financial centers worldwide. Smith reports on his activities every day in an e-mail message to his manager, Becky Lewis. Here, for example, is one week of messages:

**October 31, Monday.** I attended a meeting that lasted about 45 minutes (Pete Rancowitz led it) about the new 3.0 release that will be available on our in-house systems by the end of this week. It will be used for Front Office Interest Rates and has lots of new features.

I also started verifying the FinanFact documentation between the programmer's report sheets and the User Manual. I checked that the column names and descriptions matched. For the most part they did but I will have to add some sections.

**November 1, Tuesday.** More FinanFact verification. Two major sections that needed to be added. One was a new column about the status of the code. Then a one-sentence disclaimer that describes what the codes mean. I worked those new columns through each of the applications. It went pretty fast because I could cut and paste many of the descriptions from one application to another.

**November 2, Wednesday.** I completed my work on FinanFact by noon. Then I sent one copy of each verified manual to New York and Australia. I think the cover should say that this is an update. In other words, if it is the first revision for version 2.0, then the cover should say 2.0(a), for example. We should keep all previous versions and keep a record of what changes were made in that version. That way, it protects us (and the company) in a liability suit.

I also cleaned up my q, t, and k directories to delete all the backup and support copies and I rearranged my current files so that the proper ones are in the right folders.

**November 3, Thursday.** Today I began my work on the Interest Rate Derivatives manual by reading through the draft documentation from the programming team. I e-mailed London and Singapore to ask them to please send me their versions. I know they have made some changes

5. **If the context requires, write a discursive report**
   Briefly summarize the project in an introduction
   Record the work accomplished task by task or component by component
   Aid managers by noting any problems and likely solutions
   Look ahead to future work needed to keep the project on track
   Compare your accomplishments to date with the tasks and schedule you proposed
   Compare your current expenses to those proposed in your budget
   Establish a persona that is honest and trustworthy

we should incorporate in our work. The manual has some nice screen shots that are superimposed on each other and a good icon of a hand that helps draw attention to a specific area of the screen. But we might want to keep the screens apart and give one instruction before each screen. Maybe we could also use numbers to point out each field in the sample screens. It's sometimes hard to tell what information goes into each.

**November 4, Friday.** I plan to do two chapters of the manual every day (10 chapters total). At that rate, I should finish next Thursday and be able to print and distribute a master copy to the programming team and to London and Singapore next Friday. I'll also need to make sure style and format are consistent.

The first two chapters went well today. I had trouble near the end of Chapter 2 when I tried to capture the last screen but our system kept crashing. Steve was able to fix it. He also set up the terminal so that I can run both our own software and the dtp program together. It's great, but it took about an hour before I could start again with the manual.

Such messages *record* information adequately but in a random, scattershot way. Based on your reading of these messages, and your understanding of progress reports, design a better way to report such information.

1. Should the reports continue to be sent daily? Or would weekly reports be better?
2. Should Smith continue to use e-mail—or paper?
3. What categories does the information fit into?
4. What headings should Smith use to arrange information?
5. How can Smith convert these reports from being descriptive of what he has done to persuasive for management? *Should* he make such a conversion?
6. What is Smith's persona in these messages, and is it appropriate?

Design a template, that is, a framework of categories of information and headings, for Smith's "new and improved" progress reports.

(Thanks to Tom Tigani, SunGard Capital Markets, Inc., for his assistance in developing this case.)

## EXERCISES

1. Send an e-mail message to your technical communication instructor assessing your progress in the course. Measure that progress against your own personal goals for the course as well as the instructor's goals and procedures as outlined in the course syllabus.
2. Assume you are the president of a campus activity such as Engineering Week, Homecoming Week, or Greek Week. Prepare a progress report on programming, publicity, selection of meeting places, speakers, judges, and financial arrangements. Include a statement of the future work needed to complete the plans.

Memos A, B, and C (below) all deal with the same subject—the development of an acoustical tile for Soprano Suppressors, Inc. The writer of each is the project director, and the reader is the director's supervisor within the research organization. But the voices in the text differ, and each recommends a different strategy for dealing with the problem. Who do you believe? Why?

**Memo A**

This project, on the development of acoustical tile for the selective absorption of sound in the frequency range of 800 to 1,150 cycles per second, has been with us for almost eleven months, and the original contract is for a one-year period. Although the problem has been one of the more interesting, the particular requirements have complicated the possible solutions. From our work to date, we feel that there should be a reasonable solution to the problem of finding a material or material design arrangement for the selective absorption of audio energies in the 800 to 1,150 cps range.

A large portion of the time has been consumed in locating and setting up the equipment suitable for this problem. Although first estimates were based on converting our XYZ area and equipment to use on this work, we soon found that several problems arose in the adaptation of our existing equipment to the requirements of this problem These difficulties, which were mostly ones of response sensitivities of the instruments at various frequency combinations, are now under control.

At present, several unforeseen difficulties have been surmounted, but we are now 11 months into our year contract. Even with the best of luck, our remaining month doesn't allow us much time to get data. We should probably try to get an extension on this contract for a period of at least six months. A year's extension would be preferable as the Sponsor can use this material in several product lines.

**Memo B**

The contract on this project for Soprano Suppressors, Inc., on the development of acoustical tile for the selective absorption of sound in the frequency range of 800 to 1150 cycles per second, expires next month. In the past 11 months we have designed and constructed equipment that will measure sound absorption in this frequency range with the extreme accuracy required. Many difficulties were encountered in arriving at this workable system so that we have, unfortunately, been unable to devote much time to the study that we had planned of materials for the tile.

We are now ready to test these materials. However, we have only one month left on the present contract and, obviously, cannot test many materials, write the final report, and have it edited and processed in the time remaining. The experimental study of materials and construction methods that we proposed still looks as attractive as it did when we started on the problem. The need for the tile in suppressing sopranos is still as great as ever. Hence, we should recommend that this project be continued for another six months to enable us to obtain the data we are now prepared to collect. If work is stopped now, the $50,000 already spent will be nearly a total loss to the Sponsor. They still won't know if tile of the desired selective absorptivity can be made. With the extra six months' work, we should be able to determine either how to make them or that there is little chance of developing such tile.

**Memo C**

The contract on this project for Soprano Suppressors, Inc., on the development of acoustical tile for the selective absorption of sound in the frequency range of 800 to 1150 cycles per second, expires next month. In the past 11 months we have had one trouble after another with the equipment used in measuring the sound absorption at these frequencies. In fact, we have been able to test only three experimental tiles in this period and we are somewhat uncertain of the results we obtained on them. Although the equipment seems to be operating satisfactorily now, we have no assurance that it will continue to do so since we are not sure that we have found the basic causes for our instrumentation difficulties.

We now have one more month in which to make and test tile, write a final report, and have it edited and processed. Obviously, we cannot expect to do more than scratch the surface of the extensive development program that we originally planned. Another six months of work, at least, would be required to make an adequate study of materials and methods of construction. In view of our previous equipment difficulties, we could easily spend the six months in getting rid of new bugs and at the end of the period be in the same position as we are now.

After spending $50,000 we still don't know whether tile of such selective absorptivity can be made, and we cannot even predict whether we will have an answer after another $50,000 is spent. In all honesty, we should point this out to the Sponsor and recommend that the project be dropped.

## FOR COLLABORATION

If you are working on a team report, then interrupt that work briefly to write a progress report to your instructor. Follow the guidelines in this chapter. Include a schedule updated, if necessary, to show any changes in tasks or deadlines from what you had earlier proposed. Briefly summarize what you have learned. Note any problems you may have encountered and indicate how you plan to solve them. Be sure to include bibliographical material.

CHAPTER 17

# INSTRUCTIONS



"IF ALL ELSE FAILS, READ THE INSTRUCTIONS."

As a technical professional, you'll spend a good deal of time reading and writing instructions for making things work. As a visual, as text, or as both visual and text, such instructions appear

- On a device.
- On the box containing a device.
- On a tag.
- In a memo.
- In a printed brochure or manual.
- Online, in a window or on one or many screens.

Instructions pertain to products (like toaster ovens), systematic approaches to behavior (like conduct in class), and methods (like conducting a test in a laboratory). They may be called by a variety of names, for example, *procedures, directions,* and *directives.* The document that contains them is often called a *manual.* All instructions divide an action to be performed into steps the reader can manage. In this chapter, you will learn guidelines for instructing your reader.

## PURPOSE

You have two related purposes in writing instructions. First, your instructions enable the reader to use the product, behave appropriately, or follow the method to achieve the desired result. That's the obvious purpose. In addition, instructions *persuade* the reader to buy or behave or follow. This second purpose is one writers forget at their peril.

### Enabling Use

To achieve the first purpose, think of the reader as a *user* of the instructions. What problem is the product or system designed to solve for them? How will they measure success? Here is the goal statement for the DuPont accounting system whose users are described in Figure 17.1.

> The information products will be used to train new GTDB users and to support them when they are on their own. User support is the most important because the users
>
> - Are already under time constraints and dealing with a lot of variables (time zones, languages, currencies, etc.). They need fast solutions to their usability problems to meet their deadlines.
> - Depend on one another for quality financial information from the database. Bad data costs time at the very least and can potentially cause significant misstatements of financial information.

The following statement from an instructional insert clarifies who should use a blood sugar monitor and what that use should achieve:

> Blood sugar (serum glucose) monitors are for people with diabetes who need to check the level of sugar in their blood. The amount of insulin they will use depends on the reading this monitor gives them. By taking a sample of blood from a finger and placing it on a special strip, diabetics can tell what their current blood sugar level is. They can use this information to follow the doctor's advice for insulin dosage or other medications (Backinger and Kingsley 7).

### Persuading the Customer

To achieve the second purpose, persuasion design your instructions to fit a marketing strategy. Listen to your customers:

- Gather questions asked on a product or service help line.
- Record comments or questions submitted to a product or service Web site.
- Talk with any gatekeepers between you and the customer, for example, health care professionals who can talk about patients' needs.

Learn the answers to the following questions:

1. *How homogeneous is the audience?* Can you assume all your readers share major characteristics—motivation, expectations, prior knowledge, and the like? Are you addressing a small group of experts? A large group? How diverse is the group?
2. *How widespread are your users geographically?* Will the manual supplement personal instruction or must it stand on its own for readers who will need to use it at a distance from your company's support system?
3. *What constraints pertain to your document?* For example, what legal and commercial standards and codes apply in the markets where you'd like to sell?

**Primary Audience**

The primary audience for the GTDB information products are the financial professionals (about 90 people) who currently use the COGS Transfer DataBase. These people are responsible for managing and communicating financial information for individual DuPont subsidiaries around the world. They have individual accounting responsibilities to their own companies and governments, but they also work together as part of the DuPont global accounting network to collectively report the financial position of the DuPont businesses and consolidated DuPont Company each month.

These users work in a high-pressure environment which is imposed by a monthly series of reporting deadlines. In dealing with their accounting responsibilities they must also deal with multiple time zones, languages, cultural factors (holidays, religions, attitudes), levels of technology, types of accounting systems and equipment, and currencies with fluctuating exchange rates. Their work is very interdependent. If one subsidiary is late in providing data, it could hold up other subsidiaries and the entire corporation. If users have difficulty using a system, it will slow them down and cause greater anxiety for meeting the established deadlines.

A large percentage of this audience is not in the United States. Although they all have some understanding of the English language, English is not the native language for many. We have mailed out a survey which is aimed at learning more about their cultural, educational, and language backgrounds, as well as other information. We plan to conduct some interviews to get more information. In addition, the Business Information group has a file of help line requests from the COGS Transfer DataBase.

The people who support the GTDB Help line are another primary audience. They work under the same time constraints as the GTDB users, acting as a problem-solving resource for them. There are currently two help line people, both located in Wilmington.

**Secondary Audience**

A secondary audience will comprise the following:

- Managers who need to know how the accounting systems work
- Auditors who need to confirm system procedures
- Business analysts

**FIGURE 17.1 Audience Profile: Financial Professionals**

The term *primary audience* refers to those readers whose needs the document must serve most directly. They come first. *Secondary audiences* are readers who must be aware of the instructions but who will not specifically implement them (see Chapter 6). In the profile, GTDB is the new program; COGS is the program being replaced. (Courtesy of Susan Gallagher and the DuPont Global Business Services.)

4. *What is the user's technical level?* Will the document need to teach technical concepts or principles before it can describe the tasks at hand?
5. *What is the user's first language?* If the answer is a language other than English—and you are writing in English—you'll need to build in special features. For example, readers may be more comfortable with technical terms in English than with colloquial English.
6. *Will the manual be translated?* Prepare the document for possible translation. Such preparation helps companies release products simultaneously throughout the world even if the manual has to be localized, that is, translated for a particular country or language group. Translating is expensive, and that expense can usually be justified only if the market in a particular country is large. For most technical products, instructions in an international language like English are adequate, so long as that English is simple and precise.

For more about writing for translation, see Chapter 12

7. *How much detail do users need?* Are your readers already familiar with similar devices or systems so you only need to update them on differences? What expectations do they have about the extent of discussion in a manual?
8. *What technology do users share with you?* Can they download your manual if you send it via satellite or on the Web? Can they open a file sent as an e-mail attachment? Is a print manual the only form they can access? Is it their preferred form?

Finally, adjust your approach to overcome any resistance you anticipate from the reader. Some people readily adopt new ideas or new technologies; others resist. Because of their high motivation, adopters make fewer demands as readers than resisters. For resisters, be imaginative in the use of visuals, design elements, and interest-getting devices. The following examples summarize briefly the answers to some of these questions for a particular set of instructions. The first is for users of home medical devices and the second, for users of a software help system. Figures 17.1 and 17.2 show more extensive profiles.

- Users of home medical devices range widely from the sophisticated to the nonliterate. Many are elderly; for many, English is not their primary language; some are on medication that impedes attention span and memory; some have poor vision; some have poor hand-eye coordination (Backinger and Kingsley).
- Customers for the help system probably fall into at least three categories: new users, experienced users who need guidelines for a seldom-used operation, and advanced users who need reference information and programming details (McLaren 6). New users need basic information to get them started, perhaps with a comparison to a program or activity they already know. Any user may need a reminder about something forgotten, methods for correcting an action that led to trouble, orientation when they are lost, for example, in a hypertext program, and a quick way out.

| | |
|---|---|
| What is our objective? | Show potential users how the system can meet their needs and encourage them to try it out |
| Who are the main target readers? | Scientists who could use the system to gather hydrology data remotely |
| Who else might read the guide? | The scientists' assistants<br>Administrators who want to know how much it will cost<br>Data processing security people who want to check that any confidential data will not leak out from their sites |
| What are their main questions? | Will the system offer easy access to data?<br>What are its unique capabilities?<br>What will guarantee the continuity of the system over the next ten years?<br>Who will help if they have operational problems?<br>Which members of their peer group are already using the system and what do they think of it? |
| What else may readers have seen about the Toulouse System? | The Toulouse newsletter<br>Scientific papers mentioning the Toulouse System<br>Advertisements in professional journals<br>WWW site |
| Likely reading pattern? | Quickly skim through introductory segment on theory of operation and history<br>Read list of hydrology capabilities more closely<br>Look up specific issues in table of contents or index |
| Lessons to draw from their reading pattern? | Make each capability a separate, self-contained segment<br>Use smaller than A4 page, brochure format<br>Use plenty of tables, lists, and illustrations |

**FIGURE 17.2 Audience Profile: Scientists**

A technical writer in France created this profile of the audience for the hydrology version of a commercial geographic information system, the Toulouse System.

(Courtesy of Nigel Greenwood.)

A CLOSER LOOK

## Making Things Work

Such organizations as the American Society for Testing and Materials (ASTM) and medical boards have long codified highly technical processes like testing the strength of engineering materials and performing surgical interventions. Such documentation both establishes the best practices and instructs practitioners in performing the process. They address technical experts.

But especially since World War II, a vast new range of products has appeared that serve consumers, not experts, both at home and at the office. Such items as personal computers, exercise equipment, kitchen appliances, entertainment devices, and programmable telephones represent sophisticated technology whose users need simple directions for making them work.

In the best of all worlds, the operation of this sophisticated technology would be obvious, embedded in the good design of the product or system. Good design indeed goes a long way toward making products usable—but not all the way. The next level of help to compensate for the gap between the technology and the user consists of simple instructions at the point of use. The top of most coffee makers and the paper tray of most printers provide examples of such instructions.

Point-of-use instructions often represent a refinement over other forms. For example, instructions for operating the emergency exit on a plane have often appeared on a foldout card placed in a seat pocket, with each airline using its own form. In an emergency, the passenger has to find the card and follow the text or pictures. Instead, it makes sense to embed those instructions above the door itself in a visual form consistent on all aircraft of that type regardless of owner.

Finally, explicit instructions, in print or online, are another bridge between the technology and the user. Particularly in the United States and Europe, these guidelines are often extensive. Until recently, however, such instructions were less prevalent in Asia, where a tradition of learning from others in the group, juniors from seniors, made them less necessary.

## STRUCTURING THE INSTRUCTIONS

Creating clear instructions is demanding, made more so because, if the instructions are good, they *appear* to be simple. Keep in mind this principle: if you *can* be misunderstood, you *will* be misunderstood. It's particularly easy to be misunderstood if your readers vary in skill level, differ from you in corporate or national culture, or represent a variety of different cultures. Design your instructions to fit your profile of the readers. In doing so, include some or all of the segments described in this section.

### Introduction

An introductory unit to such instructions as package inserts, manuals, or sets of guidelines welcomes the reader and overcomes any reluctance to read. A brief introduction to instructions for assembling a device may, for example, list pieces included in the box, overview the major steps ("You can assemble the bed frame in three easy steps."), and establish a time frame ("This assembly should take one hour."). More detailed instructions for using a device or system may include a longer introduction, as in Figure 17.3. In such a segment, often headed "Read Me First" in U.S. publications, you achieve some of the following purposes:

- Reiterate the purpose and importance of the device or system.
- Explain how the manual is organized.

## Finding the Information You Need

(1) 

(2) Thank you for choosing Norsonic! The RTA840 has been designed to give you many years of safe, reliable operation.

Your approach to the RTA840 documentation depends on what you want to do and how much you already know.

The manual has been divided into ten sections plus index. Each section provides different information. In some places information has been (3) copied from other sections (but adapted) to let you have all the relevant information there and then and thus avoid unnecessary "page-riding."

Depending on your requirements and your familiarity with acoustics *per se*, you may find that you use some parts of this manual often and (4) others not at all. Note that the manual describes a fully equipped instrument. Your version may not have all the described extensions installed.

Readers of previous editions of this manual will note a change in layout. We believe the change is to the better as it provides more information in fewer pages and above all, in a much more structured form.

The present structure is well-suited for reference purposes, but beginners should note that the order of appearance of the book's topics is neither arbitrary nor alphabetic.

(5) Instead we have sorted the topics in an order reflecting the natural flow of work when dealing with a specific task. For example, the section *Making Level Measurements* starts with a discussion of general measurement aspects, then proceeds through a complete measurement preparation procedure (in the recommended order), before we outline for you how to actually make the measurement and what display tools you have at your disposal.

There is an extensive index — use it!

We consider *Basic Concepts* as the most important part of the manual, because it explains all the fundamentals , the use of menus, how to set the measurement parameters, the features and tools available inside a menu etc.

(6) Consequently, we recommend that you read through this part of the manual before you start to use the analyser.

The rest of the manual can then be consulted whenever required.

Our main objective with this manual was to address your needs and goals. Please let us know how well we succeeded!

*Using the Real Time Analyser RTA840* *iii*

1. Icon resembling a basketball repeats image elsewhere in manual and gains reader interest
2. Welcomes the reader and reiterates that the reader made the right choice
3. Overviews the benefits of the manual's organization
4. Establishes the level of special knowledge or skill required and intended user
5. Provides an example of the device's applications
6. Instructs the user on how to read the manual

**FIGURE 17.3 "Finding the Information You Need"**
An example of an effective introduction. The manual accompanies an instrument, called a *Real Time Analyser,* which performs acoustic analyses.
(Courtesy of Norsonic AS and Gustav B. Ese.)

- Establish the level of special knowledge or skill required.
- Identify the intended user of the device or system.
- Provide an example or scenario of the device's applications.
- Instruct the user on how to read the manual.

A leading technical professional in China notes that Chinese readers expect the introduction to help build a trusting relationship between producer and customer in a culture where such personal relationships are an essential part of doing business. The introduction should "establish the image of the enterprise, emphasizing the company history, business activity, range of products, and their application" (Zhu 11).

### Theory of Operation

Your audience may also expect a unit providing the theoretical underpinnings for the process. German and Dutch readers, for example, like to see an extensive discussion on the theory of operation at the beginning of a manual. They tend to read manuals carefully before performing tasks and need theory as context as in the Basic Concepts section noted in Figure 17.3.

### Safety

You'll probably want to address safety concerns in a special unit near the beginning of a manual. Government regulations in Japan, for example, require manuals to begin with extensive warnings and cautions aimed at keeping operators safe while performing the process (Burnett). In the United States, too, manufacturers devote attention to identifying hazards, in part out of fear that they may be liable for damages. Note such safety issues at the beginning of the manual and at pertinent points throughout. Include a generally recognized icon that signals the problem, briefly explain the problem and the consequences if your instruction is not followed, and instruct the user in avoiding the hazard. The two major safety notations are *warnings* and *cautions*.

**Warnings.** Write a *warning* to draw attention to something that threatens the user's personal safety, including serious injury or even death. You may also use the term *danger*.

Instructions for some medical devices have to warn the user not to operate the device if certain symptoms or problems occur. They might also warn about backup procedures necessary if life support systems fail. Most warnings are set off from the text in boxes or accompanied by icons that grab the reader's eye. Standards for the icons, however, vary. The International

Standards Organization (ISO) dictates the use of a triangle with an exclamation point for a chemical warning, a triangle with a lightning bolt for an electrical one. Some documents use a red stop sign with a raised palm. The color disappears, however, if the manual is photocopied in black and white, so relying on color as a warning may be ineffective.

**Cautions.** Use a *caution* to alert readers to potential malfunctions or failures of the system itself and any resulting damage including, for example, a loss of data if a computer system goes down. A safety precaution aims to guard both the equipment and the user against any misuse:

> Protect the printer against extreme cold or hot weather as well as dampness and wet conditions.

Cautions are sometimes indicated by the color yellow and by an exclamation point in a triangle with its base at the top.

### Conditions Affecting Use

If they are significant, explain the conditions required for operating a device or conducting a procedure. For example

- The need for a source of water and electricity
- The need to unplug the device before testing or cleaning

Similarly, clearly state any conditions in which the procedure should not be conducted or the device not operated:

> A telephone should not be used near water—near a bathtub, sink, wet basement, or swimming pool.

Tell the reader how to vary the procedure if the device operates differently

- At various altitudes or temperatures.
- In different geographic regions.
- In transit between locations.

Prepare the customer to anticipate and correct any problems.

### Setup

If appropriate, clarify whether readers should set up the device on their own or seek professional help for that step. If the readers will perform the setup, note any special requirements (Backinger and Kingsley 9):

- Parts list
- List of materials and tools needed for setup
- Unpacking instructions

- Directions for locating the device in the home, such as on a tabletop or on the floor (also state whether the device should remain in one place after setup)
- Any warnings or safety instructions specifically related to setup, placed right before the corresponding task or instruction
- Description of what happens if setup is wrong
- Setup instructions in steps numbered in logical and chronological order
- Any special preparation before first use of the device, such as cleaning or disinfection
- Space to write in user-specific instructions
- Who to call if there is a problem

In addition, explain, if necessary, any checks readers should run to ensure that the device is working and is properly calibrated.

### Step-by-Step Description

At core, the instructions describe a series of steps readers must take to achieve the desired goal. Sometimes there is only one way to do something. But often, you can reach the same goal by different routes. Find the route that best matches your profile of the readers so you reduce any initial resistance and uncertainty. Look for trouble spots as well as shortcuts. Establish the outcome of each stage.

One method for developing the sequence is to videotape an expert performing the task. Then rerun the tape to identify discrete activities. In addition, interview experts who use the product, for example, the designers who developed it or major customers. In an interview, however, an expert may not think of some action that's now second nature. For example, an expert may think the first step is simple: Turn the machine on. But look for the possibility that more than one action is required, as in the following description of turning on a machine (Backinger and Kingsley 23):

To turn the machine on,

1. Plug the power cord into an AC outlet.
2. Facing the front of the machine, find the black power switch on the right side.
3. Turn the power switch to the "ON" position.

To avoid the bias of an expert, you can work with the product or perform the method yourself, assuming you are not an expert. See what it takes to achieve the right outcome. You might also ask some novices to help you determine what steps are necessary. Suggest guidelines and then watch the novices perform the activity. Note, for example, if any materials or equipment are needed beyond what you originally planned. In addition, observe the following:

- How difficult the reader finds each step
- Whether any steps are significantly less or more difficult than others

- Where novices seem to go wrong in using the product or understanding the procedure
- How you can help the reader avoid those wrong turns

Figure 17.4 shows one expert's recommendations for determining how much information—and what information—you need in a procedure. The method is also expressed as a recipe, a common form of procedure.

## Recipe

### Lean Documents

Let users write the document. To create a concise but effective document, use our exclusive no-cost usability testing method.

#### Ingredients

**Two users** (or reasonable surrogates)
**Observer** (that's you)
**Video camera or tape recorder** (These are optional. If the users object to being taped, just take notes).
**Expert** (someone who knows the product, technology, or system you are writing about).
**Prototype** of the product (If a prototype is not available, have a subject-matter expert simulate the product).
**Realistic task** for the users to perform

#### Procedure

Give the users a task to perform, but **no documentation or training**. Tell them if they have questions they can ask them of the expert. Also explain that they do not have to ask questions. They can figure things out on their own if they prefer. They can experiment with the prototype. They can meditate and wait for psychic inspiration for all you care.

The expert can answer their questions but cannot volunteer information. The answers do not have to be "yes" or "no." Answers can be complete explanations. The expert can answer in words, draw pictures, and demonstrate actions. The one thing the subject-matter expert should not do is to volunteer information the users have not asked for. This means that you, the observer, must intervene if the expert tries to help the users without being asked or if you feel the answer to a question strays beyond the scope of the question.

#### Here's the basic cycle:

1. The users try to do a step. If they get stuck they ask a question.
2. The expert answers the question.
3. You prompt the users to paraphrase the answer.
4. If the expert agrees that the paraphrase is essentially correct:

   A. You write down the paraphrase.
   B. The users perform the step and continue the process.

At the end of the test, you string together the paraphrases and you have a concise first draft.

Now repeat the test with another two test subjects. But this time, give the test subjects the first draft you derived from the first test. If the first draft is not sufficient, that is, if the new users have additional questions, add the answers to the draft before the next test. After a few rounds of testing you have a draft that is written in the user's terms and concepts, expressed at their level of experience, containing just what they needed to know with nothing wasted.

**FIGURE 17.4 Recipe for a Lean Procedure**
Compare this procedure with the description of usability testing in Chapter 12.
(Used with permission from *Secrets of User-Seductive Documents,* 2nd ed, by William Horton. Reprinted with permission from *Technical Communication,* the journal of the Society for Technical Communication.)

## MAINTENANCE

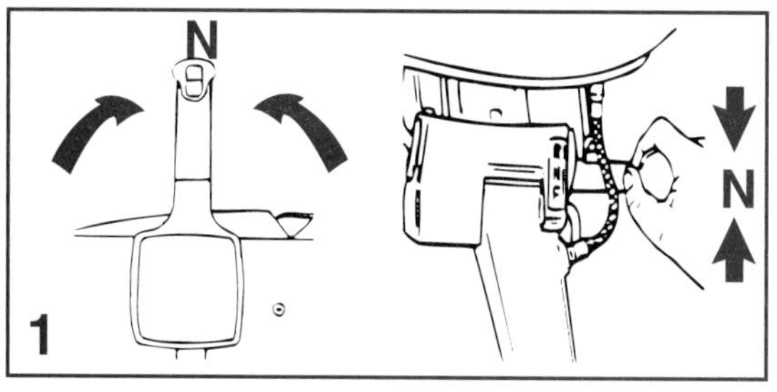


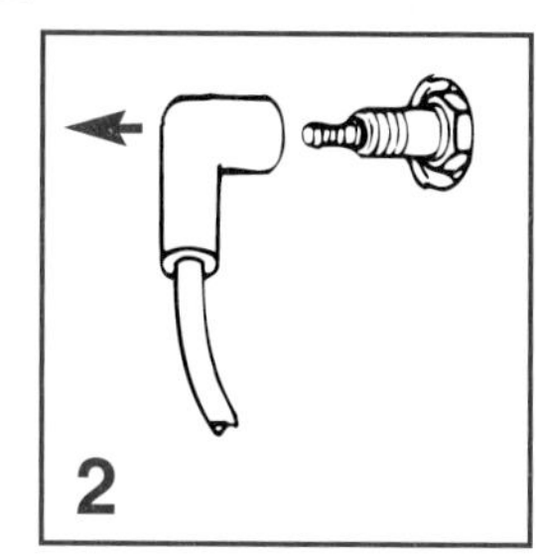


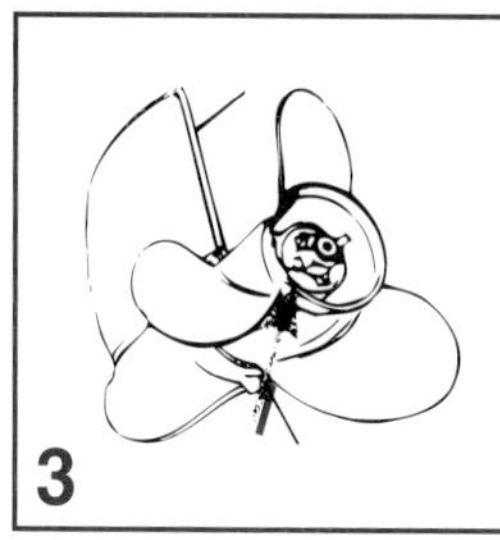


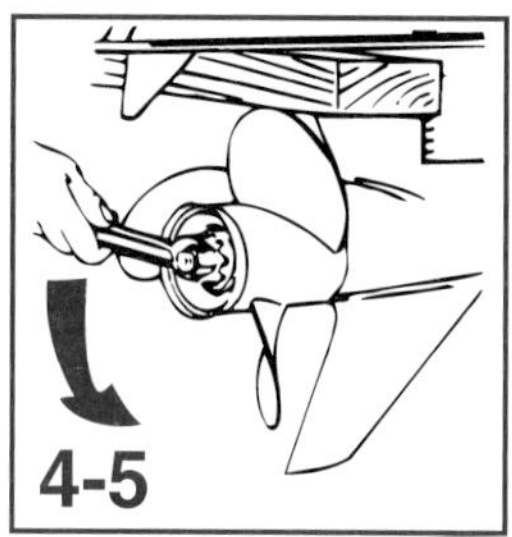


### Propeller Replacement

Safety warning

**⚠ WARNING**

**If the propeller shaft is rotated while the engine is in gear, there is the possibility that the engine will crank over and start. To prevent this type of accidental engine starting and possible serious injury caused from being struck by a rotatting propeller, always shift outboard to neutral position and reremove spark plug leads when you are servicing the propeller.**

Parallel descriptions in words and pictures with numbers keying one to the other

**1** Shift outboard to neutral (N) position.

**2** Remove spark plug leads to prevent engine from starting.

**3** Straighten the bent tabs on the propeller nut retainer.

**4** Place a block of wood between gear case and propeller to hold propeller and remove propeller nut.

**5** Place propeller straight off shaft. If propeller is seized to the shaft and cannot be removed, have the propeller removed by an authorized dealer.

FIGURE 17.5 **Maintenance**
How to replace a propeller.
(Courtesy of Mercury Marine.)

## Cleaning and Maintenance

In instructions for using devices, you may need to tell users how and how often to clean the equipment. Provide pull-out or laminated maintenance charts that can be hung on the wall for easy reference. In addition, provide guidelines for storage (for example, "must be stored in a cool place" or "keep out of sunlight"). Figure 17.5 shows how to replace a propeller on an outboard engine as part of a maintenance check.

## Troubleshooting

In your unit describing the steps in the process, you show what happens when everything goes right. In the troubleshooting section, you show what to do when the process goes wrong. Trial runs with potential users help you anticipate problems and prepare solutions. In addition, design your troubleshooting section for rapid reference:

- Provide a table or flow chart that includes signs of trouble and actions to counteract the difficulty (Figures 17.6 and 17.7).

### Troubleshooting Guide

The following checklist will assist in the correction of most problems which you may encounter with your unit.
Before going through the checklist below, refer to the connection and operating procedures.

#### General

| Problem | Cause/Solution |
|---|---|
| **No sound** | • Adjust the volume with the + button<br>• Set the fader control to the center position for 2-speaker systems |
| **Indications do not appear in the display window.** | Remove the front panel and clean the connectors. See "Cleaning the Connectors" for details. |

#### Tape playback

| Problem | Cause/Solution |
|---|---|
| **Playback sound is distorted.** | Contamination of the tape head. ⟶ Clean the head. |

#### Radio reception

| Problem | Cause/Solution |
|---|---|
| **Preset tuning is not possible.** | • Memorize the correct frequency.<br>• The broadcast is too weak. |
| **Automatic tuning is not possible.** | The broadcast is too weak. ⟶ Use the manual tuning. |
| **The "ST" indication flashes.** | • Tune in precisely<br>• The broadcast is too weak. ⟶ Press the SENS button to enter the MONO mode |

Table incorporates common problems and their resolution

FIGURE 17.6 **Troubleshooting Guidelines for a Car Stereo**
(Courtesy of Sony Electronics Inc.)

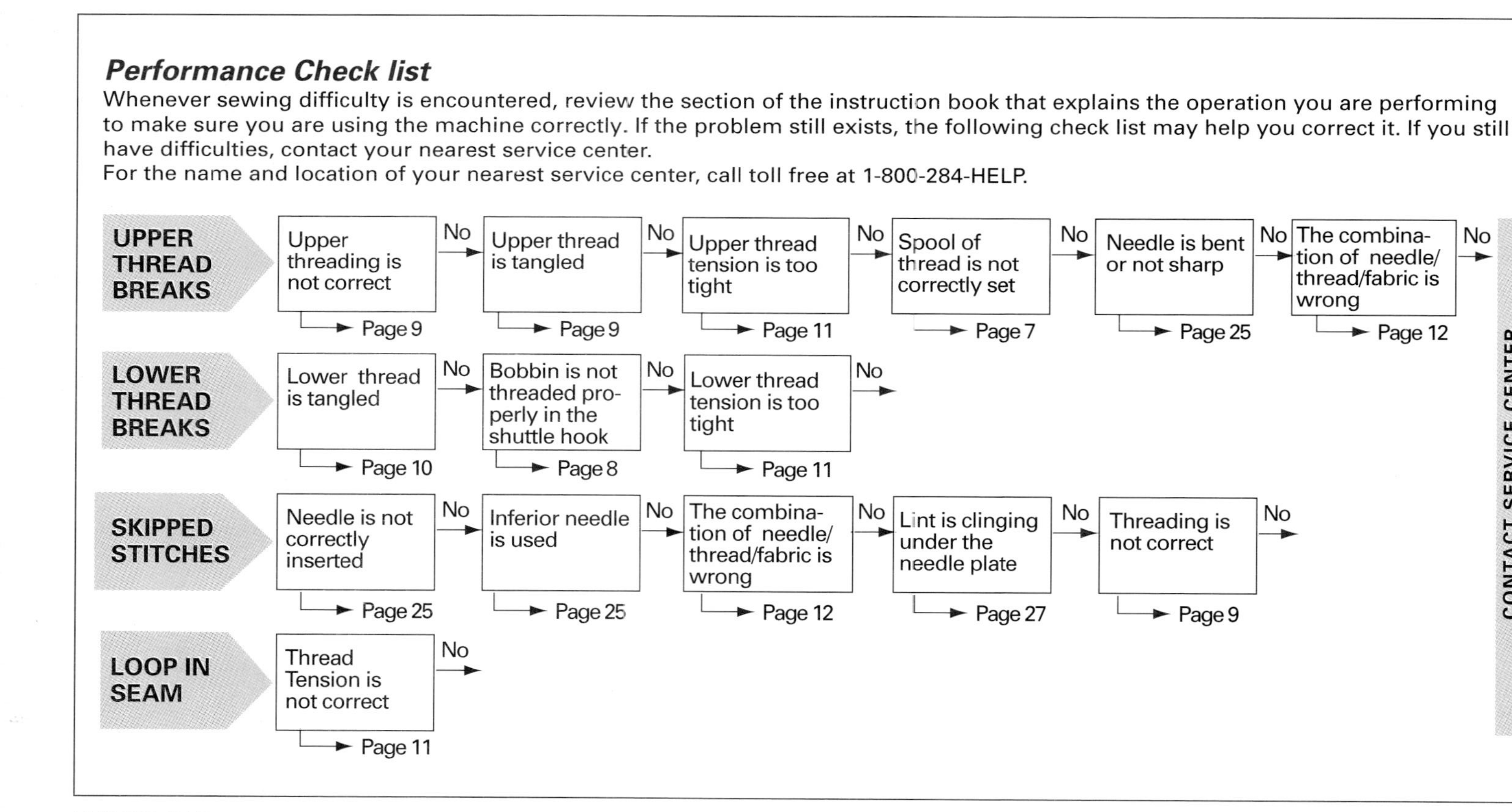


FIGURE 17.7 **Troubleshooting Guidelines for a Sewing Machine**
Flow chart form or "decision tree" shows steps in solving the problem and provides references to manual pages.
(Reprinted by permission of Brother International Corporation, © 1993 Brother International Corporation. All rights reserved.)

- List error messages if the device provides these, and then show how to correct the error. In any error messages, differentiate between what the system or product did and what the user did.
- Group related problems and use typography to highlight especially serious or common ones.
- Avoid lengthy technical explanations. As one veteran sailor noted in another context, "When the boat is taking on water, don't talk about navigation. Bail."
- Clarify what troubleshooting steps the user can complete, what problems the user should not attempt to remedy—and when and how to call a serviceperson, as in Figure 17.7.

### Other Items

In addition to these major units, you may also include other items. For example, you may create a quick reference card for seasoned users. The card provides brief reminders of key steps or functions, along with cautions and warnings. Use a single page or a narrow column that can fit on the device itself. In addition, list prominently (preferably on the device itself) an 800 number, Web site URL, or e-mail or postal address for customer service.

You may also include *notes* that qualify information in the text, provide suggestions for further information, add technical fine points, or offer hints to make the procedure easier. For example:

> Note: If your screen doesn't look like the one in Figure 2, hold down the CTRL key and press T twice. Now the layout of the screen should resemble that in the figure.

If your readers may not be familiar with the terminology for a complex device or system, include a glossary of key terms. Preparing a glossary of terms early in the writing also helps you use terms consistently, coordinate a team effort, and ease translation if your manual addresses different language groups.

## MANAGING A TEAM EFFORT

Writing the procedures for today's complex equipment, systems, and techniques usually requires the joint expertise of many people in a team endeavor. The team may include product designers, engineers, writers, artists, marketing specialists, potential users of the product, media specialists, and translators. That approach also allows insights from the process of writing instructions to aid in designing the product. In one situation, for example, mechanics working on the wing of a fighter plane often stepped on a hydraulic line running under their preferred work area. Writers preparing instructions to warn them against stepping there talked with engineers who had placed the line and found that a simple rerouting solved everyone's problem (Lippincott).

## Planning

To coordinate the team, whether it is writing instructions for building a fighter plane or maintaining a software system, prepare a plan for the documentation. Such planning helps even when you write as a single author, but it is particularly important when you write on a team. Make explicit the answers to these and other questions you have been learning how to ask in this chapter:

What is your goal in writing the instructions?
Who will use your instructions?
What is their goal in using them?
What tasks do you need to perform to complete the instructions?
What tasks do the users need to perform to use the system or device?
What is the environment in which they will use the instructions?
When do you need to deliver the instructions to the user?
What schedule do you need to follow to make that delivery?
How much money will it cost to prepare the instructions?
What component documents are needed (for example, tutorials, reference materials, help screens, help line)?
What is the format of the pages or screens?

In Figure 17.1 you read an audience profile and a purpose statement from such a plan at the DuPont Company. The plan focuses the work of the team members and provides concrete information that managers and potential users may review. Figure 17.8 reproduces a list of all the units of the DuPont plan.

## Testing

Including representatives of your users on the development team helps you troubleshoot the draft of your work. In addition, you may set up more formal tests of the product's usability, as you read in Chapter 12. In those tests you measure the users' responses to such elements as

Warnings and cautions
Extent of coverage for each step
Order of the steps
Level of difficulty in vocabulary and concepts
Graphics

Here are five techniques for such user testing recommended by the U.S. Food and Drug Administration (Backinger and Kingsley).

**For more about focus groups, see Chapters 4 and 6**

**Focus Group Interview.** Select some six to eight potential users who will meet to discuss the instructions at the outline stage or when you have completed a preliminary draft. Discuss format, problems, and solutions.

Product description and purpose

Audience profile

Purpose of the information product [the instructions]

Task description

Usability goals for the information product

Design implications

(for example, "Make the reference materials graphical wherever possible. This will help the non-U.S. users who may have limited English capabilities and who need more support due to their distance and time difference from the help line.")

Media selection (specifications for a notebook, for example)

Development strategies and concerns

(for example, information about translating, distribution, testing, management review)

Constraints (anticipated problems)

Roles and responsibilities

(a matrix outlining what each team member will do, by name)

Deadlines and project schedule for the writing team

FIGURE 17.8 **Segments of a Plan for Collaborative Writing**
(Courtesy of Susan Gallagher and the DuPont Global Business Services.)

**In-Depth Interview.** Interview potential users to determine what they want to see in the instructions or to discuss their recommendations after reviewing a draft.

**Questionnaire.** Circulate a draft and a set of questions to potential users.

**Gatekeeper Review.** Interview or circulate a questionnaire to people who are knowledgeable about the users. Manufacturers of home health care devices, for example, should interview health care professionals.

**Operator Performance Study.** Ask potential users to perform the procedure. Watch them as they work and ask for their comments about any discrepancies between the manual and what actually needs to be done.

In addition to verifying the usability of instructions, you also need to confirm their accuracy, that is, to validate them. For example, suppliers of military aircraft use one or more of three techniques: comparison, simulation, and demonstration (Lippincott). A comparison test matches the manual against the engineering documents and blueprints and makes sure there are no discrepancies. In a simulation, "you walk out to the airplane, look at the engine, and compare the manuscript to the hardware ('Yes, I can see that if you disconnect these lines and remove these bolts, the engine

CROSSING CULTURES

## User Friendliness

Companies are increasingly recognizing that manuals must do more than instruct the customer or client; they have to *sell the process* and keep the user happy. Manuals are marketing tools. To market well, U.S. companies in particular try to create manuals that simplify the process, attract reader attention through good design, and speak to the reader in a friendly voice. That voice can be chatty, casual, colloquial, even funny. The author draws the reader into a conversation.

Such a light-hearted approach, however, is not always welcome—or even understood—by customers, especially those outside the United States. Colloquial English is often hard for non-native speakers to understand. It can also be hard to translate. And it can set readers' teeth on edge. One British observer complained, "This friendly approach is part of the pernicious 'Have a nice day' insincerity from across the water and the . . . need to convince ourselves that despite the fact we're all trying to make money we really are nice and kind to each other" (Knight 18). A French reader may expect more detachment and formality in a manual, measuring its worth in part by how difficult it is. A German might want more details in an explanation.

To be friendly to users *internationally,* find out what makes readers comfortable in the culture you're addressing. For example, several researchers point out that Japanese readers find imperatives jarring because they connote aggressive behavior and confrontation. Avoid such verbs in manuals to be sold in Japan. Some Europeans find the warnings and cautions mandated by U.S. legal standards overwhelming. Place such warnings in the background for the European market. Use analogies that are friendly to your readers. For example, a manual for a British database system used an extended analogy to a Christmas card list. Recognizing that Germans pay far less attention to Christmas cards than the British, the translator used a different analogy in the German edition: a mailing list for an association (*verein*), like a singing group or club, something most of his readers would recognize. Friendly manuals also customize people's names in examples so that different users find names that are familiar and comfortable to pronounce. And they respect (and reflect) what readers value. An airline that emphasizes its *punctuality* to Western audiences translated its appeal to *hospitality* for Arabic readers (Weiss 417). Visuals, too, should show familiar forms of objects. For example, recognize the wide variety of electrical plugs and outlets internationally and make sure your instructions show the kind your reader sees everyday (see frame 5 in Figure 17.10).

*User friendly,* then, means that the document speaks in the user's terms, whatever the language of the transaction.

will come out as the manual describes.')." In a demonstration, "You walk out to the airplane and physically remove the jet engine following the tech manual steps exactly as written" (Lippincott).

## DELIVERING THE INSTRUCTIONS

Because writers think of them as workhorse documents, focus only on their content, and often write them on teams, instructions are frequently dull. The text is dense and uninviting. The voice that comes across the page or screen may be at best deliberately neutral. Sometimes, it scolds or talks down to the reader. Such sober seriousness often fails to engage the reader or build confidence for the task at hand. Increasingly, U.S. instructional documents attempt to make reading comfortable in the belief that such comfort improves the instruction's effectiveness (see *Crossing Cultures: User Friendliness*). The following guidelines will help you deliver instructions effectively.

### Visuals

Either with text or on their own, visuals are essential elements in delivering instructions (Fig. 17.9). Visuals without labels, as in Figure 17.10, adapt themselves well to manuals that serve different language groups. In addition, visuals help *show* readers something rather than *telling* them. They

(1) Install solid door with gaskets
(2) Use vibration isolators & mounts
(3) Isolate appliance from cabinet with rubber gasket
(4) Insert rubber gaskets behind cabinets and appliances to avoid wall contact
(5) Place rubber pads under small units, dish racks and in sink basins
(6) Install rubber or cork tile on backs and shelves of cabinets
(7) Apply vibration damping material
(8) Install acoustic tile
(9) Install exhuast fan on rubber mounts
(10) Install acoustic ceiling
(11) Install carpect or foam backed tile

**Steps are keyed to numbers on drawing**

**FIGURE 17.9 Reducing Kitchen Noise**
This checklist is embedded in a more general discussion of sources of noise in a home. It appears in a report issued by the U.S. Environmental Protection Agency.

1. Pictures are sequenced left to right but also numbered for an international audience that may read in a different direction

2. "Generic" electrical outlet and plug shown rather than one geared to a specific country

3. Arrow indicates direction of action

**FIGURE 17.10 Instructions for Using a German Food Processor**
(Courtesy of Krups North America.)

thus reduce ambiguity, especially for readers who are more visually than verbally minded (Figure 17.11). Instead of saying, for example,

> Click on the "Open file" icon on the tool bar

*show*

> Click 

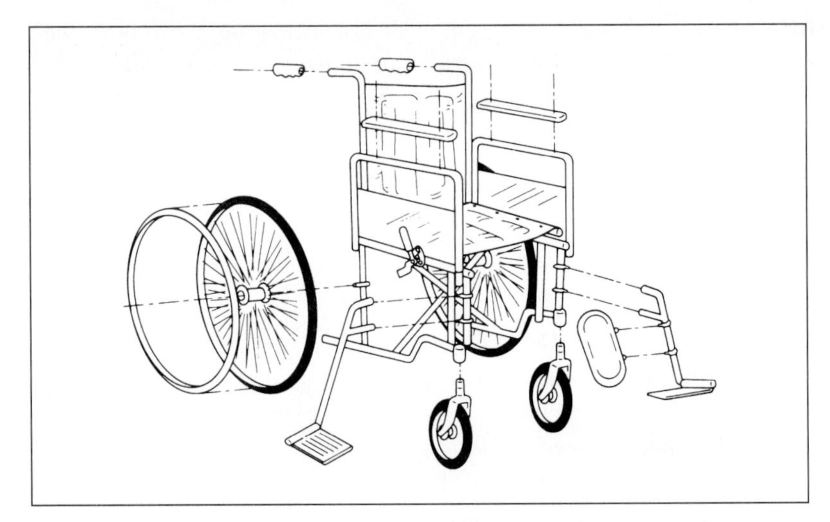

View describes both the components and how to put them together

FIGURE 17.11 **Exploded Diagram of a Wheelchair**
(Backinger and Kingsley 41.)

## Paragraphs

Some readers prefer to see instructions in units of paragraphs (Figure 17.12). Using paragraphs also helps you tuck away explanations and nuances and emphasize the flow of the process. With paragraphs of explanation, you can develop the context for the process and include brief scenarios or stories about how others have used the process successfully. But paragraphs are harder to skim than lists and pictures and usually take up more space. The following paragraph from a brief pamphlet effectively guides campers in dealing with bears in the wilderness:

> A black bear in camp requires caution but is not cause for great alarm. Most are timid enough to be scared away by yelling, waving, and banging pans. But a few are too accustomed to people to be bothered. Many people have lost their food and vacation by being timid. Campers experienced with black bears simply chase them away before the bears settle in to eating a week's supply of vacation food. They make sure the bear has a clear escape route and then yell, wave, and rush to no nearer than 15 feet of the bear. This is especially effective when several people do it together. If alone, a person might create the illusion of numbers by throwing sticks through the underbrush. Campers should not feed the bears or try to pet them. Touching a wild bear can elicit a nip or cuff. (North Central Forest Experiment Station)

## Lists

Reading at their leisure, campers might like to see bear activity explained in paragraphs. Such a description is intended for advance reading, not while they look at a bear. When you intend your instructions to be used during the process, think about using a list. Lists emphasize individual tasks better. A checklist, for example,

- Reinforces the serious consequences of not following a safety procedure, as in an aircraft pilot's checklist of actions before takeoff.
- Helps readers skim a procedure to see the big picture.
- Provides a sense of accomplishment as users check off each step.
- Reminds users of lessons learned in a tutorial.
- Encourages compliance.

As an example, consider this checklist that guides telephone operators in answering customer complaints:

1. Listen to the complaint. Do not interrupt.
2. Summarize the complaint for the customer to assure your understanding of the main point.
3. If appropriate, apologize.
4. Offer a solution or alternatives.
5. Thank the customer for calling attention to the problem.

Number the steps for ease of reference and note in the introduction the total number of steps. Use Arabic numbers, not Roman numerals or letters. Use bullets if the list indicates steps whose sequence is not significant or for a list of items like equipment, materials, or requirements. To see how checklists serve readers in different contexts, examine Figures 17.9 and 17.12.

### Verbs and Other Words

Whether you use paragraph or list form, express the action at the heart of the instructions with verbs chosen from the following options:

#### Active Imperative

*Measure* the elapsed time.
*Prepare* the solution.

#### Active Indicative

The technician *measures* the elapsed time.
The technician *prepares* the solution.

#### Passive

The elapsed time *is measured.*
The solution *is prepared.*

#### Conditional

The elapsed time should (or shall) be measured.
The technician should (or shall) prepare the solution.

Which should you choose? Choose the form that reflects your corporate style guide, personal taste, customer preference, legal mandates, or

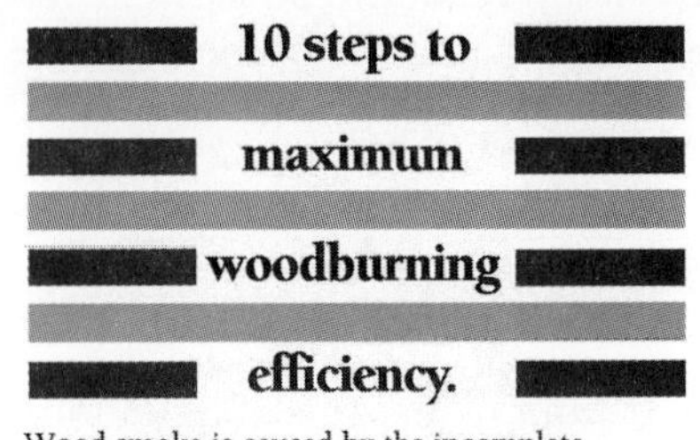


Wood smoke is caused by the incomplete combustion of wood. This can pollute the air — indoors and outdoors — as well as contribute to higher heating costs.

Fortunately, the cure for cutting down on pollution and waste also cuts the costs by burning wood with safety and efficiency.

**1** Choose the proper size stove. A properly sized wood stove will do its job efficiently even on the coldest days. One that's too big will have to be damped down, which increases creosote production. The insulation in your home is a factor as well. To be sure you select the right-size stove, take along to your dealer the number of square feet to be heated, and the amount of insulation surrounding the area to be heated.

**2** Buy the most efficient design you can afford. It'll pay for itself in the long run. Research has made great strides in designing fireboxes, drafts, catalytic combustors and other devices that improve combustion and reduce smoke. Maybe it's time to retire that old "smoker" and modernize.

**3** Burn only the fuel your stove was designed for. Don't burn coal in a wood stove, for example, unless your stove was designed to handle both wood and coal. Trash shouldn't be burned in your stove either — besides increasing the chance of starting a chimney fire, some plastics and other trash emit harmful gases, and can ruin your catalytic combustor. Driftwood, treated wood, artificial logs, or anything containing plastics, lead, zinc or sulfur will damage your catalytic combustor.

**4** Burn seasoned wood. Up to 50% of the weight of green wood can be moisture, which has to be burned off before heat can be released into your house. Seasoned wood burns hotter and more efficiently, helps decrease the amount of creosote buildup in your stovepipe, and saves you money.

**5** Make your fires small and hot. This burns volatile gases more quickly, producing fewer safety hazards and air quality problems than a fire that is over-damped. Smaller, hotter fires mean more frequent loading and tending the stove...but the improved efficiency and air quality are worth the effort.

**6** Install a stack thermometer on the stove flue. This will help you monitor the temperature of the gases as they leave the stove. Optimum range for most efficiency and least pollution: about 300° to 400° F.

**7** Remove excess ashes. Too much can clog your stove's air-intake vents and cut down on the amount of oxygen needed for efficient woodburning.

**8** Tighten up your house. Insulation, weather stripping, storm windows and caulking can all reduce the amount of wood required to heat your home, which in turn helps decrease the amount of air pollution.

**9** Check your "smokestack." Burn your stove at different rates, then go outside and check the emissions. The absence of smoke indicates that your stove is burning cleanly and effectively.

**10** Inspect your stove. Once or twice a year, depending on how often it's used, your stove and chimney should be inspected. Look for warping, check the baffle to make sure there are no gaps, check for creosote. Your dealer can make regular inspections, and so can a chimney sweep.

**Large step numbers provide good design**

**Each step is expressed in a short phrase with parallel imperative verbs**

FIGURE 17.12 **10 Steps to Maximum Woodburning Efficiency**
This checklist constitutes the centerfold of a brochure for homeowners. A state agency distributes the brochure to improve safety and reduce pollution in a state where many residents use woodstoves as a major source of heat.
(Courtesy of CONEG Policy Research Center, Inc., Northeast Regional Biomass Program.)

tradition. The simplest is the active imperative, which is clear, energetic, and direct. The active indicative is appropriate for some standards and other documents that record a procedure, but the need to repeat the noun (or use a pronoun, which may cause problems in avoiding discriminatory language) makes that usage wordy. Passives often fail to name the agent or actor. The conditional may help emphasize a particular step or action, but overused, it sounds like scolding. Wherever you can, choose the active imperative.

Once you've chosen a verb form, use it consistently. That consistent usage is a form of *parallelism*. Parallelism helps you reinforce your message and make your instructions comfortable for the reader. If some item you'd like to include resists such consistent expression, that resistance may be a sign that it isn't a step. For example,

1. Clean the beaker.
2. Place 2 mg of the reagent in the beaker.
3. The temperature in the room should be no more than 25°C.

That third item is a condition, not a step. It can't be expressed as an imperative verb (except, perhaps, in the form: "Make sure the temperature in the room is no more than 25°C.").

For more about parallelism, see Chapters 7 and 9

At a larger level, maintain parallelism among segments of the instructions. For example, if you write one segment in paragraphs, use paragraphing in all similar segments or explain explicitly why you are changing form. Keep highlighting techniques parallel, for example, box all warnings and accompany them with the same icon.

Users often hang on your every word in a procedure. Make sure those words are exactly right: unambiguous, precise, and standard. One often-told anecdote points out the disastrous effects of ambiguity. After a plane crash, inspectors reviewed the maintenance manual and found this sentence: "Unmount the bolt and replace it." Replace could mean either "put in a new bolt" or "put the old bolt back in." Unfortunately, the ground crew put in the old bolt, and the component—and the plane—failed.

In addition to avoiding ambiguity, be precise. Use concrete, operational terms (Backinger and Kingsley 28, 29):

***Poor:*** Respond quickly.
***Better:*** Respond within 1 minute.

***Poor:*** Device operates poorly in a cool room.
***Better:*** Device will not operate below 60° F.

***Poor:*** Buy lots of strawberries.
***Better:*** Buy 4 quarts of strawberries.

Use terms, too, that are standard and accepted in the field. Review dictionaries and thesauruses; scout out the terms your competitors use. Within the document, identify components, devices, and systems by the same term and avoid synonyms. A component labeled as a "dial" in one place should not be called a "knob" someplace else. Avoid acronyms or explain them on the first use when they are necessary.

## Print Design

For more about design, see Chapter 7

Design pages to accommodate readers who may face your instructions with a knot in their stomachs. Most want to get the reading over with—and get on with their work. Follow these guidelines to design effective pages.

**Separate Units of Text and Visuals Graphically on the Page.** Each step, for example, may constitute a clear unit. Sort through the tasks the user needs to perform to achieve such distinct units. Some researchers recommend a two-page spread in which one page provides a graphic depiction of a task and the other page describes that task in words.

**Use Headings and Make Them Work.** For example, use questions a reader might ask, and then answer them. Distinguish the headings from the text by highlighting or a change in font. Perhaps use a rule to separate the units. Avoid simply listing capabilities from the point of view of the system or product. Instead, indicate actions from the user's point of view:

| Capability | Action |
|---|---|
| Copy utility | Copying files |
| Calculator | Adding, subtracting, multiplying, dividing |
| Special characters | Selecting a font |

**Differentiate Levels of Information.** Different readers ask different questions of a document. Some want to read only the minimum information, perhaps just one or two key points. They'll figure out the rest on their own. Some, too, find a diagram enough; words seem confusing. Other readers seek more background, examples, scenarios, maybe a test to see if they are working right, perhaps alternative strategies for doing the same task. You could customize documents for each reader need, but that's expensive, although Web delivery helps, as you'll see. Instead, good design can help you accommodate those differences in one document. Represent the levels of information either by sequence through the document or as a hierarchy.

To follow a sequence

- Place the most important or simplest functions first, then move into less common—or more complex—applications.
- Provide an overview before segments arranged alphabetically.
- Boldface key terms or cross-referenced items to indicate that more extensive discussion is contained elsewhere in the document.

To show different levels within one unit

- Color-code items in a table of contents and use that color (ink or paper) throughout the document for items at that level. The coding, for example, might represent audiences: white pages for all users, green for mechanics interested in the details, blue for systems operators, and so on.
- Indicate levels through typography, for example, all illustrative examples in Times Roman 10 point, all notes indented five spaces, all background statements in Arial 10 point, all safety warnings in capital letters.
- Use a graphic symbol to represent information at a certain level.

**Accommodate Conditions of Use.** Determine where the user is likely to read and incorporate design features that help overcome adverse conditions, for example, large print in a manual to be used in poor light and waterproof pages for manuals in a marine environment.

**Keep a Family Look.** Make sure all manuals in a series *look* alike and reflect the image of your organization. In particular, documents for foreign markets should not look second rate. Companies sometimes print glossy manuals in their native language and provide only poorly photocopied versions of translated text. For ease of printing and reproduction, use standard paper size (U.S. 8½ × 11 inches or the European A sizes) and select a

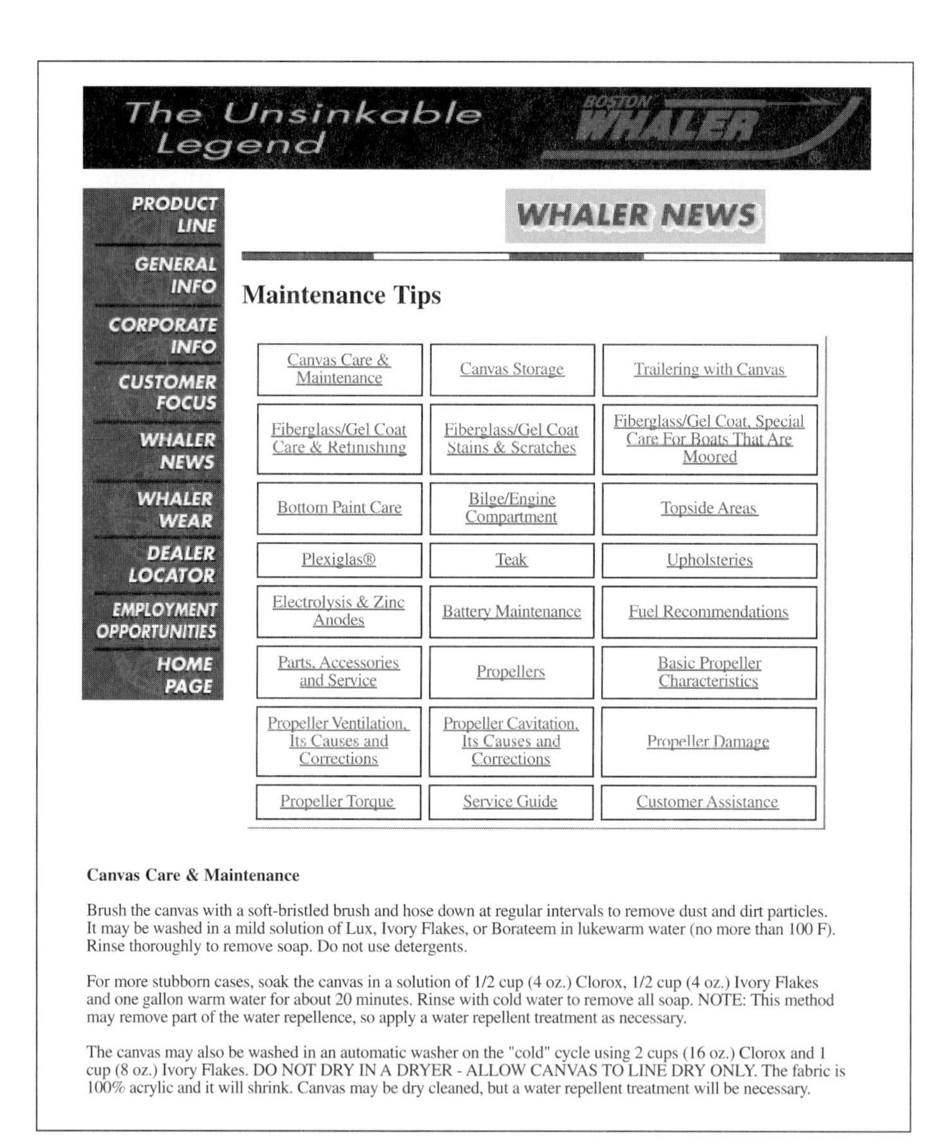


**FIGURE 17.13 Instructions on the Web**
This page from the Web site of a boat manufacturer provides a matrix of topics linked to specific advice in each category. It also offers links to other areas of the site, a quick return to the site's homepage, and a direct e-mail connection to appropriate personnel at the company (under "Customer Focus").
(Courtesy of Boston Whaler Inc.)

thickness that keeps the print from showing through. The manual should also fit into the box or carrying case that transports the equipment as well as on a standard office shelf. If the manual will reside in a remote location, consider developing a quick reference guide for nearby use. A ring binder is often appropriate for manuals that will require frequent updates because you can easily add pages.

### Screen Design

The units of your discussion may need only minor adjustment to work on a screen. Companies are increasingly using electronic delivery, especially over the Web, to reduce printing and storage costs and foster easy and speedy updating. Customers also gain the advantages of keyword searching so they can rapidly find what they need (see Figure 17.13). Electronic delivery also reduces paper use and is thus environmentally friendly.

Electronic instructions play an important role in fostering globalization. Electronic delivery not only eliminates the high cost of shipping but also reduces the difficulties and delays that still accompany the movement of paper across national borders.

As a writer of online documentation, you will face new challenges in overcoming some user resistance to reading online. But you will also help users overcome their fear in opening a hefty print manual. In an online system, you can hide the extent of the procedure while the user gets just what is needed, when it's needed. You can also embed links to other instructions.

The test of good instructions, in print or online, is that they work. The reader can use them to successfully complete the task at hand. If at the same time your instructions can make it as easy as possible for the reader and address the reader in a voice that is comfortable and engaging, so much the better.

## CHECKLIST: Instructions

1. **Think of the reader as a *user* of the instructions: Make them *easy to use***
2. **Think of the reader as a *customer:* Make the instructions *persuasive***
3. **To achieve these purposes, structure information**
   Introduction: background, materials, benefits, theory
   Step-by-step description in the easiest steps
   Adequate warnings and cautions
   Maintenance and troubleshooting suggestions
4. **Manage a team effort through an effective document plan**
   Prepare an audience profile and purpose statement
   Determine tasks, assignments, schedule, and costs
   Build in tests for usability and accuracy to make sure the instructions *work*
5. **Design the instructions to *sound* comfortable**
   Meet the readers' preference in text and visuals
   Select verbs carefully

Keep elements parallel
Use precise and appropriate diction

6. **Design the instructions to *look* welcoming**
   Divide steps into easy-to-read-and-follow units
   Clearly identify paths and priorities
   Deliver the instructions in the best physical form, in print or online
7. **Design the instructions to *work***
   Collaborate with product designers and users
   Test for usability and accuracy

## EXERCISES

1. Obtain a copy of a set of instructions that you think are either excellent—or very bad. Develop a list of criteria for good instructions based on your reading of this chapter and apply them to your sample. Then write an evaluation of the instructions in a memo to your instructor. At your instructor's suggestion, rewrite the instructions to improve their utility. Attach a copy of the instructions (or the original) to your memo.

2. For a novice audience (a freshman engineering major, for example), describe how to operate a device that measures or calibrates something. First, describe the device (see Chapter 10) and general theory of operation. Justify the need to measure whatever it measures. Then move to a step-by-step procedure. Include at least one visual.

3. Write a memo to a fellow student in your major that explains how to research the leading journals and other literature sources in your college library, on the Internet, or on the Web.

4. Revise the following instructions for "Differential Leveling" that appear in a handout advising civil engineering students on techniques for surveying. Change the instructions from paragraph form to list form. In addition, note any missing steps or guidelines. What visuals should accompany the instructions?

   Set up your tripod firmly on the ground and then take the level out of its box. When you are screwing the level onto the tripod head, always hold the level with one hand while the other hand is screwing the base onto the tripod head. Never take your hand off the level until the base is firmly screwed onto the tripod head and you have visually checked it.

   After you have determined the location of the benchmark, pace approximately 205' to 300' in the direction you are going to run your levels. The length of sights depends on the quality of your instruments (optics) and the weather conditions. In cold weather with little or no heat shimmer, your sights can be lengthened.

   As you are pacing ahead for the instrument setup, the rod person should always hold the rod on the benchmark or turnpoint. This should prevent you from setting up the instrument in a position where there will be an obstruction between the instrument and the rod.

Spread the legs of the tripod in a stable position at a convenient height, and position the legs so you will be able to stand between them and not straddle a tripod leg with your legs. Visually level the instrument plate with your eye before firmly placing the legs into the ground. As you are leveling the instrument remember thumbs in thumbs out, the bubble always moves in the direction your left thumb is moving.

5. Write a brief set of instructions for a novice (for example, a college freshman living away from home for the first time) that describes how to plan meals and shop for groceries. Assume that the audience has "ordered out" or eaten at home until now and knows nothing about shopping or basic cooking.

## FOR COLLABORATION

On a team of three students, write a document plan that you would follow in creating a set of instructions to accompany some product, system, or process on campus. At this point, you do not need to create the instructions themselves, but create a *plan* for assessing each of the elements you read about in this chapter. Such plans resemble a proposal (see Chapter 14) in that you are detailing a procedure for approval by someone else in the organization.

For example, assume that your plan details how you will create instructions for using a computer site on campus. In your document plan, cover these topics:

- Goal and scope of the project in general and the instructions in particular
- Project schedule
- Project budget
- Components of the document set (e.g., tutorials, reference materials, help screens, help line)
- Profile of the users
- Profile of the environment for use
- Profile of the user tasks (frequency and sequence)
- Format of the pages or screens

Note, too, how you will develop training materials that users can follow on their own at the site. The tutorials should engage users in doing their own work at the site. To that end, you might include a representative of the users as a consultant to your team.

# PART FIVE

# Communicating as a Professional

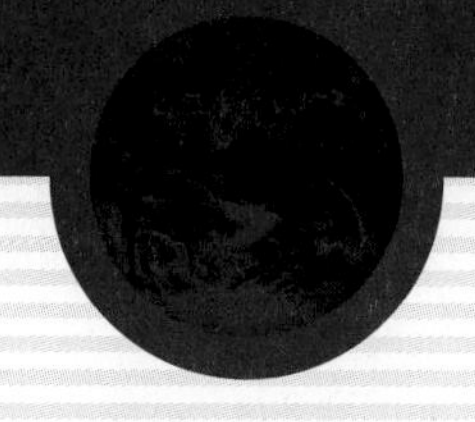

CHAPTER 18

# LETTERS



"THIS LETTER IS LONGER THAN USUAL ONLY BECAUSE I HAVEN'T HAD TIME TO MAKE IT SHORTER."

BLAISE PASCAL

Letters play an important role in getting the world's work done. They furnish information, encourage action, influence people, create a permanent record, and maintain relationships. The convenience and speed of voice mail and e-mail have meant that an ever-smaller percentage of day-to-day technical communication finds its way into traditional letters, but as a technical professional, you need to know how to write effectively in this well respected genre. This chapter and Chapter 19 will show you how.

## PURPOSE

Letters are formal documents, reserved mainly for correspondence between individuals or between an organization and an individual (messages within an organization usually take other forms, as you'll see in Chapter 20). A traditional letter, on paper, has a certain weight and authority to it. It suits the delivery of information that is complex or subject to misunderstanding, especially in discussions of legal or financial issues. It is also *private,* unlike e-mail messages, which are easily monitored and less secure. In addition, a letter seems more sensitive when you convey messages tinged with emotion, as in expressions of thanks, condolence, or congratulations in a professional setting.

Sometimes, especially in an international context, a letter is the only appropriate form to achieve your purpose. Over half the people in the world have not yet made their first phone call. Not everyone has voice mail. Not everyone receives e-mail. Even where electronic systems are in place, a letter may be the expected medium for delivery of professional messages. It negotiates time zones and provides a text for translation, if needed.

## DESIGN

For more about design, see Chapter 7

Letters work best when they are no more than two pages long. So you have a limited space in which to gain attention, promote understanding of your message, and encourage the reader to comply with any request you may be making.

### Structure

When you write as the spokesperson for an organization, follow that organization's recommended practices (Figure 18.1). Review copies of letters written in situations similar to the one you find yourself in and imitate their approach. Some companies also maintain form letters or templates in networked word processing systems. To compose the letter, select the appropriate form; choose, insert, and arrange paragraphs from the system's

**Reader:** ____________ (fill in the name of the person you are writing to)

**Purpose:** Circle the appropriate items.

| | | |
|---|---|---|
| To ask | To explain | To fulfill obligations of courtesy |
| To inform | To complain | To persuade |
| To answer | To confirm | To sell the company |

**Occasion:** Identify carefully what has prompted your correspondence (the problem)

**Facts:** What does the reader need to know? List all the items. If you have good news, put it up front, and organize the facts for a smooth flow of information.

**Action:** What do you want the reader to do? Be explicit. Any methods or means you may wish to suggest? Do you want the reader to reply, to send, to authorize, to act?

**Closing:** What will most clearly indicate the next step in the correspondence? What is the appropriate or most courteous closing?

FIGURE 18.1 **Correspondence Planner**
(Courtesy of Black & Veatch.)

memory; insert specific reader information; and then print and send. This technique's many advantages include

- Reinforcing a corporate image.
- Speeding up the writing process.
- Controlling the quality of the prose.
- Minimizing discrepancies in the message when several writers are responsible for creating letters.

Customers and other readers of such letters, however, may object to their off-the-shelf quality and lack of personal attention. Instead, customize the letter:

Paragraph 1: Briefly acknowledge any previous correspondence and explain why you are writing.

Middle paragraphs: Detail your message. If you're responding to a request, match the order of the original letter in discussing issues raised or questions asked.

Last paragraph: Note any action or decision the recipient should take and show your willingness to answer questions.

Adjust your approach, however, to the expectations of the reader, especially when you write for readers abroad (see *Crossing Cultures: French and Japanese Business Letters*). Figure 18.2 shows how a salesperson from an agricultural equipment company revised a standard letter for U.S. readers to reflect the expectations of a different audience: a Chinese agriculturalist. In centering on his audience, the writer addresses him personally, encourages trust by providing more background on the writer's company, and is less direct.

## Conventional Units

To make your letter look like a letter, include the units conventional in your organization or expected by your reader. Such conventions differ from country to country and, indeed, are changing in the global economy (Figure 18.3). This section reviews current practice for letters in the United States.

**Heading.** The heading gives your address and the date. Use letterhead stationery or design your own letterhead.

**Inside Address.** The inside address identifies the name and address of the recipient. Use a fax number instead of a postal address if you are sending the letter by fax. Avoid abbreviations except for the postal code (see Figure 18.4). Note carefully the order of elements, especially in non-U.S. addresses.

**Subject Line.** Less common on letters than memos, a subject line identifies the topic of the letter. If appropriate, refer to the number of an account, the name of a project, or prior correspondence:

Re: Frequent Traveler Account 654689201

Standard

Dear Sir:

Your name and address were referred to me by the Illinois Department of Agriculture—Far East Office. They stated that you had expressed an interest in our products and requested further information.

I am therefore enclosing a brochure which itemizes our products and services. Please let me know your exact requirements. I will be happy to provide you with further details.

Thank you for your participation at the Illinois Slide and Catalog Show. I look forward to your reply.

Centered on Reader

Dear Mr. Yen Zen-jiu:

I hope you have had a safe journey home and that you have found your family in good health. The Midwestern part of our country where you graciously visited continues to have wet weather, but I am thankful for the rain after our two years of drought.

Ag-World wishes to thank you for your participation at the state Agricultural Convention and for stopping by our booth.

Our firm is situated in Bloomington, Illinois, the heart of grain and cattle country. It has a history of 10 years' experience in selling livestock and livestock equipment. It has trade relations with more than 45 countries in the world. Our firm is well known for its excellent service and good quality products.

In 1988 we sold 168 hogs to China. We wish to establish relations with China on a regular basis. We would like to know whether our breeding livestock and livestock equipment, such as Pork-Preg, Pork-alert, and Beef-o-meter, could benefit you in any way. I will be very happy to provide you with further information.

I am also enclosing two price lists of our equipment; one is the regular price, and the other is the pricing for demonstrators.

May your seasons be fruitful and plentiful.

FIGURE 18.2 **Adapting a Standard Letter to Center on the Reader**

(From *Technical Communication Quarterly,* Vol. 4, No. 3, Summer 1995, pages 245–59, by Carolyn Boiarsky. Reprinted by permission of *Technical Communication Quarterly* and Carolyn Boiarsky.)

CROSSING CULTURES

## French and Japanese Business Letters

The increasingly global economy has muted some cultural distinctions in the structure of letters. A Japanese reading a letter from an American, for example, is probably comfortable with an American approach, even if the letter is translated into Japanese. But differences in approach still remain—and should be recognized. Two researchers characterize one set of differences as follows: American business letters are reader oriented, French business letters are writer oriented, and Japanese letters are "nonperson" oriented, reflecting the *relationship* between people more than the people themselves (Hinds and Jenkins 331).

French business letters aim at extreme precision of content in case of future litigation between buyers and sellers, who are seen as potential adversaries. The writer seeks to create an iron-clad case. The letters often follow a tight six-zone format prescribed by L'Association Française de Normalisation (see Figure 18.3). They are brief and may seem abrupt, especially in their listing of facts, although their language follows the seemingly deferential, if rigid and formal, *règles de politesse* (rules of politeness). A typical closing to a sales letter, for example, is

> En souhaitant que cette proposition retienne favorablement votre attention, et dans l'attente de vous lire très prochainement à ce sujet, Nous vous prions d'agréer, Messieurs, l'expression de nos sentiments les meilleurs.

Which, translated literally, means

> In the hope that this proposed sale continues to favorably attract your attention, and in anticipation of reading you very soon on this subject, we pray, gentlemen, that you will accept the expression of our best sentiments.

According to one account, the chief virtues of such letters are "prudence, conciseness, and precision" (Hinds and Jenkins 333).

By contrast, traditional Japanese business letters are often essentially ceremonial. Most Japanese prefer to conduct business in person or over the telephone, and letters are considered less polite than such calls. The ceremonial nature of letters is shown, for example, in the stock phrases writers use. A traditional letter opens gently with a salutation whose translation may read, "Allow us to open with all reverence to you." The letter continues with a formulaic remark about the seasons or the weather and congratulations on prosperity:

> The season for cherry blossoms is here with us and everybody is beginning to feel refreshed. We sincerely congratulate you on becoming more prosperous in your business.

It then thanks the reader for some personal gift or kindness received or for business patronage—all before stating the main message (Haneda and Shima 21). The writer also selects from a wide range of alternative expressions in Japanese that reflect the perceived relationship between writer and reader. In many companies, formal letters are the responsibility of only a few people devoted to that task. Technical professionals exchange information through conversations or in faxes that record data in visual or briefly verbal form. Japanese letters befit the "high-context" setting of Japan—as opposed to the "low-context" setting in the United States, for example, that fosters greater expression of individualism and more attention to the idiosyncratic needs of the reader.

1. Sender's address
2. Reader's address
3. Our reference number
4. Your reference: a letter you sent on 10 August
5. Salutation
6. Subject line: price list
7. Title and signature of the sender

① Societe MACQUART
12, rue de la République

75010 PARIS

② Société PRINCIPAUX
A l'attention de Mme DENEUVE
8, rue de France

68100 MULHOUSE

Paris, le 2 septembre 2001

③ N/REF: 0218
④ V/REF: Courrier du 10 août 2001

⑤ Messieurs,

⑥ Object: TARIF

Paragraph one

Paragraph two

Paragraph three

⑦ Société MACQUART

*André Dutoit*

ANDRE DUTOIT
Direction Commerciale

**FIGURE 18.3 Conventional Format for a French Business Letter**
Note that the salutation is "gentleman," although the letter addresses a woman.

**Salutation or Attention Line.** The typical salutation in the United States is "Dear," although this may seem old-fashioned, followed by the name of the recipient of the letter. Use the name as it appears in the inside address. If you are writing to a role rather than an individual, use an attention line:

Attention: Customer Relations

Or use a simplified format (Figure 18.5) that omits both the salutation and the attention line.

**Body.** The body contains the text of the letter's message.

**Complimentary Close.** Most formal U.S. letters include a single line of closure, either "Sincerely" or "Yours truly." If you omit the salutation line,

**United States**

AL Alabama
AK Alaska
AZ Arizona
AR Arkansas
CA California
CO Colorado
CT Connecticut
DE Delaware
DC District of Columbia
FL Florida
GA Georgia
HI Hawaii
ID Idaho
IL Illinois
IN Indiana
IA Iowa
KS Kansas
KY Kentucky
LA Louisiana
ME Maine
MD Maryland
MA Massachusetts
MI Michigan
MN Minnesota
MS Mississippi
MO Missouri
MT Montana
NC North Carolina
ND North Dakota
NE Nebraska
NH New Hampshire
NJ New Jersey
NM New Mexico
NV Nevada
NY New York
OH Ohio
OK Oklahoma
OR Oregon
PA Pennsylvania
PR Puerto Rico
RI Rhode Island
SC South Carolina
SD South Dakota
TN Tennessee
TX Texas
UT Utah
VA Virginia
VT Vermont
WA Washington
WI Wisconsin
WV West Virginia
WY Wyoming

**Canada**

AB Alberta
BC British Columbia
LB Labrador
MB Manitoba
NB New Brunswick
NF New Foundland
NS Nova Scotia
NT Northwest Territories
ON Ontario
PE Prince Edward Island
PQ Quebec
SK Saskatchewan
YT Yukon Territory

**Selected International Domain Codes**

AE United Arab Emirates
AF Afghanistan
AR Argentina
AT Austria
AU Australia
BE Belgium
BR Brazil
CA Canada
CH Switzerland
CL Chile
DE Germany
DK Denmark
EC Ecuador
EE Estonia
ES Spain
FR France
GB Great Britain
IT Italy
NL Netherlands
PT Portugal
US United States

FIGURE 18.4 **Postal Codes**

then omit the closing line. Professional colleagues may use a variant like "Best wishes" or "Kind regards"; "Respectfully yours" is appropriate to show deference to the reader, as in submitting a proposal or final report to a client or supervisor.

**Signature.** The signature block provides the final authority for the letter as a legal document. You sign the letter, and that signature may appear above your typed name and title or below it. You may also include the company name. Personalize an otherwise formal letter by signing your nickname if your relationship with the reader warrants.

**Notations.** The space below the typed signature may include such notations as the initials of a typist, the number of enclosures, or the names of people to whom copies were sent. The abbreviation for copies is traditionally "cc" ("carbon copies"), but some organizations use "pc" ("photocopies").

**Subsequent Pages.** Try to keep a letter to one page. But if you need another page, use nonletterhead stationery that matches the first page. You

**Every element begins at the left margin.**

1. Heading: sender's address (without name) and date
2. Inside address (with name)
3. Salutation (followed by: if formal or, if informal)
4. Complimentary close
5. Typed signature
6. Notations

① P.O. Box 3000
Park City UT 84060

25 March 2_ _ _

② John Lynch
Partridge and Smith
Surveyors and Estate Agents
300 Great Portland Street
London W1N 6AL
England

③ Dear Mr. Lynch:

Paragr ph one . . .begins at left margin and is single spaced with double spacing between paragraphs

Paragraph two . . .

Paragraph three . . . and so on.

④ Sincerely yours,

⑤ Anita Sterling-Foster

Enc. (1)
⑥ cc: Peter Smythson

FIGURE 18.5 **Three Common U.S. Letter Formats: Block**

may simply number the pages, or head the following pages with the recipient's name, the page number, and the date. Avoid having only a line or two, plus the complimentary close and signature, on a continuation page.

John Myers
29 January 2001
p. 2

## Format

Standard formats for U.S. letters are the block, modified block, and simplified (Figure 18.5) Some researchers suggest that the block form seems impersonal and advise writers to use the modified block for addressing people who expect a more personal approach in business communication. But the block form is probably the most prevalent in U.S. correspondence. For all forms, adjust the top and bottom margin to center the image area of the letter on the page. Avoid scrunching a short letter at the top or squeezing the closing of a long letter. The left margin is justified, meaning all lines beginning at the left margin line up vertically. But for ease of reading, leave the right margin ragged.

① P.O. Box 3000
Park City UT 84060

25 March 2_ _ _

② John Lynch
Partridge and Smith
Surveyors and Estate Agents
300 Great Portland Street
London W1N 6AL
England

③ Dear Mr. Lynch:

Paragraph one is indented. The text is then single spaced with double spacing between paragraphs

Paragraph two (also indented)

Paragraph three . . . and so on.

④ Sincerely yours,

⑤ Anita Sterling-Foster

⑥ Enc. (1)
cc: Peter Smythson

**Modified Block. Paragraphs are indented. Sender's address, date, complimentary close, and signature begin about seven tab stops across the page.**

1. Heading
2. Inside address
3. Salutation
4. Complimentary close
5. Typed signature
6. Notation

① P.O. Box 3000
Park City UT 84060

25 March 2_ _ _

② Partridge and Smith
Surveyors and Estate Agents
300 Great Portland Street
London W1N 6AL
England

③ Re: Rental of Pavilion Road property
④

Paragraph one begins at left margin and is single spaced with double spacing between paragraphs

Paragraph two . . .

Paragraph three . . and so on.

⑤

⑥ Anita Sterling-Foster
Enc. (1)
cc: Peter Smythson

**Simplified. No salutation or complimentary close. All elements begin at the left margin. A subject line is optional but common with this format.**

1. Heading
2. Inside address
3. Subject line—optional but common
4. No salutation
5. No complimentary close
6. Typed signature

FIGURE 18.5 **Three Common U.S. Letter Formats (continued)**

### Voice

A letter is a form of dialogue. Write in a way that respects and engages the reader, confirms your professionalism, and sounds like you at your best. The wrong word can easily alienate the recipient ("This letter corrects your mistake in calculating the payment you sent to us." "In your letter you claim that we failed to . . ."). If two or three such words appear in the same letter, the cumulative effect can be a powerful disincentive to read.

Avoid *clichés,* that is, expressions past their prime. Writers often come to their task with outdated notions of how a letter should sound, and thus they use phrases like "to whom it may concern" or "I am in receipt of your letter of recent date" or "attached hereto please find," phrases no longer current in any other context. A student who asks for "thoughtful consideration" of her resume is probably revealing a writing process based on copying an outmoded model. A British writer may use a formal closing, for example, "Should you have any queries, please do not hesitate to contact the undersigned." But such terminology would seem inappropriate in a U.S. letter.

## REQUESTING

The rest of this chapter addresses common situations in which you will need to write a letter, either as a student or as a professional. Most fall into the category of making requests and responding to requests.

Many requests are routine matters, responded to in phone calls or e-mail messages. But as part of a research project, you may need to construct a letter that conveys a more complex request. The letter may also address formally someone you don't know as opposed to the more informal approach of an e-mail message.

If you write to someone you don't know, your likelihood of receiving a response is probably directly related to the reader's likelihood of selling you something. When you write to inquire about a product, you'll probably get a response. When you write to a source you don't know to receive information to advance your own work, a response is much less likely. Some people do respond, of course. Treasure them. But be prepared to have many or even all of your requests go unanswered. In a research project, factor into your plan enough time for responses and for follow-up requests. Have an alternative plan ready just in case you receive no responses at all.

To increase the likelihood of a response, make sure you answer the following questions any reader would have in reviewing your request:

- Who are you?
- Why did you send the request to me?
- What do you want?

Here is the text of an effective letter requesting information:

Who you are

> As a student at Eastern University, I am writing a research report on the different concentrations of sulfur dioxide emitted in stack gases. Because you are the leading producer of equipment to remove this

pollutant, I would particularly welcome your responses to the following questions as they pertain to your devices:

Why are you writing to them

1. What percentage of sulfur dioxide is removed by the equipment you produce?
2. What is the cost per pound of the processed sulfur dioxide removed during the operation?
3. Is adsorption or absorption applied in the extraction process?

Questions focus the information you seek

Any further information in the form of published reports or brochures would be greatly appreciated.

I hope to complete my report by November 21. I'd be happy to send you a copy of the report, if you're interested.

Time frame for response

### Who?

If you are not already known to the reader, identify yourself. The reader will want to know who you are before deciding how to respond. Note how you will use whatever information you are asking the reader to give you. If you are requesting information for a class project, for example, identify the class and your specific requirement.

Sometimes convincing the reader that you're worth responding to is easy, especially if you're a potential customer. It's harder to ask a favor. In such situations, your clear identification of your need, precise request, knowledge of the reader, and well-crafted prose establish your credibility and foster a response.

### Why?

Choose the right person to write to. That choice may be easy and obvious (as in Figure 18.2, a letter to someone who stopped by a booth at an exposition), or it may require some research as you identify just the person who can answer your questions. If you don't know the reader, include a statement that will predispose the person to respond. A well-phrased compliment may do the trick:

I read with great interest your comments in the 2 June *Wall Street Journal* article, "Label Law Stirs Up Food Companies." I don't want to interrupt your label-every-45-minutes pace, but I would like to ask you a few questions about how your company is responding to the law. Your response will help me prepare a report on frozen food labeling for my senior nutrition class at Utah State University.

If someone suggested the source to you, include that person's name in your opening:

Professor Peter Jones suggested that I write to you concerning your company's response to the food labeling law.

### What?

For more about literature searches, see Chapter 5

Decide in advance of writing what you specifically seek. For example, develop a list of reasonable questions and keep the list to a minimum. Only request information that is not readily available in the literature on your topic. Request, for example

- Amplifying or clarifying details
- Personal approaches, opinions, or anecdotes
- Local instances from a specific organization

Provide background that shows you are not writing blind. Some inquiry letters encompass so many questions or ask such a large question that the reader may think you're saying, "Please write my report for me." Don't ask the reader to do your work.

The following effective questions appeared in a letter to the head of the technical writing department of a major corporation. The letter sought the reader's help in compiling guidelines for subject matter experts (SMEs) who provide information to technical writers:

1. What guidelines do you provide to SMEs that help them know how long a project will take and how much it will cost?
2. What control devices do you use to integrate documents produced by different people at different times? How do you ensure consistency in the way names are spelled, for example, and avoid repetition of explanations in different locations in the document?
3. Do you provide checklists for SMEs that help ensure a level of quality in their writing even before it is edited? If so, may I have a copy?

Note a time frame for the reader's response. Give the reader enough time to show your recognition that they can't drop everything to answer your request—but not so much time that they'll forget.

## RESPONDING

As a student and as a professional, you need skills at writing persuasive letters requesting information. You may also need to respond to someone else's request. Structure your response to the questions asked. Although it may strike the reader as inappropriate and even rude, jotting some notes on the original letter and sending that back may be an efficient response. If you can be sure the writer of a series of numbered questions has kept a copy of those questions, then in your letter simply number your responses in order. But people do forget, so it's best to repeat or paraphrase the questions. Enclose with your response any printed material that might be pertinent or cite references to such material available publicly.

When you write a letter of response that denies someone's request, that is, when you have to say "no" to your reader, other factors come into play. The reader will probably not welcome your message, so you need to express it in a way that is both clear and diplomatic. Researchers disagree on the best approach. Some recommend indirection: Build up to the bad news or denial with explanations that might win the reader to your side. Other researchers suggest that readers will simply skip over such text to find the *no* wherever it is located, and so a direct approach is best. As with any writing, your knowledge of the audience is critical. What would your reader expect? In addition, recognize that you have an ethical obligation to provide an explanation. In general, consider these elements as you structure your letter:

- Begin on common ground with the reader. At a minimum, acknowledge the request.
- Include a detailed and accurate explanation of why you cannot accommodate what was asked for.
- Say no—kindly but unambiguously. This statement may precede or follow your explanation depending on your sense of the reader.
- In your closing, perhaps offer a substitute for the requested solution, if that's appropriate.

The following is a polite but negative response to a student's request for information about laboratory accidents at a biotechnology company:

> You have certainly chosen an important topic for your technical writing class report. Your interest in the number and types of accidents in biotechnology companies reflects an important issue today, and we at Genetatek share your concern.

Common ground

> Because of the high risks of contamination and other severe consequences for our workers, for the environment, and for the human population at large, we take safety seriously at Genetatek. Controls and policies are in place to prevent accidents before they happen. This is a business that requires precision and careful conduct at all times. To ensure such conduct, we monitor practices closely and keep records of any kind of accident, defined at different levels from minor workplace mishaps, like paper cuts and scratches, through more serious injuries. This record is, however, proprietary.

Rationale explained and negative "accidents" turned to positive "safety"

> Thus, I cannot share it with you.

The denial

> I have enclosed, however, a copy of our policy statement on safe practices in the laboratory, as well as our annual report, which demonstrates that our safety record exceeds those of the industry as a whole. I hope this information will help you as you prepare your report and learn about the biotechnology industry.

Substitute information offered

## COMPLAINING AND CLAIMING

A second common situation for requesting and responding develops when a product or service proves unsatisfactory. Sometimes the problem can be resolved with a phone call to an 800 number. But often, you have to write a letter that makes a claim and seeks a resolution of the problem. Someone responsible for taking action on the claim then writes a letter detailing how the manufacturer or service provider will respond. The context for such letters is sometimes emotional. Somebody is dissatisfied and unhappy, but both the claimant and the responder must control their emotions and establish a suitable voice in their letters.

You are more likely to receive a favorable and speedy response if your letter is reasonable than if it is merely angry. Establish your case, as in a letter requesting information, by establishing who you are, why you are making the request to this person or organization, and what solution or compensation you are requesting:

- Identify yourself through your connection to the product or service.
- Explain why the product or service failed.
- Provide the details: exact dates, amounts, model numbers, sizes, colors, and so on.
- Describe the resulting inconvenience, if applicable.
- State what you request as a fair solution or compensation.

The amount of explanation you need—and the placement of that explanation—depends on the certainty of the claim. You can be direct when the correction is a matter of course: something covered under a warranty, a problem with some equipment, an obvious mistake, and the like (Figure 18.6). For claims a company might find less certain, consider placing the explanation first to win the reader to your side with the evidence before hitting her or him with the conclusion, as in the following effective complaint:

> The other day I heard a squeaking noise at peculiar intervals in my apartment. I searched all over to find the source and finally traced it to the smoke detector.
>
> Last Christmas, my son gave me the device, which he bought at Marshall Fields in Chicago. It is manufactured by your company, Model 23, with a warranty of five years. Since it doesn't seem to be working, I'd like return it to you so you can send me another one. Since I'm arthritic and slow-moving, I feel much safer in the apartment with a smoke detector that works.

If you are responsible for responding to a claim, first read the claim letter carefully. Take the claim seriously as an opportunity to find out about your company's products and performance and to regain the confidence of your client or customer. In your response, address the specific claim, which may mean reading between the lines to find a core issue buried within.

Your response will probably be either a "yes" or a "no": Either you will give the claimant what is wanted or you will deny the request. If the problem is an easy one to fix, as in some repairs to a faulty instrument, then you

*In reply please quote:*
NDQ JW:
106/3/2

19 April 2001

Dear Mr Smith

As you suggested in our phone call on 17 April, I have enclosed the refractometer that does not seem to be operating properly.

There appear to be two problems with this instrument:

1. The lead glass prism is not level and slants slightly to the left. This may cause problems in obtaining readings from larger stones.
2. Shadow edges observed while using this instrument are cutting the scale at an angle, making accurate readings difficult.

Would you please advise as to the possibility of repairing this instrument and the approximate costs involved.

As funds for maintenance of equipment are severely limited this year, it may be necessary to postpone any costly repairs until appropriate funds are available.

The contact office at this laboratory is Mr Stephen Spence.
Phone: 62017788

Yours faithfully

*Mary Petersen*

**FIGURE 18.6 Letter Requesting a Solution to a Problem**
The form follows conventions for letters in Australia, where this was written.
(Courtesy of Kaaren Blom and Laurie Hallam.)

may do so, at minimum cost or even for free as a goodwill effort. If the issue is big enough, your company may try to negotiate—that is, say "maybe"—a course of action that can be time consuming, legalistic, and argumentative. Here is how one company spokesperson negotiated the smoke detector owner's complaint with diplomacy and tact:

> You are right. Our smoke detectors are guaranteed for five years but should last you much longer.

We are sorry to hear that you are having trouble with yours, but we certainly intend to give you the satisfaction we have guaranteed. Enclosed is a postage-paid label you may use to send it back to us at our expense. We will then either repair it or ship you a new one.

However, before you send it back, please check the battery. The detectors are designed to beep for seven days if the battery has run down. This, of course, is for the owner's protection. Consult page 6 in the instruction manual for directions. In case you do not have a manual, we are enclosing one. If the trouble does not come from the battery, then please send the detector along to us and we will repair or replace it.

The voice in this letter is both professional and pleasant. It notes immediately that the claim is accepted and offers a possible explanation. It never attacks the writer ("If you had read the instructions you would know that the beeps indicate a dying battery."). It anticipates a problem—that the writer may no longer have the manual—and solves it without requiring another set of letters.

For more about persuading and proving, see Chapter 11

Saying "yes" to a claim is generally easy as long as you avoid sounding begrudging. You then reap the goodwill your solution should dispense. Negotiating or saying "no" taps more deeply into your skills in argumentation and persuasion. An authoritative persona and voice are critical, which means that sometimes "no" letters bear the signature of a higher authority than "yes" letters (even though they may be written by the same person, of course).

## REQUESTING APPROVAL

A third common situation for letters is requesting approval of some action or plan. As with all requests, establish who you are, why you are writing to this reader, and what you are requesting. Be polite as you build an argument to support your request, as in the following, a request from a professor in Australia to a power company to use the land on which its power lines run (courtesy of Kaaren Blom and Laurie Hallam).

**Subject:** Access to Easement for the 66 KV Power Line, Burrinjuck to Bogolara Switching Station

Following a phone conversation with Mr David Johnson today 24th June, I am writing to ask your approval for our Department to gain access to the power easement track following the 66 KV power line that runs from Burrinjuck Dam to Bogolara switching station.

The Department of Geology and Gemmology is teaching courses in Four WD Driving, Recovery, and Bush Survival, in the Childowla area. We have been in contact with the South West Slopes Superintendent, Mr Turner, and he has no objections to our using this particular track. We estimate that our usage will be approximately four times per year and will be restricted to the track area only. We would undertake repairs to any sections of track that may be required, eg washouts, so that the track will be in a passable condition.

If your approval is granted, we will also follow Mr Johnson's suggestion that our Department put locks on the chained gates.

Yours faithfully,

## LETTER QUALITY

In writing any letter, get the information right, of course. But more than that, sound like the professional you are or will become. Engage the reader's attention by good design and by good text: crisp and tailored to the reader's expectations and needs. Your letters establish or maintain a relationship that may last a few hours—or a lifetime.

### CHECKLIST: Letters

1. **Compose a letter to an individual when your message is**
   - Relatively formal and professional
   - Private
   - Complex, perhaps with legal implications
   - Sensitive
   - Addressed to readers abroad who might prefer a letter
2. **In one page or two**
   - Clarify how this letter fits the context of other correspondence
   - Gain the reader's attention by creating common ground
   - Promote understanding
   - Encourage compliance with your request or decision
3. **Make any request easy to respond to by**
   - Identifying who you are
   - Establishing why you are writing to this reader
   - Detailing specifically what information or action you seek
4. **In responding to a request**
   - Answer any direct questions
   - Offer a correction or solution to a complaint, if appropriate
   - Be clear but diplomatic in denying a request

## EXERCISES

1. Write a letter requesting information from someone who could provide useful opinions or examples for you to include in a formal report. Match your structure of the letter to the guidelines presented in this chapter.
2. As an officer of a student organization, write a letter inviting someone to speak at one of your programs.
3. Assuming that the speaker accepted your invitation and participated in your program, write a letter thanking her or him.
4. You work for a surveying company that needs access to several properties in a rural area that has recently been subdivided. To gain that access, you have to write a letter to current property owners explaining your purpose and seeking their permission. Write the form letter. (This exercise was suggested by Kaaren Blom and Laurie Hallam.)
5. One of the owners (see Exercise 4) phones you a few days after receiving your letter with some questions and objections. He's not happy to let you have access to the property. Take the role of the property owner, think of the objections he might raise, and then write a letter that responds to those objections and asks, once more, for permission to conduct the survey.
6. Write a letter of complaint about a product or service that you found defective. Determine the appropriate person to receive the letter and address that person directly.
7. A professor has asked you to participate in a research program she is conducting to test the effectiveness of a new formula shampoo. She needs subjects with long, straight, red hair, and you fit the bill. But you would prefer not to participate, even though the professor has assured you that the formula is harmless (maybe even highly beneficial) and has offered you a stipend of $100 for your agreement. She has provided you with a written consent form; you feel, in returning the form unsigned, you should also explain your refusal in a letter. Write the letter.

## FOR COLLABORATION

1. On a team of three or four, develop a set of questions and then interview people who write letters on behalf of organizations. What is their process for composing the letters? How much time do they spend on letters? Do letters go through an approval route before being mailed? At your instructor's request, write a brief memo summarizing your findings.
2. Write a letter inviting someone to speak to a meeting of an organization you belong to, asking for information, or providing directions to someone who will be visiting your campus. Then ask a person who is fluent in another language to translate the letter into that language. As the person translates, or when the translation is finished, interview the translator. How would such a message ordinarily be conveyed in that language and culture? Were there any difficulties in translating the message? What is the standard format for letters in the culture? Provide your instructor with the original letter, the translation, and a brief commentary. Then discuss these translations in general in class.

CHAPTER 19

# THE JOB SEARCH



"LOOKING FOR A JOB IS A JOB."

STUDENT COMPLAINT

You probably don't need to be convinced that the letters and resume you write as you look for a job constitute some of the most important writing of your career. This chapter provides guidelines for such writing. You can also adapt these techniques if you are seeking an internship or applying to graduate school before entering the job market.

## THE JOB SEARCH

Many people comment that searching for a job or applying to graduate school is a job in itself. It requires careful planning and accurate record keeping:

- Job leads
- Graduate school program descriptions
- Recommendations solicited
- Letters sent and received
- Interview appointments
- Notes on interviews

In a college or university setting, it's probably wise to establish a dossier with the placement office, which can then send it to potential employers. Compiling a dossier that includes a resume, recommendations, and perhaps other pertinent documents is also a good way to focus your search. In addition, gather information about potential employers, about the kinds of jobs available in your field, and about your own career interests.

### Career Information

Start your search for information about employers and jobs with the placement office, which maintains its own database and can provide advice on where else to look. You'll also find abundant information on the Web (see *Electronic Edge: The Electronic Job Search*). In addition, consult

- The classified advertisements in newspapers published in the area where you'd like to work (newspapers for large cities worldwide are available online).
- Annual reports of companies that interest you. The primary purpose of the reports is to comply with financial disclosure requirements of the U.S. Securities and Exchange Commission, but many companies see them as major public relations and recruiting tools too.
- State and federal employment offices.
- Private for-profit recruiters.
- The professional society that serves your discipline. Attend meetings of local chapters, a good place to network with people about jobs. Read their journals and newsletters, which may publish job listings.

### Your Interests and Strengths

In addition to finding out about the market, find out about yourself. Analyze your accomplishments, activities, skills, strengths, and weaknesses. Several Web sites offer self-analysis techniques. Ask questions like the following:

- What do the decisions I've made in the last year say about me?
- How do I feel about working on teams?
- Am I a risk taker or do I prefer security over surprise?
- Where do I like to work—office? Field? Lab?
- How much pressure do I want on the job?
- What's my definition of success?

### References

Develop a list of people who can testify about your abilities, accomplishments, and potential as a graduate student or professional. In letters of reference, they comment on what you have done and thus what you can do. Because their own credibility enhances the force of their recommendation, solicit letters from people whose word will be taken seriously:

- Your adviser or major professor
- A supervisor who can describe your work habits

- An adviser to a campus organization or sports team in which you have been active
- People you know who also know the university, organization, or company to which you are applying

In asking for a letter of recommendation, keep the following guidelines in mind.

**Ask First.** Never mention a recommender to a potential employer without checking first with the person you've named. That step is both courteous and self-serving. You don't want an employer to call someone who may not remember you or who provides only a lukewarm statement. Similarly, don't push your request if a potential recommender seems hesitant. She or he may be too busy to write at that time or may feel unable to recommend you. Some job candidates ask directly if the person would be able to write a strongly positive recommendation.

**Brief the Recommender.** Convincing letters are concrete and detailed. To encourage such letters, talk with your recommenders about your intentions and provide them with a resume, a statement of your goals, and perhaps some background on the company or graduate program for which you are applying.

**Do Your Homework.** Complete sections of recommendation forms that require your own information or signature. When you are applying to several graduate schools, you'll probably find that each has slightly different requirements. Do whatever you can to make the logistics of filling out the forms easy for the recommender. Figure 19.1 shows a control sheet one student prepared for a professor writing recommendations to support his application to law schools. Each column indicates a peculiarity of the application form for each school.

## COMPOSING A RESUME

A resume sometimes takes on mythic proportions in students' minds. This need not be. The following guidelines will help you compose an effective resume.

### Purpose

A resume has a simple purpose: to list what you have done as a predictor of what you can do. As such, it provides a showroom of your experience and abilities. Although readers expect certain units of information, content and design are not universally dictated and differ between print and electronic formats. Two approaches, however, are common: chronological and functional.

**Chronological Resume.** In a *chronological* resume, you list your education and experience by order of time. Figures 19.2. and 19.3 provide examples of this type. Most students find this approach the most comfortable one.

| School | Release | Questionnaire | To School | To Scott | Complete? |
|---|---|---|---|---|---|
| Widener University School of Law | | | | ✓ | ✓ |
| Roger Williams University School of Law | | | | ✓ | ✓ |
| Franklin Piece Law Center | ✓ | ✓ | ✓ | | ✓ |
| Seton Hall University School of Law | | | | ✓ | ✓ |
| New England School of Law | ✓ | ✓ | ✓ | | ✓ |
| Western New England College School of Law | ✓ | | ✓ | | ✓ |
| Rutgers School of Law – Newark | ✓ | | | ✓ | ✓ |
| Rutgers, The State University of New Jersey School of Law – Camden | ✓ | | | ✓ | ✓ |
| Vermont Law School | | | | ✓ | ✓ |
| University of Dayton School of Law | | | | ✓ | ✓ |
| Ohio Northern University Claude W. Pettit College of Law | | | | ✓ | ✓ |
| Quinnipiac College School of Law | | | | ✓ | ✓ |

**FIGURE 19.1** **Control Sheet for Letters of Recommendation**
(Courtesy of Scott Collins.)

**Functional Resume.** In a *functional* resume, you categorize information about education and experience under headings that indicate your skills and achievements (Figure 19.4). Those categories usually reflect requirements stated in a job description, for example, "team building skills," "management experience," " information technology." Use this approach when

- You have little experience.
- Your experience is scattered in type and amount of time.
- You are shifting careers.
- You are seeking a narrowly defined job, internship, or training program for whose specific requirements you need to show a match.

## Julie Anne Fine

School: 314 Sypherd Hall. Newark DE 19717. Home: 12 Bristol Ct. Hartford CT 06106.
e-mail:jaf@udel.edu. (302) 837-8969

**Education**

University of Delaware — September 1996–May 2001
B.S. Degree with Distinction. GPA 4.0/4.0
Major: Entomology (Wildlife Conservation Concentration) and Plant Science
All expenses financed by scholarship, loan, and part-time job

**Experience**

Petersen Environmental Consultants — November 1997–present
*Technician*
Perform forest surveys, wetland delineations, audits, and comprehensive studies
Prepare final reports and interpret blueprints, topographic maps, and aerial photos
Inspect debris removal operations with the DE Dept. of Natural Resources

University of Delaware Academic Services — September 1998–present
*Student Tutor*
Tutor groups and individuals in courses involving science and math
Help students with learning disabilities develop study skills

**Awards and Activities**

Dean's List 1996–2001. Delaware State Carvel Agriculture Scholars Scholarship 1997–2001. Roger Williams College Marine Biology Program 1998. Boston University Science Scholars Program 1999. Alpha Zeta Honors Fraternity; inducted 1997. Alpha Lambda Delta Honors Fraternity; inducted 1997. University Orchestra. Wildlife Conservation Club.

**Affiliations**

Wildlife Conservation Society. American Museum of Natural History. World Wildlife Fund. National Wildlife Federation. Audubon Society. National Parks and Conservation Association. Wilderness Society. Nature Conservancy.

**Special Skills and Interests**

*Environmental Techniques*
Forest surveys. Audits. Social studies. Hydrology determinations. Wetland delineations. Well monitoring. Map reading. Site inspections
*Computer Skills*
Atlas and Idrisi GIS. Data analysis. WordPerfect and Word. Excel. PowerPoint
*Interests*
Music (violin and piano). Writing (creative and technical). Sports (tennis, softball, biking). Hiking. Camping. Boating. Gardening.
*Willing to Relocate*

References furnished upon request

**FIGURE 19.2 Chronological Resume**

Traditional

1/18/01 Alvina L. Diaz

**Campus Address (until May)**
35 Northgate, Apt. 5
Houston TX 77236
713 777 4024
e-mail: 24316@ricevm1. rice.edu

**Home Address**
2234 Johnston St.
Vancouver BC V6R 3R1 Canada
604 221 2843

JOB OBJECTIVE

To obtain an entry-level position in a civil engineering firm specializing in environmental services

EDUCATION

**Bachelor of Civil Engineering.** Rice University, Houston TX. May 2001

CERTIFICATION

Passed the FE/EIT Exam October 2000

EXPERIENCE

**Rangley Engineers.** Tempe AZ, *Summer Intern,* 2000

- Designed applications to test new versions of process simulation programs
- Developed computer models to investigate contaminants in wastewater systems
- Participated in team briefings to explain the modeling procedures
- Created a preliminary brochure on the new system

**Rice University, College of Engineering.** *Undergraduate Researcher* (December 1999 to present)

- Experiment with microorganisms for the biodegradation of Toluene, a toxic waste
- Operate an electrolytic respirometer and record and graph daily results

**Department of Transportation.** *Provincial Inspector.* Rossland BC (Summer 1999)

- Supervised roadwork on a provincial transportation team
- Analyzed weak concrete bridge columns

**Lickety Split.** Vancouver BC. *Manager of an ice cream store* (Summers 1996 to 1998)

- Worked with employee scheduling, inventory, machine operation, and banking
- Attended to customer complaints and supervised store

COMPUTER SKILLS

WordPerfect. Microsoft Word. AutoCAD. BASIC. Excel. PowerPoint

LANGUAGE SKILLS

Fluent in Spanish

ACTIVITIES AND AWARDS

Member of RISE—Resource to Insure Successful Engineers
Editor of Chi Epsilon—National Civil Engineering Honor Society
Intramural volleyball, flag football, and softball
NACME (National Action Council for Minorities in Engineering)
Davidson Fielding Scholar (Texas scholarship in CE)

FIGURE 19.3 **Chronological Resume in Two Forms: Traditional**

Scannable

**Alvina L. Diaz**
35 Northgate, Apt. 5
Houston TX 77236
713 777 4024
e-mail: 24316@ricevm1. rice.edu

**Key Words:** Internship and courses in wastewater treatment and other environmental issues. Hold FE/EIT Certification. Canadian Citizen. Willing to relocate. Fluent in Spanish. Strong teamwork skills as well as field and laboratory experience.

**Objective:** Entry-level position in a civil engineering firm specializing in environmental services.

**Education:** Bachelor of Civil Engineering, Rice University, Houston TX. May 2001. Courses in computer modeling of engineering systems, wastewater treatment, ecology, and geotechnical applications.

**Experience**

**Rangley Engineers.** Tempe AZ. Summer Intern. 2000. Designed applications to test new versions of process simulation programs. Developed computer models to investigate contaminants in wastewater systems. Participated in team briefings to explain the modeling procedures. Created a preliminary brochure on the new system.

**Rice University, College of Engineering.** Undergraduate Researcher. December 1999 to present. Experiment with microorganisms for the biodegradation of Toluene, a toxic waste. Operate an electrolytic respirometer and record and graph daily results.

**Department of Transportation.** Provincial Inspector. Rossland BC summer 1999. Supervised roadwork on a provincial transportation team. Analyzed weak concrete bridge columns.

**Lickety Split.** Vancouver BC. Manager of an ice cream store. Summers 1996 to 1998. Worked with employee scheduling, inventory, machine operation, and banking. Attended to customer complaints and supervised store.

**Computer Skills.** WordPerfect. Microsoft Word. AutoCAD. BASIC. Excel. PowerPoint.

**Activities and Awards**

Member of RISE—Resource to Insure Successful Engineers
Editor of Chi Epsilon—National Civil Engineering Honor Society
Intramural volleyball, flag football, and softball
NACME (National Action Council for Minorities in Engineering)
Davidson Fielding Scholar (Texas scholarship in CE)

FIGURE 19.3 **Chronological Resume in Two Forms: Scannable (continued)**

## Peter Silverman

P. O. Box 3321 Park City, Utah 84060 phone: 801 645 8118 e-mail: psi@utah.edu

### Keywords

Sports medicine. Physical therapy. Intern with the University of Utah Hospital physical therapy staff. Volunteer as PT aide in a nursing home. Experience with Cybex, Kinetron, and Universal equipment. Laboratory research.

### Objective

To serve on the trainer/therapist staff of the U.S. ski team

### Education

University of Utah. 1997 to 2001
Bachelor of Science in Physical Therapy expected June 1998
Independent study: Honors project, "Techniques for Rehabilitation After Knee Injuries: A Case Study of Ten Recreational Skiers"
Dean's List
Northeastern University. Boston MA. 1994 to 1995

### Teamwork Skills

*Volunteer,* Skyline Adult Care Center (1998–present) and Intern, the University of Utah Hospital (1999)

- Helped to free the therapists' time for the tasks only a licensed therapist can do
- Worked effectively without close supervision
- Participated on clinical teams with therapists and nursing staff
- Led patients in group exercises

*Rehabilitation team member,* Cripple Creek Sports Medicine Clinic (Park City UT, 1999–present)

- Escort and monitor clients
- Set clients up on exercise and diagnostic equipment
- Prepare team modalities
- Share responsibility for maintaining equipment

### Athletic Skills

Competitive skier for four years of high school. Continue to ski as well as mountain biking and surfing

### Communication Skills

Work well with clients of all ages. Strong ability to explain exercises and motivate even reluctant patients

FIGURE 19.4 **Functional Resume**

## Units of Information

Whichever approach you decide on, include some or all of the units discussed in this section.

**Heading.** Standard information in the heading allows the potential employer to file your resume and to get in touch with you:

- Your name
- Mailing address
- Phone number (and cell phone and fax numbers, if appropriate)
- E-mail address

Design this unit to attract attention and set your style, as in Figure 19.2. At either the top or the bottom of the resume, you might indicate the date on which you completed it to show its currency to both the potential employer—and you. The date line will remind you to send the current version of a document you may revise many times.

**Keywords.** Whether you use a chronological or a functional approach, thinking about categories of skills will help you develop a list of keywords about yourself. Such terms are increasingly common on resumes, especially ones that will be scanned or read online and thus entered into a database that can be searched to fill job openings. Those keywords should match the terms an employer uses to describe its search. Here are some examples of keywords:

Information systems
Information technology
Ability to learn quickly
Handle complex conceptual problems
Effective team player
Mature, experienced sales executive
Strategic planning
Detailed implementation
Proven track record in [name of software or method]
Key areas of responsibility:
  Managing support services
  Reviewing and enhancing systems and procedures
  Monitoring budgets and costs
Challenging role

**Objective.** Resumes often include a statement of objective, which many recruiters say they like to see (Figures 19.3 and 19.4). In creating such a statement, be concrete: Note what you seek to do, not what you want to be.

> Objective: to serve as a support person or aide on a physical therapy team

Create different statements, if necessary, for different jobs. Computer text processing allows you to switch them as necessary. Avoid using complete sentences in stating the objective. If you can't define your objective in concrete terms, omit the statement.

**Summary.** Include a brief statement summarizing your skills and achievements if that helps you consolidate extensive amounts of experience, especially if you are changing careers:

> Experience as a teacher of secondary mathematics for over 15 years. Completing a degree in accounting (career change) with the ultimate goal of qualifying as a CPA.

A summary statement is also useful on a resume that will enter an electronic database.

**Education.** Be thorough in describing your education. As Figures 19.2 through 19.4 show, list colleges attended, beginning with the current one, and degrees expected or attained. You may also include high school if you feel identifying the school will comment strongly on your own credentials. But generally speaking, high school experience is assumed, not expressed.

Other items you may include:

- Brief list of important courses
- Grade point average (in the major or in general)
- Scholarship aid (not including dollar amounts)
- Self-financing (in percentage of expenses earned, not dollar amounts)
- Special awards (like dean's list or honorary societies)
- Computer or language skills
- Undergraduate research experience
- Peer tutoring

Be imaginative as well as honest. Use this information to differentiate yourself from the pack. Indicate activities and knowledge, not just an ability to show up. For example, rather than noting that you have had six semesters of French courses, indicate fluency in French, if indeed that's what you achieved. When you state that, of course, be prepared to respond *en français* if an interviewer greets you with *Bonjour.* List courses by their substantive titles ("Thermodynamics") rather than the university's internal code ("Eng 421"). Avoid acronyms that a potential reader might not know. A civil engineering student applying for jobs in civil engineering can feel comfortable listing the EIT (Engineer in Training) test but should explain the acronym for another reader.

**Experience.** Describe your work experience in detail, using either the chronological or functional approach. Be specific about the jobs you have held, including your title, the name of the employer, and your responsibilities.

Assess the skills you demonstrated, for example, writing reports, preparing financial statements, operating equipment, or monitoring processes:

Davidson Laboratories
Salt Lake City, Utah
June to September 200X
*Job title:* Writing Intern
*Scope of Responsibilities:*

- Designed five hypertext documents for the laboratory's Web site
- Researched medical topics and published two articles for a general audience
- Worked closely with the medical illustrator
- Delivered final product formatted in HTML

Note any increase of responsibility that shows success in your role. Where possible, quantify the results of your work, for example, an increase in sales or a decrease in the amount of organizational paperwork attributable to a computer system you installed. Describe any management or supervisory role you played. Note both independent work and teamwork. If you participated in an internship as part of your academic program, you may categorize that work as either experience or education, depending, in part, on which unit of information needs supporting entries. Even if the content of the jobs you've held doesn't resemble what you want to do in your career, your ability to work effectively in an organizational context speaks well of you to a future employer. In your unit on experience, or in a separate category, note any special skills you possess or levels of certification you have achieved. For example, if you are interested in a career in golf course management, note your golfing handicap.

**Personal Background.** Somewhere in the resume you may want to place information about your interests, activities (including sports and hobbies), and perhaps travel. In an increasingly global economy with electronic job searches that span nations and continents, indicating you are willing to relocate can also be a selling point. Weave such information into the education and experience segments or set up a catchall category at the end of the resume. You don't need to include such details as your birthdate, marital status, ethnicity, and religious preference. U.S. affirmative action guidelines prevent employers from asking for this information. But you may find that employers in other countries expect to see such facts on a resume.

**References.** To save space, most people simply indicate on their resume that letters of recommendation are available. But an impressive list of recommenders may have persuasive value. In addition, conventions differ from discipline to discipline. If the job announcement requests, include the names of recommenders in your letter of application. Otherwise, select particular recommenders to match particular jobs; indicate their names in a cover letter or note the availability of a dossier at your school.

## Design

How you package these units depends on the conventions of your discipline as well as your own design persona. As you draft the resume, look at the resumes of others who were successful at seeking the kind of job you seek. Emphasize the units that best showcase your abilities. Start, for example, with education if that's your strong suit, or with experience, if that's more persuasive.

**For more about persona, see Chapters 7 and 21**

The resume stands for you—on paper or on a screen. It should convey an appropriate persona, and it should also be easy to read, two purposes that are sometimes in conflict. Employers give resumes a notoriously short amount of time to make their point, so appropriate emphasis on key credentials and readability is essential. You want to make sure your assets as a candidate are obvious, so spend time to achieve an attractive and efficient design. Spend time, too, to ensure that your text is letter perfect. A spelling error or simple error in grammar provides an easy way for a reader to eliminate your resume; if you are not careful with a resume, the thinking goes, you will not be a careful and precise technical professional on the job.

**Print.** Figures 19.2 through 19.4 provide good models of printed resumes that take advantage of the options available in desktop publishing software. Keep such resumes short, generally one page, sometimes two. In your attempt at brevity, don't be lured, however, below a 10-point type size for major text. If the type is too small, it becomes difficult to read and you've lost the advantage of brevity. Condense the text instead. Use graphics and layout to enhance a personal statement. Create an original resume with a high-quality printer and reproduce it on white or off-white bond paper.

**Scannable Print.** The personal flourishes of a print resume may be lost if your resume meets a computer before it meets a person at a potential employer's office. In fact, some design elements reduce readability when a resume is scanned. A growing number of companies, including AT&T, Texas Instruments, Hewlett-Packard, and Ford Motor Company, use scanning devices to screen resumes for further consideration. They say scanning technology saves them millions of dollars in recruiting costs and improves their search strategies. Recruiters can spend more time with candidates and keep databases current so that people who don't fit one job can be easily searched for another.

The scanner sorts resumes by keywords that match the credentials needed for the position. To prepare a resume for electronic processing (Quible; Rifkin 12), follow these guidelines:

- In a section headed "Keywords," "Keyword Profile," or "Keyword Summary," feature your credentials in the terms requested in the job description (Figures 19.3 and 19.4). Include your job titles, skills, major, tasks performed, perhaps your university. Maximize the use of your industry's jargon and acronyms.
- Emphasize nouns in an electronic resume (the traditional resume emphasizes verbs).

- Use a standard typeface and size (Courier or Times Roman, in, for example, 12 point).
- Avoid borders and graphics that computers can't read.
- Use white paper (colored papers provide less contrast with type).
- Use one column; scanners scramble tabular material and columns.
- Separate elements with white space.
- Send the resume unfolded to avoid losing text in the fold—and to avoid having your folded sheet jam the scanner.
- Use as many pages as you need, but print on one side only because computers can't turn a sheet over. Place your name at the top of each page.

If you are uncertain about the method a potential employer will use to read your resume, consider sending two copies: one that looks good in print and one that looks good to a scanner.

**Online.** When you create a resume for delivery online, you can take advantage of the flexibility, interactivity, and speed of that delivery system and its ability to link your resume to other databases or sites on the Web. Emphasize your keywords, which are the basis of the search strategy that makes online resumes such a powerful tool for both employers and candidates worldwide. The easiest approach is to upload the text of a print resume into an e-mail message. In doing so, however,

- Display text in only one column.
- Compensate for the inability of most e-mail systems to display many design features.
- Include a brief (fewer than 45 spaces) title in the "subject line" position: Chemical Engr/5 Yrs Exp/Oil Industry-NY

A more complex, but potentially more rewarding, approach is to rethink and restyle your resume for online delivery (Krause 119). The first screen of a student resume linked to a personal home page on the Web provides what a firsttime reader would expect: name and e-mail address, objective statement, and headings (buttons) for key information like education and experience. The resume can then use links to other screens that provide further information as the reader needs it, for example,

- A more detailed description of the curriculum you studied or honorary societies you belong to
- The full text of an honors thesis or other writing samples from your portfolio
- A video taken while you performed some task
- Other Web sites you've contributed to or built
- Sites that explain your work, like the site of an employer

For privacy in such a public medium, however, you may omit your postal address or home phone number. Figure 19.5 shows the home page for a creative Web resume.

First screen reproduces his business card

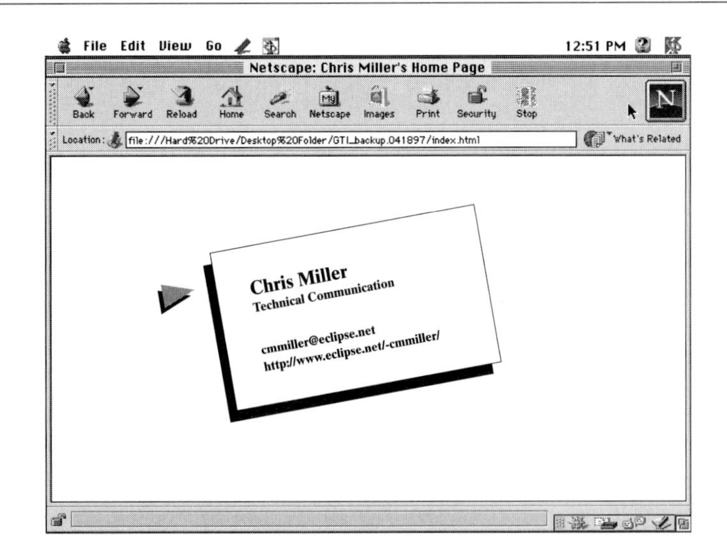


Second screen organizes active links to his resume and portfolio

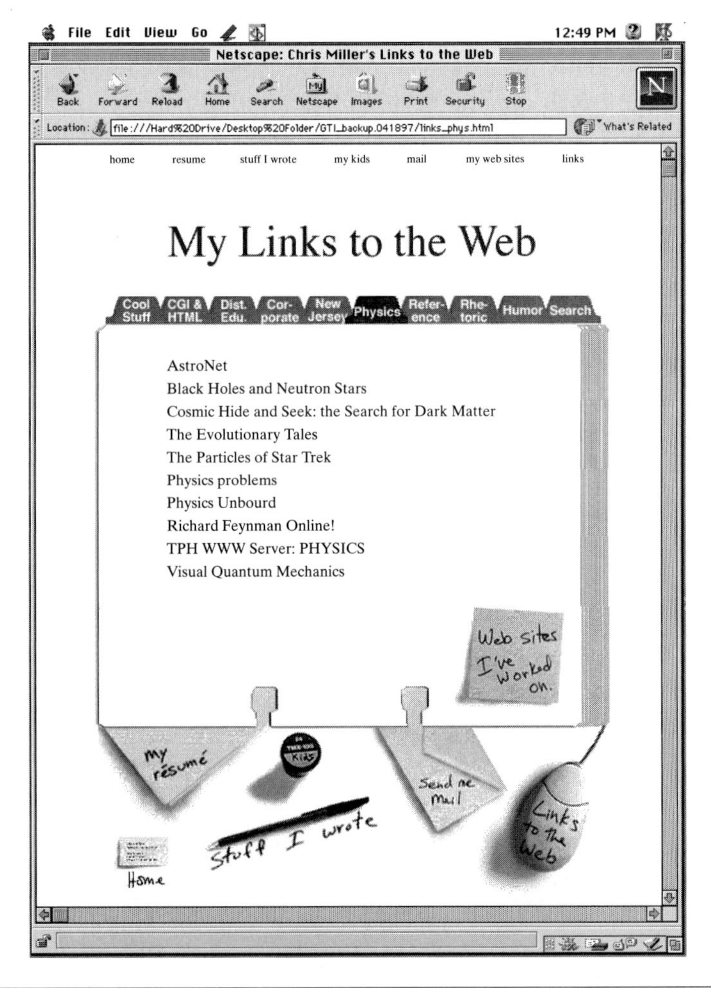


FIGURE 19.5 **A Resume on the Web**
(Courtesy of Chris Miller.)

ELECTRONIC EDGE

## The Electronic Job Search

Job candidates and potential employers are meeting each other more and more over the Web. The following list of URLs suggests some of the sites that provide advice on the job search as well as postings of both jobs and resumes (Hemby 120). Because Web sites come and go, keep checking the Web for new services.

*www.monster.com*
Monster Board. Extensive site devoted to job postings.

*www.occ.com*
Online Career Center. Charges a fee for corporate job listings but provides free posting for students (if you enter your resume electronically yourself).

*www. collegegrad.com/jobs/usenet.html*
Overview of job-related newsgroups.

*www.jobdirect.com*
Job Direct. Provides job postings and a resume template.

*www.jobweb.org*
JobWeb. Posting service maintained by the National Association of Colleges and Employers.

*www.careermosaic.com*
CareerMosaic. Allows searching of newsgroup postings. You can also post your resume directly to a newsgroup.

*www.sgx.com/cw/*
Career Network.

*www.collegegrad.com*
College Grad Job Hunter. Provides advice on the job search as well as postings.

*www.pathfinder.com/fortune/careers/*
Career Resource Center of Fortune Magazine.

*www.careermag.com*
Career Magazine.

*www.employmentspot.com*
EmploymentSpot.

*www.vault.com*
Vault. A career network.

Some experts worry, however, that students will lose control over their personal information once it is distributed over the Internet. In addition, the anonymity of the Internet means that some job listings may not be valid, and it is hard to verify sources and employers (Levenson A17). Another service being developed is an online interviewing system that allows video signals to be transmitted on telephone lines and enables companies to screen candidates for personal interviews or set up summer positions less expensively than with on-campus visits.

Because the range of available materials may be wide, you'll need to identify carefully what is appropriate and avoid a tendency to overwhelm the reader. To encourage employers to visit your Web site, note its address on a traditional print resume. You might also list it with search engine services. If you are seeking a job that requires computer skills, posting an electronic resume helps convince a prospective employer that you have those skills and are comfortable with electronic media.

## WRITING A LETTER OF APPLICATION

The resume provides the facts backing your assertion that you are the best person for a particular job. In a letter you send with the resume, called a

For more about argument, see Chapter 11

*letter of application* or a *cover letter,* you argue the truth of that assertion. You also establish your character and persona. As an example of the importance placed on character in hiring decisions, tradition in France dictates that letters be handwritten, even by those who are computer literate. Handwriting is considered an important clue to character, and handwriting analysts routinely consult with corporations on recruiting matters.

For more about proposals, see Chapter 14

Just as you comb the Request for Proposal when you write a proposal, comb the published job description when you write your letter to make sure you have responded to each requested credential (Figures 19.6 and 19.7). Keep in mind the *who, why,* and *what* approach to create an effective letter (Figure 19.8).

Date: Mon, 24 Jul 2001 10:36 : 18–1000
From: Margaret Gerard <margaret@apotheosis.com.au
To: Multiple recipients of list
Subject: Job vacancy- Sydney, Australia

Apotheosis Pty. Ltd. has a requirement for an ENTRY LEVEL technical writer to join the Computer Systems Development Centre (CSDC).

CSDC is a software engineering group which specialises in data acquisition and control systems. We are located in a pleasant working environment in Lane Cove. The development environment is Unix based and the main documentation tool is FrameMaker. Training is provided and employees are encouraged to increase their skills.

The person we are seeking will be:

- keen to pursue a career in technical writing
- an excellent written and verbal communicator
- flexible and enthusiastic
- thorough and detail-oriented
- eager to learn to use new tools and methodologies.

The duties of the vacant position cover a broad range of activities including

- producing and editing system, user, and procedural documentation
- maintaining existing documentation
- internal auditing of the CSDC Quality System
- usability testing
- administrative duties: maintaining records.

Applications to:

- Margaret Gerard
  margaret@apotheosis.com.au

FIGURE 19.6 **A Job Description Circulated Via E-Mail**
(Reprinted courtesy of Apotheosis Pty. Ltd.)

SMITHSONIAN INSTITUTION

# National Museum of Natural History

## Research Training Program

The National Museum of Natural History is offering ten-week summer internships exclusively for undergraduate students interested in a career in systematic biology and natural history research. Through the Research Training Program 24–30 outstanding students from around the world will be selected to participate in this intensive training opportunity at the Smithsonian Institution in Washington, D.C. Components of the Program include:

- *The Research Project.* Under the guidance of a Smithsonian research scientist, students will design and participate in all phases of a scientific study including gathering data, making observations, analyzing results, preparing information for publication in a scientific journal, and presenting the research project at a scientific meeting. Examples of research projects are: description and publication of a new species, morphological or molecular analysis of a taxonomic group, and mineralogical or geochemical study of a rock or mineral.
- *Lectures, Discussions, Workshops, and Field Trips.* Additional activities are scheduled to provide students with information and experience outside their selected research topic. Through formal and informal lectures and discussions, Smithsonian staff share their research and work experiences with Program participants., Topics include: species concepts, measuring biological diversity, biogeography, cladistics, morphometrics, molecular systematics, paleoecology, forensic anthropology, global volcanism, and mass extinctions. Field trips to local sites and workshops hosted by Smithsonian experts are also included in the active schedule of events.
- *Laboratory Demonstrations and Collection Tours.* Complementing the topics presented, behind-the-scene tours of many Smithsonian facilities are included. The hands-on segment provides students a chance to interact one-on-one with scientific illustrators, collection curators, research scientists and laboratory technicians. Participants also learn how research collections are organized and maintained and how to use museum collections in research.
- *Resources Available.* The NMNH collections include more than 121 million specimens of plants, animals, insects, fossils, materials, marine organisms, and human artifacts. The Smithsonian library system totals over 1 million volumes including rare books, current journals and reprints from around the world. More than 600 staff including 110 doctoral level research scientists and 250 support staff are available for consultation and guidance.

**FIGURE 19.7 Description of a Research Training Program**
Details concerning the application process and contacts at the museum are not reproduced here.

(Reprinted by permission of Smithsonian National Museum of Natural History.)

**Julie Anne Fine**

School: 314 Sypherd Hall. Newark, DE 19717-4616
e-mail:jaf@udel.edu.
(302) 837-8969

1 January 2001

Uses specific name in inside address

Mary Sangrey
Program Coordinator NHB 166
Smithsonian Institution
Washington, DC 20560

Dear Ms. Sangrey:

Notes source of interest

While scanning the Department of Entomology's bulletin board, I noticed a poster describing your summer Research Training Program. I immediately requested an application. I would be honored to participate in and contribute to your program.

Responds to "natural history-reseach"

Responds to "research project" as well as teamwork in lab

Through courses at the University of Delaware, I have become well acquainted with many aspects of natural history, including entomology, ornithology, botany, ecology, habitat management, and taxonomy. I have also taken eleven credits in courses requiring the design, implementation, and analysis of team research projects and three credits requiring an individual project.

Shows her ability to contribute original ideas because of related experience

As a technician with Petersen Environmental Consultants, I performed wetland delineations and other on-site field observations, and I have served as an inspector for several debris removal efforts in local residential areas. These experiences have helped me to learn how to recognize a wide variety of wetland and upland plants and their relationships to their habitats.

Shows success in a similar research internship

During the summer of 1999, I accepted a full scholarship to Boston University's Science and Engineering Scholars Program. While there, I worked for six weeks under the supervision of MIT Professor Jagadeesh S. Moodera. I was immersed in the world of academic research, and at the end, I presented my findings to a panel of professors.

Further evidence of her related interests

I have a broad range of interests and knowledge. My dual major in Entomology and Plant Science and concentration in Wildlife Conservation have provided me with a firm background in ecological sciences. As a member of the National Museum of Natural History and similar organizations, I am also familiar with current issues in natural history and biodiversity. I believe these skills and experiences would allow me to bring fresh ideas and creativity to a research project and team.

Enclosed is my resume listing my qualifications and work experience. I hope you will agree that your program could benefit from my participation. If you require additional information or would like an interview, I will be happy to meet your needs.

Respectfully Yours,

*Julie A. Fine*

Julie A. Fine
Enc. Resume



FIGURE 19.8 **Letter Responding to the Program Description**
Note that the letter is in *block* format. (Courtesy of Julie A. Fine.)

## Who?

First, identify yourself as a responder to a specific job opening, and then match your qualifications to those requested. Although the letter is about you, avoid opening every paragraph with "I." In particular, avoid a flat first sentence like, "I will graduate from the University of the West with a degree in civil engineering." Instead, interpret yourself in the reader's terms. Consider these more reader-oriented openings:

**Reference to an Advertisement**

I am writing in response to your advertisement for an auditor in the 16 May edition of the *Wall Street Journal.* I hope you will agree that my education and experience qualify me for your position.

**Reference to an Individual**

Professor Peter Prince of our Department of Chemical Engineering suggested that I apply to you for the opening in your environmental engineering project office.

**Reference to Past Experience with the Company**

Having spent the last four summers as a playground director with your department, I would like to apply now for a permanent position as director of one of your recreation centers.

## Why?

Then match yourself to the potential employer, and thus merge *who* and *why.* Address your letter to a specific person whenever possible (Figure 19.8). If you hear about a job from a friend, former employer, or professor, be sure to ask about the name of the person to write to. Then emphasize your interest in and understanding of the reader and his or her organization. Tailor the facts in the resume to the requirements of the job you are applying for. Show the reader what you have done and can do. Avoid mere generalizations. A statement like "I am conscientious and hardworking" carries no conviction, but specifics do: "I have maintained a grade point average of 3.8 (on a 4.0 scale) while working 20 hours a week as an assistant in the chemical engineering laboratory."

***General:*** I get along great with people.
***Specific:*** I am president of my senior class and have held two offices in my sorority: secretary and treasurer. Currently, I serve on the program committee for Engineers' Day, which entails working with 24 other students and 5 faculty members.

***General:*** I know a lot about designing children's furniture.
***Specific:*** My interest in designing children's furniture was triggered by my visit to the Furniture Mart in Chicago three years ago, where I saw several pieces designed by your company. I then switched my major from architecture to design. In my design classes, I focus any open assignments on issues in designing for children. I wrote one report, for example, on school playgrounds in Denmark. My design for a playbox was featured in our student exhibit last year.

### What?

Finally, request an interview, the goal of your letter, and refer to your enclosed resume:

The prospect of working for DelMart is most appealing. If my enclosed resume interests you, I would be happy to talk with you further at your convenience.

If you live some distance from the potential employer, mention that you would be willing to travel there for an interview (if you are) or would be available for a phone interview, and indicate appropriate times.

## FOLLOWING UP

When you put your letter of application and resume in the mail, or post them electronically, you begin a process that will require even more correspondence. Potential employers will respond to you by mail, e-mail, or phone. At least some of them will request an interview, which you may need to arrange through further letters or, more likely, through phone calls or voice mail messages. This section briefly discusses the letters you will write once you have successfully completed an interview.

### Thank You Letter

After the interview, send a brief note to the interviewer that expresses your appreciation, reminds the interviewer about yourself during that critical time when a decision is being made, and reiterates your interest in the position. Be brief but cogent:

I enjoyed talking with you yesterday at your Smeaton plant.

Touring your extensive facility and meeting potential coworkers only increased my interest in joining Peterborough Technologies. I was particularly impressed by your emphasis on quality control and team management.

Thank you for taking the time to arrange such an informative visit and especially for your personal attention to all my questions. I look forward to talking with you again.

### Letter of Acceptance

When you are offered a job that you had set your sights on, celebrate, and then write a letter of acceptance. Briefly express your pleasure in being offered the job, show that you are eager to work for the company, and confirm the starting date and any other conditions of employment that require such confirmation:

I am delighted to accept your offer to join Peterborough Technologies as a software designer.

As you suggested, I'll plan to be in Smeaton on Friday, June 15, so that I can take up my duties the following Monday.

Enclosed are the completed health survey and the signed statement concerning proprietary information.

If you need any more information from me, or if you have further information for me, please write or call. I'll be at this address until June 13, when I'll begin my relocation. I'm looking forward to working with you.

### Letter of Refusal

When you decide not to accept a job offer, write a brief letter conveying and explaining that decision:

I very much appreciate your offer of a position as a software designer.

For the past several days, I have given the matter serious consideration, but I have finally decided that I cannot accept your offer, principally because I am not able to move to Smeaton at this time.

Thank you for taking the time to meet with me and guide me around the plant. Peterborough Technologies is an interesting company, and I regret that I can't accept your offer.

### Letter Responding to a Refusal

Even if your interview does not result in a job offer (many don't), consider writing one more letter if you are still interested in the company:

> In response to your letter of 15 April, I, too, am sorry that you do not have a position for me at this time.
>
> I enjoyed talking with you and learning more about your company. Peterborough Technologies remains very attractive to me and I hope you'll keep me in mind for any future openings.

A polite letter that reiterates your interest lets you have the final word—and may also help you if a position does open up in the company.

## CHECKLIST: The Job Search

1. **Design your resume to**
   - Show what you have done as a predictor of what you can do
   - Make information easily accessible to the reader
   - Demonstrate your persona
2. **Structure your resume**
   - Chronological: education and experience in order of time
   - Functional: education and experience pegged to your skills and achievements
3. **Deliver your resume**
   - In print: appropriate typeface and page design, preferably one page
   - Online: appropriate headings, accessible through a range of systems
   - On the Web: interactive, with flexible units and appropriate links
4. **Tailor your resume to a persuasive argument in a letter of application**
   - Open with a statement about why you are writing to this person
   - Apply for the job
   - Provide evidence that you possess—or exceed—the qualifications for the job
   - Close by requesting an interview

## EXERCISES

1. Analyze the persona of the writer (see Chapter 7) and the informational content of the following paragraph, which constituted an entire letter of application:

   I would like to apply for a job with your company. I have had a good education and have taken practically every course available in electrical engineering. Therefore, I feel that I could handle any problem that you have in that area. I realize that it is important to get along with people, and I assure you this is one of my strong points. I'll be visiting my brother in Chicago during the month of September and will be glad to come to your office for an interview.

**2.** Comment on the persona that each writer demonstrates in the following opening lines of letters of application.

I am the right man for you as you'll see when you read this letter.

I'm sorry to take up your time, but I need a job, and I'd like to join your company. I'm sure you'd never regret making me an offer.

I have searched everywhere for a job commensurate with my abilities. You are my last resort, and for that reason, I have outlined for you the reasons I need a job.

My name is Pete Marwick.

Enclosed is a resume and application form for employment during the summer of ________.

At the suggestions of Patricia O'Neill, community relations director of the Utah chapter of the March of Dimes, I am applying for a position in your promotions department.

After having survived four months of your jokes about Delaware and your unrelenting teasing, I know I could put up with much more of it. Therefore, I would like to apply for the public relations position you said might be available this month.

When I address the Vong Family, I address Vong Ltd. as a whole. If one thing has impressed me about Vong Ltd., that one thing would be the facsimile of a family that you create with your employees. For the last seven years I have sought part-time work with your company, during periods off from school. Seven times I have found your arms wide open like a father bidding welcome to his prodigal son. If I have learned one thing of Vong Ltd., I have learned that you take care of your own, and now I would like to take care of Vong Ltd. That is why I am applying for a project management position with your company.

**3.** Comment on these closing paragraphs. Would you make any changes in them?

Thanking you very much for reading my letter and for considering my qualifications, I remain

Sincerely yours

I am anxious to obtain an interview and am available at any time.

Please answer as soon as possible because I want to make arrangements for interviews with other companies.

As you can see from my letter, I haven't had much practical experience, but I can learn fast and am not afraid of hard work. So please tell me when I can come and talk with you.

**4.** Interview a personnel manager at your college or at a nearby company about what she or he looks for in a letter of application and a resume. Here are some questions to consider:

- How important are spelling and grammar? Would one spelling error cause the manager to reject the resume?
- Should applicants include information about courses taken and grade point average?
- Does a resume have to be only one page?
- Should applicants include a statement of objectives?
- What if applicants don't have a lot of experience? What should they include?
- Should applicants highlight computer experience? Language fluency? Extracurricular activities?
- How important is the design of the resume: paper selection, font choice, white space, and the like?

5. Write a letter of application responding to the job description in Figure 19.6.

6. Comment on this letter applying for an accounting position. The job description indicated the need for "3+ years of accounting experience, preferably in cost accounting, and demonstrated computer experience."

    Your advertisement in the *Sun Times* for a cost accountant caught my eye this weekend. I feel that my qualifications would be a very good match for your needs in this position. I will complete requirements for my BS in accounting at Eastern State in May and have found cost accounting and the related issues in operations management an area in which I am confident I would enjoy working.

    I realize that I do not offer the precise experience stated in your advertisement. However, my experience as a teacher and high school coordinator has helped me develop many skills, including organizing my work, directing others in their work, solving problems, and delegating duties to others. This experience should help me quickly adapt to a job as a cost accountant at Four Diamond.

    While at Eastern State, I have taken several courses devoted to computer software and systems design. I have also used computer systems to complete much of the classwork in accounting, finance, business policy, statistics, and operations management. I feel comfortable with Excel spreadsheet programs and most office applications, including Word and PowerPoint. I feel I can also adapt to whatever programs you use.

    As I hope you will agree, I fit your requirements closely. I look forward to discussing this program with you in greater depth. I am available at 904 386 4163 after 4:00 P.M. on weekdays.

## FOR COLLABORATION

1. At your instructor's request, form teams of three students to prepare brief reports on job opportunities outside the United States. Select different areas of interest for each team. Research opportunities through print sources but especially through the Web. Find out, as well, the conventions for letters of application and resumes in that country or region.

2. One of the best ways to obtain comments on your resume—and to learn about other people in the class who may team up with you on projects—is to circulate a draft resume for class review. Early in the semester, each member of the class should bring adequate copies of the resume for everyone. Exchange copies. One easy technique for the exchange is to arrange chairs in a circle. Each student then places a stack of copies on her or his chair and walks around the circle collecting the other resumes and arranging them in alphabetical order. In class, over several sessions, discuss all the resumes in turn, with comments from everyone in the class. Later in the semester, turn in a revised resume in final format.

# MEMOS AND ELECTRONIC MAIL



"COURTESY . . . ACCURACY . . . SPEED."

EARLY TWENTIETH-CENTURY BRITISH ADVERTISEMENT FOR THE TELEPHONE

To encourage action and serve as its collective memory, organizations rely on three forms of mail, in addition to letters: memos, voice mail, and electronic mail (e-mail). Although electronic technology is blurring the distinctions among these forms and changing the very meaning of *mail,* the guidelines you'll read in this chapter will help you compose each form.

## MEMOS

Memos meet many of the purposes of letters in making and responding to requests and maintaining relationships. Like letters, memos should be brief, preferably no more than one page. But unlike letters, which are generally sent between individuals who are not in the same organization, memos are traditionally inside documents. Among other topics, memos report on research or field investigations, describe policies and procedures, and circulate the minutes of meetings, three types of memos you'll read about in this chapter.

### Structure

As letters look like letters, memos *look like* memos, and that look helps you determine the content appropriate for a memo. Memos begin with a *header,* which usually includes four items, preprinted on a memo form, most commonly in the following order:

To
From
Date
Subject

**To.** When you write a letter, you generally address one person. In a memo, you may write to one person, but you often address a group of people: all employees, all collaborators on a project, all people who attended a meeting; that is, you write to a set of individuals. In selecting that set, be sparing. You may be tempted to include many, but the more people named on a distribution list, the less likely it is that any one of them will read the memo, because each may think the other is taking care of the issue.

**From.** Identify yourself as the sender of the document. Add a title to your name if the context requires that the reader know your title as well as your name.

**Date.** Make sure your memo is timely, sent at the right time to meet a current need. Memos are fleeting documents, responding to a particular issue, at a particular time. But they are also permanent in creating a record for future reference. So if you write when you are angry, or say something indiscreet or without proper consideration, that memo may establish a record which later embarrasses you.

**Subject.** Make the specific purpose of your memo clear in the subject line. Here's a simple rule: one memo = one topic or action. Announcing that topic or action in the subject line helps the reader decide how (perhaps "if") and when to read the memo. In the stream of project documentation represented by memos, each category of information—for example, costs, components, personnel—may find a home in a different file. Moreover, each action may require a different approval. So covering several topics in one memo may make responding, acting, filing, and approving difficult.

**Opening Paragraph.** Following the header, provide the memo's main content. Use a brief opening paragraph to make clear what your subject line implies. Answer this reader question: "Why am I reading this memo from you now and what should I do about it?" Give the context of your memo by explaining

- What action or understanding you seek from the reader.
- Why you selected the reader to receive this message.
- Why you—rather than someone else—wrote the memo.
- Why you are sending the memo today—rather than yesterday or tomorrow.

If the memo responds to a request, note that request to remind the reader and engage her or his attention. The subject line and opening paragraph go a long way toward achieving the memo's purpose, so they deserve special care. Note how the revised opening in Figure 20.2 directly answers the reader's request: "What should I do about these complaints?"

You'll read more about Figures 20.1 and 20.2 shortly. As in Figure 20.2, a direct opening usually works best:

> I am writing in response to your request of 3 March for an update on our survey of retirees who have established homes in Rehoboth Beach during the last five years.

To initiate a specific activity in a memo, establish your need in the reader's mind:

> Please review the attached specifications for the proposed handicap retrofit of Building A before we meet with the architect on Friday.

**Supporting Paragraphs.** In a series of short, well labeled paragraphs and perhaps a visual or two, provide the supporting argument, as in the examples you'll read in this chapter. Use attachments, as needed, for the details.

**Closing Paragraph.** Note any action or decision the reader should take.

### Voice

Because memos circulate within organizations and should be brief, use a less formal style than in a letter. Feel free to use insider's language and abbreviations and to present information more baldly without the elaboration or regard for polite small talk necessary for outsiders. To encourage brevity, some organizations provide half-sheet memo forms because writers who face a full page may feel compelled to fill it. Moreover, some companies dictate that no memo, no matter how worthy its topic, can exceed one full sheet, so the reader can see the beginning and the ending at once. That's not a bad rule of thumb.

For more about persona, see Chapters 7 and 21

Although they are informal, bald, and brief, memos still shouldn't be rude. Pay attention to your persona and the voice that comes across the page or screen. The memo in Figure 20.3, sent by the director of a large production facility to all employees, demonstrates an ineffective voice.The memo at first confused the recipients, then angered them, and then became a source of humor. It didn't work. Why? For openers, it violated the readers' expectations. A memo on safety—nearly 2 million man-hours without a lost-time accident—should be an occasion for celebration. Instead, the director takes another tack: "have compassion" for the first person to cause an accident and thus break this record. The sentiment is worthy, but it is not exactly memo material. In addition, the memo slides in another message at the end: a safety audit will soon occur. Readers felt a bit betrayed. Was this the real message? Are there problems? Am I being investigated? Or was this all an afterthought?

The writer of Figure 20.3 also shifts his persona in midmemo. The voice waffles from the authoritarian ("it has been a while since a safety memo has been generated by me," "address that eventuality") to the folksy ("chugging along," "knock off that milestone," "chide or ride"). One word is simply wrong: "Unlucky sole" is a fish, not a person. Readers felt both embarrassed for the director and patronized. The intention was perhaps honorable, but the execution failed.

**MEMORANDUM**

To: Ann Thompson
From: Chris Kulchitski
Date: 16 January 2001
Subject: Cooktop Temperatures

Vague subject line. What about?

Descriptive heading that is more common in the middle of a document

**Experimental**

Two experiments were performed to measure the freestanding range cooktop temperature. One was to measure the effect of various burner bowl designs on cooktop temperature. The burner bowls consisted of current bowl and trim, one-piece bowl, and one-piece bowl elevated 1.5 mm. The other was to compare Stephens's cooktop temperature with our competitors'. The competitors' consisted of Fritag, Inco, and Johnson. All of our competitors have trim and bowl and coil heating elements similar to ours. Temperature measurements were taken at various locations. For the burner bowl variations, the tests were repeated to get an average value. The experiment details, including procedures, are in the data sheets attached to this memo.

The memo begins without any context or motivation for the reader to read

Details lack connection and emphasis

**Results:**

The effect of various burner bowl designs on cooktop temperatures is shown in the data sheets and summarized below:

Approach follows the style of a lab notebook rather than a memo

Cooktop Temperature

| | *Average Temp. (° C)* | *Std Deviation* |
|---|---|---|
| Current trim and bowl | 96 | 21.5 |
| One-piece bowl | 98 | 23.0 |
| One-piece elevated (1.5 mm) | 112 | 32.3 |

The current trim and bowl cooktop has the lowest average temperature, 96° C, followed by the one-piece bowl, 98° C, and the one-piece elevated bowl, 112° C. Elevating the one-piece bowl 1.5 mm not only caused the average temperature to rise by 16° C, it also caused a wider spread in temperatures as shown by the larger standard deviation. With the elevated one-piece bowl, the cooktop has a high of 184° C and a low of 73° C. The current bowl and trim have the least spread in temperature. The one-piece bowl design increased the cooktop temperature slightly, about 2 degrees.

Paragraph restates details without answering the central question in the consumer complaint

If we divide the cooktop into hot central and cooler outer regions, the current trim and bowl and one-piece bowl spread is still about 3 degrees in the hot zone, i.e., 116° C versus 119° C. However, the elevated one piece bowl increased the cooktop hot zone temperature significantly to 142° C.

Here is a summary of the comparison test results (see attached data sheets).

| | *Average Temp. (° C)* | *Std Deviation* |
|---|---|---|
| Freitag | 98 | 15.3 |
| Stephens | 96 | 21.6 |
| Inco | 90 | 21.8 |
| Johnson | 80 | 14.8 |

Johnson has the coolest cooktop, about 16° C cooler than Stephens's. It also has the most even temperature cooktop. Inco has the second coolest cooktop, but it has the highest temperature spread. Stephens and Freitag have the hottest cooktops.

Facts are hard to compare

Stephens has the hottest hot zone, 116° C, followed by Inco, 113° C, Freitag, 109° C, and Johnson, 93° C. The ranking of the cooler outer region is Freitag, 92° C, Stephens, 85° C, Inco, 77° C, and Johnson, 72° C. Overall, Johnson has the coolest cooktop, followed by Inco. Stephens and Freitag are comparable. We have a bigger spread in temperature compared to Freitag, that is, hotter in the center but cooler in the outer edge.

Paragraph lacks unity and development toward a point

**FIGURE 20.1 Original Memo Discussing a Problem with Cooktop Temperatures**
(Courtesy of Robert Kelton.)

Important recommendation is buried

2

About 3 percent of the customer instruction service calls are due to the hot cooktop. There are many calls and letters from consumers complaining about our hot cooktop as well. Our cooktop is so hot (116° C) that one customer called in to the hot line to complain that her fingers were blistered when she placed her hand on the stove top. It is not surprising since water boils at 100° C. To reduce customer complaints as well as addressing a safety issue, I strongly recommend that we redesign our range to reduce the stove top temperatures.

FIGURE 20.1 **Original Memo Discussing a Problem with Cooktop Temperatures (continued)**

**MEMORANDUM**

To: Ann Thompson
From: Chris Kulchitski
Date: 16 January 2001
Subject: Recommended Cooktop Redesign

Action-oriented subject line

**Overview**

Brief statement of context and reason for the memo

In response to many customer complaints, you asked me to evaluate the temperature of our cooktops. My evaluation consisted of two experiments whose results are summarized in this memo and the attached data sheets.

Summary of the memo's recommended action and rationale

Based on these tests, we concluded that our bowl and trim design is as cool as possible, but that our cooktop should be redesigned to offer a cooler cooktop temperature. Because of the potential for customer injury, we recommend speedy action.

Informative heading

**Tests for Bowl Design Effect on Temperature**

Structures information to meet a business need rather than merely recording the experiment

Here are the results of our tests on bowl design:

| *Bowl Type* | *Average Temp. (° C)* | *Std Deviation* |
|---|---|---|
| Current trim and bowl | 96 | 21.5 |
| One-piece bowl | 98 | 23.0 |
| One-piece elevated bowl (1.5 mm) | 112 | 32.3 |

Answers the reader's question: So what? Attachment provides details

In summary, the lowest temperatures were associated with the current design, which also offers the lowest temperature spread. The one-piece bowl increased cooktop temperatures by about two degrees and had a slightly higher temperature spread. By far, the hottest and most varied temperatures were associated with the elevated bowl design, which had a high of 184° C and a low of 73° C. This information suggests that we should maintain our current bowl design.

**Cooktop Comparison by Brand**

Here are the results of our tests to determine the average temperatures and standard deviations for our competitors' models in comparison with ours.

Boldface highlights the company's statistics

| *Model* | *Average Temp. (° C)* | *Std Deviation* |
|---|---|---|
| Freitag | 98 | 15.3 |
| **Stephens** | **96** | **21.6** |
| Inco | 90 | 21.8 |
| Johnson | 80 | 14.8 |

Action closing

These results suggest that we should alter our design to reduce cooktop temperatures and to make our heating more even. Further evidence for this recommendation is contained in the attached tables that compare hot and cool zones separately.

FIGURE 20.2 **Revised Memo**
(Courtesy of Robert Kelton.)

MEMORANDUM

Date: 1 May 2001
To: All Staff
From: R. J. Nettleson, Head
Re: Safety

Even though I have attended some recent safety meetings, it has been a while since a safety communication has been generated by me. Well, I am still very interested in the subject both on my own and because it is still a heavy-duty item in the viewpoint of the company.

We are chugging along toward two million man-hours without a lost-time accident and, if we all keep ourselves alert to safe practices, we will make it. A lot of people are looking forward to that later this year. It is my hope that we can knock off that milestone and go after several million more man-hours without a lost-time accident in the future.

The reality is, unfortunately, that some day someone will have a lost-time accident, and I want to address that eventuality today. The larger our number of man-hours gets, the worse the person who finally causes it to end will feel. That is why, before we know who that will be, I am writing what follows. Both myself and the unlucky sole will feel extremely bad when it happens, and I ask that each of you have compassion for the individual when it occurs. Please do not chide or ride or make light of the fact that they were responsible. I feel that the vast majority of people and maybe all would not do that and do not need the request. However, if I did not mention this and it happened, my opinion of me would be that I neglected to do something I should have done.

Additionally, as a continued enhancement of our safety efforts, we will be conducting an International Safety Rating System baseline audit of R&D sometime in July. It is designed to show us where we are weak so we can improve. This will be coordinated by John Jones who will communicate additional information further down the road.

Thank you for your time and have a good and safe summer.

FIGURE 20.3 **An Ineffective Memo Style**

## Brief Reports

Unlike the director, the author of the memo in Figure 20.2 speaks in a clear and professional voice. That voice, and the reader-oriented design of the final memo, illustrate the effects of good revising. In the memo, Chris Kulchitski responded to a request from his supervisor, Ann Thompson, at the Stephens Company, which produces stoves and other kitchen appliances. Ann asked Chris to investigate complaints from customers about the high temperature of Stephens's cooktops. Chris's report had to achieve several purposes: furnish information, encourage action, provide a record, and maintain his good working relationship with Ann. In revising, he had to decide *which* purpose was most important. His draft (Figure 20.1) is a flat and unemphatic record of information. It follows the structure

**MEMORANDUM**

Date: 30 October 2001
To: Jim Dunns, VP Marketing
cc: The Marketing Group
From: Steve Milquist
Re: Conference on International Technical Marketing on the Web

**Overview of the memo's purpose, in informal style**

The attached report details the information I picked up at the International Technical Marketing conference last week at Snowbird, Utah. It also strongly recommends that we develop a Web page and, more generally, an electronic marketing strategy.

The conference consisted of both lectures, including a keynote address by a Web marketing guru at the University of Utah (Lamar Summerhays), and tutorials. Those tutorials were first-rate. In sum, the presenters emphasized these guidelines:

**List of main points to alert readers who may not read the whole report**

- *Graphic distinctiveness.* Use graphics creatively but not so extensively that it takes a long time to download the page.
- *Easy response.* Set up the page with an e-mail dialog box so the visitor can get back to us easily.
- *Interactivity.* Use a contest, fill-in-the-blank exercise, etc., to get the viewer to interact with the page.
- *Easy links.* Tie in the page to popular search engines and to other relevant and popular pages.
- *Easy navigating.* Help the viewer move easily through the text by markers like page-link graphics.
- *Abundant information.* Avoid mere fluff pages; while not letting complex information get in the way of simple stuff, satisfy the viewer's need for information.
- *Frequent updates.* Keep the page fresh and enticing, inviting restless viewers to return for more. Stale pages are a real turn-off.

The report discusses these guidelines in detail and shows how we might apply them to our effort.

**Friendly sign-off, thanks—and point toward action**

Thanks for supporting my attendance at the conference. I look forward to applying what I learned in our own marketing—soon!

FIGURE 20.4 **Covering Memo That Summarizes a Longer Report**
The writer reports on what he learned at a conference, a common occasion for memo writing.

Chris used in his laboratory notebook. Figure 20.2 organizes information strategically to support a technical and business decision in response to the complaints.

Like Chris, use a memo to report—the results of an investigation, the activities of a business trip, the information derived from a conference, and the like. But you risk violating the sense of a memo if you drag the message over more than two pages. Instead, provide a covering memo and attach a longer report (as in Figure 20.4).

CROSSING CULTURES

## A Scottish Voice

Memos play an important role in organizational communication throughout the world. The format that is conventional in the United States is also common in Europe and Asia, and multinational companies often require that memos appear in English even if their home office sits in a Swedish- or German-speaking country. An exception is Québec, where political concerns and legislation add another perspective to the choice of French or English in internal correspondence.

But even if they appear in English, memos around the world *sound* different. The casual voice common in the United States may seem inappropriate in other countries. You'll recognize the document produced below as a memo—an effective one. The style, however, reflects British usage (actually, the author is a Scot). In particular, its tone is more formal and it relies more on the passive voice. (Courtesy of Malcolm McLean.)

IMPERIAL COLLEGE
DEPARTMENT OF MATERIALS

MEMORANDUM

To: All Departmental Staff
From: Malcolm McLean
Subject: Laboratory supervision and risk assessment
Date: 17 November 2001

**1.** Safety during laboratory classes is the responsibility of the member of academic staff running the particular activity. Supervision of the activity can be delegated to postgraduate students, PDRAs, technicians, or others; however, the responsibility remains that of the member of academic staff. He/she must delegate this responsibility to demonstrators in a formal way through written procedures.

*Action.* Academic staff should provide demonstrators with written instructions concerning laboratory experiments and have a copy signed by the demonstrator as an acknowledgment of having received them.

**2.** Any piece of equipment or apparatus or experimental procedure that is considered to constitute a safety hazard should be subject to risk assessment. This can be fulfilled by the obvious and ready availability of written procedures for its use or execution. In the case of commercial equipment this will normally be satisfied by the availability of an operating manual.

MEMORANDUM

To: Lab Supervisors
From: Safety
Date: 1 October 2001
Re: Bypass Lines

Flat subject line

You may recall that I recently sent out a notice stating that "we are no longer going to allow bypass lines to be used to circumvent check/surge valves." Our shop mechanics are now in the process of removing all of the bypass lines that are presently violating this safety procedure. To date 15 lines have been identified that are in violation.

Opening line lacks emphasis. So what?

The mechanics need your help to identify these lines as to who is the owner. So I am asking each person to check your gas lines, that come from cylinders, to see if you have bypass lines in them. If you do, please identify these lines by putting your name and lab # on a tag and hanging this tag on the bypass line.

Request for action is buried in midparagraph

If you don't identify your lines, when the mechanics go to remove the bypass lines your gas supply will be shut off. They will then not be able to notify you that this is going to occur. This could cause you serious interruptions. So take the time to identify your lines!

Consequence of noncompliance comes too late in memo

This identifying of lines must be completed by Friday 23 October. After this date the mechanics are going to start removing the bypass lines.

Because it is critical, date should come earlier

FIGURE 20.5 **Policy Memo: Before**

## Policies and Procedures

The director's memo in Figure 20.3 responds (rather badly) to another common memo writing occasion in addition to brief reports: conveying an organizational policy or procedure or announcing that more extensive descriptions of such procedures are available elsewhere. Writing such statements often seems simple but can land you in difficulties similar to those of the director. Figure 20.5 shows two versions (before and after revision) of a memo announcing a new procedure concerning bypass lines in a research lab.

## Minutes

You'll also use memos to circulate the minutes of meetings to members of a team or to a larger group of people (Figure 20.6). The minutes should represent the discussion and discussants fairly, stripped of the drama that may have consumed some of the time. You may find writing minutes tedious, but a little attention can make the job easy. In addition, as the compiler of the

MEMORANDUM

To: Lab Supervisors
From: Safety
Date: 1 October 2001
Re: **WARNING: All remaining Bypass Lines will be removed after 23 October**

Before 23 October, please hang a tag with your name and lab number on any bypass lines around check/surge valves in your lab's gas lines.

After that date, mechanics will start removing the lines. The work requires that we shut off the gas to any lab which has such a line.

Because of the potential for explosion, we are no longer allowing such bypasses. See my memo dated 10 September for more details. Shop mechanics have already identified 15 violations, but we need your help in tagging any others.

If you identify your lines, we can schedule their removal to fit your needs and thus avoid any interruption in service. If you wait until we identify them, then we will not be able to notify you in advance of their removal.

**Action-oriented subject line**

**Requested action simply stated**

**Consequences of noncompliance**

**Rationale**

**Benefit of requested action**

FIGURE 20.5 **Policy Memo (continued): After**

minutes, you can play an important role in controlling the meeting's outcome. Good minutes also provide excellent visibility for you as they circulate in the organization.

To prepare the minutes

- Begin with the meeting's agenda and summarize the results on each point.
- Note any conclusions arrived at in the meeting.
- List assignments (actions to be taken) with the names of those responsible for performing them.
- Confirm any deadlines.
- Streamline the record of the discussion.
- Note any subsequent meetings.
- Attach any documents circulated during the meeting as an appendix.
- Ask the reader to validate the correctness of the record.

Send the minutes to everyone who attended the meeting and to others who did not attend but who will be affected by actions discussed there.

**MEMORANDUM**

Date: 14 April 2001
To: Marketing Group (see Distribution)
From: Paula Petersen
Re: Minutes of 12 April Meeting on Web Marketing

**Notes attendees for the record**

Attending: Steve, Amy, Kesha, Paula (Quinn was on assignment)

**Purpose and overview**

The MG met to determine if, and how, we should market on the Web. Steve reported on a recent conference he attended about technical marketing on the Internet (see attached report). He strongly recommended that we develop a Web page for our photochemicals and extend the approach to other products if the response warrants.

**List form eases skimming**

**Note parallelism in the list**

Discussion centered on three issues:

1. *Cost.* In general, except for the onetime costs of start-up and development, the cost of such marketing is much lower than through traditional media.
2. *Effectiveness.* At this point, it's hard to determine how effective such marketing is. We discussed this at some length.
3. *Ease of operation.* Steve assured us that the learning curve is not all that steep; we should be able to develop and run the Web site ourselves without additional hiring or extensive additional work.

**Clearly defines each attendee's further responsibilities**

**Action**

*Steve.* Develop a prototype Web page for photochemicals

*Amy.* Survey local Web access providers to check on fees, requirements, etc.

*Kesha.* Review our competitors' Web pages (if there are any) and in general browse the Web for ideas

*Quinn.* Outline a document package that would include the information for the Web page and supporting follow-up documents

**Uses highlighting to remind readers about the next meeting**

**Next Meeting: Monday 1 May 3 P.M. in the conference room**
Agenda: Reports on the actions listed above

FIGURE 20.6 **Memo Providing Minutes of a Meeting**

## VOICE MAIL

Information on paper, like memos, is gradually being replaced by electronic information in many offices and laboratories. For years, experts have been predicting the arrival of the "paperless office." It's not here. But increasingly, you will fulfill the purposes of a memo electronically, through either voice mail or e-mail.

### Advantages and Disadvantages

Voice mail systems are sophisticated answering machines (and talking machines) that combine phones and computers to handle routine transfers of information. The story (perhaps apocryphal) of the invention of voice mail fits the pattern of a "eureka" moment of insight. An entrepreneur stuck in a rainstorm noticed a pile of pink "while you were out" message slips in a nearby dump. That sight inspired him to think of a machine that would allow people to record messages in their own voice and reduce both paper and confusion (Ramirez 9). Automated voice mail is used, for example, to take routine orders from customers, register students for classes, provide updated quotes on stocks, and record buy and sell orders that will then execute when the designated price is reached. Voice mail has become so common in organizations that many callers expect machines and are surprised to hear a person answer a phone call. Some callers even prefer machines because the call is shorter: You can leave a brief message and hang up.

These significant advantages of voice mail are matched by two disadvantages. First, some systems trap callers with excessive branching through many levels of prompts and choices in what is referred to as "voice mail jail." A second disadvantage is long outgoing messages. Well designed systems avoid branching beyond four levels and limit initial outgoing messages to ten seconds. You may not be able to change your organization's system, but you can at least make sure the outgoing message on your own phone is brief and pleasant.

### Guidelines

Choose voice mail over a memo when your message is short and not controversial, when you'd like the speed and ease of a phone call and its cost is not a consideration, or when your call activates a process, as in registering for classes through a voice mail system. When creating a message to leave in someone's voice mailbox

- Start with a plan, even if you don't normally plan phone calls.
- Gather necessary information before the call.
- Note the date and time of your call; not all systems record that information automatically.
- Speak clearly, especially when you give your name and phone number. Provide that information twice—at the beginning and at the end of your message.
- State directly the purpose of the call and the action or response you seek. Listeners need more time to prepare for such information than readers.

## ELECTRONIC MAIL

E-mail messages are rapidly replacing most paper memos in organizations around the globe. People in offices, laboratories, and field sites are spending

ELECTRONIC EDGE

## Privacy

Who owns the messages that circulate in an office's voice mail or e-mail system? Does an employer have the right to monitor such messages? Employers contend that they should be allowed to determine appropriate use of their system—and to sanction workers who send personal messages and eavesdrop on potentially disruptive conversations. Employees often see such behavior as an invasion of their privacy. In a recent survey of employers conducted by a computer magazine, 22 percent of the respondents said they monitored employee voice mail, e-mail, or computer files—many without the workers' knowledge or consent (McMorris B1).

The legal issues are murky. Those who support the privacy of voice mail argue that such messages are protected by the U.S. federal wiretap law and its 1986 amendment, the Electronic Communications Privacy Act. The wiretap law, written long before the boom in electronic communication, prohibits listening in on telephone conversations unless one of the parties has given permission or there is a warrant. It does allow employers to monitor telephone conversations for business purposes. That right, however, is limited by an appeals court ruling that required employers to separate business-related and personal calls and not eavesdrop on personal ones.

The privacy of voice mail is a more open question than that of e-mail. Recent court cases have clearly established the right of employers to read e-mail circulated on systems they own and operate. While these issues are being resolved, companies should create and circulate guidelines for fair use of electronic systems that protect everyone's interests. Employees in the meantime should not expect that anything they say or write electronically on those systems is private.

increasing amounts of time typing at one another through e-mail systems and more broadly through a wide range of network services. By one estimate, more than 1.4 billion e-mail messages circulated in mid 2000 (Labaton and Richtel 1).

### Advantages and Disadvantages

The head of information services at Aetna Life & Casualty Co. notes, "Paper in a service business is like cholesterol in the bloodstream. . . . Good paper is what you need to communicate with others—claims checks and premium notices. Bad paper is the internal stuff that clogs up the arteries" (Bulkeley B1). Some people continue to prefer seeing "internal stuff"—especially memos—on paper. Most people find print easier to read than screen text because of its higher resolution. It's also easier to skim a pile of pages, so long documents in particular may be taxing in an electronic version.

But, as you are certainly aware, e-mail has serious advantages in shuttling information around an organization or around the world. E-mail

- Encourages communication and discussion.
- Reaches its intended reader rapidly.
- Saves paper and thus contributes to a better environment.
- Reduces the cost of maintaining paper files.
- Supports teamwork both locally and globally.
- Aids telecommuting and wide dispersion of employees in networked organizations.
- Helps communicators cross political and language borders.

Some research also suggests that e-mail fosters democratic thinking. In addition, shy people may find their voice on e-mail; the form seems less intimidating than telephone calls or letters. The democracy of networked computing was emphasized in a cartoon in *The New Yorker* that showed two dogs at a terminal. One says to the other, "On the Internet, nobody knows you're a dog." Finally, in the global economy, e-mail accommodates differences in time zones or work hours. A colleague in Europe writes while the United States sleeps; the message waits for the colleague in New Mexico to start her day.

The many advantages of e-mail come at some cost, however. One is message overload, which may reduce productivity. Another is the ineffectiveness of many of these messages. Matching the speed of delivering e-mails with speedy composing, some writers send messages that are poorly thought out and sloppy in expression and rely on a series of messages, rather than one good message, to do the work. Researchers, too, point out a growing isolation among individuals tied to their computers rather than participating in a real community.

## Guidelines for Writing

To reap the advantages of e-mail, and overcome the disadvantages, *structure* your message, as you would a print memo, and create a voice appropriate for a message-on-a-screen (see *A Closer Look: Netiquette*). Figure 20.7 shows an effective e-mail message in the casual voice common in such messages. In doing so

- Keep messages short—even shorter than a memo. Readers don't like to scroll.
- Divide the message into paragraphs. Avoid the common practice of one-paragraph e-mails.
- Use a richly informative subject line that can stand for the memo in an index.
- Take advantage of abbreviations and acronyms (see Figure 20.8) when you know your reader will understand them and such informality is thus welcome.
- Use a conversational style, more like your speaking voice than your formal report voice.
- Remember that a person is reading your message. Take time to be civil.

From: Angelina Petrofsky (corky@udel.edu)

Date: Tues., 8 Nov 2001 01:41 : 40 - 500
To: dandrews@udel.edu
Subject: Rewrite of Individual Report

Well, I'm finally ready to rewrite the report. Thanks for all of your suggestions.

A couple of questions—

Oh, just in case you don't remember, I'm developing a telephone interface to be run on a computer. The phone will be embedded in the computer.

1. In my report, does it make sense to cover:
   - The steps I took in making the design
   - Decisions made for the interface
   - Improvements to be recommended for future versions
2. I have a screen capture of the interface. Should I maybe start my report with it? I could label it with markers like A, phone pad; B, conference phone number; etc., and then refer to them in my report.

Thanks

Angie

**FIGURE 20.7** **Effective E-mail Message**

- Proofread. Make sure your message is correct in grammar and punctuation.

## Guidelines for Reading

Because you may face an overload of e-mail messages, you need a strategy for sorting and prioritizing them. One strategy is automatic: You can introduce "intelligent agents" into your e-mail program to sort messages by sender or recipient, keyword, or date. If you don't have an automatic pre-reader, filter messages yourself. To do so,

- Read the entire index first to determine which messages are most important, and start your detailed reading with those.
- Avoid merely reading messages in the order (usually chronologically as they arrived) in which they appear in the index because you may waste energy replying to insignificant messages.
- Jot down on a paper pad the points from a message that you need to respond to.
- Keep a notepad list of messages you need to send to keep your session at the terminal efficient.

| | |
|---|---|
| BBL | Be back later |
| BFN | Bye for now |
| BRB | Be right back |
| BTW | By the way |
| FWIW | For what it's worth |
| FYI | For your information |
| HSIK | How should I know? |
| IMO | In my opinion |
| IMHO | In my humble opinion |
| IOW | In other words |
| LOL | Laughing out loud |
| NBD | No big deal |
| NOYB | None of your business |
| OTOH | On the other hand |
| PMFJI | Pardon me for jumping in |
| PTB | Powers that be |
| ROTF,L | Rolling on the floor, laughing |
| SNAFU | Situation normal; all fouled up |
| Sysop | Systems operator; person who manages a newsgroup |
| TIA | Thanks in advance |
| TIC | Tongue in cheek |
| TTFN | Ta ta for now |
| TTYL | Talk to you later |
| WYSIWYG | What you see is what you get |

FIGURE 20.8 **Glossary of E-mail Abbreviations**
These abbreviations appear frequently in the postings of newsgroups as well as in personal e-mail messages, although their use represents a writer's shortcut more than a formal code.

- Read the last message in a series from the same sender before you read any earlier ones, because that final message may correct or amplify partially developed earlier attempts.
- Read messages addressed only to you (most indexes automatically note this) before messages from bulletin boards and newsgroups. Such individual messages are usually more important.
- Read your e-mail at least once a day. Because the systems are generally fast and reliable, your correspondent will assume the message has gotten through.
- Print only those messages that require further processing as paper, for example, directions to a meeting location that you'll need in your car, an attachment for a print report that must be immediately mailed, or a policy statement you must keep in a file. Read your messages at the screen, as the author intended, and delete messages of only momentary interest to conserve disk space in the e-mail system.

A CLOSER LOOK

## "Netiquette"

Because e-mail is an evolving form, the norms for using it are also evolving. Here are some commonly accepted rules of "netiquette," the etiquette of e-mail, especially for messages posted to a newsgroup:

- *Send only those messages you're willing to make public.* Companies own their internal e-mail systems and may monitor messages; messages are easily forwarded and may reach people you never intended to see them (Legeros 10–11).
- *Consider layout.* Avoid using only capital letters because they are hard to read and seem to shout. Use paragraph breaks, short lines, and lists.
- *Focus the message.* Label each message and center each on one point that fits the label.
- *Paraphrase, don't just reproduce, the message you are responding to.* Most systems allow an easy response mode in which your new message includes the entire text of the message you are responding to. A series of such messages can bury the new information in layers of dead text.
- *Provide context.* It's easy to hit a reply button or key and type in "yes," but the reader, removed in time from her own request, may have forgotten what a "yes" answer means. Give enough information to clarify your answer. Make sure, too, that you respond to all questions or items in the original message.
- *Avoid trivia.* The spontaneity of e-mail and perhaps its newness as a form encourage trivial postings. Think before you write.
- *Post messages to the appropriate discussion group.* Read the FAQ (frequently asked questions) document to know the range of topics. Stay away if you just want to complain about the group.
- *Sort messages.* Determine which messages should go to individuals and which are of interest to the whole group; for example, the group won't be interested in your subscribe/unsubscribe requests.
- *Note the source of any information you quote.* Cite proprietary or copywritten material as you would in print communication.
- *Differentiate your text from what you're quoting.* Use indenting, quotation marks, and the like, to mark text that's not your own.
- *Obtain permission* before you forward any message a writer may have intended only for you.
- *Don't expect instant responses.* The speed of writing your message won't necessarily be matched by the speed of its being read. You may need to phone if you're not sure of your reader's e-mail habits.
- *Notify e-mail correspondents* if you will be "offline" for a period of time so they won't expect an immediate reply.
- *Respect privacy.* Avert your eyes when you pass by someone's screen displaying e-mail.

# CASE
## REPORTING AN INCIDENT

Mark Bernstein was pleased when his work-study position came through. That meant not only money but also the opportunity to *work* in another culture during his semester in London. His duties were varied as he assisted the program's resident director, Dan Robinson, a British professor of geography. Most work was routine: photocopying, taking phone messages, filing, running errands to the bank and the post office. He rarely had to initiate anything. On one occasion, however, Mark felt he should say something about behavior that didn't seem quite right.

The occasion was a field trip to a small village and agricultural station about 2 hours from London. Robinson hired a coach [bus] to transport the 21 students. On the way there, the coach driver stopped at a commercial rest area on the motorway. Since it was lunchtime, students hopped out, bought something to eat, and reassembled at the coach about 15 minutes later. For the return trip, Robinson decided to stay behind at the site and put Mark in charge of overseeing the group and tipping the driver when they arrived in London. Students picked up snacks near the farm, ate them right away on the coach, and many began to sleep. When the coach approached the same commercial rest stop, the driver asked Mark if he wanted to stop. Because everyone had eaten and many were sleeping, Mark said no. The driver then tried to persuade Mark to stop by saying that he had been given a coupon by the area's owners; if he presented the coupon he would receive £10. This was the restaurant's way of "saying thanks" when drivers brought in patrons. But he had to stop to get the money.

Mark knew that this coach company was one Robinson always hired, so it must be ethical. Besides, the driver had been pleasant and maybe someone did need to use a toilet. But the practice sounded like a "kickback," Mark was eager to get back to London, and by American standards, the distance was short. So Mark again turned down the driver's request. But the driver was adamant. He pulled into the rest stop, parked, told Mark he would be right back, ran in to redeem his coupon, and returned, for a total elapsed time of perhaps 10 minutes. Mark was annoyed. When they returned to London, Mark wanted to withhold the tip, which should recognize excellent service, but he felt unauthorized to do so. Reluctantly, he gave it to the driver. For the next several days, however, Mark wondered if he should tell anyone about the driver's actions.

1. Should Mark tell Robinson?
2. If so, should he convey the information orally or in a memo? Why?
3. Should Mark tell anyone at his college's International Programs office in the United States in case this small incident might be a symptom of any larger problems?

## CHECKLIST: Memos and Electronic Mail

1. **Write memos within an organization to**
   - Encourage action
   - Serve as the organization's collective memory
   - Report briefly on events or investigations
2. **Structure your memo**
   - Create an effective subject line
   - Open directly with the main point and context
   - Keep to one page
   - Write one memo for each main point
   - Anticipate reader questions and expectations
   - Write in well labeled modules
   - Use insider's code and be brief, yet polite
3. **In a memo conveying the minutes of a meeting**
   - Review the disposition of each agenda item
   - Note conclusions from the discussion and decisions taken
   - List follow-up assignments and deadlines
   - Give the date of the next meeting, if appropriate
4. **Use e-mail to send messages rapidly and inexpensively**
   - Write within or outside your organization
   - Use attachments for long texts
   - Paraphrase, don't repeat, a message you are responding to
   - Keep messages short and in discrete units
   - Avoid sensitive or complex messages
   - Assume that your message is not private

## EXERCISES

1. Using guidelines you learned in this chapter, revise the following memo that summarizes the process of testing a crude oil sample and reports test results.

INTEROFFICE MEMORANDUM

Date: 20 March 2001
From: Sherry Ortiz
Dept: Analytical Technology
To: Pete Bronski
Subject: IR RESULTS OF SAMPLE FROM HOUSTON

THE FOLLOWING SAMPLES WERE EXTRACTED WITH CHLOROFORM: TRAY #34, TRAY #32, TRAY #31, AND THE CRUDE LINE SLUDGE. THE SAMPLES WERE DILUTED WITH CHLOROFORM

AND HEATED GENTLY FOR APPROXIMATELY 5 MINUTES, REMOVED FROM THE HEAT, AND ALLOWED TO SOAK IN THE CHLOROFORM FOR APPROXIMATELY 6 HOURS. THEY WERE THEN FILTERED AND STRIPPED UNDER NITROGEN PURGE. ANY EXTRACTED MATERIAL WAS THEN SCANNED BY IR.

THE EXTRACTED MATERIAL FROM SAMPLES #31, #32, AND #34 WAS YELLOW IN COLOR AND CLEAR. IR SCANS OF ALL 3 SAMPLES INDICATED THAT A PARAFFINIC OIL WAS PRESENT. QUANTITATIVELY, THE MATERIAL EXTRACTED FROM #32 AND #34 REPRESENTED APPROXIMATELY 10 PERCENT OF EACH OF THE SAMPLES. THIS MATERIAL WAS SCANNED IN A .107 MM SEALED CELL WHICH MADE IT EASIER TO FURTHER IDENTIFY THE MATERIAL. THE OIL IS A PARAFFINIC BASE OIL WITH A MODERATE AMOUNT OF AROMATICS PRESENT. THE SPECTRA OF THE EXTRACTED MATERIALS IS VERY SIMILAR TO BASE15. THERE ARE INDICATIONS THAT A DILUENT IS PRESENT, POSSIBLY A "KEROSENE TYPE" MATERIAL. ALSO THERE IS A SMALL AMOUNT OF CARBONYL, POSSIBLY A CARBOXYLIC ACID, PRESENT IN THE OIL.

THE EXTRACTED MATERIAL FROM THE CRUDE LINE SLUDGE WAS DARK IN COLOR, PROBABLY DUE TO CARBON FINES THAT I WAS UNABLE TO REMOVE. IR SPECTRUM INDICATED THAT THE EXTRACTED MATERIAL IS A PARAFFIN OIL AND POSSIBLY THE SAME TYPE OF OIL THAT WAS SEEN IN SAMPLES #32 AND #34 BUT THIS HAS NOT BEEN CONFIRMED. I WAS UNABLE TO ESTIMATE THE QUANTITY OF OIL EXTRACTED DUE TO THE DIFFICULTY OF REMOVING THE CARBON FINES.

IF YOU HAVE ANY QUESTIONS GIVE ME A CALL (X 8788).

**2.** What is the core message in the following memo? Revise the memo to make that clear.

As you may or may not know, the weekend crew has implemented a general greenhouse maintenance program. The goal of this program is to provide a clean, safe working environment within the greenhouses. This is done by performing periodic algicide applications to the floor followed by a scrubbing of the floor to remove any excess algae.

As far as standard procedure is concerned, after the greenhouses have been treated and cleaned, I will circulate a list of treatment dates which you may keep for your own records. This list will be sent out approximately once a month, and it will indicate what services, if any, have been performed.

If your greenhouse is in special need of attention prior to our regular treatment, please feel free to contact me via e-mail, account name "WEEKENDR," or in person.

> More than ever, I feel it is very important for the weekend crew to become more involved with the ever-changing structure of this company. This program is just one way in which to help reach that goal. If you have any comments or suggestions concerning this program, or if you have any ideas which could lead to future programs, please do not hesitate to discuss them with me.

**3.** The following description of a meeting is dramatic but unstructured. First, revise the description to create a memo that provides the minutes of the meeting. Second, use the description as the basis of a memo to be sent by Peter Jackson, director of safety, to remind employees about the dress code.

> The announcement for this meeting was made possible by Peter Jackson who sent a memo to all supervisors of the loading dock. The reasons for the meeting centered around the issues of the dress code and the safety rules for employees who work on the loading dock. I played a major role in this meeting since I was the first employee who started to wear shorts and non–steel-tipped boots to work. When everyone else saw me do this they all started to wear shorts to work. This form of dress is a major violation of the dress code. The leader of the meeting was Peter Jackson, director of safety. He started the meeting out by stating the problems of the dress code and the safety rules. All of the supervisors agreed that these problems needed to be resolved. I tried to defend the issue of wearing shorts to work in the hot weather but I was told to be quiet by Sam McWilliams. When I was told to be quiet I just laughed it off with a smile. Sam and I never got along when it came to making a decision. The form for the new dress code was outlined by Peter Jackson. Afterward he asked if anyone wanted to add to the discussion. The new dress code stated that all employees working on the loading dock are required to wear steel-tipped work boots and pants covering the entire leg area.
>
> The second part of the meeting concentrated on safety procedures. Not all of the employees know the safety procedures required by state law. When Jackson started to talk about this issue, all of the supervisors and I became very bored. Everyone knows that Jackson gets overly excited about safety issues so they tend not to listen. To correct this problem, the supervisors voted to make all employees take a written exam concerning safety procedures. After this vote was taken the meeting came to an end. The meeting lasted for 20 minutes.

**4.** For one semester or quarter, monitor an electronic discussion group on a topic related to your major field. Write a report summarizing the themes of the discussion, the most active discussants, and the style and approach of most comments. Note, for example, what acronyms surface in the messages and how they affect your interpretation of the information and your view of the sender. At your instructor's request, present your report orally to the class.

## FOR COLLABORATION

1. Form a team to review some issue, real or hypothetical, concerning use of computing services at your college, for example, access to games or to pornographic sites on the Internet or misuse of the system for commercial gain. Research the issue and any applicable formal or informal computing guidelines, and then formulate a policy to solve the problem. Assume that the head of computing services has asked you to write a memo for her or his signature that describes the policy. It will circulate via the e-mail system to all students. Write the memo, keeping in mind its appropriateness for reading at a screen and for an audience of students.
2. In a team of two or three, prepare a brief report on the effects of technology—voice mail or e-mail—on internal communication at your college. You might prepare a questionnaire (see Chapter 4) to circulate to fellow students. Or interview a sample of faculty or staff to learn their opinions and examples of how such technology has changed their communication patterns.

   If your college does not currently offer such services, then prepare a review of literature on the topic of e-mail or voice mail from sources available at your library. One issue of interest, for example, is a growing tolerance for the mistakes in spelling and grammar that seem endemic in e-mail messages. You might review discussions of the tone and informality of e-mail.
3. Arrange for an e-mail discussion with students in a technical class in another country. To start that discussion, contact, for example, the Intercultural E-Mail Classroom Connections (IECC) Program at St. Olaf College in Minnesota (www.iecc.org). Three mailing lists of interest are

   - IECC-PROJECTS for teachers to announce specific projects
   - IECC-SURVEYS for students and teachers to post surveys and questionnaires and request assistance with short-term projects
   - IECC-DISCUSSION for general conversation about cross-cultural e-mail exchanges

   To subscribe to the mailing lists, send an e-mail message containing the word *subscribe* (and leaving the subject line blank) to

   iecc-projects-request@stolaf.edu
   iecc-surveys-request@stolaf.edu
   iecc-discussion-request@stolaf.edu

# ARTICLES AND REVIEWS OF LITERATURE



**"IN WRITING, A SCIENTIST EXCHANGES INFORMATION FOR RECOGNITION."**

DIANA CRANE

As a technical professional, you are likely to write articles, on your own or as a collaborative endeavor, to gain visibility in an organization and to carve out your research territory. Such writing also fulfills an ethical obligation to discuss your research and scientific issues with broad audiences when the issues are in the public interest. This chapter provides guidelines for writing three common genres of articles: a report on original research, a review of literature, and a popular science piece.

## SCIENTIFIC AND TECHNICAL PUBLICATIONS

Scientific and technical publications differ widely in the topics they cover, the breadth and depth of their coverage, their circulation and frequency of publication, and their design and voice. Most of those differences derive from differences in the audience each aims to serve. A simple way to classify those audiences is to think of their level of technical understanding. Consider three categories of audiences and, thus, of publications: scientific and professional, semitechnical, and popular.

### Scientific and Professional Publications

The most technical publications are the scientific journals published by professional societies, government agencies, and commercial presses. The first such journal, *The Philosophical Transactions of the Royal Society of London,* was founded in 1665 and published monthly. Today, at least 40,000 scientific journals publish more than a million new articles per year. They vary in prestige and coverage. Among the most prestigious are *Nature* (published in the United Kingdom); *Science, The Scientist,* and *The New England Journal of Medicine* (published in the United States); and the Japanese *Journal of Ophthalmology* and *Kenchiku Zasshi* (*Journal of Architecture and Building Science*). Such professional societies as the American Chemical Society, the American Medical Association, and the Society of American Foresters sponsor journals. The journals mainly publish original research, although those addressed to practitioners, like engineers and architects, also include tutorials that show ideas in practice.

## Semitechnical Publications

Semitechnical publications include *Scientific American* and *Science Digest,* the bulletins of agricultural and engineering experiment stations and extension services, business and financial publications (such as *Latin Finance, Business Latin America,* and *The Economist*), and corporate in-house magazines. Although focused on technical or scientific material, they are less specialized and less theoretical than scientific publications and expect their readers to have less knowledge of the topics discussed. Established researchers report on their own work and staff writers create original articles based on interviews and careful reading of the scientific literature. Computer magazines are one form of semitechnical publication that has grown extensively in the last several years, tracking the growth in ownership of personal computers.

## Popular Publications

Editors of computer-oriented magazines can usually assume their readers share a basic understanding of digital technology (increasing numbers of people do), as the editors of *Psychology Today,* for example, assume a shared context of interest and background in the social sciences. Popular science writers appeal to an even wider circle of readers who share only a general interest in scientific issues. For example, museum publications like *Natural History* (published by the American Museum of Natural History) and *Smithsonian* (published by the Smithsonian Institution) appeal to an educated audience interested in areas outside their specialties—topics like energy conservation, art history, preservation of wildlife, new discoveries about prehistoric mammals, and health care. Science articles also appear in general interest publications like *The New Yorker, Time, Newsweek,* and the weekly "Science Times" section of the *New York Times.* In addition, biotechnology and other scientific companies publish magazines to tell their story to customers, stockholders, and the public at large (Figure 21.5).

## PREPARING AN ARTICLE

Writing for publication is a complex process, especially because your article will compete with many others for highly limited publication space. To compete well, determine how what you have to say could strike a responsive chord in a reader. What in your work is newsworthy, or significant, or timely? Your article may derive from a report you wrote on the topic, although reports that have only local implications are seldom suitable for broader distribution. You'll need to rework a report to create an article suitable for publication.

If your work was sponsored by a client or organization, gain permission from the sponsor to discuss that work publicly. You may also need to submit a draft of your article to them for approval. The answer might not always be "yes." As you read in Chapter 3, companies are often understandably reluctant to reveal proprietary information. In addition, if you include visuals that were originally published somewhere else, or if you quote extensively from another text, you will need to obtain permission from the person or organization holding the copyright for that material.

### Select a Publication

For more about information searches, see Chapter 5

Determine which publications reach the readers you think would be interested in your topic and would take an approach to the topic that parallels your own. Analyze the journals or magazines you already know, and then search the literature to find others that have published information on your topic. *Writer's Market,* an annual publication, is a good starting point.

Review several issues of the publications that seem most promising and check their *Guidelines for Authors*. Those guidelines are the most direct specifications for your work, and you need to follow them exactly. For technical, scientific, and professional publications, such guidelines usually appear in each issue, or in one issue a year (Figures 21.1 and 21.2). Journal Web sites also usually include authors' guidelines (Figure 21.3). You may have to write a letter or send an e-mail request to obtain guidelines for semitechnical and popular publications. To supplement this explicit advice, examine tables of contents and skim sample articles in several issues to determine:

Range of subjects included
Editorial viewpoint
Categories of contributions (sometimes called departments)
Special features
Design and structure of each article type
Use of visuals
Level of language

It's sometimes helpful to send an e-mail message or a letter (called a "query letter") to the editor of the most promising publication. Describe the article you intend to submit to see if the editor is interested. An editor's

The *IEEE Transactions on Professional Communication* is a refereed publication with quarterly distribution of more than 4000 copies per issue. *Transactions* is indexed and abstracted worldwide. Its main goal is to help those involved in technical communication, peripherally or full-time, to do their job effectively.

**Audience**

- Practicing engineers and scientists
- Technical project and program managers
- Writers and editors
- Teachers of professional and technical writing

**Scope**

Papers need not fit special categories; they can range from theoretical treatises to pragmatic discussions, tutorials, or how-to treatments of special problems. Some broad areas of interest:

- Oral, graphic, and written communication
- Communication issues in software engineering
- Communication technology
- Communication education and training
- Book and software reviews

**Selection Criteria**

Submitted papers are reviewed by at least two referees; their comments are reported to the author. Papers are judged on the basis of the following:

- Subject interest
- Usefulness, relevance
- Good scholarship

FIGURE 21.1 **Excerpt from the "Information for Authors" Page, *IEEE Transactions on Professional Communication***


expression of interest won't mean that your article is assured publication. But an editor can at least tell you early on that, for example, your topic is not suitable for their readers. You can then look for a publisher elsewhere.

## Tailor Your Material

When you've selected a target publication, tailor your material to its design. The best method is imitation. Read one or two articles several times to thoroughly grasp their structure and voice. Then, without, of course, copying what they say, apply their approach to your message. To imitate the design, make sure you include any required supplements, for example, an abstract,

## INTERNATIONAL JOURNAL OF ENGINEERING SCIENCE

**Incorporating:** *Letters in Applied and Engineering Sciences*

*Notes for Contributors*

**1.** The Editors will be pleased to receive contributions from all parts of the world and manuscripts for publication should be sent in triplicate to the appropriate members of the Board of Editors or the Editorial Advisory Board.

**2.** Only papers not previously published will be accepted and authors must agree not to publish elsewhere a paper submitted to and accepted by the journal. In the case of rejected papers it is the policy of the journal not to return the manuscript to the authors.

**3.** The languages of the journal are English (preferred), German and French.

**4.** *Manuscripts should be typed with double spacing and submitted in triplicate with ample margins.* Pages should be numbered consecutively.

**5.** Manuscripts should begin with the title of the article, the author's name and address from which the communication comes. In the case of co-authors, respective addresses should be clearly indicated, as well as which author is to receive correspondence and proofs for correction. Where possible, the **Fax number of the corresponding author** should be supplied with the manuscript, for use by the publisher. It would be appreciated if authors would notify the publisher of any change of address which occurs whilst their paper is in the process of publication.

**6.** The essential contents of each paper should be briefly recapitulated in an abstract.

**7.** Half-tone illustrations are to be restricted to the *minimum* necessary. They should accompany the script and should not be included on manuscript pages. Line drawings should be originals and include all relevant details; no photo-prints should be sent. Photographs should be enlarged sufficiently to permit clear reproduction in half-tone. If words or numbers are to appear on a photograph two prints should be sent, the lettering being clearly indicated on one print only. Drawings, etc. should be about twice the final size required and lettering must be clear and "open," and sufficiently large to permit the necessary reduction of size for publication.

**8.** In the interest of economy and in order to avoid the introduction of errors, tables will be reproduced by photo-offset means directly from the authors' typed manuscripts. The guidelines detailed below should be followed:

(a) Page size should be $8.5 \times 11$ in. (or $21 \times 30$ cm).

(b) Large or long tables should be typed on continuing sheets.

(c) Original typed tables are required.

**9.** Figure legends, table headings and table footnotes should be typed on a separate sheet and placed at the end of the manuscript.

**10.** References to published literature should be quoted in the text in square brackets. References should be listed together at the end of the paper in numerical order. Double-spacing must be used throughout. Journal references should be arranged thus:

[1] S. A. TOBIAS, *Proc. Instn Mech. Engrs* 173, 474 (1959).

[2] A. A. ANDRONOV and C. E. CHAIKIN, *Theory of Oscillation,* Chap. 5, Princeton University Press (1959).

**11.** All mathematical symbols may be either handwritten or typewritten but no ambiguities should arise. Greek letters and unusual symbols should be identified in the margin. Distinction should be made between capital and lower case letters; between letter O and zero; between the letter l, the number one and prime; between k and kappa. A vector will be printed in bold face and to indicate this the letter should be underscored with a single wavy line. The numbers identifying mathematical expressions should be placed in parentheses.

**12.** Nomenclature should conform to that most frequently used in the engineering field concerned.

**13.** Page proofs will be sent to the author (or the corresponding author in a paper of multiple authorship) for checking. **Corrections to the proofs must be restricted to printer's errors.** Any substantial alterations other than these may be charged to the author. Authors are particularly requested to return their corrected proofs as quickly as possible in order to facilitate rapid publication. Please note that **authors are urged to check their proofs carefully before return, since late corrections cannot be guaranteed for inclusion in the printed journal.** Reprints and copies of the issue (at a specially reduced rate) can be ordered on the form which will accompany the proofs. These should be returned to: Pergamon Press Ltd, Pergamon House, Bampfylde Street, Exeter EX1 2AH, England. The corresponding author of each paper will receive 25 reprints free of charge.

**14.** The original manuscript and diagrams will be discarded 1 month after publication unless the publisher is requested to return original material to the author.

**FIGURE 21.2** ***International Journal of Engineering Science.* Notes for Contributors**

(Reprinted from the *International Journal of Engineering Science* © 1996, with permission from Elsevier Science.)

FIGURE 21.3 At its Web site, visitors can search current and past issues of the *Journal of Biological Chemistry* as well as other journals in the field and find out related information about the American Society for Biochemistry and Molecular Biology, which publishes the journal. Potential authors also must submit their manuscripts through the site. (© 2000 by Journal of Biological Chemistry.)

keywords, photograph or brief biography of the author (you!), and notes acknowledging sources of funding. Adhere to the conventions for illustrations, that is, the appropriate kind, number, and labels. If the publication's articles follow a standard structure, like the IMRAD approach for reporting empirical research, make sure your article follows that structure. If articles incorporate headings, organize your text to include them. If you cite secondary sources, make sure those citations are in the journal's standard form. Imitating the voice of a published article is a bit harder, but the

**For more about the IMRAD approach, see Chapter 15**

A CLOSER LOOK

## An Objective Persona

In a technical or scientific article in the United States, the traditional persona is that of an objective, invisible observer. The following guidelines provide the generally accepted approach to creating such a persona:

- Be open with information. Detail your method so that others can verify your results by reproducing your work or show why your results are false.
- Use technical terminology to achieve precision and accuracy.
- Write in a style that focuses the reader's attention on the object under study and does not draw attention to itself or to you.
- Avoid emotion, humor, or irony.
- Avoid personal pronouns ("I" and "we") and use the passive voice to keep the agent at a remove ("The instrument was calibrated.").

Objectivity as you gather information is an essential element of the scientific method, but current research suggests an objective persona that conforms to these guidelines can sometimes be misleading. To share knowledge honestly and accurately, make explicit your personal responsibility for what you say. In addition, recognize that writing about science and technology is often a form of argument. Writers and readers establish the truth through discussion. Within the bounds of accurate observation and honest reporting, scientists are less interested in facts in themselves than in facts as illustrations that support some interpretation. Our understanding of natural phenomena thus advances through the collaboration of people who see things differently, take responsibility for what they see, and show in their writings their personal points of view. Truth is derived and error uncovered as professionals convince each other in their discussions and publications. As one authority notes, "The physics of undergraduate textbooks is 90 percent true; the contents of the primary

examples in this chapter should help you recognize how articles in different journals *sound* different. *A Closer Look: An Objective Persona* also discusses the persona and voice of a scientific article.

Before submitting your article, make sure you have complied with any instructions for submission (Figures 21.1–21.3). In particular, note the number of copies or type of file or e-mail attachment necessary, rules for the preparation of illustrations, the form of submission (online? on disk? on paper?), and the form of the title page. Journals that send manuscripts to expert reviewers for evaluation (called *refereed journals*) usually ask authors to provide title pages that do not include the author's name. Reviewers do not know the identity of the people whose articles they are evaluating.

research journals of physics is 90 percent false" (Ziman 40). The journals provide a forum for debate before ideas are enshrined and taught to students.

Recognizing the argumentative function of reports and articles, many current authors deliberately create a persona that is not "invisible" but suggests trustworthiness by displaying some of the writer's personality and engaging the reader's attention. This new voice in American publications strikes a Russian translator as significant. She says that American scientific texts "give a more vivid picture of the problem under consideration, the result obtained or the theories developed" than articles written by their Russian counterparts, which are "academic," and "frozen" in a narrow, unemotional style. "The number of nonneutral, expressive, emotive elements in American texts is large . . . which testifies to the individuality of both the author and of the style of scientific sublanguage in general" (Serebryakova).

Recent American articles use at least mild exaggeration in terms like *crucial* or *exciting* and may introduce figures of speech and casual expressions. They also avoid impersonal constructions like "It has been decided that" or "It has been shown that" or "It has been observed that." Such expressions dilute responsibility because they avoid saying who decided, or showed, or observed. In addition, they make the writer sound stuffy and imply that the reader should simply yield to whatever unnamed authority the writer brings up.

But too much personality in a document may make a reader suspicious. You have to choose how much of yourself to show depending on your reader's expectations and the type of document you are writing. Too much informality in a scientific article may make readers from another culture suspicious of the author's credibility and authority. For such audiences, you may have to speak not only in another language, but also in another voice, keeping your personality under wraps and inviting less interaction.

## SCIENTIFIC ARTICLE

**For more about empirical research, see Chapter 4**

If you plan a career in research and development, you'll need to know how to write a scientific article, the most prestigious form of professional publication, sometimes called a scientific paper. It is a thorough, detailed account of an empirical research project, emphasizing new and significant findings that make a substantial contribution to the discipline represented by the publication. It shows how the work being described supports, refutes, or extends established theory or confirms a new experimental approach. Many scientific articles are collaborative endeavors, and most are written by the researchers themselves. Professional writers, either on the staff of the publication or independently, often write articles in semitechnical or popular magazines.

## Structure

The form of the scientific article differs somewhat from journal to journal, but most follow the IMRAD approach also common in reports. A report, however, often addresses managers, clients, or sponsors, while the article highlights what's news to colleagues. Thus the emphasis in each segment may differ. For example, a report may center on the description of a procedure, but because such information may be simply common knowledge to an audience of colleagues, the article's focus should be on how the results contribute to the profession's understanding of key phenomena. If the report you're working from presents several case studies, you may combine them to make one example in the article or highlight just one significant or interesting case.

Figure 21.4 shows the elements of a scientific article schematically. The following introduction to a scientific article shows some of the introductory moves noted in the figure. The paragraph develops by comparison.

*Stasis*—the research territory

*Disruption*—the gap in information

*Resolution*—the authors' research niche

> The creep behavior of trabecular bone may play an important role in the loosening of implanted prostheses or in the etiology of vertebral body fractures. However, the only reported study which addressed the creep of trabecular bone [1] failed to characterize creep behavior suitably for biomechanical analysis. By contrast, creep of cortical bone is well documented and is characterized by the shape of the creep strain vs. time curve [1, 2, 3], the existence of an Arrhenius relationship between applied stress, temperature, and creep strain rate [3], and a strong relationship between applied stress and time-to-failure [2]. Our overall goal with this study was to determine whether similarities exist between trabecular and cortical bone creep behaviors. Specifically, our objectives were to: a) characterize the creep strain vs. time curve for trabecular bone; b) determine if the creep strain rate at constant temperature may be described by a power law relationship; and c) investigate if the time-to-failure may be described by a power law relationship. (Bowman et al. 173. Used with permission.)

The following paragraph from the "Results and Discussion" segment shows how the argument continues. It includes a reference to a figure (not reproduced here).

**Comparison of expected and experimental results**

**Comparison of present work with published findings**

**Interpretation of a significant result**

> These data demonstrate that the creep behaviors of trabecular and cortical bone are qualitatively similar. Both the shape of the creep vs. time curve and the experimental power law relationships are similar to those obtained from earlier work on cortical and trabecular bone creep behavior [1, 2, 3]. The strong relationship between normalized stress and secondary creep strain rate suggests that an Arrhenius relationship, which is also a function of temperature, may exist for trabecular bone. Therefore, it is possible that the creep behavior of trabecular bone, like cortical bone [3], is temperature dependent. Comparison of the creep time-to-failure curves for bovine trabecular and human cortical bone [2] (Figure 2) indicates that, under similar normalized compressive stresses . . . cortical bone will fail in creep before trabecular bone. (Bowman et al.)

1. FRONT MATTER
   - Title
   - Byline
   - Abstract
2. INTRODUCTION
   - *Stasis*—define research territory
   - *Disruption*—interrupt stasis so as to create a niche within territory
   - *Resolution*—occupy or defend that niche
3. METHODS
   - Procedures used to occupy or defend niche
   - Materials used in carrying out procedures
   - Theoretical principles and assumptions behind procedures
4. RESULTS AND DISCUSSION (Occupying the Niche)
   - Experimental or calculated results in text, tables, figures
   - Comparison of results (present vs. published earlier, baseline vs. altered state, experimental vs. control group, theoretical calculations vs. experimental measurements)
   - Reference to previous research for purposes of criticism or support
   - Interpretation of significance of results and comparisons
   - Explanations for surprising or contradictory results
5. CONCLUSION
   - Main claims derived from having occupied the niche
   - Wider significance of those claims to research territory
   - Suggestions on future work to validate or expand upon claims
6. BACK MATTER
   - List of literature cited
   - Acknowledgment of assistance provided during writing or research

FIGURE 21.4 **Elements of the Scientific Article**
(Source: Joseph E. Harmon and Alan G. Gross, "The Scientific Style Manual: A Reliable Guide to Practice?," reprinted with permission from *Technical Communication,* the Journal of the Society for Technical Communication.)

The last paragraph of the article, in another typical approach, looks to practical applications of the research; it begins, "This demonstrated creep behavior of trabecular bone may have significant clinical implications."

## Voice

As high-context documents, circulated to people in the know, scientific articles tend to rely on extensive shared understanding. In the paragraphs on

ELECTRONIC EDGE

## Electronic Publishing

Increasing numbers of publishers are delivering their articles to readers online rather than in print—or in both forms. Electronic publication has many advantages, including its ability to deliver up-to-the-minute news and its interactivity. Late-breaking scientific developments reach readers faster than in the more time-consuming form of print distribution. Moreover, articles that are not read-only documents become the occasion for a dialogue between authors and readers. In addition, if articles are not just pages-on-a-screen but are structured for electronic delivery, then they can embed hypertext links to related articles or databases, and they lend themselves to immediate and easy searching by keywords, like topics, authors' names, dates, product names, and process names.

Electronic production also speeds the editing process and thus reinforces the dynamic nature of research as it saves on printing and mailing costs. Printing costs in particular can be high, given the need to reproduce illustrations and other visual material at high resolution. Electronic text is often more accessible, affordable, and global than print versions. Two other advantages deserve further discussion. First, electronic technology is transforming the publishing business from one concentrated in a few major publishers to an open field in which a small organization or even an individual can disseminate documents worldwide, rapidly and inexpensively. The Internet, and especially the Web, have shifted the playing field for publication. It's easy now to create a journal for a "niche market" of only a few specialist readers.

On the other hand, several researchers question this approach, or at least point out the need for new policies to govern electronic publication. For example, while the process of review by experts helps verify the authority and accuracy of print articles, many electronic publishers skip such reviews in their search for speedier publication, and thus errors and inaccuracies may proliferate. When articles become sites for interactivity between authors and readers, then what becomes the *definitive* article? Should the publisher label different versions? How does an author or publisher secure the right to an electronic article, especially one that embeds links to other articles? These are only a few of the issues that follow from a broadening of the publishing business.

Second, electronic technology is vastly enhancing an author's ability to help readers visualize scientific and technical information through multimedia presentation. Film clips, video, high-resolution and computer-enhanced photography, sound—these media are easily incorporated in CD-ROM or Web delivery. Where graphic elements have traditionally been supplemental to an article's text, they can now become the core, often to the great benefit of readers.

A browse through a catalog of CD-ROMs or search engines on the Web reveals a growing number of publications. Many are enhanced electronic versions of print journals, like *Wired* magazine's *Hot Wired* and other online information at URLs that begin *www.wired.com* or *Nature* at *www.nature.com.* As more authors and publishers incorporate the unique benefits of electronic delivery, they are changing the elements of an article and perhaps creating a new genre, an *e-article.* They are moving away from thinking of screens as pages. At the same time, it's interesting to note, many print magazines seem to be incorporating the windows and other graphic features of screen design into their pages.

creep behavior of bone, for example, the author uses technical terms freely (like "Arrhenius relationship"), assuming the reader is familiar with them. Researchers tend to qualify their conclusions (*strong relationship . . . suggests . . . may exist . . . it is possible that*). Qualifications are common because authors are naturally cautious about bald statements of cause and effect in their research. To bolster their authority and show their familiarity with the field, authors also cite other researchers abundantly (as reference numbers in brackets in the bone example). In addition, expression tends to be terse. For example, a guideline from the American Chemical Society (ACS) notes that "journal space is a precious resource created at considerable cost, and thus improper use of that space limits the disseminating of knowledge" (Dodd 219). That search for brevity has meant that some publications circulate only abstracts, often online, to all readers; from those, researchers determine which items to request in full. Overdone, however, such compression can create tough slogging for readers. An engaging voice in articles, as in all writing, attracts readers' attention and creates a more lasting impression of your work—and of you.

## REVIEW OF LITERATURE

A second important genre of publication is a *review of literature,* sometimes called a "critical review" or "review article." One expert notes that the goal of a review is "to give order to the past, so as to establish a shared present that will be the basis for coordinated work in the future" (Bazerman and Paradis 5). A report, proposal, or article on empirical research may also open with a brief review that

- Shows how your project fits with other projects and doesn't duplicate them.
- Highlights what's new in your approach.
- Helps confirm the authority and validity of your work.

Writing a review is a good way for you to master the literature in your field, and thus an instructor may ask you to prepare an entire document that reviews information from a variety of print and online sources and provides a coherent explanation of one issue, technical approach, event, or topic. You'll find examples of such reviews in a variety of publications, but especially in journals devoted exclusively to them, such as *Chemical Reviews* or *Nutrition Reviews.* The *ACS Style Guide* defines such a review as follows:

> Reviews integrate, correlate, and evaluate results from published literature on a particular subject. They seldom report new experimental findings. Effective review articles have a well-defined theme, are usually critical, and should present novel theoretical interpretations. Ordinarily they do not give experimental details, but in special cases (as when a technique is of central interest) experimental procedures may be included. An important function of reviews is to serve as a guide to the original literature; for this reason accuracy and completeness of references cited are essential. (Dodd 9)

In a review article, as the definitions you just read suggest, you provide a complete and authoritative account that describes which researchers are working on the topic and with what success. Because engineers, for example, are legally responsible for knowing all relevant information in a field when they select a material or complete a design, such reviews provide a valuable service. Reviews are also useful in drawing together information from various disciplines—economics, law, engineering, science, psychology—to focus on problems that necessitate broad understanding, like the design of earthquake-resistant highway overpasses or the implementation of recycling programs. But completeness in a review is not clutter. You must be selective about what you display as evidence. Irrelevant information—no matter how hard it was to come by, how interesting it may seem, or how clever a researcher it proves you to be—must be discarded.

### Structure

The writing of review articles presents special problems in fitting together materials from disparate sources so they form a coherent presentation and not a collage of clippings. As with any piece of writing, a point ("a well defined theme") and a plan are essential. Two plans are common:

1. Show how authorities either agree or disagree on key issues and findings.
2. Proceed publication by publication, in chronological or logical order.

The following paragraph shows how authors agree about methods to remediate cadmium-contaminated soils:

> Researchers are documenting serious harmful effects when such heavy metals as cadmium leach into ground water and are taken up by crops (Adriano, 1986; Pierzynski and Schwab, 1993; Sherlock, 1991). As a toxic substance, cadmium can cause bone disease and renal dysfunction. In Japan, for example, rice farmers who used irrigation water contaminated with cadmium from mining operations showed severe symptoms of cadmium poisoning (Logan, 1992). Lagerhoff and Bower (1994) also report that persons living downwind of a smelter near Galena, Kansas, may be consuming at least 50 percent more cadmium than other individuals by eating home-produced vegetables and meats.

The author of the next paragraph, June Fullmer, frames the discussion in terms of the disagreement of authorities concerning the role of craftsmen in technological advances:

> Melvin and Robinson (19) argue against the notion that most technological advances were the result of the work of anonymous "little men," craftsmen and technicians often innocent of basic scientific development and training. They think that to insist upon the importance of such craftsmen "serves to reinforce the belief in the autonomy and purity of science as a concept-generating activity. . . . Ruling out any significant interaction between technologists and scientists until the late nineteenth century [merely] protects a particular historiographic viewpoint," a viewpoint that has tied to it "a

particular kind of scientific community" (p. 78). Their opinion is in direct contrast to that of Thomas Kuhn, who sees the polarization between technology and science as springing from subterranean roots, "for almost no historical society has managed successfully to nurture both at the same time" (p. 50).

The author of an article about the use of illustrations in a scientific monograph takes a chronological approach:

> If a child were to look into *Nature* and, say, *Sociology* or the *Journal of Linguistics,* the first thing that might strike him or her as important would be that the scientific journal had pictures, while the others just had print. Those of us who study scientific texts have, until recently, ignored those pictures. But since Martin Rudwick commented on this lack of attention in 1976, a number of studies of scientific discourse have discussed the use of illustrations in scientists' communications with scientists (Latour, 1985; Shapin, 1984; Lynch, 1985c; Bastide, 1985). Illustrations are also important in communications between scientists and readers outside their specialties (Jacobi, 1985, 1986; Gilbert and Mulkay, 1984; Pickering, 1988). Indeed, the iconography of a science is more likely to have an impact on the public than the words or mathematics, which may be incomprehensible to them. If we ask, for instance, what most people would recognize from Watson and Crick's 1953 *Nature* article, it would not be the exact phrasing of the claim, it would be the picture of the double helix, with the phosphate chains like flat ribbons, the base pairs as rods between them. (Myers 231)

## Voice

These model paragraphs demonstrate successful strategies for achieving a unified voice in a review. Three matters of expression that require attention in any writing project are particularly significant in reviews: integrating quotations, selecting the appropriate verb tense, and citing references.

**Integrating Quotations.** Be sparing in quoting your sources. Use direct quotation to

- Call attention to a particularly significant interpretation.
- Summarize with special finality a line of reasoning.
- Present the accurate text of a law or regulation.
- Open the discussion on a point of familiarity with the audience when the source is well known.
- Capture a particularly vivid expression of an idea.

The following paragraph, about the pace and direction of negotiations in China, integrates quotations effectively. The author follows the guidelines of the American Psychological Association in citing sources, the standard for the journal that published the article.

> Silence is permissible, comfortable, and respected. Talking too much is not respected (Huang, Andrulis, & Chen, 1994). U.S. negotiating meetings are planned, whereas Chinese negotiating meetings unfold. Salacuse (1988) observed, "Rarely does one hear an American complain that negotiations are going too quickly" (p. 12). The discussion takes circuitous paths. Chinese

> negotiators return repeatedly to items that Americans have not yielded (Bucknall, 1994, p. 127). O'Hara-Devereaux and Johansen (1994) say that the links "needed to complete a task were like the vertebrae of a large, flexible dragon rather than a rigid-backed steer" (p. 211) and that "the quickest route from A to B may not be a straight line if there is a cliff in between" (p. 215):
>
>> To monochronic people the information flowcharts of polychronic cultures often looked like doodles gone mad. To get from point A to point B, information often travels throughout the entire alphabet. The high-context Chinese had a people-intensive work process and were polychronic. . . . Following the power/information flow was more effective than going straight to the point—because . . . tactics to shorten the flow upset the cultural logic and often landed him at zero (O'Hara-Devereaux & Johansen, 1994, p. 211).
>
> Clearly, time is a major investment. The foreign businessperson's focus should encompass not just one or a few negotiation meetings, but a series, possibly involving multiple trips to China. Indeed, Chronin (1991) writes, "Five years is not an unrealistic waiting time for a sale to China. . . . Time-wise, China is a life commitment" (p. 162). (Gilsdorf 28)

Overused, however, quotations impede the flow of the text. They indicate an author who, out of laziness or modesty, failed to tailor sources to the new occasion of the review. As an author, you need to rework others' material into a new context, for a new reader, so avoid long quotations. If you feel compelled to quote extensively, set the material off by indention from the left margin, introduce it with a line of text, and double-space above and below. No quotation marks are necessary, because the spacing and indention indicate quotation.

**Selecting the Appropriate Tense.** Adhere to the conventions of the review when you select verb tenses. Those conventions dictate that you use the present tense to discuss authors whose ideas you consider valid or current:

> Smith (1999) argues that wetland restrictions are ineffective. His complaints are also validated by Rogers (2000).

Use the past tense to argue for the lack of currency in an interpretation that has been corrected. Use the past, too, to indicate the real time in which a case study was conducted or a test performed:

> Smith's criticism is based on a 1999 study in which he surveyed 25 federally designated wetlands in the Delaware Valley.

**Citing Sources.** Third, as the definition from the ACS urges, provide a comprehensive "guide to the original literature" through "accuracy and completeness of references." Provide enough citations, that is, all the major works, and complete citations, that is, adequate bibliographic information, so the reader can find the original. In addition, include that information as noninvasively as possible within the text so the reader isn't thwarted in getting the gist of the review because of an intrusive referencing system.

## POPULAR SCIENCE ARTICLE

Popular science articles address a broad readership. They appear in newspapers (like your college newspaper), general interest magazines, publications for college or university alumni, and, increasingly, on the Web. Developers of Web sites may reproduce articles from print sources, or they may create their own. Health care–oriented sites, for example, provide abundant examples of articles for general readers about diet, disease, medical interventions, drugs, and the like.

Scientific articles have something of a captive audience, but readers browsing popular magazines or Web sites are likely to ask, "What's in it for me?" Aim to inform, convince, or advise, with large doses of entertainment. Consider what's new and significant in your work, as the guidelines from *Wired*, a computer-oriented magazine, suggest (Figure 21.3). Don't be surprised if a reader's interest centers on some small detail on the periphery of your major research effort.

Use the following guidelines to select your material:

- Wrap a scientific development in a story rather than simply reciting the facts.
- Perhaps appeal to the reader's fundamental interests: money, security, safety, health, and recreation.
- Stress the startling dimensions or conflict or human interest in your work.
- Consider using a question-and-answer format. The Q&A format is particularly effective when you offer advice—about homes, cars, plants, diet, health, exercise.
- Be concrete and specific rather than abstract and theoretical.
- Relate the unknown to what the reader is already familiar with by extended comparisons and analogies. Define keyterms.
- Introduce yourself and your work; that is, weave a narrative around yourself.
- Use visuals in each of these approaches—and as an approach, particularly in multimedia presentations.

### Structure

As these guidelines suggest, you aim to motivate the reader to enter your discussion. The beginning is critical. It's the wedge into the reader's attention. Begin with one fact, one visual, one short scene, and then broaden into a discussion of this item's larger significance. To shift the image, hook the reader on the detail and then develop that line. For example:

> At 7 P.M. on Thursday, Jamie Sanchez answered the last e-mail message, shut down his computer, checked his voice-mail, made notes on his tasks for Friday, and headed to the gym. The gym, he often told his friends, was both his

addiction and his salvation. He changed clothes, mounted the one unoccupied exercycle, and prepared to be soothed.

That opening scene led, in one writer's telling, to a lengthy discussion of the emergency response network in a large city—a network whose delays contributed to Sanchez's death after he suffered a heart attack while exercising. The article also discussed the need for special resuscitation equipment at places like gyms.

As you develop the units of your article, pay particular attention to the title, the lead (the first sentence or two), and the ending.

**Title.** Bait the reader with the title:

> ***Wordplay:*** Students Get That Sinking Feeling at a Personal Submarine School. (Massey B1)
>
> ***Question:*** How Much Is It Worth to You Not to Go to Houston? (Title of an article about how airlines set priorities for bumping on overbooked flights)

**Lead.** You'll further strengthen the appeal of the title in your lead, or first sentence or two, which builds momentum both as you write and later as the reader reads. The lead may strike a familiar chord, set up a good story, describe a myth or error that the article will correct, amuse, startle, or intrigue. Note the appeal of the following effective leads:

> ***Startling Statement:*** As you read this, about 100 million sleep-deprived Americans are driving cars and trucks, operating hazardous machinery, administering medical care, monitoring nuclear power plants, and even piloting commercial jets. (Brody C14)
>
> ***Metaphor:*** Nature herself is at the window—rattling the panes, roaring in the treetops, shouldering inside to rustle the scholarly papers on Donald Worster's desk. The gale seems bent on rearranging the very thoughts the papers contain.
>
> Another, metaphorical gale is now roaring through the fields of scholarly thought about nature . . . (Farney 1)
>
> ***Twist on a Familiar Chord:*** The purpose of technology is usually to make life easier. In the case of car security, the opposite is true: the purpose is to make life more difficult—for car thieves. (Cohen C2)
>
> ***Double Meaning:*** Jack Hidary's business is going under.
>
> Mr. Hidary's job is to persuade people to pay $200 for about 30 minutes of deep-sea diving. (Massey B1)
>
> ***Intrigue:*** In May 1953, Raymond Greene and Katharina Dalton gave the world a new disease. ("That Time of the Month" 75; about the identification of premenstrual syndrome)
>
> ***Anecdote/Humor:*** The oft-told Frank Lloyd Wright story is of the indignant client who phoned to report that the roof of his new Wright house was leaking onto his desk.
>
> "Well, why don't you move the desk?" replied the master irritably. (Patton 32)

# Novo Nordisk comes clean

*Novo Nordisk's first environmental report links up with an international trend for companies to communicate more openly about environmental matters.*

Let's look at Novo Nordisk's first environmental report in the context of some of the international trends in this area. From Alcan to Zexel, an international A-to-Z of companies is now producing new types of environmental performance reports. Some are doing so in response to new regulations, others in response to emerging business requirements or changing public expectations. Behind the scenes, the accountancy profession increasingly recognises that industry's environmental liabilities are likely to mushroom – and is beginning to think about how companies should reflect such liabilities in their annual reports and accounts.

Norway's Norsk Hydro made a huge impact when it published its first free-standing environmental report in 1990. Whether or not it meant to do so, Norsk Hydro broke a log-jam in the area of corporate environmental disclosure. Soon, a growing number of companies were producing their own reports, often competing on the basis of who felt able to disclose most. While European companies began to report on a voluntary basis, most US companies began to report on the basis that the data were already available to the public because of the Toxic Release Inventory (TRI) scheme.

SustainAbility, which began to work with Novo Nordisk in 1989, concluded that corporate environmental reporting would emerge rapidly – with major repercussions for industry. As a result, we have carried out several international surveys of environmental reporting. The first, undertaken alongside Deloitte Touche Tohmatsu and Canada's International Institute for Sustainable Development, resulted in a report – Coming Clean – published in 1993 and offering 23 recommendations for would-be-reports.

***How do companies rank?***

Coming Clean encouraged growing numbers of companies to produce their own reports. So our second project, funded by the United Nations Environment Programme (UNEP), involved an even more ambitious survey. The final report, Company Environmental Reporting, was published in April and ranks the reports produced by

*The author of this article John Elkington, a director of SustainAbility Ltd and a member of the EU Consultative Forum on the Environment.*

*Figure 1: Five stages in corporate environmental reporting.*

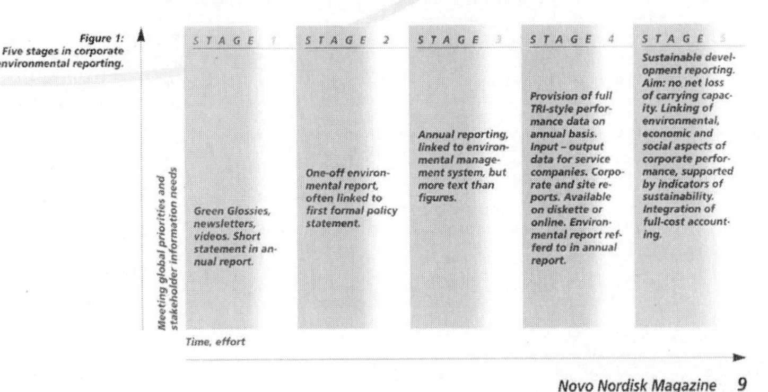


Novo Nordisk Magazine 9

**FIGURE 21.5** **"Novo Nordisk Comes Clean,"** ***Novo Nordisk Magazine***
(Courtesy of Novo Nordisk, *Novo Nordisk Magazine,* May 1994.)

The Novo Nordisk Pro

100 pioneering companies worldwide. We benchmarked the various reports against the five stages of reporting shown in Figure 1.

Companies already have to report their contingent liabilities – including environmental liabilities – in their annual reports and accounts. Ensuring that environmental liabilities are fully disclosed and appropriate provisions made represents a major challenge for companies. Business leaders in Europe, North America and Japan express concern that current pressures for corporate environmental disclosure will intensify, broaden in scope and affect industries well beyond the environmental firing line.

The voluntary environmental reporting and accounting initiatives undertaken by many of the companies we surveyed are profoundly changing society's expectations in terms of corporate disclosure. Inevitably, however, the pioneers remain divided on where these new paths will lead: some see free-standing environmental reports as a transitional stage to fuller disclosure in company annual reports, while others forecast that free-standing reports will find growing favour.

But whatever style of reporting they have adopted, growing numbers of companies think that financial analysts worldwide will soon scrutinise the environmental reports produced by leading companies at least as carefully as they currently study corporate financial statements. In doing so, and in acting on the basis of the information they glean from such reports, analysts could help wipe billions of dollars from the value of major industries – and boost the perceived value of others.

Last year, Fortune magazine published its first ratings of major US corporations, sending shock-waves through the boardrooms of those listed as "laggards". Interestingly, companies that had produced a report were viewed with particular favour, although reporting was no guarantee of a positive ranking. The "10 Laggards" included several companies that have been prominent reporters.

***Target audiences***

For anyone expecting to find that report makers were mainly targeting the noisiest stakeholders, our results may come as a surprise. The key target audience for report makers in North America (96%) and Europe (88%) turned out to be their own employees. Over three-quarters of the Japanese companies also mentioned employees. Partly as a result, many report makers laid a great deal of emphasis (but with varying degree of success) on the user-friendliness of their reports.

Local communities got a higher rating in North America and Europe than in Japan while shareholders were seen as a particularly important target audience in North America. With the growth in ethical and green investment funds in the UK and some other EU member states, it is not surprising to see shareholder pressures receiving a high rating in Europe, too.

Environmental campaigners were seen as a second order – but nonetheless critical – audience in North America and Europe, whereas they appear to be a third order priority in Japan. The media were seen as a more significant factor in Europe and Japan than in North America. Given the impact of green consumerism in Europe in the late 1980s, it is interesting to see consumers rated higher there than in North America. Green consumers have also been a factor in the growing business-to-business pressures illustrated by supplier challenges, so again we see European report makers rating trade and industry customer pressures at a higher level than do their North American counterparts. Interestingly, Japanese report makers were somewhat more likely to have responded to this type of pressure than North American companies.

***The Novo Nordisk approach***

In all three regions, a worryingly small number of report makers had carried out any real research on the needs of the various "user"



FIGURE 21.5 "Novo Nordisk Comes Clean," *Novo Nordisk Magazine* (continued)

duction Cycle

groups.

3M is one and Monsanto – which organised a series of focus groups – another. BP Chemicals also held an Environmental Forum to encourage environmental stakeholders to provide feedback on the company's first environmental report. We hope to see this more sophisticated approach to environmental communications spreading.

Novo Nordisk – as the 1993 report makes clear – has gone even further, organising a series of no-holds-barred visits by European environmentalists to its Bagsværd and Kalundborg sites. The fourth of these annual visits took place earlier this year. The idea has not been to apply a coat of "green gloss" to the company's thinking and activities, but to open them up to close, critical challenge by key environmental stakeholders. The aim: to encourage these stakeholders to consider themselves as "customers", whose needs the company is dedicated to understanding and, where possible, meeting.

Novo Nordisk, in fact, has already made a number of changes to its policies and practices as a result of recommendations made during these visits, and the company's 1993 environmental report reflects the very considerable benefits of the process. Now, as the panel confirms, it's your turn to read the report and suggest improvements that might be made in the next edition.

If you would like a copy of the report, contact the Novo Nordisk Environmental Services Department on tel (+ 45) 4444 8888 or fax (+ 45) 4444 4039.

## We do not have all the answers

*"One thing is clear," says Novo Nordisk chief executive Mads Øvlisen in the foreword to the 1993 report: "around the world, people want to know a great deal more about the industries and corporations they work for, live next to, invest in and buy from."*

*The task of preparing the report was a challenging one, involving more than 50 professionals working in 14 specialist groups. A key feature of the report, not yet seen in any other, is the introduction of an Eco-Productivity Index. But Mads Øvlisen is in no mood to sit on the company's laurels. "We know that we do not have all the answers," he concluded. "Indeed, we invite your comment on our environmental performance, on the performance indicators we have chosen for this report – and on those we might use to assess our performance in future years."*

*Novo Nordisk Magazine* 11

**FIGURE 21.5** "Novo Nordisk Comes Clean," *Novo Nordisk Magazine* (continued)

CROSSING CULTURES

## Translating Science for a General Audience

In theory, a researcher who knows a subject well should be able to present at least its main points clearly to any audience. But as a random look at journals and Web sites may show, many scientists and engineers fail that test. Their failure is often *cultural;* they haven't accommodated the cultural differences among the readers of different publications. Scientific journals, for example, are high-context documents. Writers and readers share a strong motivation to learn about important topics in the field, rely on a depth of common understanding of theories and mechanisms, and speak in code—the jargon of their discipline. Much can be left unsaid.

At the other extreme, however, general publications are low context. Writers can assume little preprogrammed motivation, understanding, or vocabulary on the part of their readers. So writers of popular accounts have to *establish* the context before they can convey their message. They have to motivate and teach their readers. They have to "translate science into English," as one expert puts it.

In addition, they have to understand what is true and accurate. In their otherwise honorable attempts to be accurate, some writers instead confuse and mislead the reader. Popular accounts must balance between "incomprehensible accuracy" and "comprehensible inaccuracy." One researcher proposes as a solution the concept of "communicative accuracy":

> This concept rests upon the fact, not always recognized, that the effective accuracy of a written statement depends primarily upon the interpretation given to it by the reader. A statement may be said to have communicative accuracy, relative to a given audience of readers or hearers, if it fulfills two conditions. First, taking into account what the audience does and does not already know, it must take the audience closer to a correct understanding. The better an example of communicative accuracy is, the more gain in understanding it will achieve—but the basic point is simply that it must gain ground in the right direction. Second, its inaccuracies (as judged at a more sophisticated level) must not mislead, must not be of a sort which will block subsequent and further progress toward the truth. Both of these criteria, moreover, are to be applied from the point of view of the audience, not from the more informed and properly more critical point of view of an expert. (Weaver 499)

"Communicative accuracy" is not lying or cutting corners. Technical professionals need to keep their ethical values intact as they take their information into their reader's culture. But they also need to do what they can to help their readers understand.

***Personification:*** After a century of fading into bed-side tables and kitchen walls, the telephone—both the instrument and its network—is on the march again. (Gleick 103)

***Setting Up a Story:*** Upon first entering the rain forest in Borneo, I was surprised that the palms I'd seen overrunning the trees had long, inch-wide stems instead of stout trunks. (Moffett 110)

*Knocking Down an Argument:* Who owns the law? If you feel a momentary temptation to raise your hand and chirp, "Hey, this is a democracy; nobody can own the law," then you should probably sit tight for a minute and wait for the impulse to pass." (Wolf 98)

*Correcting a Previous Belief:* The destruction of habitat and loss of living species as a result of deforestation in the Amazon basin of Brazil is much greater than had previously been believed, according to a new analysis of satellite photographs. (Stevens C4)

**Ending.** End with a punch—for example, a final piece of advice, a quick look to the future, or a return to the narrative that brackets a technical discussion.

## Voice

As these examples demonstrate, your voice in a popular article should engage the readers' attention. Figure 21.5 demonstrates an effective voice in a popular article about a topic—accounting practices—that general readers might approach with a pronounced yawn. It uses a pun in the title ("Novo Nordisk Comes Clean"), meaning both "owning up to" and "having clean environmental practices." It has a strong lead and ending, and throughout it uses vivid images, strong verbs, and visuals to hold reader interest. In writing, avoid the arrogance of thinking you are dumbing down your work. That attitude is fatal, as it often leads to prose that is patronizing, pretentious, and unengaging. Instead, be conversational, authoritative, maybe even humorous, if your readers would find such humor understandable and welcome.

## CHECKLIST: Articles and Reviews of Literature

1. **Write an article for publication to**
   - Enhance your visibility as a professional
   - Carve out a research territory
   - Share findings with colleagues
   - Fulfill an ethical responsibility to discuss work in the public interest
2. **Select the appropriate publication**
   - Determine which publications the readers you would like to address read
   - Determine which publications cover your topic
3. **Imitate the design of articles in the target publication as you write**
   - Use the appropriate structure, for example, the IMRAD approach
   - Conform to the expected persona and voice
   - Use the expected visual forms
   - Follow guidelines for special elements and form of submission
4. **Write a review of literature to round up the major work on a topic**
   - Align your research to a theme
   - Include key researchers and publications

Use a chronological or comparative structure
Avoid merely pasting together many quotations; establish your own voice
Keep citations correct and noninvasive

5. **Write a popular article to go public with work in the public interest**
Incorporate interest-getting devices to motivate the reader
Be concrete and specific rather than abstract
Use stories, visuals, and color, especially on the Web
Pay particular attention to the title and lead sentence

## EXERCISES

1. Find two articles, one from a technical journal and one from a semitechnical or popular one, that discuss roughly the same topic. Compare the articles in the following categories (for a model of such comparison, see Chapter 6):

   Audience
   Format (abstract, headings, design)
   Style (length of sentences and paragraphs, use of technical terms, general readability)
   Content (theoretical? practical? examples? mathematics? extent of discussion? interest-getting devices?)
   Visuals (number, purpose, style)

   Then write a memo addressed to a fellow student comparing the articles. Attach a photocopy of representative passages or of the entire work. Make sure that you provide a complete citation.

2. Write an article addressed to a general audience based on material you presented in a technical report for a writing class or a technical class. Adjust the content to the requirements of one publication. At your instructor's request, attach a sample table of contents from the publication. Think of a good title and lead sentence. Label, in the margin, the interest-getting devices you use.

3. Collect examples of analogies and other devices used to explain concepts in journal or newspaper articles about science. Review your technical textbooks, too, for such devices. Then write a brief report on figures of speech in writing about science. Chapter 10 provides explanations of metaphors and analogies. You can also check a dictionary to refresh your thinking about such terms.

4. If you wrote a report on empirical research in one of your technical or scientific classes, convert it into a scientific article. Choose a particular journal to write for. When you submit the article to your instructor, attach a copy of the guidelines for authors of the journal and a copy of one article whose structure and style you imitated in your paper.

5. Collect a sample of "guidelines for authors" pages from several publications in your field. Many publications include these on their Web sites; for example, www.sciencemag.org. Include at least one non-U.S. publication. Analyze them for similarities and differences in their style and content—and thus

what they say about the style and content of the journals they represent. Summarize your analysis in a memo to your instructor.

6. Review several Web sites that convey technical or scientific information to a general audience. Sites concerned with environmental, nutritional, or medical information make good candidates. Analyze the techniques used to design the site as an information product (see Chapter 7). Answer questions like the following. What persona comes through the site's text and visuals? What is the relationship between text and visuals? How long is each segment on the site?

7. Analyze the *voice* in Figure 21.5. What is the effect on you as a reader, for example, of the use of a contraction and the invitation to join the author in that first sentence ("Let's look . . .")? Do colloquialisms and figures of speech ("liabilities are likely to mushroom," "broke a log-jam") contribute to an informal tone? What is the relationship of the author to the company? Is this an example of *marketing*, a voice that is trying to sell the reader on something?

## FOR COLLABORATION

1. In a team of two or three students, gather statistics on some topic of interest. Then review publications that deal with that topic. In an oral report to the class, discuss how you would base an article on that information for each of the publications. How would your articles differ?
2. Collaborate with two or three other students in your class to create a newsletter for some campus organization or small business you may be associated with. Include information about the organization and its members, about activities, policies, or national news that may have local implications for you. Use a desktop publishing program to develop a good design (see Chapter 7).

CHAPTER 22

# ORAL PRESENTATIONS

**"BE BOLD. BE BRIEF. AND BE SEATED."**

**F. D. ROOSEVELT**

In addition to *writing* about their work, technical professionals need to *talk* about it in such settings as conferences, proposal reviews, and public interest meetings. You are in good company, however, if the thought of making an oral presentation terrifies you. It is a common fear. To help overcome that fear, adapt the strategies you have learned for effective writing to preparing and delivering an effective oral presentation. This chapter will show you how. Good preparation reduces anxiety and increases the chances that your presentation will meet your audience's needs—and your needs.

## PURPOSE

Like documents, presentations both inform and persuade. You might deliver an informative presentation to

- Alert other members of your project team to the status of your work.
- Review your research with colleagues at a professional conference.
- Brief members of your class on the major findings in your final report.
- Instruct employees of your company in using a software system.
- Explain a medical procedure to a patient.
- Describe a new agricultural device to a retired farmers' club.

In a persuasive presentation, you might

- Defend your company's logging practices to a public environmental oversight committee.
- Recommend the best of three alternative treatments for a patient.
- Urge employee compliance with a new safety policy.
- Show how your proposed solution to a problem is better than your competitor's.
- Ask your boss for a raise.
- Urge other members of your team to complete their work.

As these few illustrations suggest, the range of purposes is wide. Some presentations meet internal purposes within organizations, and some meet the needs of colleagues, clients, or customers outside. The range of settings also varies widely, from offices, cafeterias, or airport conference rooms to classrooms, auditoriums, and convention centers, among other locations.

## TYPES OF PRESENTATIONS

You'll accommodate these differences in purpose and setting by selecting the right type of oral presentation. Most fall into one of three types. The types differ in how much of the statement is prepared in advance and how much interactivity the presentation allows, and thus in their relative formality.

### Impromptu

The type of presentation least explicitly prepared for, and thus generally the most interactive and informal, is the impromptu talk or off-the-cuff remarks. In a meeting or other gathering, you may respond to a request for such a presentation and engage in some give-and-take of ideas. "Least explicitly prepared for" means that you can make such remarks easily because you are already an expert on the topic, you find it easy to express your information to a variety of audiences on the spot, and you don't need to do any special research to come up with convincing material. A classroom teacher may ask you to speak off the cuff about an assigned reading or exercise. Often, such talks take place in high-context settings, like team meetings, where only a few words are needed to establish a message because the audience shares a broad understanding of the issues.

One advantage of such a presentation is the lack of explicit preparation time (although all the time spent learning the technical information itself does count). In addition, its language can be fresh and its material well directed to the audience, because the presentation is composed on the spot in their presence, with their questions perhaps shaping what is said. But to the degree that a technical expert lacks skills in selecting and presenting information orally, that approach can be highly uncomfortable for both the speaker and the audience. In the wrong hands, such presentations are too long, digress ineffectively, and lack emphasis. If you are at all nervous about oral presentations, try to avoid an impromptu approach.

## Outlined

Instead, think ahead to line up the main ideas and appropriate supporting evidence in an outline and notes (Figure 22.1). The outline may appear on note cards, on one side of a sheet of paper, or in a three-ring binder that includes your notes as well as the transparencies you'll use in the presentation. Sometimes that means you prepare a presentation even

2 min — Intro. to talk content: roles of medical writers, types of writing projects

4 min — team concept for writing: internal (with other writers) and external (on team with functional groups)

skills needed: writing, design, editing, knowledge of health science, quality control

5 min — projects overview: regulatory documents compared to publication documents

- regulatory: trial reports, investigational brochures, new drug applications
- publication: manuscripts, abstracts, presentations, conference materials, advertising

5 min — life cycle of documents, initial draft to filing of paper and electronic versions

12 min — case study: the clinical trial report

FDA guidelines
ICH (International Committee on Harmonization) guidelines
company guidelines
structure

- front matter
- introduction
- methods: trial design, assessments, statistics, administrative issues
- results: demography, efficacy, pharmacokinetics/others, safety, discussion
- conclusions: main findings and alignment with objectives
- appendices

2 min — complex topic, but this talk has covered one of many documents that help bring the benefits of pharmaceutical research safely and effectively to you.

**FIGURE 22.1 Outline of a Talk in Units of Time**
Topics for a 30-minute talk introducing pharmaceutical writing to undergraduate biology majors.
(Adapted from a presentation by Robert J. Bonk.)

when you know you may not need to give one, as in a class whose instructor arbitrarily calls on students. Notes are also helpful when you need to brief an audience about a topic or situation. Briefings range from informal sessions around a conference table or in a classroom to more formal occasions when, for example, generals brief reporters about the conduct of a war or department heads brief a company president about activities in their area.

Another form of note-based presentation is a bullet report, which is often replacing written reports in fast-moving, high-tech companies (see Figure 22.2). You present your main message as bulleted points on a large-display computer screen or on an overhead transparency—perhaps as a series of such screens or transparencies. Your oral remarks expand on and support the implications of the bulleted items, although you are not limited to those items, because you adjust your presentation to the audience's questions and generally use conversational, if technical, language.

### Scripted

In the most highly structured, least interactive, and most formal presentation, you read directly from a written text, often called "prepared remarks." You work from a script, and your voice thus often reflects the formality of prose rather than the informality of conversation. The script gives you the most control as a speaker and thus may be essential, for example, when the topic is controversial and legal counsel has been sought to define exact wording. But few people read papers well. In addition, if you memorize the talk, you run the risk of forgetting the entire text if you forget one word. Prepared remarks lend themselves to a monotone presentation, unenticing for the audience and unemphatic for the material. So when you have a choice, speak from notes.

## DESIGNING THE PRESENTATION

Design your presentation to match your purpose and the appropriate type of delivery. This section provides guidelines for designing well.

### Point

Like readers, listeners ask of any presentation: What's the point? Many presentations fail because speakers don't really know the point of the presentation or because they fail to make that point obvious and compelling. What do you want the audience to know? To do? Fill in the blank:

> After hearing this presentation, my audience should ______________.

In a ten-minute talk, one main point is plenty. In an hour, you might cover three points—three is probably a maximum for any presentation (other than a classroom lecture). Deliver that point early in the talk and restate it at the end. It gives you a track to run on. It also gives the audience a

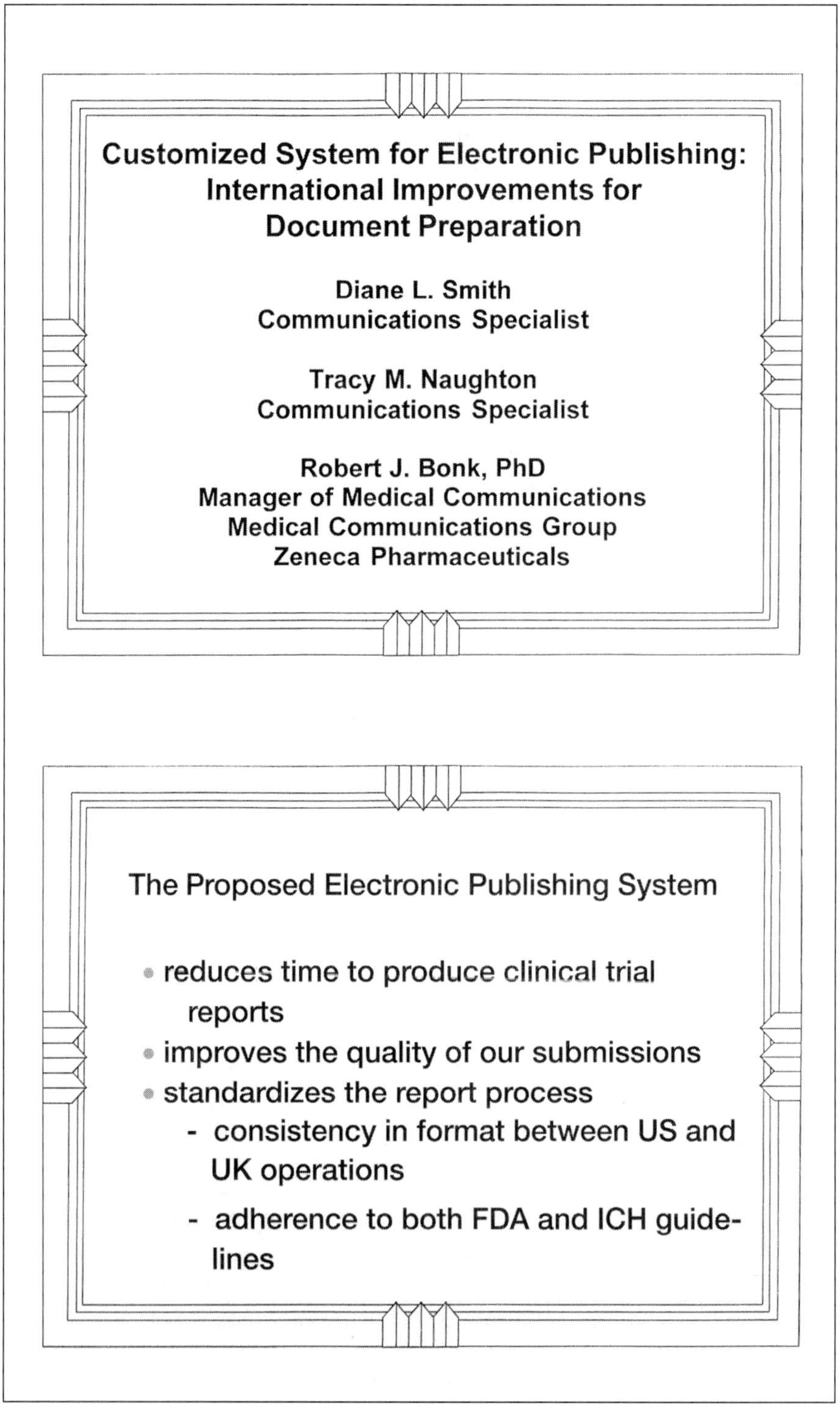

Customized System for Electronic Publishing:
International Improvements for
Document Preparation

Diane L. Smith
Communications Specialist

Tracy M. Naughton
Communications Specialist

Robert J. Bonk, PhD
Manager of Medical Communications
Medical Communications Group
Zeneca Pharmaceuticals

The Proposed Electronic Publishing System

- reduces time to produce clinical trial reports
- improves the quality of our submissions
- standardizes the report process
  - consistency in format between US and UK operations
  - adherence to both FDA and ICH guide-lines

FIGURE 22.2 Title Slide and Opening Slide from a Bullet Report
(Courtesy of Robert J. Bonk.)

framework for assembling the information you will present, for remembering and acting. Translate your purpose, then, into a point, for example,

> ***Purpose:*** update
> ***Point:*** The installation of our new workstations is now 60 percent complete, as anticipated.
>
> ***Purpose:*** research report
> ***Point:*** Our studies demonstrate that strong similarities exist between the creep behaviors of trabecular and cortical bone.
>
> ***Purpose:*** proposal defense
> ***Point:*** The Toulouse System meets or exceeds each of the requirements for flexibility, reliability, and geographic sensitivity stated in the Request for Proposal.
>
> ***Purpose:*** description of a process
> ***Point:*** In the next fifteen minutes, I'll describe briefly our procedure for designing an overcoat for a male paraplegic, from the initial interview with the client through fabrication.

## Audience

**For more about audience analysis, see Chapter 6**

To help determine your point, and the best way to develop it, consider what your audience expects and needs. Like documents that must work for readers, oral presentations must work for listeners. But keep some key differences in mind. Readers have many options in reading: skimming, returning to the beginning if they get lost, looking at all the pictures and then the text, and reading at their own pace. Listeners have none of these options. They have to depend on you to highlight a main point and impose a structure and pace that work. Unless your presentation is on tape, listeners can't catch it in the middle and then go back to the beginning. The beginning is gone. If their attention drifts, listeners have missed the information you delivered during that time, and you've lost them. To prevent such losses, make sure you understand your audience.

When someone requests that you give a presentation, ask about the audience. Here are some questions to ask:

- Why is the audience attending the presentation?
- How much do they know about the topic?
- How much do they know about the technical or scientific field as a whole?
- What do they expect you to tell them?
- How willing are they to pay attention to your presentation?
- What is the length of presentation they are accustomed to?
- How many people will attend the talk?
- How diverse—in age, nationality, technical skills, experience, and the like—is the group?
- Do all members of the audience speak English fluently?

If you will be speaking in English and your listeners will be hearing the talk in another language, that is, if your talk will be interpreted for the local audience, then

- Prepare visuals that can substitute for text.
- Write the main points on a handout, preferably in both English and the language of the audience.
- Streamline your vocabulary.
- Avoid jokes and metaphors outside the experience of the audience.
- Talk with the audience, not to the interpreter, a common problem in such settings.

Incorporate the answers to these questions in your design. For example, if the group will be small, plan to talk informally, with lots of interaction. If the audience is likely to be antagonistic, prepare counterarguments carefully to show that you take their objections seriously and to defuse their opposition. If the audience is diverse, aim at a middle level of understanding in your general remarks, but keep such comments brief to allow time for a question-and-answer period that accommodates different interests. Circulate handouts that provide background information as well as more technical details.

### Structure

To please the audience and get your point across, develop a structure for the talk. Even a relatively informal presentation profits from structure. As one British expert notes, the goal is to

> create a state of mind, or a point of view. . . . There should be one main theme, and all the subsidiary interesting points, experiments, or demonstrations should be such that they remind the hearer of the theme . . . the force of the impression depends upon a ruthless sacrifice of unnecessary detail. (Bragg)

Assembling the units of your talk in a structure helps you develop your key message (or "theme") and achieve the "ruthless sacrifice" of detail so that you provide the most telling information—and no more—in the appropriate type of presentation. Peg each unit to the time needed for its delivery (Figure 22.1). Exceeding your time limit is unfair to the audience and may cause them to stop listening before you stop talking. Individual companies or professional organizations tend to establish standard lengths to suit particular routine purposes. Presentations at professional conferences, for example, are often 20 minutes. A panel discussion at a corporate meeting may begin with a 10-minute statement from each participant. You may have only five minutes to talk with a company president. Rarely does any presentation last for more than one hour without a break. Within whatever time frame is set for your talk, develop three segments: introduction, middle, and ending.

**Introduction.** An old piece of advice about presentations runs, "First you tell 'em what you're gonna tell 'em, then you tell 'em, then you tell 'em what you've told 'em." That's a good approach.

In the opening segment, draw the audience's attention and engage their curiosity and interest. For highly motivated or technical audiences, get directly to the point, without wasting their time with preliminaries they will find frustrating. Do, however, provide background for a general audience that needs such information to understand the technical details to follow. Some of the interest-getting devices that work in the introductions of popular science articles serve speakers, too, when they address a general or diverse audience. Consider these openers:

**For more about popular articles, see Chapter 21**

- Mention a local reference point—a person, a place, or a favorite saying, for example—familiar to the audience.
- Tell a story or anecdote that relates to your main point.
- Show a pertinent cartoon.
- Open with a question you assume to be on the audience's mind.
- Deal immediately with any problem or complaint that the presentation is designed to address. Don't let people harbor their grudge and thus not listen.
- Frame the purpose of the talk in terms of what the audience will gain from listening: "By the end of this morning's session, you should be able to conduct keyword searches through at least 10 search engines on the Web."
- Introduce yourself if you have not already been introduced.
- Show a slide or screen that previews the structure of your presentation.

Here's the beginning of an effective student presentation:

> The senior design project I'd like to talk about with you today really began two years ago while I was riding my bike behind a bus I couldn't seem to escape from. The exhaust smelled terrible. I knew that smell, too, had to be a sign of all kinds of things I was breathing that couldn't be good for me—or anyone. So when the time came for a design project, I remembered that incident and decided to investigate odor-causing components in diesel exhaust.

Even if only briefly, give the context for your remarks, the "why" before the "how." Note the problem, if there is one (for example, the smell), before turning to the solution. Many speakers omit this step, plunging immediately into the technical details while the audience strains for an overview. Give them the overview, as in this statement from the presentation on diesel exhaust:

> I'll first describe the composition of diesel exhaust as we characterized it in the laboratory and then show the modifications we recommend in the design of the engine to reduce harmful emissions.

**Middle.** In the middle, you tell them. In an informative presentation, you describe or narrate, and you provide instances. In a persuasive presentation, you use examples, statistics, expert testimony, and the like to help the audience visualize and commit to the position you advocate. As you do so, "ruthlessly eliminate unnecessary details," because too much information can hinder the audience's memory. Unlike a written report, a presentation aims more at impression than depth.

CROSSING CULTURES

## The Culture of Presentations

In a presentation, you have to accommodate the values and expectations shared by people in the culture where you are speaking. That culture may be both corporate and national.

Different organizations (and even divisions within organizations) set different norms for presentations. One analyst of those norms, Rae Gorin Cook, labels two major corporate cultures in the United States: "cowboys" and "suits." In cowboy cultures, presenters dress informally and idiosyncratically, encourage interruptions and confrontations, emphasize spontaneity, and cultivate an aggressive, "no-nonsense" style. In "suits" settings, by contrast, presenters dress formally (in suits that resemble other presenters' suits), adhere to rigid standards of formal presentation that allow few questions, use visuals extensively and speak from a prepared text, and cultivate an authoritative stance. Similarly, successful presentations in some companies are low keyed, highly technical and detailed; in others, they are more general, breezier, more likely to dazzle than inform. So a "one size fits all" approach to presentations won't work.

Flexibility in approach is even more necessary in international settings, where values and expected behaviors can vary even more widely. If possible, sit in on several successful presentations before you need to give one and observe closely the elements of proper behavior. Ask someone familiar with the audience and setting about appropriate style. Cook notes, for example, that Chinese dislike boasting and guessing; they find the speculations that fill many American presentations wasteful. They prefer to speak only when they are positive about what they are saying. Hispanic professionals often welcome direct, emotional confrontation on technical issues—confrontation that may not be appropriate in an American environment where managers prefer to avoid public criticism.

How valued is emotional sensitivity in the audience's culture? Should you display your emotions—or hide them? Should you maintain eye contact, as in the United States, or is such contact considered uncomfortable and invasive? In any setting, watch your body language, and limit or avoid such sensitive behavior as touching, pointing a finger, or using other hand signs, like the American "thumbs up" for approval or "OK" sign. People from other cultures may read very different meanings in those signs (Cook).

In selecting and arranging your information, accommodate your listeners' perspective on what constitutes appropriate content. The American belief in objectivity is not universally shared. Many people draw the line very differently between facts and opinion and between what is private information and what can be said publicly. How open are your listeners to hearing negative information? How *direct* should your talk be? A good American presentation is usually direct: the main point followed by its support. Other cultures value indirection, with facts and details serving as a ramp into the main point. In some situations, the concept of a "presentation" is itself foreign. Instead, a group asks and answers questions in an associative way to build information as they build personal relationships. Accommodate differences, for example, by allowing people who don't like confrontation or public display to ask questions in private or in writing.

**Ending.** In the end, you make sure the audience knows what your presentation adds up to. You don't want to leave them thinking, "So what?" Prepare a solid ending for your talk, one that will prevent you from rambling and from looking beseechingly at an instructor or other external authority to tell you that you are done. Control the ending in your own terms:

- Clarify any action the audience is supposed to take.
- Build motivation for such action.
- Broaden the focus.
- Recap the main theme.
- Summarize the main evidence or components.
- Create a dramatic image in the listeners' minds.

Here's an example of a brief but effective ending to a presentation about how participation in college sports builds lifelong skills:

> What you have learned on the playing fields will help you participate on any kind of team in any kind of organization, manage your time effectively, compete with others and share with others, overcome defeat and be modest in success, and think quickly on your feet—skills that will stand you in good stead when the air goes out of the collegiate soccer ball.

## Making the Structure Obvious

Written reports can have an endoskeleton supporting information from within, but oral reports need an exoskeleton, obvious from the outside. One professor, for example, marks the structure of each class with jokes—one at the beginning, one in the middle, and one at the end—a pattern that helps students feel comfortable and provides a break for the professor, too, who loves to tell jokes.

Make the structure of your presentation obvious to your listeners (with or without jokes). You can do this with both visuals and words. Use a transparency or screen, for example, to provide a map for your presentation (Figure 22.3) which you then display again at key points, reminding the audience that "you are here." Open with a title slide, then show main topics with the first highlighted, then show the slide again and change the highlighting as you shift topics. In addition, use repetition and transitional words to keep listeners from falling through the cracks and to aid those whose minds might have momentarily wandered.

Repeat the main point several times—at the beginning, in the middle, and at the end. Include frequent transitional statements:

> Here are our findings on the composition of diesel exhaust . . .
>
> Now that we've seen what the exhaust is made of, let's look at what's going on in the engine that produces it.
>
> That's how currently available engines operate. Let's look at the modification we're proposing to control emissions.

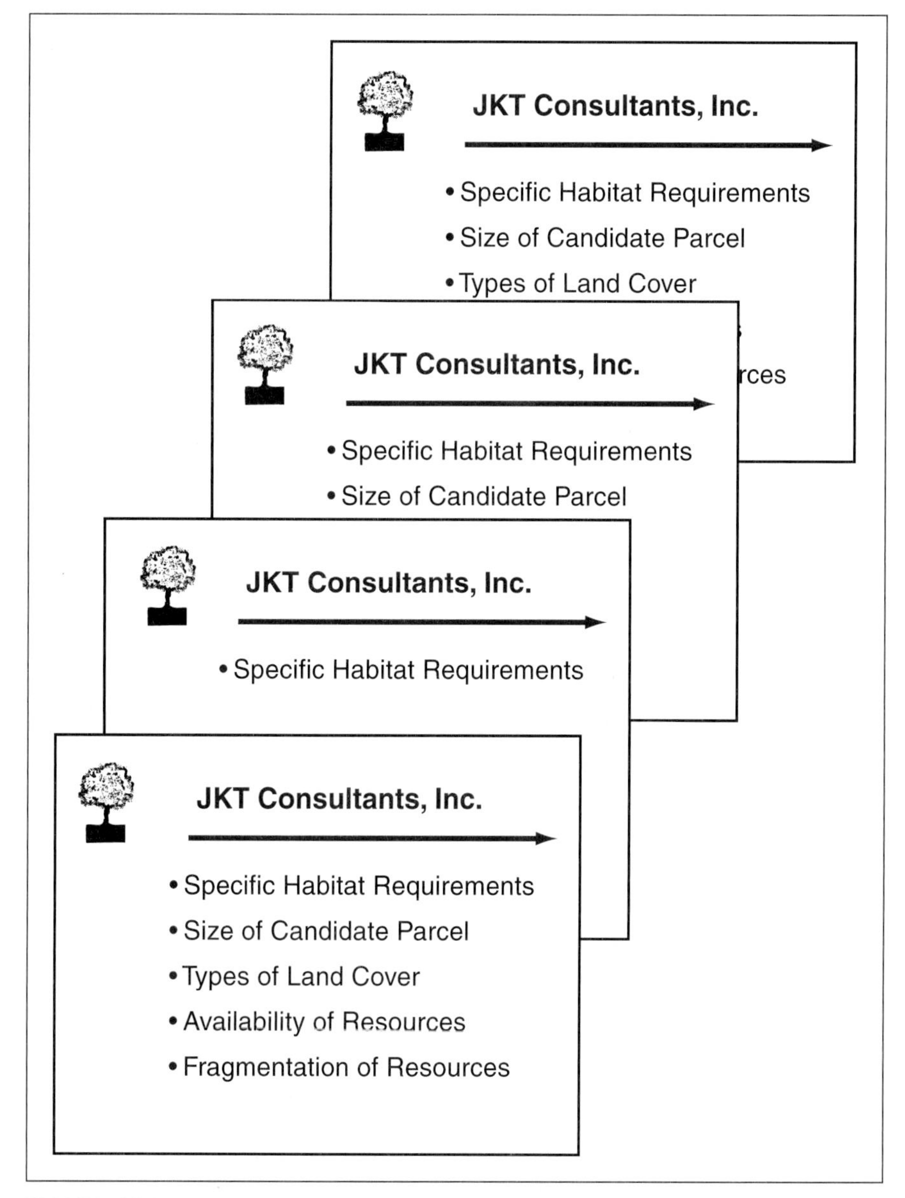


**FIGURE 22.3 Structure Slides**
In a computer-driven presentation, each line can "fly in" from the border as the speaker reaches that topic.
(Courtesy of Rebecca B. Worley.)

## DESIGNING THE VISUALS

To reinforce the structure of your talk, use visuals, as in Figure 22.3. In addition, use visuals to display and explain its content (Figure 22.4). Both general and technical audiences expect visuals, and many professionals realistically describe the act of preparing a presentation as "getting my slides together." Good visuals can make even a mediocre performer seem dynamic. So use visuals effectively to

- Aid the audience in remembering. Research shows that people remember best concepts and systems they associate with a picture.
- Help emphasize key points.
- Summarize supporting data quickly and efficiently.
- Provide graphic analogies to bring complex ideas into the experience of the audience.
- Add a dynamic and dramatic element.

Currently available presentation software helps you create professional-looking visuals easily. If the technology is available, you can plug your laptop computer directly into a projector for large-screen display of up-to-the minute slides and demonstrations. You can also weave sound, animation, and video motion into your presentation. Consider the following guidelines for determining both the appropriate technology and the appropriate content in the visuals you use.

### Appropriate Display Technology

Decide on the appropriate technology to support your presentation. Learn what technology is available where you will be talking and have a backup plan if promised equipment is not delivered. In international settings, you may find that your equipment doesn't plug into the available outlets, your videotape doesn't fit the available videocassette recorder, or electricity is rationed and not operating when you need it. In any setting, don't assume an elaborate approach is always better than a simple one. Adjust the technology to your audience's expectations and the type of presentation. For example, flip charts and blackboards help you create a presentation on the spot and encourage interactivity between you and the audience.

From simple to complex, here are the advantages and disadvantages of the most common options for displaying visuals.

**Blackboards.** Blackboards (or whiteboards) have these three major advantages as aids to presentations:

✓ Widely available.

✓ Interactive; you can write on the board as items come up.

✓ Well paced; you can record information while the audience assimilates it, so the pace is regulated better than in a transparency that may show too much at once.

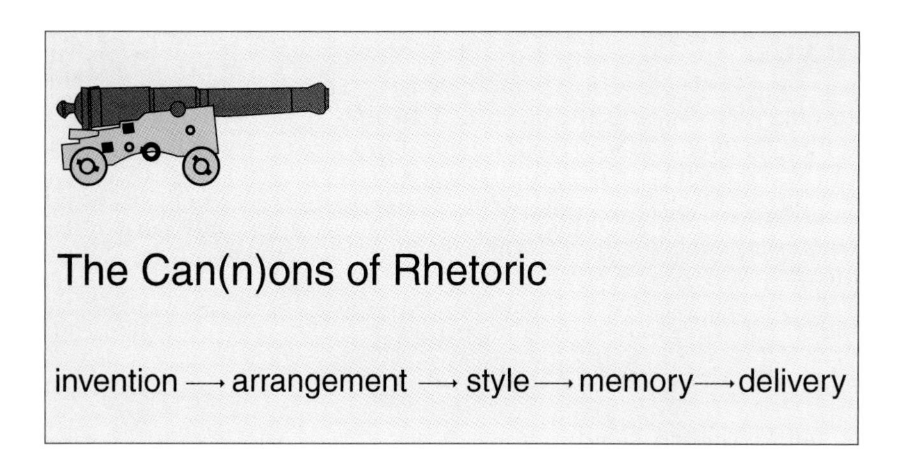


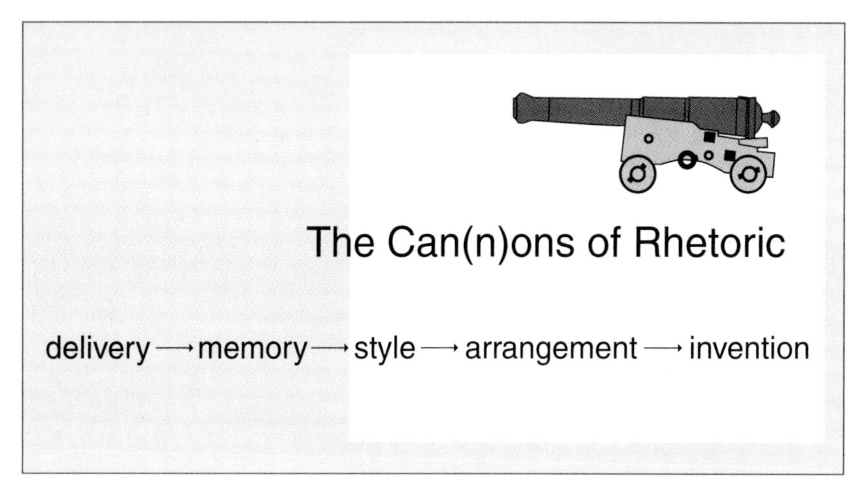


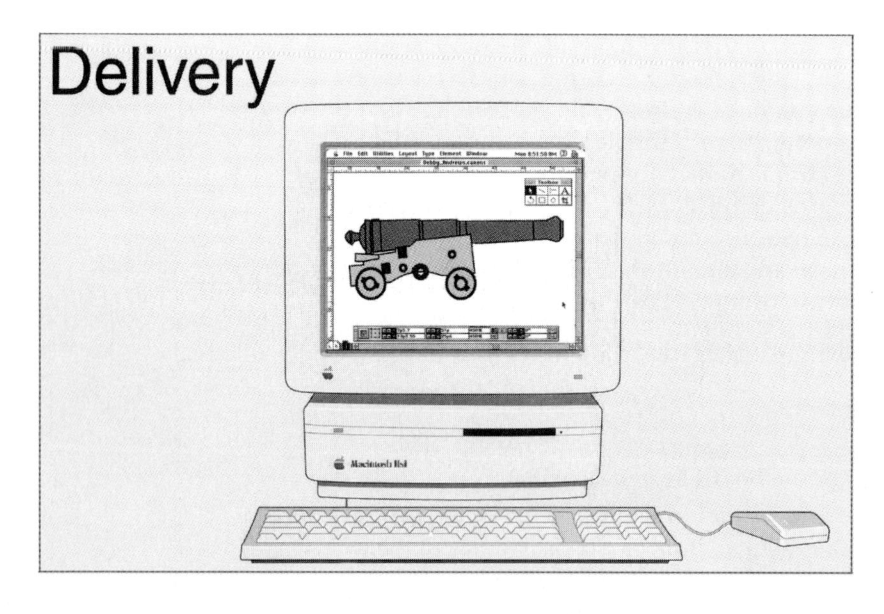


FIGURE 22.4 **Slides from an Effective Talk**

The speaker discussed how digital delivery systems are changing the concept of rhetoric. The image of the cannon is a visual pun, because the *canons* of rhetoric are its rules and elements, and *cannons* are weapons. The cannon is also appropriate because the speaker's main point is the revolutionary nature of the changes, and the talk was given on the anniversary of a political revolution. In the "Delivery" slide, the cannon appears on the screen of the force for change: a computer.

(Courtesy of Chris Miller. Graphic taken from: ClickArt Business Illustrations, © 1994 T/Maker Company, a wholly owned subsidiary of Brøderbund Software, Inc.)

However, they may

- ✗ Require you to turn your back on the audience while you write.
- ✗ Be hard to read if your handwriting is poor.
- ✗ Block the audience's view of the board as you write.
- ✗ Delay the delivery of the main point if you spend too much time drawing.

**Flip Charts.** Flip charts, large (46 × 61 cm or larger) pads of newsprint affixed to an easel, suffer the same disadvantages as blackboards and are limited to even smaller groups for clear viewing. But they have two additional advantages. They can be

- ✓ Prepared in advance.
- ✓ Taken away from the meeting as a record from which, for example, to write minutes.

**Transparencies.** Sometimes called *overheads* (because they fit on an overhead projector), these plastic sheets are a staple of business and technical presentations (Figure 22.4). Use presentation software to prepare them, reproduce a hard-copy original on an acetate sheet, or write on a blank acetate. Most bookstores and quick-copy shops stock such sheets. Transparencies are

- ✓ Easy to prepare.
- ✓ Easy to arrange and rearrange.
- ✓ Professional looking.
- ✓ Welcomed by most audiences.
- ✓ Able to be displayed while you face the audience.
- ✓ Visible in a lighted room.

You can show a sequence of ideas through overlays, with each new transparency adding a new line of text or feature of a drawing.

**35mm Slides.** For presentations to groups of all sizes, including those in multitiered lecture halls, 35mm slides have been a standard medium and remain popular. Slides can reproduce in detail and accurately paintings and buildings, protozoa and archaeological digs, the microstructure of metals and of human cells.

Slides, however,

- ✗ May show too much detail.
- ✗ Are less flexible than transparencies because they must be arranged in trays in advance.
- ✗ Are more difficult to prepare than simple transparencies.
- ✗ Are not interactive; they cannot be written on during the presentation.
- ✗ Require a darkened room.

**Large-Screen Computer Presentations.** In classrooms and conference rooms across the world, presenters are using laptop computers connected to 8′ × 8′ projection screens to display the visual support for their talks. Presentation software like Microsoft's PowerPoint facilitates the display. With such software, you can create slides like those shown in Figures 22.2 and 22.3. You can also scan in images from print sources as well as 35mm slides. You can mix sound, text, video, and animation in one easily controlled file. In the right hands, such presentations are just what you need to inform, persuade—or dazzle—an audience. Such a presentation

- ✓ Accommodates the multiple ways that people display their intelligence. A training demonstration, for example, can include a video of someone performing the task for audiences who imitate well; a series of drawings about the task for those who need to understand the process; textual instructions for those who prefer to learn through words.
- ✓ Lends itself to modular delivery, so that, for example, you can insert translations of segments into different languages.
- ✓ Demonstrates your technical abilities effectively, further corroboration of your credibility especially when the content of the presentation is highly technical.

However, the presentation

- ✗ Requires expensive hardware and software both in its preparation and its delivery.
- ✗ May overwhelm the audience so that they remember the medium more than the message.

The more technology driven the presentation, the more important it is for you to have a backup plan. If you anticipate using a laptop, take copies of your major slides as overhead transparencies, just in case the laptop fails to connect with the projection hardware available in the presentation setting. Print preview versions of the slides as a handout. Have a disk copy of your presentation available for transfer to another working laptop.

## Appropriate Visual Content

For more about visuals, see Chapter 8

For more about storyboards, see Chapter 14

Your choice of display technology helps shape the design of visuals to support your presentation. In addition, design visuals to emphasize your central message and highlight key evidence for that message. To prepare the design, develop a storyboard. Divide a sheet of paper into two columns: In one, note the points you want to convey, in sequence from top to bottom; in the other, draw a thumbnail sketch of the visual that will support that point (Figure 22.5). In preparing content slides, minimize or simplify information. If your presentation is based on a report, you may start with the report's visuals, but you'll probably need to eliminate details and enlarge the type size for a transparency. It's simple to make transparencies from figures in a document, but it is often not simple for listeners to understand these. Compressed charts of molecular weights and tables for calculating wind loads, for example, are fine for readers who have enough time to puzzle them out, but not for listeners (Figure 22.6).

| | |
|---|---|
| Intro. to talk content | title screen: my name, topic, company logo |
| Roles of medical writers | photo of medical writing group at a desk [A] |
| types of writing projects | list of 15 types of projects, 3 or 4 on each screen, text flies in from side, then fades as I discuss. Cartoon? [B] |
| Writing team concept | Screen [A] again, then photo of writer interviewing scientist |
| Projects overview | Screen [B] again, then blank screen, zoom in on each type name as I discuss<br>Photos of sample documents?<br>TOCs from sample docs? |
| Life cycle of documents | flow chart of the approval process<br>Chart of the hierarchy of regulatory agencies |

**FIGURE 22.5 Storyboard**
Here's how the speaker planned the screens he would create for a segment of the talk outlined in Figure 22.1.
(Courtesy of Robert J. Bonk.)

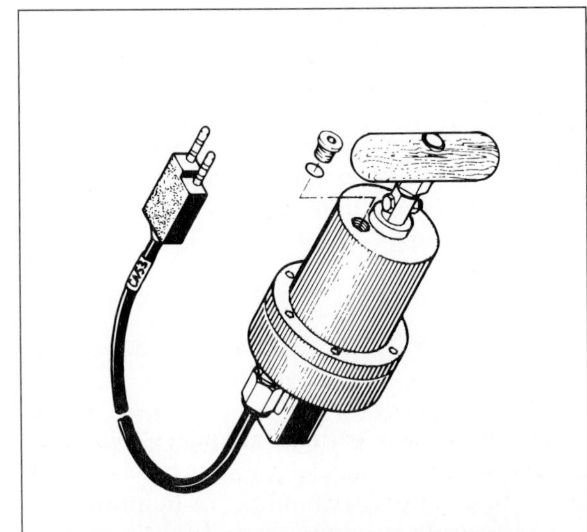

Source: Backinger, C. L., Kinglsey, P. A. *Write It Right: Recommendations for Developing User Instructions for Medical Devices Used in Home Health Care.* HHS Publication FDA 93-4258 (August 1993), pp. 39–40.

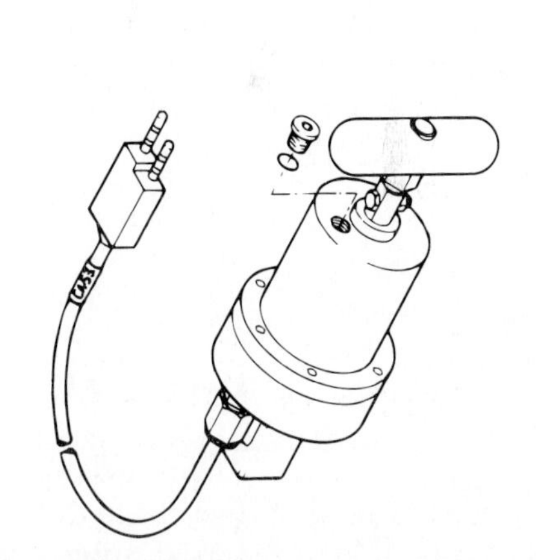

Source: Backinger, C. L., Kinglsey, P. A. *Write It Right: Recommendations for Developing User Instructions for Medical Devices Used in Home Health Care.* HHS Publication FDA 93-4258 (August 1993), pp. 39–40.

**FIGURE 22.6 Simplifying**
The drawing on the left provides shading and texture, but the one on the right might be more appropriate if your goal is to focus on the device's main components. Note the use of a source line.

Some presentations are primarily visual, in particular, poster sessions at research conferences. *A Closer Look: Poster Sessions* provides guidelines for posters that present empirical research.

The following advice will help you design visuals with the appropriate content:

- Give each visual a title.
- Use visuals to show relationships, not details. If the details are necessary, consider reproducing them in a handout that the audience can review after the presentation.
- Limit the lines of text and the total number of words on a slide or screen: no more than seven lines, no more than six or seven words per line. Create headlines, not sentences.
- Design all visuals in the same format to reinforce the coherence in your talk.
- Create a master slide that others follow. Use a consistent typeface, preferably sans serif, probably 36 points for the title, 24 points for headings, 22 points for subheadings.
- Use color cautiously, especially for multinational audiences.
- Use no more than three high-contrasting colors: one for text, one for background, one for highlighted words or numbers.
- Proofread all visuals. Standing next to a 1-foot-high spelling error is particularly embarrassing.
- If you are showing slides, concentrate them in one segment of your presentation to avoid toggling between lights on and lights off.
- Use "leftover" slides as needed to answer questions.

Students often try to make do with only one or two visuals, and professionals often include too many. There is no simple equation for determining how many visuals you need, but think in terms of at least one or two for each major segment of your presentation: introduction, middle, and ending.

## HANDOUTS

At professional conferences and business meetings, attendees sometimes feel cheated if they go away empty-handed. Especially if your talk is controversial, subject to more than ordinary misunderstanding, or in an international setting, you may want to provide handouts that allow listeners to follow along while you speak or to work out problems on their own after the talk. Use handouts, then, to

- Explain a difficult decision or political issue.
- Provide a text so the audience is likely to quote you accurately.
- Reproduce the main steps in a process for reference when people try the process on their own.

A CLOSER LOOK

# Poster Sessions

One form of presentation that is common at undergraduate, graduate, and professional research conferences is a *poster.* The poster is a large board (generally around 4′ wide by 7′ high) that provides your theme and most significant evidence. In a room filled with long rows of tables or easels, you stand by your poster for a designated period of time. Visitors pass by and stop to discuss issues with people whose topic interests them. Abstracts of all the posters and their locations usually appear in a program for visitors.

In preparing a poster, avoid the temptation to merely tape up the pages of a report, but instead use large print (at least ⅜″ high) and visuals to attract the attention of passersby. During the session, because you are probably presenting work in process, take notes when you hear questions and comments that may help you improve your approach. The Philadelphia Chapter of the Association for Women in Science offers these guidelines for entrants in its competition for student posters on scientific research (courtesy of The Philadelphia Chapter, Association for Women in Science):

- Your audience came to your poster to learn something interesting. Make a good story out of your research. Show the experiments that best allow you to make your point. Sometimes, preparing a presentation helps you see more clearly how to adjust your future research direction.
- Include the following sections: introduction, methods, results, and conclusions. Methods are described in the shortest possible fashion. Supporting and contradictory evidence is part of results, and conclusions should be brief and meaningful.
- Number the poster panels so that anyone can easily follow them in your absence.
- Use color to make your presentation more exciting.
- Choose a simple type face. Do not use all capital letters because they are difficult to read.
- Label axes in graphs clearly. Labels in poster presentations should be more extensive than in publications. Multiple traces should be identified directly on the graph rather than in the legend.
- Keep graphs and tables uncluttered. Multiple graphs are preferable to multiple kinds of data on a single graph. Use graphs to show trends in data. Uses tables for precise numerical values. It is easier to compare data in columns than in rows. Have a title above each graph or table.
- Show how reproducible and accurate your results are by including numbers of replicates, standard errors, standard deviations, or confidence intervals in graphs and tables. Include only significant figures.
- Acknowledge the source if you include someone else's data.
- Proofread and check the spelling of your presentation.
- Rehearse aloud and time yourself for an initial 5-minute presentation.

- Allow listeners who are hearing your talk in a non-native language to see as well as hear your words.
- Provide a translation of your main points and the text of your major visuals.

Circulate handouts before the talk if you will refer to them directly. Otherwise, note at the beginning of your talk that you have prepared handouts, but reserve them until after the talk so they will not compete with you for the audience's attention.

Package your handouts. Even if they are simply paper copies of your transparencies, give them coherence as a document. You may show several slides or screens in reduced form on one page. Attach a cover page with the title of the talk, your name, and the date. Number the pages of handouts if the individual transparencies don't already have numbers. Include information needed to contact you if listeners have questions: telephone and fax numbers, postal and e-mail addresses.

## PRESENTING

Before you go on stage, practice your presentation. If possible, practice in the room where you'll be speaking so you can adjust yourself to the equipment available there and become comfortable with the setting. If you are using slides, make sure they are arranged to project right side up. Know how to place a transparency on an overhead projector. Even if that setting isn't available, practice in another room. Perhaps assemble a group of friends to simulate an audience and be willing to be part of a practice audience for your friends. If your college or other organization offers such a service, videotape your practice session. Reviewing the tape will help you "see yourself as others see you" so you can make appropriate corrections.

When the day of the presentation arrives, expect to be nervous (even professionals are nervous); then take a deep breath and forge ahead. Use your nervous energy to deliver what the audience needs. You are there to achieve something with the audience, at that time and in that space, that can't be done in writing. Don't waste that occasion. You'll probably be especially nervous during the first few minutes—and especially thankful that you prepared well. As you move through the presentation, let your good preparation help you adjust the talk's structure and pace. In addition, make sure you talk *with* the audience and control the visuals.

### Talk with the Audience

From the beginning of the talk, develop rapport and common ground with the audience. Talk with them, not just to them, in a cordial voice that demonstrates your role as part of the group gathered in the room:

- Be flexible. Look for any signs, like restlessness or puzzled facial expressions, that the audience hasn't understood what you've said.

You may need to back up, define a term, or give another example or analogy.

- Slow down. Don't let nervousness make you rush through your presentation or raise the pitch of your voice.
- Avoid slang, jargon not familiar to everyone in the audience, and acronyms. Especially when you address an international audience, use formal English.
- Make sure all pronouns are clear. When you say, "They've developed a promising approach," you may know who "they" are, but your audience may be left wondering.
- Watch for distracting mannerisms. If you've videotaped yourself, you'll see in the playback what these are. Don't push your glasses up on your nose, or tug at a sweater, or move your hands around in your pockets, or push a pointer in and out, or clutch the lectern.
- If you are speaking in front of a blackboard, erase any leftover messages that might deflect the audience's attention from your presentation.
- Maintain eye contact with the audience in cultures like the United States where that is valued.
- Pause when you need to think and before you say something that is complex or potentially confusing.

In general, be confident and show your enthusiasm for the topic. If you don't demonstrate strong interest in what you are saying, it's highly unlikely the audience will be anything but bored.

### Control the Visuals

Audiences have a right to revolt, too, if your presentation is simply a series of comments beginning "This slide shows . . ." Don't let your visuals override the message you are delivering, and take charge of the devices necessary to display them.

- Remove a visual whose message competes with what you are saying.
- Reveal each part of a structure slide as appropriate, covering the rest.
- Turn off the machine when you are not using it; don't force the audience to stare at a lighted but empty screen.
- Resist the temptation to talk to the screen instead of the audience.
- Don't insult your audience by reading the slides to them. As the audience reads, tie the slide into the point it is proving.

### Coordinate a Team Presentation

When the situation requires a team presentation, preparation is the key. First, agree on the purpose, the genre, and the plan. Make sure everyone knows how long the entire presentation should last and buys into the main point and the best approach. An outline is essential. Storyboarding can be

Date:
Speaker:
Timing: started . . .
stopped . . .
total elapsed time . . .
over/under/at required time . . .

**The Point**

Summarize the main point of the presentation in one sentence:

Comment on the point:

Clear?
Appropriate?
Innovative?
Compelling?

**The Structure**

Was the introduction effective in giving an overview and establishing the presentation's purpose?

Did the speaker provide evidence to support the main point?

Adequate?
Clear?
Logically sequenced?
Any gaps in development?

Did the presentation end well?

FIGURE 22.7 **Checklist for an Effective Presentation**

particularly effective, because it allows the team the opportunity to see how all the visuals, and thus each person's presentation, will be integrated. It also helps you set a common format and style, for example, in border design and typeface, for the visuals. Agree on how—or if—you will answer questions. For example, will questions be appropriate at the end of each presentation? At the end of the entire presentation? Will one person handle all questions? Will you ask people to see you after the presentation if they have questions? The opening speaker should announce the agreed-on approach, and you should stick to that decision.

Second, assign speakers carefully. Remember that a presentation is a performance, requiring a dynamic interaction with the audience, especially at the beginning and the end. As on a relay team of runners or swimmers, place your strongest presenters first and last. Determine who on the project team should speak. A student team presentation usually requires the participation of everyone. A professional presentation may fall to only

**The Visuals**

Number: Too many Too few Just right

Effectiveness:

Structure slides:

Content slides:

Overall:

Legible?
Correct?
Memorable?
Clever?
Dramatic?
A help? Or a hindrance?

**The Performance: individual (answer yes or no)**

The speaker held my attention.
The speaker was easy to hear and understand.
The speaker convinced me.
The speaker was energetic and vigorous.

**The Performance: team (yes or no)**

The team was well prepared.
The presentation was unified around a central message.
The speakers seemed interested in each other.
The speakers made an easy and effective transition from one to another.

FIGURE 22.7 **Checklist for an Effective Presentation (continued)**

one speaker who summarizes the collaboration or may include the expert testimony of people representing different disciplines.

Third, balance a unity of impression against a diversity of voice. The message should be unified around a single point and plan, but the expression of that message should reflect the individual differences among team members. Unlike a collaborative document, which generally requires a consistent style, a presentation leaves room for individuality. In doing so, however, make sure the performance is under control. Practice where team members will stand when they are not talking, how you will move back and forth to the podium, how you will pass the baton, so to speak, from one speaker to the next. Practice transitions for a smooth handoff, and ruthlessly govern the timing of each person's talk. No one should run over. Make sure that presenters who are not speaking show interest in their teammate on stage. If you are not interested in each other, you send a signal to the audience that they should not be interested either. One good technique for balancing unity and diversity is for the opening speaker to announce the plan, as in this introduction to a team presentation:

> Over the last two months, our team reviewed the operation of the three university computer sites where we serve as monitors. Our goal was to determine how we could improve site operation, in particular, the training of first-time users, who require extensive hands-on help and dominate the machines so that they are not available to others. I'll introduce the problem, then Alvina will talk about a survey we conducted among users and site monitors, followed by Rick, who will describe a training course we are proposing. We'll talk for about 15 minutes total, and then we'd be happy to answer any questions you have.

### Answer Any Questions

In either a team or individual presentation, if appropriate, encourage questions from the audience. Those questions are often central to impromptu occasions and bullet reports. When you deliver prepared remarks, you may offer a question-and-answer session at the end. You or another person who is in charge of the presentation should announce a time limit for questions, just so the session doesn't simply dribble out, and enforce that limit with a remark like "We have time for one more question" at the end. Here are some guidelines for answering questions:

- Answer the question asked.
- If you don't understand the question, request that it be rephrased or explained.
- If you don't know the answer, say that and perhaps suggest where the answer may be found.
- If you are addressing a large group, repeat (or restate if the original was roundabout) the question before you answer so the audience is clear about what you're discussing.

Because most questions occur toward the end of the talk, your answers leave an important impression.

Figure 22.7 provides a checklist for evaluating a speaker's performance. It's probably scarier to think about speaking in public than to do it. The key ingredients are proper preparation and respect for the audience. In each presentation, display enough self-confidence to carry through positively and in a spirit of goodwill, even if your host mispronounces your name or fails to provide the equipment you specifically requested. You do have something to say, and saying that, to an appreciative audience, can mark a high point in your education or your career.

## CHECKLIST: Oral Presentations

**1. Determine the purpose and thus the type of presentation needed**

Impromptu: light preparation, lots of interactivity, informal

Outlined: more preparation, less interactivity, more formal

Scripted: written in advance, little or no interactivity, formal

2. **Determine the central message that reflects the purpose, setting, and audience**
3. **Profile the audience: needs, expectations, level of understanding, values**
4. **Structure the talk**
   Divide it into units pegged to time
   Create a clear introduction, middle, and, especially, ending
   Make the structure obvious
   Ruthlessly eliminate details
5. **Make the talk *visible***
   Select the appropriate display technology
   Use both content and structure visuals
   Limit the text on a visual
   Motivate but don't overload or merely dazzle
6. **Use handouts**
   Back up controversial information
   Provide translation of main points if needed
   Give a reference for a complex process
7. **Practice**
   Coordinate the roles, order, and transitions of a team presentation
   Make sure the display technology works in the setting
   Become comfortable with the technology and with speaking out loud
8. **Talk with the audience—not at them**

## EXERCISES

1. Present an 8-minute talk based on a report you prepared for your technical writing class. Don't read the report. Instead, *adapt* it for oral presentation. If the report addressed a client or supervisor, for example, then direct the oral report to the class. You'll need to change emphasis and content to match the new audience. For example,

   *Report:* An International Program for Improving Veterinary Practices in Estonia

   *Presentation:* Commentary on one brief video used in the Estonian training program

   *Report:* The development and design of new signs for the intersection of Interstates 95, 495, and 295

   *Presentation:* The process for determining the text and location of interstate highway signs

   *Report:* A survey of documentation practices in a Swiss software firm

   *Presentation:* Differences in teamwork techniques between Swiss and U.S. software engineers

2. Develop an outline for your presentation. Attach it to a memo addressed to your instructor that covers the following topics as you learned about them in this chapter:

   a. Audience profile
   b. Presentation genre
   c. Purpose
   d. Main point or theme
   e. Visuals to be used—note the equipment necessary and briefly characterize what you will show on the visuals and how you will create them
   f. Plan of the presentation
      - Opening line
      - Overview statement
      - Supporting evidence or components
        (1)
        (2)
        (3)
      - Closing line
   g. Anticipated audience questions—and responses.

3. As part of a class exercise in oral presentations, develop and deliver a brief (3-minute) introduction to a classmate's talk. Discuss the content of the talk and the speaker's background with the speaker, and select information that will set the context for the talk and warm up the audience.

## FOR COLLABORATION

In a team of two or three, prepare a review of literature (see Chapter 21) on some aspect of oral presentations, for example, advice from master presenters, a review of software for presentation graphics, advice on the best use of graphics in presentations, or techniques for structuring the presentation. Then *brief* the class informally about your findings. The briefings will help you prepare for a more formal team or individual presentation.

# Works Cited

## Chapter 1

Bronowski, Jacob. *Science and Human Values.* New York: Harper & Row, 1956.

Brooks, David. "Why Nice Folks Build Bigger Businesses." *Wall Street Journal* 11 Aug. 1995: A7.

Cowell, Alan. "Europe Plays Internet Catch-Up." *New York Times* 11 March 2000: B1.

Deal, Terrence E., and Allen A. Kennedy. *Corporate Cultures.* Reading: Addison-Wesley, 1982.

Freudenreich, L. Ben. Personal interview. 21 Mar. 1980.

Hunter, Margaret. "Best Practice Teamwork." *Novo Nordisk Magazine* Dec. 1995: 6–8.

Malone, Michael S. "Translating Diversity into High-Tech Gains: The Executive Life." *New York Times* 18 July 1993: 29F.

Perlez, Jane. "Clinton Lauds Technology as Key to India's Economy." *New York Times* 25 March 2000: A4.

Petersen, Andrea. "Opening a Portal: E-commerce Apostles Target Latin America, but It's a Tough Sell." *Wall Street Journal* 25 January 2000: 1.

Talbert, Tonya. "Skinner Conquers World's Climbing Challenges." *Alumnews* [U Wyoming] Dec. 1993/Jan. 1994: 4.

Ting-Toomey, Stella. "Toward a Theory of Conflict and Culture." *Communication, Culture, and Organizational Process.* Ed. Stewart Gudykunst and Ting-Toomey. Beverly Hills: Sage, 1985. 71–86.

"Web WorldWide: The Revolution Goes Global." *Global Reach Express* 8 March 2000. Available: http://glreach.com.

Wurman, Richard S. *Information Anxiety.* New York: Doubleday, 1989.

## Chapter 2

Bigness, Jon. "Here Today, There Tomorrow: Offices Add Mobile Furniture to Increase Flexibility." *Wall Street Journal* 24 July 1995: B1.

Bosley, Deborah S. "Cross-Cultural Collaboration: Whose Culture Is It, Anyway?" *Technical Communication Quarterly* 2.1 (1993): 51–62.

Bulkeley, William M. "'Computerizing' Dull Meetings Is Touted as an Antidote to the Mouth That Bored." *Wall Street Journal* 28 Jan. 1992: B1.

———. "The World's a Lab." *Wall Street Journal* 13 Nov. 1995: R16.

Dennis, Christopher, and Amy Brown. *Elements of the Animal Science Discourse Community: A Case Study.* Unpublished report submitted to English 411, University of Delaware, Spring 1994.

MacLeod, Laura. "Computer-aided Peer Review of Writing." *Business Communication Quarterly* 62.3 (1999): 87–94.

"Management Focus: The Trouble with Teams." *Economist* 14 Jan. 1995: 61.

Solomon, Nancy B. "Laboratory Innovations." *Architecture* Mar. 1993: 123–27.

Stern, Aimee L. "Managing by Team Is Not Always as Easy as It Looks." *New York Times* 18 July 1993: F5.

*Wall Street Journal,* 1 July 1984: 1.

Watkins, Beverly T. "A Far-Flung Collaboration by Scientists." *Chronicle of Higher Education* 8 June 1994: A15–17.

## Chapter 3

Ashbery, John. *And the Stars Were Shining.* New York: Farrar Straus & Giroux. © 1994 by John Ashbery. Reprinted by permission Farrar Straus & Giroux, Inc.

Bernhart, Michael H. "Preparation of Technology Transfer Agents." *International Communication of Technology.* Ed. Richard D. Robinson. New York: Taylor and Francis, 1991. 131–45.

"Cracking the China Market." *Wall Street Journal* 10 Dec. 1993: R1.

Daniel, Lucille. "Paul Petzoldt: The View from the Summit." *AMC Outdoors* Jan.–Feb. 1995: 10–11.

"Ethical Guidelines to Publication of Chemical Research." *The ACS Style Guide: A Manual for Authors and Editors.* Ed. Janet S. Dodd. Washington: American Chemical Society, 1986. 217–22.

Hays, Laurie, "Working It Out." *Wall Street Journal* 14 Nov. 1994: R22.

Hunt, Morton. "A Fraud That Shook the World of Science." *New York Times Magazine* 1 Nov. 1981: 42–75.

Leonhardt, David. "Management: In Language You Can Understand, SEC Edict Revives Push to Keep Documents Simple." *New York Times* 8 Dec. 1999: C1.

Lyonnaise des Eaux Group. *Code of Ethics.*

Monmonier, Mark. *How to Lie with Maps.* Chicago: U of Chicago P, 1991.

Morgan, Bruce. "Transferring Soft Technology." *The International Communication of Technology.* Ed. Richard D. Robinson. New York: Taylor and Francis, 1991. 149–66.

"Plain English Campaign" (on-line). Available: wysiwyg://13http://www.plainenglish.co.uk.

Slossen, Edward, as quoted by Robert H. Grant, and Kenneth D. Fisher. "Scientists and Science Writers: Concerns and Proposed Solutions." *Federation Proceedings* (Federation of American Societies for Experimental Biology) 30 (1971): 819.

[U.S.] Securities and Exchange Commission. *A Plain English Handbook.* Available: www.sec.gov/consumer/plaine.htm#A4.

## Chapter 4

Beveridge, W. I. B. *The Art of Scientific Investigation.* New York: Vintage, 1958.

Black, Alex (originator) (1993 Dec. 11). *Surveyors Are Watching YOU,* e-mail to multiple recipients of list MBU-L, from *Composition Digest,* Robert Royar, moderator. al@debra.dgbt.doc.ca.

Carlton, Jim. "Tub Toys Are Ducky Ocean Researchers." *Wall Street Journal* Oct. 1994: B1.

Flower, Linda. "Rhetorical Problem Solving: Cognition and Professional Writing." *Writing in the Business Professions.* Ed. Myra Kogen. Urbana IL: National Council of Teachers of English and the Association for Business Communication, 1989. 3–36.

Goleman, Daniel. "Pollsters Enlist Psychologists in Quest for Unbiased Results." *New York Times* 7 Sept. 1993: C1.

Mangan, James. "Cultural Conventions of Pictorial Representation: Iconic Literacy and Education." *Educational Communication and Technology,* Fall 1978: 245–67.

Miller, Christopher. "Building Illusions: Culture Determines What We See." *Business Communication Quarterly* 59.1: 87–90.

## Chapter 5

Arnzen, Michael A. "Cyber Citations." *Internet World* Sept. 1996: 72–74.

Broad, William J. "Doing Science on the Network: A Long Way from Gutenberg." *New York Times* 18 May 1993: C1.

Goad, Meredith. "Doctors Adapt to Stay Current." *Maine Sunday Telegram* 13 Aug. 1993: 1B.

Jameson, Daphne A. "The Ethics of Plagiarism: How Genre Affects Writers' Use of Source Material." *The Bulletin of the Association for Business Communication* 56.2 (1993): 18–28.

Miller, Michael W. "U.S. Spies Help Scientists Pierce Data Jungle." *Wall Street Journal* 27 July 1993: B1.

Pitman, Joanna. "Lost in the Language of Cyberspace." *The Times* (London) 18 Sept. 1995: 17.

Specter, Michael. "World, Wide, Web: 3 English Words." *New York Times* 14 Apr. 1996: 1E, 5E.

Specter, Michael. "Postcard from Silicon Valley: Search and Deploy: The Race to Build a Better Search Engine." *New Yorker* May 29 2000: 88–100.

"The U.S. No Longer Has The Monopoly on the Internet . . . and Even Less in E-commerce." *Global Reach Express* 23 June 2000.

Weiss, Timothy. "Translation in a Borderless World." *Technical Communication Quarterly* 4.4 (1995): 407–23.

"Who Can Measure the Net?" *Economist* 22 July 1995: 61.

Wilson, David L. "Teaching a Computer to Find and Retrieve Stored Images." *Chronicle of Higher Education* 12 Oct. 1994: A20–21.

Winslow, Ron. "More Doctors Are Adding On-Line Tools to Their Kits." *Wall Street Journal* 7 Oct. 1994: B1.

Wurman, Richard S. *Information Anxiety.* New York: Doubleday, 1989.

"You Say Tomato, She Says Jitomate." *Global Reach Express* 23 November 1999. Available: http://glreach.com.

## Chapter 6

Browne, Malcolm W. "Scientist at Work: Roald Hoffmann: Seeking Beauty in Atoms." *New York Times* 6 July 1993: C1.

Forslund, Charlene Johnson. "Analyzing Pictorial Messages Across Cultures." *International Dimensions of Technical Communication.* Ed. Deborah C. Andrews. Arlington: Society for Technical Communication. 1996. 45–58.

Lombard, Catherine. "Let's Get Visual: Revelations After Six Days with Japanese Customers." *Technical Communication* 39.4 (1992): 689–91.

Mathes, J. C., and Dwight W. Stevenson. *Designing Technical Reports.* 2nd ed. New York: Macmillan, 1991.

Miller, Christopher. "Building Illusions: Culture Determines What We See." *Business Communication Quarterly* 59.1 (1996): 87–90.

Sobel, Dava. "Scientists Discover Answer to Color Perception." *New York Times* Nov 28 1979.

Stevenson, Dwight W. "Audience Analysis Across Cultures." *Journal of Technical Writing and Communication* 13. 4: 319–30.

Williams, Mike. "Consumer Product Instructions—An Artist's Perspective." *Communicator* 5.1 (Mar. 1995): 4–6.

Wurman, Richard S., Ed. *Information Architects.* New York: Graphis, 1996.

Zimmerman, Margot, Nancy Newton, Lena Frumin, and Scott Wittett. *Developing Health and Family Planning Materials for Low-Literate Audiences: A Guide.* Rev. ed. Washington: PATH (Program for Appropriate Technology in Health), 1996.

## Chapter 7

Backinger, C.L., and P.A. Kingsley. *Write It Right: Recommendations for Developing User Instructions for Medical Devices Used in Home Health Care.* HHS Publication FDA 93-4258. Washington: FDA, Aug. 1993.

Horton, William. "The Almost Universal Language: Graphics for International Documents." *Technical Communication* 40.4 (1993): 682–93.

Jones, Gerald E. *How to Lie with Charts.* San Francisco: Sybex, 1995. 205.

Levinson, Mark. (1994, June 26). *A4 page size.* e-mail to multiple recipients of list TECHWR-L [online]. Available e-mail:LISTSERV@ TECHWR-L@osuvm1.bitnet.

White, Jan V. *Graphic Design for the Electronic Age.* New York: Watson-Guptill, 1988.

## Chapter 8

American Institute of Graphic Arts (AIGA). *Symbol Signs,* 2nd ed. New York: AIGA, 1993.

Bolter, Jay David. *Writing Space: The Computer, Hypertext, and the History of Writing.* Hillsdale: Lawrence Erlbaum, 1991.

Ferguson, Eugene S. *Engineering and the Mind's Eye.* Cambridge: MIT P, 1992.

Greenough, G. B., as quoted in Wm G. Dean, "Foreword." *Historical Atlas of Canada.* Volume III: *Addressing the Twentieth Century 1891–1961.* Ed. Donald Kerr and Deryck W. Holdsworth. Toronto: U of Toronto P, 1990.

Helfman, Elizabeth S. *Signs and Symbols around the World.* New York: Lothrop, Lee & Shepard, 1967.

Horton, William. "The Almost Universal Language: Graphics for International Documents." *Technical Communication* 40.4 (1993): 682–93.

Ireland, Peter. "Top of the Shopping Lists." *Financial Times* 4 Oct. 1995: 11.

Monmonier, Mark. *How to Lie with Maps.* Chicago: U of Chicago P, 1991.

Moore, Bob. "Forging the Information Link." *Habitat: The Journal of the Maine Audubon Society* Winter 1994: 29–31.

Tufte, Edward. *Envisioning Information.* Cheshire: Graphics Press, 1990.

Wheeler, David L. " 'At the Helm of the Most Important Scientific Enterprise of the Century,' the NIH's Human Genome Project." *Chronicle of Higher Education* 4 Aug. 1993: A8.

Williams, Thomas R. "What's So Different About Visuals?" *Technical Communication* 40.4: 669–76.

## Chapter 9

Beveridge, W. I. B. *The Art of Scientific Investigation.* New York: Vintage, 1958. 11.

Dennett, Joann Temple. "Not to Say Is Better Than to Say: How Rhetorical Structure Reflects Cultural Context in Japanese-English Technical Writing." *IEEE Transactions of Professional Communication* 31.3 (1988): 116–19.

Kaplan, Robert B. "Cultural Thought Patterns in Inter-Cultural Education." *Readings on English as a Second Language,* 2nd ed. Ed. Kenneth Croft. Cambridge: Winthrop, 1980. 399–418.

Kirkman, John. In a seminar at the University of Delaware London Centre, 11 Oct. 1995.

National Science Board (NSB). *Science & Engineering Indicators—1993.* Washington: U.S. Government Printing Office, 1993 (NSB93-1).

Stevenson, Dwight W. "Audience Analysis Across Cultures." *Journal of Technical Writing and Communication* 13.4 (1983): 319–30.

Turkle, Sherry, and Seymour Papert. "Epistemological Pluralism: Styles and Voices within the Computer Culture." *Signs* 16.1 (1990): 136.

## Chapter 10

Austin, Phyllis. "A Clearcut Case." *Maine Times* Nov. 1994: 2–6.

"Brewing." *Enzymes at Work.* Bagsvaerd: Novo Nordisk A/S, 1992. 26.

Coletta, W. John. "The Ideologically Biased Use of Language in Scientific and Technical Writing." *Technical Communication Quarterly* 1.1 (1992): 59–70.

Fellman, Bruce. "Forestry Goes Global." *Yale* Mar. 1993: 48–52.

Gifford Pinchot National Forest. "Mount St. Helens National Volcanic Monument." Draft Environmental Impact Statement. Comprehensive Management Plan A13.92/2 M865. Vancouver, WA: U.S. Department of Agriculture, Forest Service, Pacific Northwest Region, 1984.

Novo Nordisk A/S. *Novo Nordisk's Little Book on Genetic Engineering.* 3rd ed. Bagsvaerd: Novo Nordisk A/S, 1992.

*Proposed Coastal Management Program for the State of Delaware* EIS, C55.34 D5:1. Washington: Department of Commerce, NOAA Office of Coastal Zone Management, 1979.

Riggi, Andrew, Anthony Rossi, Stacy Shulley, and Tesfaye N. Tesfaye. "Skeleton Construction." Paper submitted to English 410, University of Delaware, Spring 1994.

Scott/S. D. Warren Company. "Clearcutting Can Be a Productive Tool for Intensive Forest Management." *Maine News* 1.1 (1994): 2.

Subbiah, Mahalingam. "Adding a New Dimension to the Teaching of Audience Analysis: Cultural Awareness." *IEEE Transactions on Professional Communication* 35 (1992): 14–18.

U.S. Committee for Energy Awareness. *Energy Options: The Practical and the Promising.* Washington: The Committee, n.d.

U.S. Environmental Protection Agency, Office of Noise Abatement and Control (AW471). *Quieting in the Home.* Washington: EPA, 1978.

Woudstra, Egbert. "Asking the Right Visuals Questions." *Business Communication Quarterly* 59.1 (1996): 93–95.

## Chapter 11

Berreby, David. "Get on Line for Plato's Cave." *New York Times* 25 June 1995: E5.

Beveridge, W. I. B. *The Art of Scientific Investigation.* New York: Vintage, 1950.

Broad, William J. "Cancer Fear Is Unfounded, Physicists Say." *New York Times* 14 May 1995: 19.

Brody, Jane E. "Personal Health. Health Alarm for a Sleep-Deprived Society." *New York Times* 19 Jan. 1994: C14.

Burros, Marian. "Eating Well: Testing of Food Pyramid Comes Full Circle," *New York Times* 25 Mar. 1992: C1, C4.

Carnevale, Mary Lu, "Ultrasensitive, Cheap Test Sought for Breast Cancer." *Wall Street Journal* 2 Aug. 1994: B1+.

Crossen, Cynthia. "Fright by the Numbers: Alarming Disease Data Are Frequently Flawed." *Wall Street Journal* 11 Apr. 1996: B1.

Frankel, Mark S. "Multicultural Science." *Chronicle of Higher Education* 10 Nov. 1993: B1–B2.

Gall, Gwen. (13 Sept. 1994). E-mail message to multiple recipients of list TECH-WRL. Available on e-mail: LISTSERV@TECHWR-L@osuvm1.bitnet.

Gladwell, Malcolm. "Dept. of Disputation: The Tipping Point." *New Yorker* 3 June 1996: 32–38.

Huff, Darrell, and Irving Geis. *How to Lie with Statistics.* New York: Norton, 1954.

Koch, Barbara Johnstone. "Presentation as Proof: The Language of Arabic Rhetoric." *Anthropological Linguistics* 25.1: 47–60.

Krugman, Paul. "Heisenberg Certainty Lecture #4," as quoted in *The Mini-Journal of Irreproducible Results* Nov. 1993 [online].

Miller, Michael W. "Call for Daily Dose of Wine Ferments Critics." *Wall Street Journal* 17 June 1994: B1.

Passell, Peter. "How Much for a Life? Try $3 Million to $5 Million." *New York Times* 29 Jan. 1995: F3.

Ross, John F. "Risk: Where Do Real Dangers Lie?" *Smithsonian* Nov. 1995: 43–53.

## Chapter 12

Bell, Alice. "Writing Standards for Translation." *News & Views* [Delaware Valley Chapter, Society for Technical Communication] May 1994: 9.

Borghi, Vincent. (1994 Sept. 23). *Preparing documents for translation.* E-mail to multiple recipients of list TECHWR-L [online]. Available on e-mail: LISTSERV@TECHWR-L@osuvm1.bitnet.

Elliott, Keith, "A Layered Approach to Translating Online Documentation." *1993 Proceedings, 40th Annual Conference, STC:* 115–18.

Franklin, Benjamin. *Pennsylvania Gazette* 2 Aug. 1733.

Gilmore, Elizabeth. "What Is SGML Anyway? Should You Care?" *Intercom* Sept. 1993: 3–4.

Grayling, Trevor. 18 Oct. 1994. E-mail to the multiple recipients of the list TECHWR-L. Subject: Usability testing—our experience. Available from: Trevor@mdli.com.

Miller, L. Chris. "Transborder Tips and Traps." *BYTE* June 1994: 93–102.

Norman, Rose, and Daryl Grider. "Structured Document Processors: Implications for Technical Writing." *Technical Communication Quarterly* 1.3 (1992): 5–21.

## Chapter 13

Borko, Harold, and Charles L. Bernier. *Abstracting Concepts and Methods.* New York: Academic P, 1975.

Evans, Mary. "Structured Abstracts: Rationale and Construction." *European Science Editing* Oct. 1995: 4–5.

Grinbergs, Arvid, and Albert H. Rubenstein. "Software-Engineering Management: A Comparison of Methods in Switzerland and the United States." *IEEE Transactions on Engineering Management* 40.1 (Feb. 1993): 22–29.

Myer-Rochow, V.B. "Inuit or Eskimo: Politically Correct Terminology in Scientific Writing." *European Science Editing* 25.2 (1999): 55–57.

"The Use of Models: Nineteenth-Century Church Architecture in Quebec," from p. 1 of the Centre Canadien d'Architecture/Canadian Centre for Architecture Exhibition Programme 1994–1997.

Wong, F.H., and Lai Phooi-Ching. "Chinese Cultural Values and Performance at Job Interviews: A Singapore Perspective." *Business Communication Quarterly* 63.1 (2000): 9–22.

## Chapter 14

Hamilton, Chris. "Proposal for Kenaf as an Alternative Crop for the Atlantic Coastal Plain." Report submitted to English 410, University of Delaware, 2 Nov. 1993.

Mihm, Stephen. "1000 Crates of Sprinkles with That?" *New York Times* 29 March 2000: 28.

Oppel, Richard. A., Jr. "The Higher Stakes of Business-to-Business Trade." *New York Times* 5 March 2000: BU3.

Stein, Harley. In conversation with the author in person and in e-mail, 5 March 1997.

Wiese, W. C. "Managing and Editing Multinational Joint Venture Proposals." *International Technical Communications Conference Proceedings* (1990), MG 85–87.

## Chapter 15

Boulton, Leyla. "The Trouble with Green Audits." *Financial Times* 4 Oct. 1995: 14.

Capitol Region Council of Governments (CRCOG). *Preliminary Alternatives Review: Identification of Alternatives for Inclusion in Full Analysis.* Hartford: CRCOG, Apr. 19, 1993.

Global Climate Coalition. *Issues & Options: Potential Global Climate Change.* Washington: The Coalition, 1994.

Harmon, Joseph E., and Alan G. Gross. "The Scientific Style Manual: A Reliable Guide to Practice?" *Technical Communication* 43:1 (1996): 61–72.

"Marketplace: A World of Information at Your Fingertips," *The Times* (London) 1 Nov. 1995: Interface 6.

"Novo Nordisk Comes Clean." *Novo Nordisk Magazine* May 1994: 9–11.

Plamondon, Robert, to multiple recipients of list TECHWR-L. Re: White Papers 29 April 1995. Available on e-mail: LISTSERV@TECHWR-L@osuvm1.bitnet.

Ramsey, Kelvin W., John H. Talley, and Darlene V. Wells. Delaware Geological Survey (DGS). *Summary Report, The Coastal Storm of December* 10–14, 1992, Delaware and Maryland, Open File Report no. 37, Feb. 1993.

Warden, Richard E., and W. Tim Dagodag. *A Guide to the Preparation and Review of Environmental Impact Reports.* Los Angeles: Security World Pub Co., 1996.

## Chapter 16

Graham, C. D., Jr. "A Glossary for Research Reports." *Metal Progress* May 1957.

## Chapter 17

Backinger, C. L., and P. A. Kingsley. *Write It Right: Recommendations for Developing User Instructions for Medical Devices Used in Home Health Care.* HHS Publication FDA 93-4258. Washington: FDA, Aug. 1993.

Bulkeley, William M. "Information Age: Advances in Networking and Software Push Firms Closer to Paperless Office." *Wall Street Journal* 5 Aug. 1993: B1+.

Burnett, Rebecca E. "Readers' Presumed Preferences in Japanese-Language and English-Language Manuals." Presentation at the 61st Annual Convention of the Association for Business Communication. Chicago: 8 Nov. 1997.

Knight, Kathryn. "Sign of the Times: Don't Call Us . . . Please." *The Times* (London) 14 June 1996: 18. Available on e-mail: LISTSERV@TECHWR-L@osuvm1.bitnet.

Lippincott, Rick. *Validation vs Verification.* E-mail to the multiple recipients of the list TECHWR-L. 16 Sept. 1994.

Mankins, David, as quoted by harding@us.net, 11 June 1995.

North Central Forest Experiment Station, USDA Forest Service. *How to Live with Black Bears* HT-66. Washington: GPO, 1991.

McLaren, Thomas A. "Help Systems Today—New Domain for Technical Communication." *Intercom* Mar. 1993: 5–6.

Weiss, Timothy. "Translation in a Borderless World." *Technical Communication Quarterly* 4.3 (1995): 407–23.

Zhu Chang Gen. Statement in the announcement of the FORUM95 Video Conferences, 13–15 Nov. 1995, Dortmund, Germany.

## Chapter 18

Boiarsky, Carolyn. "The Relationship between Cultural and Rhetorical Conventions: Engaging in International Communication." *Technical Communication Quarterly* 4.3 (1995): 245–59.

Haneda, Saburo, and Hirosuke Shima. "Japanese Communication Behavior as Reflected in Letter Writing." *The Journal of Business Communication* 19.1 (1982): 19–32.

Hinds, John, and Susan Jenkins. "Business Letter Writing: English, French, and Japanese." *TESOL Quarterly* 21.2 (1987): 327–50.

## Chapter 19

Hemby, K. Virginia. "Using the World Wide Web to Teach Employment Communication." *Business Communication Quarterly* 60.1 (Mar. 1997): 120–21.

Krause, Tim. "Preparing an On-Line Resume." *Business Communication Quarterly* 60.1 (1997): 119–20.

Levenson, Lisa. "High-Tech Job Searching." *Chronicle of Higher Education* 14 July 1995: A16–17.

Quible, Zane K. "Electronic Resumes: Their Time Is Coming." *Business Communication Quarterly* 58.3 (1995): 5–7.

Rifkin, Glenn. "Virtual Recruiter: Software That Reads Resumes." *New York Times* 24 Mar. 1996: 12.

## Chapter 20

Bulkeley, William M. "Information Age: Advances in Networking and Software Push Firms Closer to Paperless Office." *Wall Street Journal* 5 Aug. 1993: B1.

Labaton, Stephen, and Matt Richtel. "Proposal Offers Surveillance Rules for the Internet." *New York Times* 18 July 2000:1.

Legeros, Michael J. "Etiquette and Email: Rules for Online Behavior." *Intercom* July/Aug. 1995: 10–11.

McMorris, Frances A. "Is Office Voice Mail Private? Don't Bet on It." *Wall Street Journal* 28 Feb. 1995: B1.

Ramirez, Anthony. "From the Voice-Mail Acorn, a Still-Spreading Oak." *New York Times* 3 May 1992: F9.

## Chapter 21

Bazerman, Charles, and James Paradis. *Textual Dynamics of the Professions.* Madison: U of Wisconsin P, 1991.

Bowman, Steven M., Tony M. Keaveny, and Thomas A. McMahon. "Compressive Creep Behavior in Bovine Trabecular Bone." *Transactions of the 39th Annual Meeting of the Orthopaedic Research Society,* 1993. 173.

Brody, Jane E. "Personal Health: Health Alarm for a Sleep-Deprived Society." *New York Times* 19 Jan. 1994: C14.

Cohen, Jeffrey. "Making Car Theft More Difficult." *New York Times* 28 Oct. 1993: C2.

Crane, Diana. *Invisible Colleges: Diffusion of Knowledge in Scientific Communities.* Chicago: U of Chicago P: 1972.

Dodd, Janet S., ed. *The ACS Style Guide: A Manual for Authors and Editors.* Washington: American Chemical Society, 1986.

Farney, Dennis. "Natural Questions: Chaos Theory Seeps into Ecology Debate, Stirring Up a Tempest." *Wall Street Journal* 11 July 1994: 1.

Gilsdorf, Jeanette. "Metacommunication Effects on International Business Negotiating in China." *Business Communication Quarterly* 60.2 (1997): 20–37.

Gleick, James. "The Net." "The Seven Wired Wonders." Ed. Jeff Greenwald. *Wired* Dec. 1993: 103.

Harmon, Joseph E., and Alan G. Gross. "The Scientific Style Manual: A Reliable Guide to Practice?" *Technical Communication* 43.1 (1996): 61–72.

Massey, Sharon. "Students Get That Sinking Feeling at a Personal Submarine School." *Wall Street Journal* 27 July 1994: B1.

Moffett, Mark. "These Plants Scratch, Claw and Strangle Their Way to the Top." *Smithsonian* Sept. 1993: 110–19.

Myers, Greg. "Every Picture Tells a Story: Illustrations in E. O. Wilson's Sociobiology." *Representation in Scientific Practice.* Ed. Michael Lynch and Steve Woolgar. Cambridge: MIT P, 1990.

Patton, Phil. "From the Desktop of Frank Lloyd Wright." *Esquire* Nov. 1994: 32.

Serebryakova, Irina. "Translation Problem of English Metaphoric Terms into Russian." 1993. Unpublished.

Stevens, William K. "Loss of Species Is Worse Than Thought in Brazil's Amazon." *New York Times* 29 June 1993: C4.

"That Time of the Month." *Economist* 31 July 1993: 75–76.

Weaver, Warren. "Communicative Accuracy." *Science* 127 (1958): 499.

Wolf, Gary. "Who Owns the Law?" *Wired* May 1994: 98–101+.

## Chapter 22

Bragg, Lawrence. "The Art of Talking About Science." *Science* 154 (1985).

Cook, Rae Gorin. "Enhancing the Participation of Foreign-Born Professionals in U.S. Business and Technology." *International Dimensions of Technical Communication.* Ed. Deborah C. Andrews. Arlington: The Society for Technical Communication, 1996. 5–22.

# INDEX

## F



## G

## J, K

**N**



**O**



**P**

## S

## T



## U and V



## W, X, Y, Z